Nanofluids and Their Engineering Applications

Nanofluids and Their Engineering Applications

Edited by
K.R.V. Subramanian
Tubati Nageswara Rao
Avinash Balakrishnan

CRC Press
Taylor & Francis Group
Boca Raton London New York

CRC Press is an imprint of the
Taylor & Francis Group, an **informa** business

CRC Press
Taylor & Francis Group
6000 Broken Sound Parkway NW, Suite 300
Boca Raton, FL 33487-2742

First issued in paperback 2020

© 2020 by Taylor & Francis Group, LLC
CRC Press is an imprint of Taylor & Francis Group, an Informa business

No claim to original U.S. Government works

ISBN-13: 978-1-138-60526-8 (hbk)
ISBN-13: 978-0-367-72754-3 (pbk)

Visit the Taylor & Francis Web site at
http://www.taylorandfrancis.com

and the CRC Press Web site at
http://www.crcpress.com

Contents

Section IV   Nanofluids in Solar Applications

Section V   Nanofluids in Oil and Gas Industry and Carbon Sequestration

Preface

Nanofluids are dilute liquid suspensions of nanoparticles with at least one of their principal dimensions smaller than 100 nm. Nanofluids are a new class of nanomaterials that offer significant advantages of superior thermophysical properties such as thermal conductivity, viscosity, stability, and specific heat with higher values of base numbers (Nusselt, Prandtl, Darcy, etc.) when compared to base fluids. Owing to their enhanced properties as thermal transfer fluids for instance, nanofluids can be used in a plethora of engineering applications ranging from use in the automotive industry to the medical arena to use in power plant cooling systems as well as computers.

Phase change materials as nanoparticles have been used in nanofluids to simultaneously enhance the effective thermal conductivity and specific heat of the fluids. Nanofluids have been demonstrated to be able to handle their role in some instances as a smart fluid. In the sub-area of power conversion technology, improving heat-transfer performance for lower-temperature nanofluids, and developing plant designs for higher resource temperatures to the supercritical water region would lead to an order of magnitude (or more) gain in both reservoir performance and heat-to power conversion efficiency. Engine oils, automatic transmission fluids, coolants, lubricants, and other synthetic high-temperature heat transfer fluids found in conventional truck thermal systems—radiators, engines, heating, ventilation, and air-conditioning (HVAC)—have inherently poor heat transfer properties. These could benefit from the high thermal conductivity offered by nanofluids. A principal limitation on developing smaller microchips is the rapid heat dissipation. However, nanofluids can be used for liquid cooling of computer processors due to their high thermal conductivity. There is a new initiative which takes advantage of several properties of certain nanofluids to use in cancer imaging and drug delivery. This initiative involves the use of iron-based nanoparticles as delivery vehicles for drugs or radiation in cancer patients. Magnetic nanofluids are to be used to guide the particles up the bloodstream to a tumor with magnets.

Looking at the possible uses of nanofluids, a variety of different possibilities open: transportation (engine cooling/vehicle thermal management), electronics cooling, defense and space, nuclear systems cooling, heat exchangers, biomedicine and other biomedical applications, heat pipes, oil recovery, fuel cell, concentrated solar technology, solar water heating, chillers, domestic refrigerator, diesel combustion, drilling, lubrications, thermal storage, and many others.

This book deals with some of the technologically most important engineering applications of nanofluids as well as cutting edge research areas. It also attempts to deal with some practical engineering challenges facing nanofluids. One of the most crucial issues related to nanofluids is the stability of the nanoparticles suspension. The agglomeration of nanoparticles results in the settlement and clogging of microchannels and also the decreasing of thermal conductivity of nanofluids. The nanofluids research has seen a large amount of experimental works, it is still in initial phase and there are several open issues, like the lack of agreement regarding the results obtained by different researchers, and the lack of theoretical understanding of the mechanisms responsible for changes in the properties. In fact, there are many important variables and issues related to the production and the usage of nanofluids, which may cause significant discrepancy in the acquired experimental data. The type of nanoparticle, its size, its shape, and distribution are important properties that cannot be easily measured nor well-defined or properly reported in the publications. The type of base fluids used, the method for the nanofluid production, use of surfactants and stabilization additives, including pH adjusters, etc. are other important factors.

We, the editors, Subramanian, Rao, and Avinash have drawn upon our own research thrusts, work, and proposals in nanofluid technology and strived to bring the application areas to the forefront with the challenges. We would like to express our sincere thanks to our management, viz.,

Gandhi Institute of Technology and Management (GITAM) School of Technology, Bangalore, India (Prof. P.V. Sivapullaiah, Pro Vice Chancellor, Prof. Vijaya Bhaskar Raju, Director) as well as Suzlon Energy Ltd., Gujarat, India, for supporting this endeavor. Our thanks are also due to the distinguished authors who have contributed their might. We finally thank our family members for their support.

Prof. K.R.V. Subramanian

Prof. Tubati Nageswara Rao

Prof. Avinash Balakrishnan

Editors

Prof. K.R.V. Subramanian is working as Associate Professor with Mechanical Engineering Department, GITAM University, Bangalore, India. His research interests are solar technology, nanotechnology, nanofabrication, energy storage devices, carbon nanotubes, nanofluids, hydrogen storage, fuel cells, sensors. He earned his PhD from Cambridge University in 2006 specializing in nanotechnology. His bachelor's and master's degree in Materials Engineering were from NIT Trichy and IISc Bangalore. He has over 20 years of academic and industrial work experience. He has published over 100 journal and conference papers. He has also edited two books. He regularly reviews for prestigious RSC, American Chemical Society (ACS), and Elsevier journals. He is an affiliate member of RSC (Royal Society of Chemistry, United Kingdom) and Fellow of Cambridge Commonwealth Society, United Kingdom. He is a Marquis Who's Who of the World and International Biographical Centre (IBC) Top 100 professionals. He has been co-principal investigator for many government-funded research projects.

Prof. Tubati Nageswara Rao is working as Head of Department and Professor with Mechanical Engineering Department, GITAM University, Bangalore, India. His research interests are heat transfer and thermal engineering. He earned his PhD from Indian Institute of Technology, Chennai, in 2001 specializing in thermal engineering. He has over 25 years of academic and industrial work experience. He has published over 30 journal and conference papers.

Dr. Avinash Balakrishnan joined as Manager at Suzlon Energy Limited in 2016, where he is heading the Materials Laboratory for Suzlon Blade Technology vertical. He did his PhD and MS in Materials Engineering from Paichai University, S. Korea. He is an alumnus of National Institute of Technology, Karnataka, where he completed his bachelors in Metallurgical Engineering. His past professional experience also includes being research scientist at Korea Research Institute of Standards and Science (KRISS), South Korea, where he extensively worked on ceramic materials for structural and high temperature applications. He was a post-doctoral fellow at Grenoble Institute of Technology (Grenoble-INP), France. He was also heading the Research and Development (R&D) division for English Indian Clay Limited, India, from 2015 to 2016. His current research interests include developing nanostructured materials for variety of applications including energy storage, composites, and ceramics. He has co-authored over 85 research publications, 2 books, and filed 1 patent.

Contributors

Teresa Aguilar
Departamento de Química Física
Facultad de Ciencias
Universidad de Cádiz
Puerto Real (Cádiz), Spain

Rodrigo Alcántara
Departamento de Química Física
Facultad de Ciencias
Universidad de Cádiz
Puerto Real (Cádiz), Spain

Mohamed Kamal Ahmed Ali
Automotive and Tractors Engineering
    Department
Faculty of Engineering
Minia University
El-Minia, Egypt

Amine Allouhi
Ecole Supérieure de Technologie de Fès
U.S.M.B.A, Route d'Imouzzer
Fez, Morocco

Avinash Balakrishnan
Advanced Materials Division
Suzlon Energy Ltd.
Gujarat, India

Divya P. Barai
Department of Chemical Engineering
Laxminarayan Institute of Technology
Rashtrasant Tukadoji Maharaj Nagpur University
Nagpur, India

Bharat A. Bhanvase
Department of Chemical Engineering
Laxminarayan Institute of Technology
Rashtrasant Tukadoji Maharaj Nagpur University
Nagpur, India

Mahmut Sami Buker
Department of Aeronautical Engineering
Konya NEU University
Konya, Turkey

Jianchao Cai
Hubei Subsurface Multi-scale Imaging Key
    Laboratory
Institute of Geophysics and Geomatics
China University of Geosciences
Wuhan, China

N. Fedal Castro
Department of Mechanical Engineering
Aarupadai Veedu Institute of Technology
Paiyanoor, India

Feng Chen
Fujian Province University Key Laboratory of
    Green Energy and Environment Catalysis
Fujian Provincial Key Laboratory of Featured
    Materials in Biochemical Industry
Ningde Normal University
Ningde, China

Kalyani K. Chichghare
Department of Chemical Engineering
Laxminarayan Institute of Technology
Rashtrasant Tukadoji Maharaj Nagpur University
Nagpur, India

B. Chitra
Department of Chemical Engineering
SSN College of Engineering
Kalavakkam, India

Nur Çobanoğlu
İzmir Katip Çelebi University
Graduate School of Natural and Applied Sciences
Department of Mechanical Engineering
İzmir, Turkey

Vincent Cregan
Department of Mathematics and Statistics
University of Limerick
Co. Limerick, Ireland

Mahidzal Dahari
Department of Electrical Engineering
University of Malaya
Kuala Lumpur, Malaysia

Amit Rai Dixit
Department of Mechanical Engineering
Indian Institute of Technology (ISM)
Dhanbad, India

Chun Du
State Key Laboratory of Materials Processing and
    Die & Mould Technology
School of Materials Science and Engineering
Huazhong University of Science and Technology
Wuhan, People's Republic of China

Elbager M.A. Edreis
Department of Mechanical Engineering
Faculty of Engineering
University of Blue Nile
Ad-Damazin, Sudan

Muftah H. El-Naas
Gas Processing Center
College of Engineering
Qatar University
Doha, Qatar

Ammar H. Elsheikh
Production Engineering and Mechanical Design
    Department
Faculty of Engineering
Tanta University
Tanta, Egypt

X. Fang
Key Laboratory of Air-driven Equipment
    Technology of Zhejiang Province
Quzhou University
Quzhou, China

Concha Fernández-Lorenzo
Departamento de Química Física
Facultad de Ciencias
Universidad de Cádiz
Puerto Real (Cádiz), Spain

Juan Jesús Gallardo
Departamento de Química Física
Facultad de Ciencias
Universidad de Cádiz
Puerto Real (Cádiz), Spain

Roberto Gómez-Villarejo
Departamento de Química Física
Facultad de Ciencias
Universidad de Cádiz
Puerto Real (Cádiz), Spain

F.A. Hammad
Mechanical Power Engineering Department
Faculty of Engineering
Tanta University
Tanta, Egypt

M. Hatami
International Research Center for Renewable
    Energy
State Key Laboratory of Multiphase Flow in
    Power Engineering
Xi'an Jiaotong University
Xi'an, China

and

Department of Mechanical Engineering
Esfarayen University of Technology
Esfarayen, Iran

Yurong He
School of Energy Science and Engineering
Harbin Institute of Technology
and
Heilongjiang Key Laboratory of New Energy
    Storage Materials and Processes
Harbin, People's Republic of China

Abdolhossein Hemmati-Sarapardeh
Department of Petroleum Engineering
Shahid Bahonar University of Kerman
Kerman, Iran

Yanwei Hu
School of Energy Science and Engineering
Harbin Institute of Technology
and
Heilongjiang Key Laboratory of New Energy
    Storage Materials and Processes
Harbin, People's Republic of China

Angel Huminic
Aerodynamics Laboratory
Transilvania University of Brasov
Brasov, Romania

Gabriela Huminic
Applied Thermodynamics Laboratory
Transilvania University of Brasov
Brasov, Romania

Dengwei Jing Jing
International Research Center for Renewable
Energy
State Key Laboratory of Multiphase Flow in
Power Engineering
Xi'an Jiaotong University
Xi'an, China

Ziya Haktan Karadeniz
İzmir Katip Çelebi University
Department of Mechanical Engineering
Çiğli, Turkey

S.N. Kazi
Department of Mechanical Engineering
University of Malaya
Kuala Lumpur, Malaysia

Khalil Khanafer
Mechanical Engineering Department
Australian College of Kuwait
Safat, Kuwait

and

Advanced Manufacturing Lab (AML)
School of Engineering
University of Guelph
Guelph, Ontario, Canada

Ravi Shankar Kumar
Enhanced Oil Recovery Laboratory
Rajiv Gandhi Institute of Petroleum Technology
Jais, Amethi (UP), India

Elisa I. Martín
Departamento de Ingeniería Química
Facultad de Química
Universidad de Sevilla
Sevilla, Spain

Alina Adriana Minea
Faculty of Materials Science and Engineering
"Gheorghe Asachi" Technical University of Iasi
Iasi, Romania

Sarah Mitchell
Department of Mathematics and Statistics
University of Limerick
Co. Limerick, Ireland

Majid Mohammadi
Department of Energy Engineering
Faculty of Science
Qom University of Technology
Qom, Iran

Tim Myers
Centre de Recerca Matemàtica
Barcelona, Spain

Javier Navas
Departamento de Química Física
Facultad de Ciencias
Universidad de Cádiz
Puerto Real (Cádiz), Spain

Gary O'Keeffe
Department of Mathematics and Statistics
University of Limerick
Co. Limerick, Ireland

Hakan F. Oztop
Department of Mechanical Engineering,
Technology Faculty
Fırat University
Elazığ, Turkey

P. Padmaja
Department of Chemistry
Faculty of Science
The Maharaja Sayajirao University of Baroda
Vadodara, India

Zhen Pan
College of Petroleum Engineering
Liaoning Shihua University
Fushun, China

K.K. Salman
Mechanical Department
Kut Technical Institute
Middle Technical University
Baghdad, Iraq

Antonio Sánchez-Coronilla
Departamento de Química Física
Facultad de Farmacia
Universidad de Sevilla
Sevilla, Spain

Sajjad Sargana
Key Laboratory of Efficient Utilization of Low
    and Medium Grade Energy
Tianjin University
Ministry of Education
Tianjin, China

K. Sathishkumar
Department of Chemical Engineering
SSN College of Engineering
Kalavakkam, India

Mehdi Sedighi
Department of Chemical Engineering
University of Qom
Qom, Iran

Fatih Selimefendigil
Department of Mechanical Engineering
Celal Bayar University
Manisa, Turkey

J. Shaibo
Department of Physics
Al-Fashir University
Al-Fashir, Sudan

Anuj Kumar Sharma
Centre for Advanced Studies
Dr. A.P.J. Abdul Kalam Technical University
Uttar Pradesh, Lucknow, India

Tushar Sharma
Enhanced Oil Recovery Laboratory
Rajiv Gandhi Institute of Petroleum Technology
Jais, Amethi (UP) India

S.W. Sharshir
Mechanical Engineering Department
Faculty of Engineering
Kafrelsheikh University
Kafrelsheikh, Egypt

Rabesh Kumar Singh
Centre for Advanced Studies
Dr. A.P.J. Abdul Kalam Technical University
Uttar Pradesh, Lucknow, India

and

Department of Mechanical Engineering
Indian Institute of Technology (ISM)
Dhanbad, India

S.M. Sohel Murshed
Center for Innovation, Technology and Policy
    Research
Department of Mechanical Engineering
Instituto Superior Técnico
Universidade de Lisboa
Lisboa, Portugal

Harnish Soni
Department of Chemistry
Faculty of Science
The Maharaja Sayajirao University of Baroda
Vadodara, India

S. Srinivas
Department of Chemical Engineering
SSN College of Engineering
Kalavakkam, India

K.R.V. Subramanian
Department of Mechanical Engineering
GITAM School of Technology
GITAM (Deemed to be University)
Bangalore, India

L. Sun
International Research Center for Renewable
    Energy
State Key Laboratory of Multiphase Flow in
    Power Engineering
Xi'an Jiaotong University
Xi'an, China

A. Radhakrishna
ChemBioTech Research Ltd.
Bangalore, India

Tubati Nageswara Rao
Department of Mechanical Engineering
GITAM School of Technology
GITAM (Deemed to be University)
Bangalore, India

Miriam Teruel
Departamento de Química Física
Facultad de Ciencias
Universidad de Cádiz
Puerto Real (Cádiz), Spain

Arun Kumar Tiwari
Department of Mechanical Engineering
Institute of Engineering and Technology
Dr. A.P.J. Abdul Kalam Technical University
Uttar Pradesh, Lucknow, India

Alpaslan Turgut
Department of Mechanical Engineering
Dokuz Eylül University
Buca, Turkey

Kambiz Vafai
Mechanical Engineering Department
University of California
Riverside, California

Amir Varamesh
Institute of Petroleum Engineering
School of Chemical Engineering
College of Engineering
University of Tehran
Tehran, Iran

Xinzhi Wang
School of Energy Science and Engineering
Harbin Institute of Technology
and
Heilongjiang Key Laboratory of New Energy
    Storage Materials and Processes
Harbin, People's Republic of China

Z.B. Xing
Zhongtong Bus Holding Co., Ltd.
Liaocheng, China

and

College of Pipeline and Civil Engineering
China University of Petroleum
Qingdao, China

H.J. Xu
Research Center of Renewable Energy and
    Energy Storage
China-UK Low Carbon College
Shanghai Jiao Tong University
Shanghai, China

Hooman Yarmand
Centre for Energy Sciences
Department of Mechanical Engineering
University of Malaya
Kuala Lumpur, Malaysia

M.E. Zayed
Key Laboratory of Efficient Utilization of Low
    and Medium Grade Energy
Tianjin University
Ministry of Education
Tianjin, China

and

Mechanical Power Engineering Department
Faculty of Engineering
Tanta University
Tanta, Egypt

W. Zhang
College of pipeline and Civil Engineering
China University of Petroleum
Qingdao, China

Na Zhang
College of Petroleum Engineering
Liaoning Shihua University
Fushun, China

Zhien Zhang
Fujian Province University Key Laboratory of
    Green Energy and Environment Catalysis
and
Fujian Provincial Key Laboratory of Featured
    Materials in Biochemical Industry
Ningde Normal University
Ningde, China

Jun Zhao
Key Laboratory of Efficient Utilization of Low
    and Medium Grade Energy
Tianjin University
Ministry of Education
Tianjin, China

Z.Z. Zhou
Key Laboratory of Air-Driven Equipment
    Technology of Zhejiang Province
Quzhou University
Quzhou, China

Nurin Wahidah Binti Mohd Zulkifli
Centre for Energy Sciences
Department of Mechanical Engineering
University of Malaya
Kuala Lumpur, Malaysia

Section I

Understanding Nanofluids

1

Nanofluids: Preparation Methods and Challenges in Stability

P. Padmaja and Harnish Soni

CONTENTS

1.1 Introduction

A stable dispersion of nanometer-sized materials in base fluids have been termed as "nanofluids" by Choi in 1995 [1]. They are fabricated by dispersing nanometer-sized materials (nanoparticles [NPs], nanofibers, nanotubes, nanowires, nanorods, or nanosheets) into the base fluids [2]. The nanoparticles can be metallic (copper, silver, gold), nonmetallic, oxide (copper oxide, alumina, and titania), carbide, ceramics, carbonics, mixture of different nanoparticles (hybrid nanoparticles), and even nanoscale liquid droplets. Carbonics include carbon nanomaterials such as graphite [3], carbon nanotube (CNT) [4], graphene [5,6], graphene oxide (GO) [7,8], and other carbon allotropes [9].

The base fluid may be a low viscous liquid like water, refrigerant, or a high viscous liquid like ethylene glycol, mineral oil, or a mixture of different types of liquids (e.g. water, water/propylene glycol, etc.), ionic liquids [10], or eutectic fluids (eutectic mixture of diphenyl oxide and biphenyl) [11]. Unlike water-based nanofluids, which have a relative narrow operation temperature range due to the freezing and vaporization of water, oil-based nanofluids have much broader application temperatures.

Nanofluids have been found to possess enhanced thermophysical properties such as thermal conductivity, thermal diffusivity, viscosity, and convective heat transfer coefficients compared to those of pristine base fluids like oil or water. Nanofluids have demonstrated great potential applications in many fields as coolants due to their enhanced thermal properties.

However, when a single nanomaterial does not possess all the requisite characteristics for a particular application (it may either have only good thermal properties or rheological properties), a hybrid nanofluid is used to take advantage of the synergistic effect of both constituent nanomaterials. In this chapter, we will discuss the progress in the methods for preparing stable nanofluids.

1.2 Preparation Methods for Nanofluids

Two different techniques have been applied to produce nanofluids. One is a single-step method, and the other is a two-step method. Further, there are two ways to produce hybrid nanofluids: (a) addition of two different types of nanoparticles into base fluid and (b) fabrication of nanocomposites and dispersion into base fluid. The former method is less popular among the researchers, although the preparation process is very simple. By simply adding two different types of nanoparticles into a base fluid, one may not able to fully utilize improved synergistic physicochemical properties of the individual materials [12]. In the latter method, fabrication of nanocomposites is a tedious and complex process. The two different nanoparticles that are bound together must be stable enough to sustain its dispersion into base fluid. If they break apart during dispersion, they will become similar to adding two different nanoparticles into base fluid.

1.2.1 One-Step Method

The one-step method is a process wherein nanoparticle fabrication is combined with nanofluid synthesis. Several single-step methods have been arrived for nanofluids preparation.

1.2.1.1 *Vapor Deposition*

Direct evaporation and condensation of the nanoparticulate materials in the base liquid are done to produce stable nanofluids. A more complex method is evaporating bulk metal precursors which can be in the form of wires in an inert-gas atmosphere, which is then followed by vapor condensation. The size of nanoparticles depends on degree of heat applied for evaporation, as with increase in super heat, the particle size reduces. The particle size decreases with increase in pressure of the inert gas and also with decrease in wire diameter [13]. During vapor condensation, nanoparticle sizes and the particle size distribution are controlled by regulating the condensation conditions [14]. Other techniques that use metal evaporation include cathodic spraying or sputtering [13], plasma arc [15,16], laser ablation [17], lithography, radiolysis, and resistive heating [18–20].

The nanoparticles can also be produced by their condensation from the vapor phase into a flowing low vapor pressure fluid such as oil which is called "vacuum evaporation onto a running oil substrate" [21]. In this method, drying, storage, transportation, and dispersion of nanoparticles are avoided, so the agglomeration of nanoparticles is minimized and the stability of the nanofluids is increased [22]. A disadvantage of this method is that it is impossible to scale it up for industrial applications and is applicable only for low vapor pressure fluids.

Another approach is chemical vapor deposition, wherein vapor phase precursors are brought into a hot-wall reactor under conditions that favor nucleation of particles in the vapor phase rather than deposition of a film on the wall. M. A. Akhavan used chemical vapor deposition process to produce multi-walled carbon nanotubes (MWCNTs) nanofluids with deionized water [23].

1.2.1.2 Electrical Explosion of Wire

A promising one-step method is the electrical explosion of wire (EEW) in the desired base fluid. EEW technique uses high electric voltage and current to cause explosion of a thin metal wire in a container of base fluid to produce the nanofluid (Figure 1.1). Any type of thin conductive wire can be transformed into nanoparticles by this method. Sadegh Aberoumand and Amin Jafarimoghaddam prepared a colloidal suspension consisting of Cu NPs and engine oil by EEW in three different weight concentrations of 0.2%, 0.5%, and 1% [24].

A novel approach to EEW is μ-micro-electrical discharge machining technique wherein material is removed from both the electrodes through melting and evaporation. Sahu et al. synthesized copper nanofluids by μ-micro-electrical discharge machining technique [25]. Lo et al. introduced a novel system to prepare nanofluid using submerged arc nanoparticle synthesis system technique in which a pure metal rod is heated by a submerged arc formed by electrodes as depicted in Figure 1.2 [26].

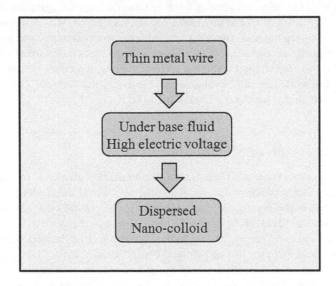

FIGURE 1.1 Schematic of the EEW method.

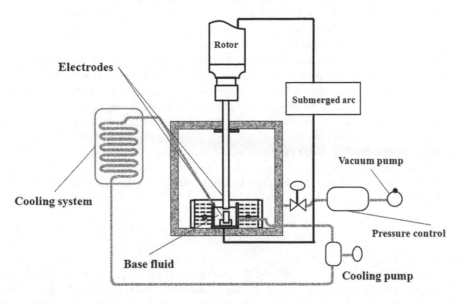

FIGURE 1.2 Submerged arc technique for nanofluid synthesis. (Adapted from Aberoumand, S. and Jafarimoghaddam, A., *J. Taiwan Inst. Chem. Eng.*, 71, 315–322, 2017.)

The main advantage of one-step method is low agglomeration of NPs, which leads to improved stability of nanofluids. Cost of drying and dispersion can be avoided using this method. However, there are some disadvantages for one-step method. Major disadvantage is that it is impossible to scale it up due to high cost of production, and it is compatible with only low vapor pressure base fluids [27]. Nine et al. reported the preparation of Cu_2O-water hybrid nanofluid by using wet ball milling process [28]. The ball milling accelerated hydrolysis of Cu particles in the presence of de-ionized water and several Cu particles transformed to Cu_2O nanoparticles at a temperature slightly higher than ambient temperature.

1.2.2 Two-Step Method

Two-step method is the most widely used method for preparing nanofluids. The nanomaterials are first produced as dry powders by chemical or physical methods and then dispersed into a fluid in the second processing step with the help of magnetic force agitation, ultrasonic agitation, high-shear mixing, homogenizing, and ball milling [29]. The process of drying, transportation, and storage of nanoparticles cannot be avoided in two-step method. Two-step method is the most economic method to produce nanofluids in large scale, because several nanopowder synthesis techniques have already been scaled up. The major disadvantage of two-step method is that the formation of agglomerates is easy due to high surface energy of nanoparticles, while this can be minimized in one-step method [30]. On the other hand, two-step method can be used almost with every kind of fluid [31].

1.3 Stability Mechanisms of Nanofluids

Preparing a stable and durable nanofluid is a prerequisite to make it amenable for various heat transfer applications. The main factor which makes the nanofluid unstable is the tendency of nanoparticles to coagulate/aggregate due to the presence of high surface charge present on them [32]. Aggregation of nanoparticles within the nanofluid can block the channels of heat exchanger used for heat transfer. Also, instability of the nanofluid can alter its thermophysical properties like thermal conductivity, viscosity, density, etc. with time, leading to the loss of potential benefits of nanofluids [33].

The stability of the nanofluids depends on the preparation methods, nanoparticle characteristics, type of base fluids, pH, and electrolyte [34]. The electrolytes (e.g. NaCl) can dramatically reduce the repulsive forces between NPs accelerating their flocculation and coagulation due to the increased rate of collision and coalescences of NPs in the suspension, leading to phase separation. Nanofluids have tremendous potential in a broad range of applications that use saline fluids, including pharmacy, medicine, water treatment, soil decontamination, or oil recovery and CO_2.

Stability of a nanofluid is directly associated with its electrokinetic properties determined by different forces such as Brownian, thermophoresis, drag, and repulsive forces that include steric and electrostatic repulsion. Nanoparticles suspended in a base fluid with a temperature gradient experience a force in the direction opposite to that of the temperature gradient, known as thermophoresis, while "drag" is the force generated in opposition to the direction of motion of a particle in a fluid and is proportional to the relative velocity between the base fluid and NP.

The Brownian motion arises as result of the collisions of molecules of the base fluid with the NPs. The Brownian motion counterbalances the gravitational sedimentation and helps disperse the NP into base fluids. In the stationary state, the sedimentation velocity of small spherical particles in a liquid follows the Stokes law [35]:

$$V = \frac{2R^2}{9\mu}\left(\rho_p - \rho_L\right)\cdot g \qquad (1.1)$$

where V is the particle's sedimentation velocity; R is the spherical particle's radius; μ is the dynamic viscosity of the fluid; ρ_p and ρ_L are the particle and the liquid medium density, respectively, and g is the acceleration of gravity. This equation reveals a balance of the gravity, buoyancy force, and viscous drag that are acting on the suspended nanoparticles. According to Eq. (1.1), the following measures can be taken to improve the stability of the nanofluids: (1) reducing R, the NP size; (2) increasing the base fluid viscosity; and (3) lessening the difference of density between the nanoparticles and the base fluid $(\rho_p - \rho_L)$. We can see from Eq. (1.1) that reducing the particle size should remarkably decrease the sedimentation speed of the nanoparticles and improve the stability of nanofluids as V is proportional to the R^2. According to the theory in colloid chemistry, when the size of particle decreases to a critical size, Rc, no sedimentation will take place because of the Brownian motion of nanoparticles (diffusion). However, smaller nanoparticles have a higher surface energy, resulting in an overall thermodynamically unstable colloid amenable for nanoparticle aggregation with time, and this is initiated by van der Waals forces [36,37].

Derjaguin, Landau, Verwey and OverbeeK (DLVO) developed a theory which dealt with colloidal stability [38]. DLVO theory suggests that the stability of a particle in solution is determined by the sum of van der Waals attractive and electrical double layer repulsive forces that exist between particles as they approach each other due to the Brownian motion they are undergoing. This interaction depends on the distance between particles and the total interface energy F_{DLVO} that is the sum of the van der Waals attraction and the electrostatic repulsion between them.

$$F_{DLVO} = k\pi\varepsilon\varepsilon_0 d_p \varphi_0^2 e^{-k(L)} - \frac{Ad_p}{24(L)^2} \qquad (1.2)$$

where ψ_0 is the surface potential (created by the surface charge), ε is the relative permittivity of base fluid, L is the separation distance between the particles, A is the Hamaker constant, d_p is the diameter of nanoparticle, k is inverse of Debye length ($k = 3.29 \times \sqrt{(c)}$ (nm^{-1}), (c) is molar concentration of monovalent electrolyte, and ε_0 is permittivity of free space ($\varepsilon_0 = 8.854 \times 10^{-12}$ c^2.m^{-1}) [39]. However, each of these forces has a particular range of effectiveness and some limitations to be effective. If the attractive force is larger than the repulsive force, the two particles will collide, and the suspension is not stable. For stable nanofluids, the repulsive forces between particles must be predominant. The attractive interaction energy between the nanoparticles depends on only the volume fraction, regardless of shape. However, the repulsive interaction energy depends strongly on the surface energy (zeta potential) and surface area, which vary with the shape of the particles. For instance, Kim et al. reported that alumina nanofluids with brick-shaped nanoparticles had the best suspension stability and showed the highest enhancement of thermal conductivity, while the nanofluids with blade-shaped nanoparticles were least stable and showed the lowest enhancement of thermal conductivity despite having the same volume faction and thermal conductivity [40]. To prepare a stable homogeneous suspension of nanofluids, the challenge is to overcome the strong van der Waals interactions between nanoparticles that induce the formation of aggregates [41].

The most widely applied methods for nanofluid stabilization are mechanical dispersion, electrostatic stabilization, and steric stabilization. In mechanical dispersion, a high energy force is applied within the nanofluid to break the clusters of nanoparticles. Electrostatic stabilization occurs when nanoparticles present in nanofluids attain some charge due to adsorption of ions. This results in an electrical double layer around nanoparticles which creates repulsive forces between nanoparticles. This is a pH sensitive method [33]. Chemical treatment such as addition of surfactant, polymers, surface treatment of nanoparticles, and change in pH of nanofluid are adopted for electrostatic and steric stabilization. Some of the researchers applied all of these methods to gain better stability.

1.4 Stabilization Techniques for Nanofluids

1.4.1 Mechanical Methods of Dispersion

Mechanical means of dispersion include physical techniques involving dispersing devices such as magnetic stirrer [42], homogenizer [43], or use of ultrasonic devices like ultrasonic bath, ultrasonic processor, etc. [44].

1.4.1.1 Ultrasonication

Mechanical mixing such as ultrasonication is effective in dispersing agglomerated nanoparticles into the base fluid [45]. Cavitation and bubbles are formed and collapsed during the process of ultrasonication. However, heat generated during the ultrasonication process increased the temperature of the nanofluid affecting various parameters like density, viscosity, thermal conductivity, stability, etc. [46]. Ideally, ultrasonication time must be optimized because excessive ultrasonication might introduce some defects [47] or may reduce the size of the nanoparticles. Ultrasonication bath- or probe-based ultrasonic devices are most commonly used. The probe-based ultrasonic devices operate at very high frequency. So, there may be the probability of contamination of nanofluids due to the detachment of very minute metal particles from the surface of metal probe affecting the stability of nanofluids. Such particles which are present in nanofluids may erode the heat transfer equipment surface during its application. CuO-water nanofluids have been prepared by ultrasonication [48].

Hybrid nanofluids which included the combinations of Ag/hydrogen exfoliated graphene /MWCNTs were prepared by Theres Baby et al. Catalytic chemical vapor deposition was used to prepare MWNTs and graphene from GO. Graphene was exfoliated by hydrogen and MWNTs were made to functionalize in an acidic solution of H_2SO_4, and HNO_3 followed by addition of $AgNO_3$ with stirring to form of Ag/hydrogen exfoliated graphene/MWCNT hybrids. The hybrid nanoparticles were dispersed in polar base fluids with the help of ultrasonicator [49].

A hybrid nano fluid suspension of phase change material (PCM) and Al_2O_3 was prepared by Ho et al. [50]. They used interfacial poly condensation together with emulsion technique to prepare PCM suspension. The core material in the micro encapsulated PCM particles was n-eicosane which was emulsified in water-soluble urea-formaldehyde per-polymer solution. Appropriate quantities of micro encapsulated PCM particles were mixed with ultra-pure Milli-Q water in a flask and ultrasonicated. Water-based Al_2O_3 nanofluid was prepared by dispersing Al_2O_3 nanoparticles in ultra-pure Milli-Q water using a magnetic stirrer and further mixed with PCM suspension in an ultrasonic vibration bath to give hybrid water-based nanofluid.

Han et al. fabricated a alumina/iron oxide hybrid sphere/CNT particle comprising of numerous CNTs attached to the hybrid sphere for applications in nanofluids. The hybrid sphere of alumina/iron oxide particles was synthesized by spray pyrolysis followed by catalytic growth of CNTs. These hybrid nanoparticles were dispersed in poly alpha-olefin with sonication and a small amount of surfactant was added to form stable nanofluids. They proposed that in such hybrid nanoparticles heat might transport rapidly from one CNT to another through the center sphere and thus lead to less thermal contact resistance between CNTs when compared to pristine CNTs dispersed in fluids [51].

1.4.1.2 Homogenization

High pressure/shear homogenization method is more efficient to break the agglomerates of nanoparticles in the nanofluid due to the combined effects of cavitation and high shear force [13] than ultrasonication.

Filho et al. prepared a silver/deionized water nanofluid which was found to be stable for more than 3 months [52]. A homogenizer consists of two micro-channels which divided the feed stream into two parts that again combined in the mixing chamber. There was a huge increase in velocity inside the mixing chamber because the diameter of the feed stream was very less which resulted in high shearing, large impact onto the walls of chamber, and formation of strong cavitations within the nanofluid. All these effects were responsible for the breaking of agglomerates present within the nanofluids.

Mechanical or sonic agitation is only effective in temporarily dispersing dilute water-based dispersion, so it is used as a secondary technique for preparation of nanofluids with long-term dispersion stability.

1.4.2 Chemical Treatment

Chemical methods include using surface modification, functionalization [53], acid treatment [54] to provide electrostatic/steric stabilization or change in polarity to the nanoparticles in the nanofluid and the aqueous organic process involving phase transfer [55] for synthesis of nanofluids.

Certain additives like surfactants/dispersants or polymers possess the ability to prevent the aggregation of nanoparticles dispersed in nanofluids. The surfactants cover the surface of nanoparticles with a long loop and tail which extend out into the nanofluids stabilizing them sterically and electrostatically for a long period of time either in polar or nonpolar solvents as the need arises [56,57]. The interactions between surfactants and nanoparticles are generally noncovalent interactions such as van der Waals, hydrogen bonding, hydrophobic, and $\pi-\pi$ interactions [58]. Figure 1.3 depicts the metal oxide nanoparticle functionalized with hydroxyl group enabling hydrogen bond formation with surfactant to form a stable nanofluid in nonpolar base fluid.

The surfactants are divided into four classes: nonionic surfactants without charge groups on its head (include polyethylene oxide, alcohols, and other polar groups), anionic surfactants with negatively charged head groups (include long-chain fatty acids, sulfosuccinates, alkyl sulfates, phosphates, and sulfonates), cationic surfactants with positively charged head groups (may include protonated long-chain amines and long-chain quaternary ammonium compounds), and amphoteric surfactants with zwitterionic head groups wherein charge depends on pH (such as betaines and certain lecithins). Gemini surfactants are a new class of surfactants comprising of two amphiphilic moieties connected with a spacer group at or near the head groups. Compared with conventional surfactants with an alkyl chain, gemini surfactants have the advantages of strong solubilization and adsorbability. In general, when the base fluid is polar, water-soluble surfactants are used, while for nonpolar base fluids, nonpolar surfactants are used. On the other hand, the solubility of nonionic surfactants is described by the hydrophilic/lipophilic balance value. The lower the hydrophilic/lipophilic balance number, the more oil soluble the surfactants, while a higher hydrophilic/lipophilic balance number surfactant will solubilize polar base fluids.

However, care should be taken to use appropriate amount of surfactant since use of less surfactant will not produce an efficient coating which creates an electrostatic repulsion and compensate the van der Waals attractions. Generally, a lot of surfactant is required to completely cover the particles, which affects the thermophysical properties of the nanofluids [33]. Further, surfactants may produce foam during heating in heat exchange systems. The functionality of the surfactants under high temperature is also a big concern, especially for high-temperature applications as bonding between surfactant and

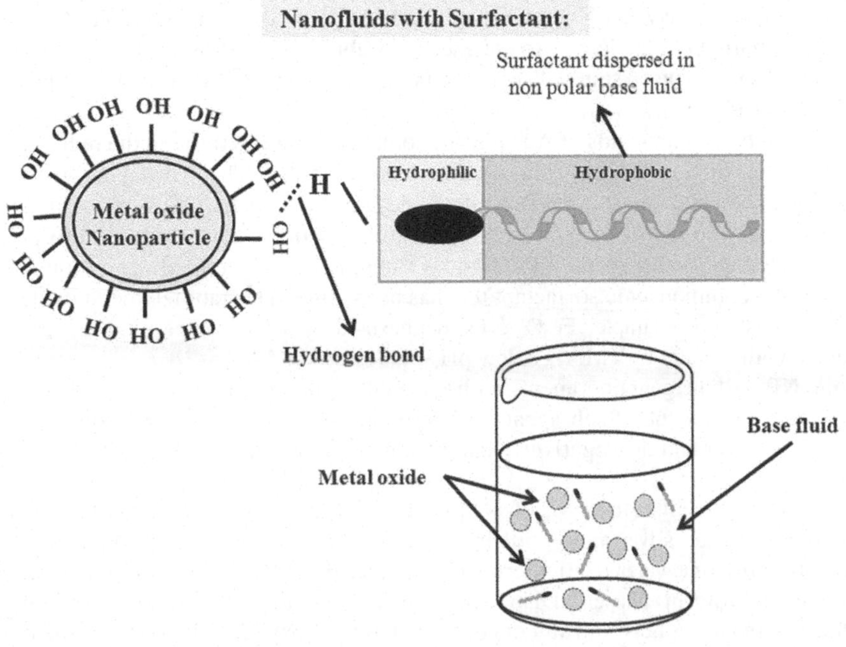

FIGURE 1.3 Stabilization of functionalized metal oxide nanoparticle in nonpolar base fluid by hydrogen bonding with a surfactant.

nanoparticles can be damaged, and they may contaminate the heat transfer media or fluids [52,59]. Therefore, the nanofluid will lose its stability and sedimentation of nanoparticles will occur.

In order to overcome the problems encountered during noncovalent functionalization especially with addition of surfactants, functional nanoparticles can be injected directly into the base fluids [60].

1.4.2.1 Steric Stabilization

For steric stabilization, polymers such as polymethylmethacrylate, Polyvinylpyrrolidone, and poly-acrylamide are added into the suspension system, and they will adsorb onto the particles surface, producing an additional steric repulsive force [61,62]. The steric effect of polymer dispersant is determined by the concentration of the dispersant. Grafting water miscible polymer chains onto the surface of NPs has also been proved to be able to suppress the inter-particle attraction. The grafted long chains could physically separate neighboring NPs, thus minimizing the van der Waals attraction force [63]. Generally, there are two grafting technologies: "grafting-to" and "grafting-from." The former one directly attaches polymer chains onto NP surface. The latter approach first anchors initiators, and then starts polymerization process to grow polymer chains out from the surface of NPs. The "grafting to" approach may hold better potential for the preparation of stably dispersed nanofluids at large scale because of its simplicity. The "grafting-from" approach is more complicated than "grafting-to" method, but it has better control over the grafting density and grafted chain length. The Fe_3O_4 NPs were modified with (3-mercaptopropyl) trimethoxysilane to obtain the Fe_3O_4-SH NPs. The attached initiators then started the polymerization of MMA monomers. The steric repulsion between the grafted polymethylmethacrylate chains rendered the grafted NPs excellent dispersion stability in organic solvents. However, the densely grafted polymer chains would also reduce the thermal energy transfer efficiency among different particles [64].

Nonionic surfactant mixtures of sorbitan trioleate and polysorbate have shown to be effective stabilizers of n-decane/water emulsions due to steric effects of their large, polar head groups [65].

1.4.2.2 Electrostatic Stabilization

Since the repulsive force between NPs is strongly influenced by the concentration of surface charge, electrostatic stabilization is a facile way to enhance the stability of nanofluids. This technique has been proved to be an effective route to stably disperse metal, metal oxide, and ceramic NPs in base fluids with relatively high polarities such as water.

To prepare water-based nanofluids of Au NPs, the common method is through the reduction of the corresponding salt of $HAuCl_4$ with sodium citrate solution under reflux. The hydroxyl groups in the citrate ions could efficiently reduce Au^{3+} to Au NPs while the negatively charged COO– groups of the citrate ions electrostatically bound to the surface of NPs [66]. Similar methods have been used to prepare stable nanofluids dispersed with other metal NPs. Besides sodium citrate, tetramethyl ammonium hydroxide (TMAH) is another common ionic surfactant that has been utilized to stabilize metal oxide NPs during the synthesis process. For example, Fe_3O_4 NPs synthesized by a co-precipitation technique could be stably dispersed within water by TMAH below pH ~ 10 [67].

For ceramic NPs, silane coupling agents with ionizable amino group were used to electrostatically stabilize the NPs in water [68]. Such agents can form silanol groups that can tightly bind to the NP surfaces and at the other end the negatively charged amino group can electrostatically repel the neighboring NPs.

Cationic gemini surfactants (16-6-16, 14-6-14, and 12-6-12) were used as capped stabilizers to synthesize stable water-based silver nanofluids by Dan Li et al. [69]. The prepared silver nanoparticles could be transferred from water to dichloromethane, and the NPs were stably dispersed in dichloromethane and chloroform phases. Dispersing agents including 4-benzenediazonium sulfonate, 1-pyrenebutyrate, and tetrapotassium salt of coronene tetracarboxylic acid could significantly improve the dispersion of graphene within water. These treating agents have an aromatic ring to attach onto the basal plane of graphene and a water-interacting end group to enable dispersing graphene within water [70].

A mixture of high concentration of acids (nitric, sulfuric, or hydrochloric acid or combination of them) has been used to covalently functionalize carbon nanomaterials such as carbon nanofibers (CNFs) and CNTs. During the acid treatment, carboxyl (–COOH) functional groups are attached on CNFs as well as CNTs and are broken into shorter and less twisted shreds resulting in higher stability and better dispersion. The van der Waals forces among the nanoparticles were eliminated as the ionization of acidic surface group on the CNFs and CNTs increased the electrostatic repulsion among the particles and made the CNFs and CNTs hydrophilic [71,72].

1.4.2.3 Change of pH

Controlling pH can lead to electrostatic stabilization through a high surface charge density due to strong repulsive forces. At the isoelectric point pH, the surface charge density equals the charge density of the bulk solution and the zeta potential (the potential difference between the surface of a solid particle immersed in a base fluid and the bulk of the fluid) is zero (zero point charge,). Therefore, the charge density in this layer is zero. At this pH the nanofluids become unstable as the repulsive forces between NPs suspended in base fluid are zero [73]. As the pH of the solution departs from the isoelectric point of particles, zeta potential becomes large and the nano particles get more stable. For instance, it was observed that Cu NPs modified with sodium dodecyl benzene sulfonate have improved dispersion stability within water when the pH of the solution was increased from 2 to 9.5 [74]. Peterson et al. investigated the effect of pH on the stability of Al_2O_3 nanofluids. The experiments indicated that the nanofluid was more stable at a pH of 1.7 than at 7.66. When Al_2O_3 particles are immersed in water, hydroxyl groups (–OH) are produced at the surface of the Al_2O_3 particle. When the pH of the solution is lower than the point of zero charge (PZC), the hydroxyl groups react with H^+ from water, which leads to a positively charged surface. On the other hand, when the pH of the solution was higher than the PZC, the hydroxyl groups reacted with OH^- from water and created a negatively charged surface [75].

1.4.2.4 Electrosteric Stabilization

Some researchers have adapted both electrostatic and steric mechanisms for stabilization of nanofluids which can be termed as electrosteric stabilization. Sen et al. developed a one-step surface modification technique to functionalize the TiO_2 NPs with 3-(trihydroxysilyl)-1-propanesulfonic acid [64]. The silanol groups could bind to the TiO_2 NP surface, and the propyl sulfonate groups ensured good dispersion stability of the modified NPs through steric and charge-based electrostatic stabilization.

Electrosteric stabilization can be also achieved with the use of polyelectrolytes. Kiruba et al prepared gamma Al_2O_3 nanofluids with cationic polyelectrolyte additives [76].

In another case, Zhang et al. transferred hydrophobic Fe_2O_3 and TiO_2 NPs into water by a ligand exchange approach. The original hydrophobic oleic acid ligands were replaced by polyelectrolytes such as poly (acrylic acid) and poly (allylamine) at a high temperature in a polyol solvent. The polyelectrolytes have high charge density as well as multiple strong binding sites facilitating the attachment of polymer chains onto NP surface. The water-based nanofluids prepared from this route have demonstrated excellent dispersion stability owing to the dual electrostatic and steric stabilization mechanisms from the anchored polyelectrolyte ligands [77]. While chemical surface modification offers general and effective tools to enhance the dispersion stability of nanofluids, this approach has some limitations in particular for scalable production because the treatment process is often time-consuming and is usually only effective for small scale preparations.

In another example, phosphate-terminated polydimethylsiloxane chains were grafted onto Fe_3O_4 NP surfaces that were synthesized in benzyl alcohol to provide steric stabilization against the strong van der Waals attraction between NPs within the silicone oil base fluids. Meanwhile, the same chemical composition of the grafted chains as the base silicone oil fluid ensured both excellent miscibility with the base fluid and good thermal stability of the grafted chains. The synthesized NPs were weakly capped by benzyl alcohol, the synthetic solvent, and thus could be easily replaced with phosphate-terminated chains [78].

1.4.2.5 Change of Polarity by Functionalization

Through plasma treatment using gas mixtures of methane and oxygen, various polar groups were imparted on the surface of the diamond nanoparticles, improving their dispersion property in water [79].

A stable dispersion of titania nanoparticles was prepared by surface modification with silane coupling agents, (3-acryl-oxypropyl) trimethoxysilane and trimethoxypropylsilane in an organic solvent of diethylene glycol dimethylether using a ball milling process [80].

To avoid the aggregation of silica nanoparticles, surface modification was attempted with vinyltriethoxysilane and 2-mercaptobenzimidazole by Caili Dai et al. in order to fabricate a new kind of self-dispersing silica nanoparticle for enhanced oil recovery applications [81].

Oil phase ligands such as fatty acids or amines are often employed to mediate the growth of NPs and cap the synthesized NPs, resulting in uniform dispersion within organic solvents. To disperse the nanoparticles back in water, exchanging the synthetic oil phase ligands with ionizable ligands of stronger binding capability could effectively disperse the synthesized colloidal NPs within water. For instance, oleylamine-capped Au and Ag NPs were post-treated with 11-mercaptoundecanoic acid, which has an aliphatic thiol group to replace the oleylamine ligands and carboxylic acid groups to help disperse the modified NPs within water [82].

Alternatively, the NPs synthesized within oil phase could be re-dispersed in water by adding a layer of surfactant to form a double layer structure on the surface of NPs [83]. It has been observed that surfactants of the same chemistry (e.g. lauric acid or myristic acid) favored dense interdigitation between chains within two layers and longer chain favored the formation of thicker bilayer providing better steric stabilization of the nano fluids. For instance, oleylamine-capped Au and Ag NPs have been stabilized by adding a second surfactant layer that has a hydrophilic end and a hydrophobic chain to form a double layer structure on the surface of nanoparticles. The hydrophilic component extends into water aiding in dispersibility of the nanoparticles in water [84].

1.4.2.6 Phase-Transfer

The method is also a facile way to obtain monodisperse nanofluids [85]. In a water cyclohexane two-phase system, aqueous formaldehyde is transferred to cyclohexane phase via reaction with dodecylamine to form reductive intermediates in cyclohexane that are capable of reducing silver or gold ions in aqueous solution to form dodecylamine-protected silver and gold nanoparticles in cyclohexane. Similarly, homogeneous and stable graphene oxide colloids were prepared by transferring from water to n-octane after modification by oleylamine. Schematic illustration of the phase transfer process is shown in Figure 1.4 [86].

1.4.2.7 Combination of Methods

Researchers have adopted both chemical and physical techniques to prepare stable nanofluids. Batmunkh et al. applied two-step method to prepare TiO_2/Ag-based water nanofluids. Initially, a TiO_2 composite was prepared comprising of 70% small (15 nm) and 30% large (300 nm) TiO_2 nanoparticles under magnetic stirring [87]. Flat Ag nanoparticles were fabricated by ball milling of Ag nanoparticles and were then ultrasonically dispersed into the TiO_2 composite suspension.

In another study, the in-situ growth and chemical co-precipitation method was used for the synthesis of uniform dispersion of Co_3O_4 nanoparticles on GO nanosheets. The reduction of aqueous $CoCl_2$ in the presence of GO with sodium borohydrate resulted in the formation of hybrid GO/Co_3O_4 nanoparticles. The hybrid nanofluids were prepared by dispersing synthesized GO/Co_3O_4 nanoparticles in water, ethylene glycol, and ethylene glycol/water mixtures of different weight ratios [88].

Soner Gumus prepared diesel-based nanofluids by direct mixing of CuO, Al_2O_3 nanoparticles, and diesel fuel as base fluid and ultrasonicating for 1 hour at 40 Hz [89]. Ammonium polymethacrylate and sodium silicate ($Na_2O_7Si_3$) were used as dispersants for preventing particle agglomeration during formation of nanofluids.

In another instance, GO surfaces were functionalized with hydroxyl and epoxy groups on the basal plane and carboxyl groups mainly at the edges enabling a subsequent chemical reaction between these

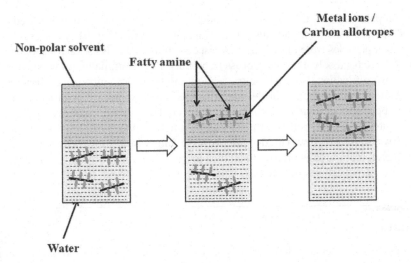

FIGURE 1.4   Phase transfer technique for preparation of nanofluids. (Adapted from Williams, O. A. et al., *ACS Nano*, 4, 4824–4830, 2010.)

functional groups and surface modifier. After surface modification, the GO nanosheets could be reduced into graphene. Further, various functional moieties and polymer chains could be covalently anchored onto the surface of graphene, and the resultant graphene nanosheets could be readily dispersed within a variety of solvents [90].

Williams et al. performed gas annealing to modify the surface of particles with hydrogen, which prevented cluster formation in buffer solution and was amenable for chemical surface modifications. Dispersion of these particles into water via high power ultrasound and high speed centrifugation resulted in a monodisperse nanodiamond colloid high positive zeta potential (>60 mV), with exceptional stability in a wide range of pH for a long period of time [91].

A. Shahsavar prepared hybrid nanofluids containing Fe_3O_4 and carbon nanotube nanoparticles by ultrasonication and further emulsification and dispersion using gum arabic and TMAH [92]. Physisorption between the TMAH-coated magnetic nanoparticles and gum arabic-coated nanotubes resulted in the formation of nanofluids. Gaweł Zyła prepared yttrium aluminium garnet ($Y_3Al_5O_{12}$– ethylene glycol) nanofluids by two-step method with mechanical stirring and ultrasonication [93]. Paul et al. synthesized hybrid Al-Zn nanofluid in ethylene glycol by two-step method. The Al-Zn nanoparticles were prepared by mechanical alloying, wherein they blended powders of elemental aluminium (95%) and zinc (5%) at room temperature using a high energy planetary ball mill [94] for sufficient time to get a homogeneous blend. The hybrid nanofluids were then prepared by adding ultra-fine Al-Zn nanoparticles in appropriate quantity to ethylene glycol and subjecting the resultant mixture to ultrasonic vibration followed by magnetic stirring.

Megatif et al. prepared water-based TiO_2-CNT hybrid NPs, wherein TiO_2 NPs were prepared by sol gel technique [95]. Firstly, the CNTs were treated with acid and dispersed into water at pH 1.5 by ultrasonication. Next, ethylene glycol and 2-propanol were added to the dispersion under stirring followed by the addition of $Ti(OBu)_4$. The resulting precipitates were then vacuum-filtered, washed, and dried in air. Lastly, the powder was calcined to produce the desired CNT/TiO_2 hybrid nanoparticles which were dispersed in water.

Among various oil fluids, silicone oil is one of the technologically important base fluids that can stably operate from −50°C to more than 200°C. However, the chemical inertness and low surface energy of the silicone oil base fluids pose great challenge to disperse NPs within them, in particular under high operation temperatures.

Yogo et al. synthesized polymer hybrid by mixing $PbTiO_3$ with silicone oil. $PbTiO_3$ was synthesized from titanium isopropoxide, lead acetate, and 2-(methacryloyloxy) ethyl acetoacetate by in situ hydrolysis using de-ionized water [96].

Jeong et al. prepared Ag/MWCNT hybrid nanofluids. The wet grinding method was used for the synthesis of MWCNTs/water nanofluids and pulsed wire evaporation method for the synthesis of silver nanoparticles. The Ag nanoparticles were further made to disperse in the MWCNTs/water nanofluids [97].

Abbasi et al. have used solvothermal process to produce CNT/gamma alumina nanofluids. The pristine MWCNTs were functionalized with nitric acid or sulphuric acid-nitric acid mixture and added to aluminium acetate powder dispersed in absolute ethanol and ultrasonicated. Ammonia solution was then added to obtain boehmite particles and was further subjected to solvothermal treatment in a Teflon lined stainless steel auto-clave chamber. Finally, the obtained hybrid nanoparticles were washed, dried, and calcined [98]. Two-step method was used to prepare nanofluids where gum arabic was used as surfactant. The oleic acid-coated alumina is prepared and then dispersed in the oil by ultrasonication to overcome the aggregation of nanoparticles in base fluid [99].

1.5 Summary

Recent advances in nanotechnology have led to the development of a variety of methods to synthesize nanofluids from a diverse range of NPs that include metals, ceramics, metal oxides, etc. The stability of the nanofluids depends on the morphology and surface properties of the NPs, the nature of base fluid, and synthesis methods of the nanofluids. The different techniques adopted and discussed are summarized in Figure 1.5.

Chemical and mechanical treatments have been shown to be effective to disentangle the NP aggregates and to enhance the dispersibility of NPs for preparing homogeneous and stable nanofluids.

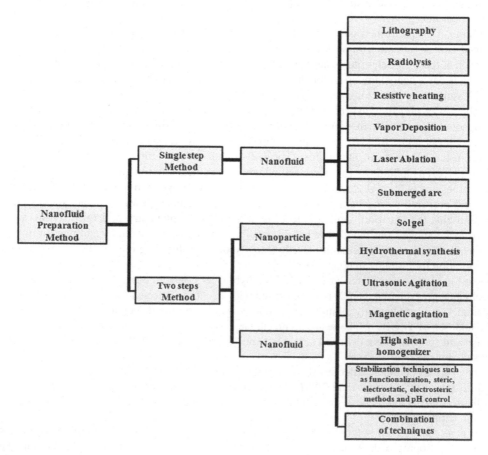

FIGURE 1.5 Techniques for the synthesis of nanofluids.

Functional groups (carbonyl group, hydroxyl group, polymer, or surfactant molecule) attached covalently or noncovalently onto the nanoparticles after these treatments have proved to be beneficial to the dispersibility of the nanoparticles in the base fluids. The functional groups changed the surface characteristics of the nanoparticles and increased the repulsive forces between them resulting in stable nanofluids. From the literature reviews on preparation of nanofluids, it was found that the stability of suspension of nanoparticles is better when the nanofluid is prepared by the one-step method as compared to the two-step method. However, the preparation of nanofluids by one-step method is difficult and expensive relative than two-step method. To get a stable nanofluid pH is also an important parameter. When pH values results are far away from the isoelectric points of the prepared nanofluid, the prepared nanoparticles were observed to be well dispersed. *Nanofluids* with *zeta potential* values greater than +25 mV or less than −25 mV typically have *high* degrees of stability.

Nanofluids have a number of applications in a number of industrial processes, wherein the stability of nanofluids should be at least equivalent to the life span of the equipment or the process cycle. The preparation of such a stable nanofluid is still a challenge. Readers can refer to the reviews that have critically discussed the recent advances in preparation, stabilization, and applications of different categories of nanofluids. Babita Sharma et al. have reviewed the advances and modifications suggested by various researchers in the preparation and analytical techniques used for formation of stable nanofluids during the last two decades [100]. Sadeghinezhad et al. summarized the recent progress on preparation and evaluation methods for the preparation of stable graphene nanofluids [5].

Yu et al. gave a comprehensive overview of the recent efforts and progresses in improving the dispersion stability of thermal nanofluids including the mechanistic understanding of dispersion behavior of nanofluids, strategies to stabilize nanofluids, and characterization techniques for dispersion behavior of nanofluids [84]. Several researchers have reviewed the latest developments in the synthesis and stabilization of hybrid nanofluids [101,102].

REFERENCES

1. Choi, S. U. S. (2009). Nanofluids: From vision to reality through research. *Journal of Heat Transfer* 131, 33106–33109.
2. Moghadassi, A., Ghomi, E., & Parvizian, F. (2015). A numerical study of water based Al₂O₃ and Al₂O₃–Cu hybrid nanofluid effect on forced convective heat transfer. *International Journal of Thermal Sciences* 92, 50–57.
3. Ladjevardi, S. M., Asnaghi, A., Izadkhast, P. S., & Kashani, A. H. (2013). Applicability of graphite nanofluids in direct solar energy absorption. *Solar Energy* 94, 327–334.
4. Aravind, S. S. J., Baskar, P., Baby, T. T., Sabareesh, R. K., Das, S., & Ramaprabhu, S. (2011). Investigation of structural stability, dispersion, viscosity, and conductive heat transfer properties of functionalized carbon nanotube based nanofluids. *The Journal of Physical Chemistry C* 115, 16737–16744.
5. Sadeghinezhad, E., Mehrali, M., Saidur, R., Mehrali, M., Latibari, S. T., Akhiani, A. R., & Metselaar, H. S. C. (2016). A comprehensive review on graphene nanofluids: Recent research, development and applications. *Energy Conversion and Management* 111, 466–487.
6. Agarwal, D. K., Vaidyanathan, A., & Sunil Kumar, S. (2016). Experimental investigation on thermal performance of kerosene–graphene nanofluid. *Experimental Thermal and Fluid Science* 71, 126–137.
7. Ijam, A., Saidur, R., Ganesan, P., & Golsheikh, A. M. (2015). Stability, thermo-physical properties, and electrical conductivity of graphene oxide-deionized water/ethylene glycol based nanofluid. *International Journal of Heat and Mass Transfer* 87, 92–103.
8. Yu, W., Xie, H., & Bao, D. (2010). Enhanced thermal conductivities of nanofluids containing graphene oxide nanosheets. *Nanotechnology* 21, 55705.
9. Han, D., Meng, Z., Wu, D., Zhang, C., & Zhu, H. (2011). Thermal properties of carbon black aqueous nanofluids for solar absorption. *Nanoscale Research Letters* 6, 457.
10. Liu, J., Xu, C., Chen, L., Fang, X., & Zhang, Z. (2017). Preparation and photo-thermal conversion performance of modified graphene/ionic liquid nanofluids with excellent dispersion stability. *Solar Energy Materials and Solar Cells* 170, 219–232.

11. Sánchez-Coronilla, A., Martín, E. I., Navas, J., Aguilar, T., Gómez-Villarejo, R., Alcántara, R. et al. (2018). Experimental and theoretical analysis of NiO nanofluids in presence of surfactants. *Journal of Molecular Liquids* 252, 211–217.

12. Kumar, M. S., Vasu, V., & Gopal, A. V. (2016). Thermal conductivity and rheological studies for Cu–Zn hybrid nanofluids with various basefluids. *Journal of the Taiwan Institute of Chemical Engineers* 66, 321–327.

13. Hwang, Y., Lee, J.-K., Lee, J.-K., Jeong, Y.-M., Cheong, S., Ahn, Y.-C., & Kim, S. H. (2008). Production and dispersion stability of nanoparticles in nanofluids. *Powder Technology* 186, 145–153.

14. Chamsa-ard, W., Brundavanam, S., Fung, C. C., Fawcett, D., & Poinern, G. (2017). Nanofluid types, Their synthesis, properties and incorporation in direct solar thermal collectors: A review. *Nanomaterials* 7, 131.

15. Chang, H., & Chang, Y.-C. (2008). Fabrication of Al$_2$O$_3$ nanofluid by a plasma arc nanoparticles synthesis system. *Journal of Materials Processing Technology* 207, 193–199.

16. Teng, T.-P., Cheng, C.-M., & Pai, F.-Y. (2011). Preparation and characterization of carbon nanofluid by a plasma arc nanoparticles synthesis system. *Nanoscale Research Letters* 6, 293.

17. Kim, H. J., Bang, I. C., & Onoe, J. (2009). Characteristic stability of bare Au-water nanofluids fabricated by pulsed laser ablation in liquids. *Optics and Lasers in Engineering* 47, 532–538.

18. Zhang, G., & Wang, D. (2008). Fabrication of heterogeneous binary arrays of nanoparticles via colloidal lithography. *Journal of the American Chemical Society* 130, 5616–5617.

19. Treguer, M., de Cointet, C., Remita, H., Khatouri, J., Mostafavi, M., Amblard, J. et al. (1998). Dose rate effects on radiolytic synthesis of gold–silver bimetallic clusters in solution. *The Journal of Physical Chemistry B* 102, 4310–4321.

20. Hummelgård, M., Zhang, R., Nilsson, H.-E., & Olin, H. (2011). Electrical sintering of silver nanoparticle ink studied by in-situ TEM probing. *PLoS ONE* 6, e17209.

21. Akoh, H., Tsukasaki, Y., Yatsuya, S., & Tasaki, A. (1978). Magnetic properties of ferromagnetic ultrafine particles prepared by vacuum evaporation on running oil substrate. *Journal of Crystal Growth* 45, 495–500.

22. Li, Y., Zhou, J., Tung, S., Schneider, E., & Xi, S. (2009). A review on development of nanofluid preparation and characterization. *Powder Technology* 196, 89–101.

23. Akhavan-Behabadi, M. A., Shahidi, M., & Aligoodarz, M. R. (2015). An experimental study on heat transfer and pressure drop of MWCNT–water nano-fluid inside horizontal coiled wire inserted tube. *International Communications in Heat and Mass Transfer* 63, 62–72.

24. Aberoumand, S., & Jafarimoghaddam, A. (2017). Experimental study on synthesis, stability, thermal conductivity and viscosity of Cu–engine oil nanofluid. *Journal of the Taiwan Institute of Chemical Engineers* 71, 315–322.

25. Sahu, R. K., Hiremath, S. S., & Manivannan, P. V. (2015). Ultrasonic technique for concentration characterization of copper nanofluids synthesized using μ-EDM: A novel experimental approach. *Powder Technology* 284, 429–436.

26. Lo, C., Tsung, T., Chen, L., Su, C., & Lin, H. (2005). Fabrication of copper oxide nanofluid using submerged arc nanoparticle synthesis system (SANSS). *Journal of Nanoparticle Research* 7, 313–320.

27. Prakash, S. B., Kotin, K. N., & Praveen Kumar, M. (2015). Preparation and characterization of Nanofluid (CuO – Water, TiO2 – Water). *International Journal of Science and Engineering* 1, 14–20.

28. Nine, M. J., Munkhbayar, B., Rahman, M. S., Chung, H., & Jeong, H. (2013). Highly productive synthesis process of well dispersed Cu$_2$O and Cu/Cu$_2$O nanoparticles and its thermal characterization. *Materials Chemistry and Physics* 141, 636–642.

29. Sreelakshmy, K. R., Ashwathy, S. N., Vidhya, K. M., Saranya, T. R., & Sreeja, C. N. (2014). An overview of recent nanofluid research. *International Research Journal of Pharmacy* 5, 239–243.

30. Mohammed, H. A., Al-aswadi, A. A., Shuaib, N. H., & Saidur, R. (2011). Convective heat transfer and fluid flow study over a step using nanofluids: A review. *Renewable and Sustainable Energy Reviews* 15, 2921–2939.

31. Wang, X.-Q., & Mujumdar, A. S. (2007). Heat transfer characteristics of nanofluids: A review. *International Journal of Thermal Sciences* 46, 1–19.

32. Ghadimi, A., Saidur, R., & Metselaar, H. S. C. (2011). A review of nanofluid stability properties and characterization in stationary conditions. *International Journal of Heat and Mass Transfer* 54, 4051–4068.

33. Mukherjee, S., & Paria, S. (2013). Preparation and stability of nanofluids: A review. *IOSR Journal of Mechanical and Civil Engineering* 9, 63–69.

34. Chen, W., Zou, C., Li, X., & Li, L. (2017). Experimental investigation of SiC nanofluids for solar distillation system: Stability, optical properties and thermal conductivity with saline water-based fluid. *International Journal of Heat and Mass Transfer* 107, 264–270.

35. Hiemenz, P. C., & Dekker, M. (1986). *Principles of Colloid and Surface Chemistry*, 2nd ed. New York: Dekker.

36. Liu, L., Wang, L., Zhu, H., & Wu, D. (2009). Critical issues in nanofluids preparation, characterization and thermal conductivity. *Current Nanoscience* 5, 103–112.

37. Azizian, R., Doroodchi, E., & Moghtaderi, B. (2015). Influence of controlled aggregation on thermal conductivity of nanofluids. *Journal of Heat Transfer* 138, 21301–21306.

38. Popa, I., Gillies, G., Papastavrou, G., & Borkovec, M. (2010). Attractive and repulsive electrostatic forces between positively charged latex particles in the presence of anionic linear polyelectrolytes. *The Journal of Physical Chemistry B* 114, 3170–3177.

39. Rhodes M. J. (2008). Colloids and fine particles. In: *Introduction to Particle Technology*, 2nd ed. John Wiley & Sons.

40. Kim, H. J., Lee, S.-H., Lee, J.-H., & Jang, S. P. (2015). Effect of particle shape on suspension stability and thermal conductivities of water-based bohemite alumina nanofluids. *Energy* 90, 1290–1297.

41. Goodwin, J. (2009). *Colloids and Interfaces with Surfactants and Polymers.* John Wiley & Sons.

42. Kavitha, T., Rajendran, A., & Durairajan, A. (2012). Synthesis, characterization of TiO_2 nano powder and water based nanofluids using two step method. *European Journal of Applied Engineering and Scientific Research* 1, 235–240.

43. Fontes, D. H., Ribatski, G., & Filho, E. P. B. (2015). Experimental evaluation of thermal conductivity, viscosity and breakdown voltage AC of nanofluids of carbon nanotubes and diamond in transformer oil. *Diamond and Related Materials* 58, 115–121.

44. Ahammed, N., Asirvatham, L. G., & Wongwises, S. (2016). Thermoelectric cooling of electronic devices with nanofluid in a multiport minichannel heat exchanger. *Experimental Thermal and Fluid Science* 74, 81–90.

45. Martínez, V. A., Vasco, D. A., & García–Herrera, C. M. (2018). Transient measurement of the thermal conductivity as a tool for the evaluation of the stability of nanofluids subjected to a pressure treatment. *International Communications in Heat and Mass Transfer* 91, 234–238.

46. Sadri, R., Ahmadi, G., Togun, H., Dahari, M., Kazi, S. N., Sadeghinezhad, E., & Zubir, N. (2014). An experimental study on thermal conductivity and viscosity of nanofluids containing carbon nanotubes. *Nanoscale Research Letters* 9, 151.

47. Amrollahi, A., Hamidi, A. A., & Rashidi, A. M. (2008). The effects of temperature, volume fraction and vibration time on the thermo-physical properties of a carbon nanotube suspension (carbon nanofluid). *Nanotechnology* 19, 315701.

48. Venkatachalapathy, S., Kumaresan, G., & Suresh, S. (2015). Performance analysis of cylindrical heat pipe using nanofluids: An experimental study. *International Journal of Multiphase Flow* 72, 188–197.

49. Theres Baby, T., & Sundara, R. (2013). Synthesis of silver nanoparticle decorated multiwalled carbon nanotubes-graphene mixture and its heat transfer studies in nanofluid. *AIP Advances* 3, 012111.

50. Ho, C. J., Huang, J. B., Tsai, P. S., & Yang, Y. M. (2011). On laminar convective cooling performance of hybrid water-based suspensions of Al_2O_3 nanoparticles and MEPCM particles in a circular tube. *International Journal of Heat and Mass Transfer* 54, 2397–2407.

51. Han, Z. H., Yang, B., Kim, S. H., & Zachariah, M. R. (2007). Application of hybrid sphere/carbon nanotube particles in nanofluids. *Nanotechnology* 18, 105701.

52. Filho, E. P. B., Mendoza, O. S. H., Beicker, C. L. L., Menezes, A., & Wen, D. (2014). Experimental investigation of a silver nanoparticle-based direct absorption solar thermal system. *Energy Conversion and Management* 84, 261–267.

53. Hwang, Y.-J., Lee, J. K., Lee, C. H., Jung, Y. M., Cheong, S. I., Lee, C. G. et al. (2007). Stability and thermal conductivity characteristics of nanofluids. *Thermochimica Acta* 455, 70–74.

54. Yarmand, H., Gharehkhani, S., Shirazi, S. F. S., Amiri, A., Alehashem, M. S., Dahari, M., & Kazi, S. N. (2016). Experimental investigation of thermo-physical properties, convective heat transfer and pressure drop of functionalized graphene nanoplatelets aqueous nanofluid in a square heated pipe. *Energy Conversion and Management* 114, 38–49.

55. Feng, X., Ma, H., Huang, S., Pan, W., Zhang, X., Tian, F. et al. (2006). Aqueous–organic phase-transfer of highly stable gold, silver, and platinum nanoparticles and new route for fabrication of gold nanofilms at the oil/water interface and on solid supports. *The Journal of Physical Chemistry B* 110, 12311–12317.

56. Das, P. K., Mallik, A. K., Ganguly, R., & Santra, A. K. (2018). Stability and thermophysical measurements of TiO2 (anatase) nanofluids with different surfactants. *Journal of Molecular Liquids* 254, 98–107.

57. Choi, T. J., Jang, S. P., & Kedzierski, M. A. (2018). Effect of surfactants on the stability and solar thermal absorption characteristics of water-based nanofluids with multi-walled carbon nanotubes. *International Journal of Heat and Mass Transfer* 122, 483–490.

58. Sarsam, W. S., Amiri, A., Kazi, S. N., & Badarudin, A. (2016). Stability and thermophysical properties of non-covalently functionalized graphene nanoplatelets nanofluids. *Energy Conversion and Management* 116, 101–111.

59. Nasiri, A., Shariaty-Niasar, M., Rashidi, A., Amrollahi, A., & Khodafarin, R. (2011). Effect of dispersion method on thermal conductivity and stability of nanofluid. *Experimental Thermal and Fluid Science* 35, 717–723.

60. Amiri, A., Sadri, R., Shanbedi, M., Ahmadi, G., Chew, B. T., Kazi, S. N., & Dahari, M. (2015). Performance dependence of thermosyphon on the functionalization approaches: An experimental study on thermo-physical properties of graphene nanoplatelet-based water nanofluids. *Energy Conversion and Management* 92, 322–330.

61. Kumar, R. S., & Sharma, T. (2018). Stability and rheological properties of nanofluids stabilized by SiO_2 nanoparticles and SiO_2-TiO_2 nanocomposites for oilfield applications. *Colloids and Surfaces A: Physicochemical and Engineering Aspects* 539, 171–183.

62. Zhu, H., Zhang, C., Tang, Y., Wang, J., Ren, B., & Yin, Y. (2007). Preparation and thermal conductivity of suspensions of graphite nanoparticles. *Carbon* 45, 226–228.

63. Li, Y., Krentz, T. M., Wang, L., Benicewicz, B. C., & Schadler, L. S. (2014). Ligand engineering of polymer nanocomposites: From the simple to the complex. *ACS Applied Materials & Interfaces* 6, 6005–6021.

64. Sen, S., Govindarajan, V., Pelliccione, C. J., Wang, J., Miller, D. J., & Timofeeva, E. V. (2015). Surface modification approach to TiO_2 nanofluids with high particle concentration, low viscosity, and electrochemical activity. *ACS Applied Materials & Interfaces* 7, 20538–20547.

65. Saarinen, S., Puupponen, S., Meriläinen, A., Joneidi, A., Seppälä, A., Saari, K., & Ala-Nissila, T. (2015). Turbulent heat transfer characteristics in a circular tube and thermal properties of n-decane-in-water nanoemulsion fluids and micelles-in-water fluids. *International Journal of Heat and Mass Transfer* 81, 246–251.

66. Leng, W., Pati, P., & Vikesland, P. J. (2015). Room temperature seed mediated growth of gold nanoparticles: Mechanistic investigations and life cycle assesment. *Environmental Science: Nano* 2, 440–453.

67. Abareshi, M., Goharshadi, E. K., Zebarjad, S. M., Fadafan, H. K., & Youssefi, A. (2010). Fabrication, characterization and measurement of thermal conductivity of Fe_3O_4 nanofluids. *Journal of Magnetism and Magnetic Materials* 322, 3895–3901.

68. Plueddemann, E. P. (1991). *Chemistry of Silane Coupling Agents*. New York: Silane Coupling Agents Springer.

69. Li, D., Fang, W., Zhang, Y., Wang, X., Guo, M., & Qin, X. (2017). Stability and thermal conductivity enhancement of silver nanofluids with gemini surfactants. *Industrial & Engineering Chemistry Research* 56, 12369–12375.

70. Su, Q., Pang, S., Alijani, V., Li, C., Feng, X., & Müllen, K. (n.d.). Composites of graphene with large aromatic molecules. *Advanced Materials* 21, 3191–3195.

71. Farbod, M., Ahangarpour, A., & Etemad, S. G. (2015). Stability and thermal conductivity of water-based carbon nanotube nanofluids. *Particuology* 22, 59–65.

72. Said, Z., Allagui, A., Abdelkareem, M. A., Alawadhi, H., & Elsaid, K. (2018). Acid-functionalized carbon nanofibers for high stability, thermoelectrical and electrochemical properties of nanofluids. *Journal of Colloid and Interface Science* 520, 50–57.

73. Zawrah, M. F., Khattab, R. M., Girgis, L. G., Daidamony, H. El, & Aziz, R. E. A. (2016). Stability and electrical conductivity of water-base Al_2O_3 nanofluids for different applications. *HBRC Journal*, 12, 227–234.

74. Li, X. F., Zhu, D. S., Wang, X. J., Wang, N., Gao, J. W., & Li, H. (2008). Thermal conductivity enhancement dependent pH and chemical surfactant for Cu-H$_2$O nanofluids. *Thermochimica Acta* 469, 98–103.

75. Peterson, G. P., & Li, C. H. (2006). Heat and mass transfer in fluids with nanoparticle suspensions. *Advances in Heat Transfer* 39, 257–376.

76. Kiruba, R., Vinod, S., Zaibudeen, A. W., Solomon, R. V., & Philip, J. (2018). Stability and rheological properties of hybrid γ-Al$_2$O$_3$ nanofluids with cationic polyelectrolyte additives. *Colloids and Surfaces A: Physicochemical and Engineering Aspects* 555, 63–71.

77. Zhang, T., Ge, J., Hu, Y., & Yin, Y. (2007). A general approach for transferring hydrophobic nanocrystals into water. *Nano Letters* 7, 3203–3207.

78. Chen, Y., Quan, X., Wang, Z., Lee, C., Wang, Z., Tao, P. et al. (2016). Stably dispersed high-temperature Fe$_3$O$_4$/silicone-oil nanofluids for direct solar thermal energy harvesting. *Journal of Materials Chemistry A* 4, 17503–17511.

79. Yu, Q., Kim, Y. J., & Ma, H. (2008). Nanofluids with plasma treated diamond nanoparticles. *Applied Physics Letters* 92, 103111.

80. Joni, I. M., Purwanto, A., Iskandar, F., & Okuyama, K. (2009). Dispersion stability enhancement of titania nanoparticles in organic solvent using a bead mill process. *Industrial & Engineering Chemistry Research* 48, 6916–6922.

81. Dai, C., Wang, X., Li, Y., Lv, W., Zou, C., Gao, M., & Zhao, M. (2017). Spontaneous imbibition investigation of self-dispersing silica nanofluids for enhanced oil recovery in low-permeability cores. *Energy & Fuels* 31, 2663–2668.

82. Hiramatsu, H., & Osterloh, F. E. (2004). A simple large-scale synthesis of nearly monodisperse gold and silver nanoparticles with adjustable sizes and with exchangeable surfactants. *Chemistry of Materials* 16, 2509–2511.

83. Bica, D., Vékás, L., Avdeev, M. V., Marinică, O., Socoliuc, V., Bălăsoiu, M., & Garamus, V. M. (2007). Sterically stabilized water based magnetic fluids: Synthesis, structure and properties. *Journal of Magnetism and Magnetic Materials* 311, 17–21.

84. Yu, F., Chen, Y., Liang, X., Xu, J., Lee, C., Liang, Q. et al. (2017). Dispersion stability of thermal nanofluids. *Progress in Natural Science: Materials International* 27, 531–542.

85. Chen, Y., & Wang, X. (2008). Novel phase-transfer preparation of monodisperse silver and gold nanoparticles at room temperature. *Materials Letters* 62, 2215–2218.

86. Yu, W., Xie, H., Wang, X., & Wang, X. (2011). Highly efficient method for preparing homogeneous and stable colloids containing graphene oxide. *Nanoscale Research Letters* 6, 47.

87. Batmunkh, M., Tanshen, M. R., Nine, M. J., Myekhlai, M., Choi, H., Chung, H., & Jeong, H. (2014). Thermal conductivity of TiO$_2$ nanoparticles based aqueous nanofluids with an addition of a modified silver particle. *Industrial & Engineering Chemistry Research* 53, 8445–8451.

88. Sundar, L. S., Singh, M. K., Ferro, M. C., & Sousa, A. C. M. (2017). Experimental investigation of the thermal transport properties of graphene oxide/Co$_3$O$_4$ hybrid nanofluids. *International Communications in Heat and Mass Transfer* 84, 1–10.

89. Gumus, S., Ozcan, H., Ozbey, M., & Topaloglu, B. (2016). Aluminum oxide and copper oxide nano-diesel fuel properties and usage in a compression ignition engine. *Fuel* 163, 80–87.

90. Dreyer, D. R., Park, S., Bielawski, C. W., & Ruoff, R. S. (2010). The chemistry of graphene oxide. *Chemical Society Reviews* 39, 228–240.

91. Williams, O. A., Hees, J., Dieker, C., Jäger, W., Kirste, L., & Nebel, C. E. (2010). Size-dependent reactivity of diamond nanoparticles. *ACS Nano* 4, 4824–4830.

92. Shahsavar, A., Salimpour, M. R., Saghafian, M., & Shafii, M. B. (2015). An experimental study on the effect of ultrasonication on thermal conductivity of ferrofluid loaded with carbon nanotubes. *Thermochimica Acta* 617, 102–110.

93. Żyła, G. (2016). Thermophysical properties of ethylene glycol based yttrium aluminum garnet (Y$_3$Al$_5$O$_{12}$–EG) nanofluids. *International Journal of Heat and Mass Transfer* 92, 751–756.

94. Paul, G., Philip, J., Raj, B., Das, P. K., & Manna, I. (2011). Synthesis, characterization, and thermal property measurement of nano-Al(95)Zn(05) dispersed nanofluid prepared by a two-step process. *International Journal of Heat and Mass Transfer* 54, 3783–3788.

95. Megatif, L., Ghozatloo, A., Arimi, A., & Shariati-Niasar, M. (2016). Investigation of laminar convective heat transfer of a novel TiO$_2$–carbon nanotube hybrid water-based nanofluid. *Experimental Heat Transfer* 29, 124–138.

96. Kato, Y., Mimura, K., Umeda, J., Sakamoto, W., & Yogo, T. (2012). Synthesis of field-responsive PbTiO$_3$ particle/polymer hybrids from metal-organics. *Colloids and Surfaces A: Physicochemical and Engineering Aspects* 408, 57–63.

97. Munkhbayar, B., Tanshen, M. R., Jeoun, J., Chung, H., & Jeong, H. (2013). Surfactant-free dispersion of silver nanoparticles into MWCNT-aqueous nanofluids prepared by one-step technique and their thermal characteristics. *Ceramics International* 39, 6415–6425.

98. Abbasi, S. M., Rashidi, A., Nemati, A., & Arzani, K. (2013). The effect of functionalisation method on the stability and the thermal conductivity of nanofluid hybrids of carbon nanotubes/gamma alumina. *Ceramics International* 39, 3885–3891.

99. Ilyas, S. U., Pendyala, R., Narahari, M., & Susin, L. (2017). Stability, rheology and thermal analysis of functionalized alumina- thermal oil-based nanofluids for advanced cooling systems. *Energy Conversion and Management* 142, 215–229.

100. Babita, Sharma, S. K., & Gupta, S. M. (2016). Preparation and evaluation of stable nanofluids for heat transfer application: A review. *Experimental Thermal and Fluid Science* 79, 202–212.

101. Che Sidik, N. A., Mahmud Jamil, M., Aziz Japar, W. M. A., & Muhammad Adamu, I. (2017). A review on preparation methods, stability and applications of hybrid nanofluids. *Renewable and Sustainable Energy Reviews* 80, 1112–1122.

102. Kumar, D. D., & Arasu, A. V. (2018). A comprehensive review of preparation, characterization, properties and stability of hybrid nanofluids. *Renewable and Sustainable Energy Reviews* 81, 1669–1689.

2

Thermophysical Properties Enhancement Using Hybrid Nanofluids and Their Synthesis and Applications

B. Chitra, K. Sathishkumar, N. Fedal Castro, and S. Srinivas

CONTENTS

2.1 Introduction

Heat transfer plays a vital role in many engineering applications, heat is transferred by devices like heat exchangers, condensers, evaporators, and sinks. The demand for enhancement of heat transfer rate rises to make these devices more efficient, compact in design, and minimize the cost of energy and material. Several techniques can be adopted to increase the heat transfer rate such as using extended surfaces (fins), jet impingement, and application of magnetic field. Heat transfer efficiency can also be improved by increasing the thermal conductivity of the working fluid. Nanotechnology has been accepted universally as an efficient technique for enhancement of heat transfer rates. The high thermal conductivities of solids are employed to improve the thermal conductivities of the fluids by adding solid particles (1 ~ 100 nm) to the respective fluid. The resulting fluids are colloidal suspensions of nano-particles in a base fluid.

The usual method for the improvement in heat transfer involves: (i) active methods and (ii) passive methods. Active methods include forced pulsation by cams and reciprocating plungers, the usage of electrical or magnetic fields as suggested by Ahuja [1] and Bergles [2] to interfere with the seeded light particles in a flowing stream. Passive heat transfer intensification does not need external power input, i.e., an increase in the surface area available for heat exchange and the residence time of the heat transfer fluids which is impractical or unacceptable to increase the size of the thermal management system.

Alternatively, the size of the heat transfer equipment can be minimized by adding solid crystalline particles. These particles have high thermal conductivity when compared to the conventional heat transfer fluids like water or ethylene glycol [3]. If the particles are in the range of millimeter or micron size are used, serious problems occur in heat transfer equipment. The problems include large particles rapidly settle down and, therefore, more pressure drop can happen in the equipment. Corrosion in pipelines might occur because of the abrasive action of the particles. Because of these drawbacks, Milli size or micron size particles are not used widely. Choi introduced the particles in nanometer dimensions as a suspended solution [4]. The small particle size and their low volume fractions prevent particles clogging and also reduce the pressure drop. Stability is increased, and sedimentation is reduced because of the large surface area of nanoparticles. The heat transfer efficiency is improved because of the reduction in particle size since heat transfer takes place at the particle surface [5]. Large savings in pumping power can be accomplished if a high thermal conductivity increase can be achieved with a small volume fraction of nanoparticles. Moreover, reduced particle clogging occurs when compared to conventional slurries, which promote system miniaturization. The suspension of these nanoparticles in base fluids like water, ethylene glycol, or oil was named as nanofluids by Choi in 1995 [4].

Cooling can maintain the desired performance and reliability of various products like car engines, power electronics, computers, and high-powered lasers or x-rays. With a rise in heat loads and heat fluxes, cooling is one of the most important challenges for these products, which is caused by higher power and smaller size, faced by several industries like microelectronics, defense, transportation, manufacturing, and metrology [6].

In the transportation industry, cooling is a major concern since the fashion for very high engine power and exhaust gas regulation for hybrid vehicles ultimately led to larger radiators and increased frontal areas, which cause a significant increase in fuel consumption and aerodynamic drag [6].

The usual way to increase heat transfer is either to increase the flow velocity or to enlarge the available surface area of cooling devices or to disperse solid particles in heat transfer fluids [7].

In the recent years, modelling and analysis of nanofluids has become an interesting area of research. Researchers have explained in their work about the various properties of nanofluids such as density, specific heat capacity, and electrical conductivity [8–11]. The high thermal properties of nanofluids are the major reason for its potential application in heat transfer enhancement, this specific advantage of nanofluids are explained in several works [12–14]. In addition to this, the heat transfer properties of nanofluids were found to be greatly influenced by other factors like particle diameter, volume concentrations, temperature, and velocities. Eastman et al. [15] reported in their research the successful enhancement of thermal conductivity with the aluminium (Al_2O_3)/ water nanofluids of volume concentration of 5% and particle diameter of 30 mm. The modelling of nanofluids is carried out in two ways, single-phase model and two-phase model. In the single-phase model slip motion exists, whereas in the latter model, this impact is not considered to be negligible. Sheikholeslami and Rokni [16] in their review paper have clearly explained about the various modelling methods of nanofluids. There are various forces that have a major influence in the treatment of nanofluids, these forces include Coulomb's forces, Lorentz forces, etc. Sheikholeslami [16] studied the nanofluid forced convection in a porous enclosure in the existence of Lorentz forces. In this work, Darcy and Koo–Kleinstreuer–Li models were taken into account for porous media and the nanofluid, respectively. There were several other works establishing the effect of such forces on the nanofluid [18–21].

The research based on convective heat transfer using nanofluids in order to enhance the thermal activity of heat transfer fluids has been a key area of interest for the past few years. Eastman et al. [15] carried out experiments in order to surmount the basic restriction of low thermal conductivity through the growth of nanofluids and were found to be successful in achieving enhancement of up to 30% in thermal conductivity. In addition to heat transfer enhancement with the development of nanofluids, various research investigations employing hybrid nanofluids (a colloidal dispersion of two different nano particles into a conventional heat transfer fluid) were found to provide more satisfying results for the purpose of heat transfer enhancement. Madhesh et al. [20] experimentally investigated the convective heat transfer and rheological characteristics of copper (Cu)-TiO_2 hybrid nanofluid. They found that the convective heat transfer coefficient (h) increased by about 52% for the hybrid nanofluid which was higher in comparison. There were several other research works too that proved the better efficiency of hybrid nanofluids for enhancing heat transfer [23–26].

This book chapter highlights the synthesis and properties of mono and hybrid nanofluids. Further, the theoretical models involved in calculation of thermophysical properties were analyzed along with its optimization technique. Finally, nanofluids applications and challenges were outlined.

2.2 Various Techniques of Heat Transfer Enhancement

2.2.1 Suspension of Micro or Millimeter Size Solids in Liquid and Their Limitations

One of the old techniques used to increase the heat rejection rate is the dispersion of millimeter or micron size particles in conventional heat transfer fluids [7]. Some of the drawbacks of this suspension are: (i) rapid settling of particles, (ii) wear out problems in pipes, pumps, and bearings if the fluid is kept on circulating, and (iii) high pressure drop and pumping power because of the addition of large quantity of particles [27].

2.2.2 Use of Extended Surfaces and Their Drawbacks

For air or liquid cooling, the other way to increase the cooling rate is to use extended surfaces like fins or micro channels. These micro channels have the ability to reduce the size and effectiveness of various heat exchange devices. Microscale heat transfer equipment has numerous advantages like high thermal efficiency, high heat transfer surface/volume ratio, small size, low weight, relatively low fluid consumption and, design flexibility [6,28]. The cost of materials and manufacturing costs could be lowered because of its size. Though micro channels are advantageous, they have certain drawbacks. Some of the drawbacks are: (i) the pressure drop through the system requires high mechanical power for running the system, (ii) they are sensitive to corrosion, roughness, and fouling of the surfaces, and (iii) very clean fluids could be used with protective filters, mechanical cleaning, and maintenance, which are impossible in a continuous process. The drawbacks in using large-sized particles and micro channels lead to an evolution of nanofluids [28].

Nanofluid is a sol colloid involving nanoscale less than 100 nm-sized particles dispersed in a base fluid. Because of the nano-sized particles, these nanofluids retain their Newtonian behavior at low volume concentrations, and hence these nanofluids can be applied as heat transfer fluids in industries as an effective substitute for conventional heat transfer fluids because of their enhanced thermophysical properties [29]. Research on hydrodynamic studies showed an increase in viscosity, which is important to determine the pumping power and convective heat transfer coefficient. Because of these hydrodynamic and thermophysical properties, nanofluids are suitable for applications in micro channel flow passages with smaller and lighter heat exchangers, reduced heat transfer fluid inventory, and reduced emissions. They also find use in drag reduction, the binding ability for sand consolidation, gel formation, wettability alteration, and corrosion control and fields like tribology, chemistry, surfactants and coatings, pharmaceutical, and medical applications [30]. Nanofluids have the ability to achieve ultra-high performance cooling, and thus they have the potential to be considered as next-generation coolants.

2.2.3 Use of Nano-Sized Particles

The thermal conductivity can be enhanced by dispersing the solids in heat transfer liquids as suggested by Maxwell [60]. Other problems include abrasion and clogging. These problems can be overcome by dispersing nano-sized particles in fluids [31]. Nanoparticles remain suspended for a longer time when compared to large-sized particles and also have a high surface area. The surface to volume ratio is 1000 times greater when compared to micro particles. Since heat transfer occurs on the surface of the particles, the high surface area enhances the heat conduction of nanoparticles. Because of its stability and high thermal conductivity, nanofluids are highly desired in heat transfer operations. Further, nanofluids reduce erosion and clogging because of their smaller particle size. Other advantages include a decrease in pumping power, reduced consumption of heat transfer fluid, and significant energy savings [32].

For the synthesis of nanofluids, there are several factors to be considered. They are: (i) thermal stability, (ii) dispersibility in a fluid medium, and (iii) chemical compatibility and ease of chemical manipulation.

2.3 Synthesis and Stability of Nanofluids

Stable nanofluids can be produced by one- and two-step method. Both one-step and two-step methods can create nanoparticle suspension and agglomeration of particles. Thus, synthesis and suspension of non-agglomerated and nearly monodispersed nanoparticles in the liquid are the key steps for the enhancement of thermal properties of nanofluids. Nanostructured materials are made of nanometer-sized substances on the atomic or molecular scale, and these materials exhibit new or enhanced physical properties when compared to conventional bulk solids. These remarkable properties of nanophase materials occur because of the high surface area/volume ratio, which is because of the large proportion of atoms present within the grain boundaries. Therefore, properties like thermal, mechanical, optical, magnetic, and electrical properties of nanomaterials are superior to conventional bulk solids [33].

Different types of nanomaterials can be used for the synthesis of nanofluids. Nanomaterials used are oxide ceramics, nitride ceramics, metals, semiconductors, carbon nanotubes, and composite materials.

Liquids like water, ethylene glycol, propylene glycol, and oil are widely used as the base fluid for the synthesis of nanofluid.

2.3.1 Nanoparticle Synthesis

Nanoparticles can be synthesized either by physical or chemical process [34–36]. Inert gas condensation and mechanical grinding are the physical processes. Chemical processes include chemical vapor deposition, chemical precipitation, micro emulsions, thermal spray, and spray pyrolysis.

The current methods to the synthesis of metal nanoparticles include inert gas condensation, mechanical milling, chemical precipitation, thermal spray, and spray pyrolysis.

2.3.2 Dispersion of Nanoparticles in Liquids

Stable nanofluids are produced by two methods, namely: (i) the one/single-step technique and (ii) the two-step technique.

2.3.2.1 Synthesis of Nanofluids by the One-Step Method

The single-step method simultaneously makes and disperses nanoparticles directly into base fluids. Synthesis of nanofluids by the one-step method produces stable nanofluids because of its combined operations like drying, storage, transportation, and dispersion of particles are being involved in liquid [37] and is shown in Figure 2.1.

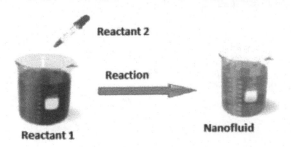

FIGURE 2.1 Synthesis of nanofluids by one step method. (From Mukherjee, S. and Paria, S. *IOSR J. Mech. Civil Eng.*, 9, 63–69, 2013.)

2.3.2.1.1 Advantages, Disadvantages, and Challenges in One-Step Method

The single-step physical method is preferable than two-step method for high conductivity metals, but it is hard to scale up. The production of nanoparticles and nanofluids will be slow since it requires vacuum, and the production cost is high. Alternatively, one-step chemical method can be used in the synthesis of nanofluids [38] with same thermal conductivity enhancement as that of one-step physical method. Though one-step chemical method produces small quantities of nanofluids, this approach has the potential to effectively produce significant amounts of nanofluids faster than the one-step physical process. One-step chemical method also minimizes agglomeration of nanoparticles and produce dispersed nanofluids. The major limitation of one-step chemical method is that the volume fraction and the quantity of nanoparticles and nanofluids produced are much more limited than the two-step process. The one-step process is not commercially available like that of the two-step process.

The major challenge in the one-step process is to adopt innovative technologies to produce large quantities of nanofluids at low cost. The one-step process runs in batch mode with control on particle size, and if the same process is allowed to operate in the continuous mode, commercial application is possible [39].

2.3.2.2 Synthesis of Nanofluids by the Two-Step Method

In two-step method, nanoparticles are first synthesized by either physical or chemical process and then dispersed in base fluids. Dispersion techniques used in two-step method are high shear and ultrasound for the creation of different particle fluid combination. Most of the nanofluids are synthesized by the two-step method. Some agglomeration of individual nanoparticles might occur by van der Waals forces if the nanoparticles are produced in powder form. Because of this agglomeration property, producing nanofluids using the two-step method is a tough task since these particles settle down in liquids. Stable nanofluids could be achieved by separating the nanoparticles by high shear or by ultrasonication. Stable nanofluids are necessary since unstable nanofluids reduce the conductivity of the nanofluid. The first ever nanofluid was carbon nanotubes with synthetic oil (poly alpha olefins) which was synthesized by two-step method [40]. Figure 2.2 represents the synthesis of nanofluid by two-step method.

2.3.2.2.1 Advantages, Disadvantages, and Limitations in Two-Step Method

The two-step process has commercial application, and it can produce bulk quantities of nanofluids economically. Since the nanoparticles which are produced in bulk are available at low prices, these nanoparticles are used in the synthesis of nanofluids in two-step method. Because of the commercial availability and relatively low cost of nanoparticles, the two-step method is used to produce nanofluids for industrial applications. However, the nanofluids produced in this method form aggregate of nanoparticles, and it requires high volume concentration of oxide nanoparticles to achieve a significant enhancement in thermal conductivity [41].

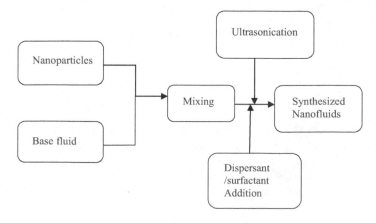

FIGURE 2.2 Synthesis of nanofluids using two step method.

The major challenge in this two-step method is to produce well dispersed nanofluids to a large volume with the help of new technologies.

2.3.3 Stability of Nanofluids

One of the key features for any nanofluid system in heat transfer application is its stability. There lies a strong tendency of the nanofluids to form aggregates in the liquid medium, and this aggregation clogs micro channels and reduces nanofluid thermal properties. Hence, the study of stability, factors influencing the stability, and the techniques in the evaluation of stability are critical.

2.3.3.1 Factors Affecting Stability

The factors which influence the stability of nanofluids are surface modifiers like surfactants, pH adjustment, nanofluid preparation method, mixing or homogenization method, and nanoparticles loading. For the preparation of each nanofluid system, a particular dispersion method is used to stabilize the nanoparticles in the base liquid [42].

2.3.3.1.1 pH of Nanofluids

When nanoparticles are dispersed throughout the base liquid, surface charge on nanoparticles is important to achieve stability. Surface charge emits an electric field resulting in attraction or repulsion of nanoparticles. This surface charge depends upon the pH value of the suspension. To achieve a stable nanofluid, the pH value of the nanofluid should be far from the iso electric point of the particles. When the pH value of the nanofluid is close to an iso electric point, because of the weak repulsive forces between the nanoparticles, they tend to agglomerate, and finally, precipitation occurs. Hence, the pH value is adjusted in such a way that it is far from the iso electric point, the absolute electrical charge gets increased, which in turn gradually increases the repulsive interaction between the nanoparticles. So, the stability of the nanofluids can be controlled by adjusting the pH value of the nanofluids. Optimum stability can be achieved at optimum pH [43].

2.3.3.1.2 Surface Modifiers

The modification to the surface of nanoparticles in the liquid medium can be accomplished by the addition of surface modifiers. Addition of proper surface modifiers like surfactants or other kind of modifiers like polymers or zwitterionic molecules is used to avoid sedimentation of nanoparticles. Surfactants consist of the hydrophilic head and hydrophobic tails, and they are used to improve the stability of nanofluids. Depending on the type of nanoparticles and the base liquid, the proper surfactant should be selected to modify the surface of nanoparticles which results in excellent electrostatic repulsion among nanoparticles. By using surfactants, hydrophilic surfaces of nanoparticles are modified into hydrophobic and vice versa. It is important to apply sufficient surfactant to the suspension and depending upon the suspension system, proper surfactant should be chosen to overcome van der Waals attractions [42]. Four types of surfactants like cationic, anionic, non-ionic or zwitterionic, or other types of surfactants like polymers can be added to the enhancement of stability of nanofluids. Temperature range of the surfactant should also be taken into account when nanofluids are used in higher temperature applications since it might damage the bonding between nanoparticles and surface modifiers.

2.3.3.1.3 Fabrication Method

Stability of the nanofluids can also be affected by their preparation method. One-step method shows a higher stability than two-step method, since in one-step method stages like drying, storage, and dispersion of nanoparticles are avoided and nanoparticle preparation and nanofluid synthesis co-occur [29].

2.3.3.1.4 Nanoparticle Concentration

If the solution is very concentrated, the nanoparticles agglomerate in the liquid medium, and so it is important to control the concentration of nanoparticle in the nanofluid to achieve better stability [44].

2.3.3.1.5 Homogenization Methods

To disperse the nanoparticles well in the base fluid, mixing or homogenization method can be used. Different methods for mixing like ultrasonic bath, homogenizer, and processors are employed by the different researcher to achieve well dispersed nanofluids [45]. All this mixing equipment is energetically powerful to break down the nanoparticle agglomerates.

2.3.3.2 Methods to Evaluate the Stability of Nanofluids

The most commonly used technique to assess the stability of nanofluids is zeta potential analysis. The optimum value of pH to stabilize nanoparticles in base liquid has been identified by using this method. The significance of zeta potential relates to the stability of the nanofluid and is shown in Figure 2.3. When the zeta potential value is in the range of −30 mV to +30 mV, the repulsion forces among nanoparticles are high, and these kind of values are said to have excellent stability. If the value is in the range of 60 mV to −60 mV, the stability is good [46].

The other methods in the evaluation of stability are sediment photography, ultra violet-visible spectrophotometry, and the sediment balance method [47].

One of the simplest methods to evaluate the stability of nanofluids is sediment photography where pictures of settling nanoparticles are recorded and compared with time. This method cannot be used for nanofluids with slow sedimentation rate [47].

In ultra violet-visible spectrophotometer, light is passed through the nanofluid, and the graph of absorbance against wavelength provides the absorption spectrum of the sample. This method cannot be used for the nanofluids with high concentration since it is hard to pass the light through the nanofluids with high nanoparticle loading [47].

The rate of change of nanoparticles sediments on a tray which is immersed into a suspension is recorded in sedimentation balance method. Since this method is simple, cheap, and it does not have any limitation of other methods, this method can be a good choice in the evaluation of stability of nanofluids [47].

The present focus is on the application of hybrid nanofluids for heat transfer enhancement in progressive cooling systems, internal combustion engines, and so on. In order to apply nanofluids for heat transfer enhancement, the laws of enhanced heat transfer must be met. The enhancement of heat transfer is due to the nanoparticles suspended in the base fluids because of which the thermal conductivity increases. Hence, a model using artificial neural network (ANN) in MATLAB to determine the optimum thermal conductivity model of the corresponding hybrid nanofluid from the experimentally measured thermal conductivity values and other parameters such as specific heat capacity and density is used. After measuring the optimum model of thermal conductivities from the several experimental values,

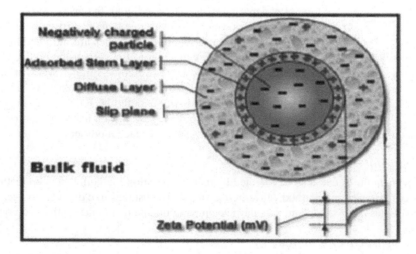

FIGURE 2.3   Zeta potential analysis of nanofluids. (From Mukherjee, S. and Paria, S., *IOSR J. Mech. Civil Eng.*, 9, 63–69, 2013.)

this selected model was used to enhance the heat transfer in system cooling and in internal combustion engines and the results obtained were comparatively much more efficient for the hybrid systems than the mono nanofluid system.

2.4 Hybrid Nanofluids: Synthesis and Properties

Hybrid nanofluids are synthesized by suspending dissimilar nanoparticles either in a mixture or composite form in the base fluid. By using hybrid nanofluids, heat transfer was furthermore increased, and good pressure drop characteristics could be achieved. This section summarizes some of the research work on synthesis, thermophysical properties, heat transfer, and pressure drop characteristics and their possible applications. Hybrid nanofluids have the properties of their constituent materials whereas a single material possesses particular property, which may or may not show good thermal or rheological properties. Hybrid nanofluids can be prepared either by dispersing two or more nanoparticles in base fluid or dispersing composite nanoparticles in base fluid.

2.4.1 Experimental Methodology for Hybrid Nanofluids

Silica and silver nanofluids were synthesized by one-step method [38]. Silica was first mixed with base fluid by magnetic stirring at 13°C, followed by introducing silver nitrate to silica and base fluid. At 130°C, reduction of Ag^+ ions to Ag by electron transfer reaction occurs.

Gold and carbon nanotube (CNT) nanofluids were prepared by suspending in water. CNT nanofluids and gold nanofluids dispersed in water were developed on an individual basis. Gold nanoparticle suspension was then added to CNT suspension to achieve gold and CNT suspension. Copper and CNT suspension can also be prepared using deionized water as the base fluid. Laurate salt was added to achieve stability of copper nanofluids. An ultrasonic cleaner was used as a low power sonication to disperse the nanoparticles in water. The prepared gold nanofluids were then added to CNT suspension to produce hybrid nanofluids [48].

Alumina-copper nanofluids can also be synthesized by first synthesizing alumina-copper composite powder by thermochemical synthesis. The steps involved in thermochemical synthesis methods are spray drying, oxidation of precursor powder, reduction of hydrogen, and homogenization. Nitrates of copper and aluminium soluble in water, $Cu(NO_3)_2 \cdot 3H_2O$ and $Al(NO_3)_3 \cdot 9H_2O$, were first prepared followed by spray drying to obtain the precursor powder. Heating precursor powder formed a mixture of copper oxide and stable Al_2O_3 in an open atmosphere. The alumina-copper oxide mixture was heated using tubular furnace in a hydrogen atmosphere and kept in an alumina boat to reduce the copper oxide to metallic copper. The mixture was ball milled to obtain a homogeneous alumina-copper nano composite powder. Alumina-copper nanofluids were prepared by dispersing specific amount of alumina-copper nanoparticles in water with sodium lauryl sulfate as surfactant using an ultrasonic vibrator [49].

Hybrid nanofluids of silica-multi walled nanotube (MWNT) in water could be synthesized using wet chemical method. COOH functional groups have been used to functionalize MWNTs. Sodium silicate was suspended in water followed by dispersion of functionalized MWNT to the solution using ultrasonic bath. Cetyl trimethyl ammonium bromide and dimethyl formamide were added to the above mixture by magnetic stirring. The products were subsequently filtered, washed, and dried to form gray powder, and these powders are later suspended in water to form hybrid nanofluids [50].

Alumina-MWNT nanofluids were synthesized by solvothermal process using ethanol as base fluid. Aluminium acetate powder was first diffused in ethanol, followed by the addition of pure MWNT and functionalized MWNT using the ultrasonic water bath. Ammonia solution was then added and transferred to Teflon-lined stainless steel autoclave mixture to perform solvothermal synthesis and hybrid γ-Al_2O_3/MWNT nanomaterials were prepared. Hybrid nanofluids were prepared by dispersing hybrid nanomaterial in deionized water using gum arabic as dispersant in ultrasonic water bath [51].

Cu-cuprous oxide in water nanofluids were prepared using wet ball milling process. Ball milling is used to hydrolyze copper nanoparticle to convert to cupric oxide nanoparticle at a temperature higher than ambient temperature. Three steps are involved in the synthesis process, namely: (i) oxidation of copper to cupric oxide, (ii) formation of cu-cuprous oxide on Cu particle, and (iii) breaking cupric oxide layer into smaller particle by 20 nm size [52].

Aluminium and zinc nanofluids were synthesized by two-step method. Al and zinc (Zn) nanoparticles were prepared by mechanical alloying. Elemental powder of aluminium (95%) and zinc (5%) were blended at room temperature by mechanical alloying using planetary ball mill for a desired period to reach to steady state. Hybrid nanofluids were prepared by the two-step process by adding Al-Zn nanoparticles in ethylene glycol under ultrasonic vibration, followed by magnetic stirring [53].

Al_2O_3 and cupric oxide nanoparticles are used as co-dispersoids in an ethylene glycol-water dispersion medium. Mono nanofluids of alumina and cupric oxide were prepared and then dispersed together to form hybrid nanofluids using ultrasonication [58]. The thermophysical properties were measured and compared with calculated values using theoretical models. Hybrid nanofluids were characterized using X-ray diffraction and transmission electron microscopy analysis and are shown in Figure 2.4.

2.4.2 Thermophysical Properties of Hybrid Nanofluids

The thermal conductivity of water-based CNT-gold nanoparticles and CNT-copper nanoparticles was compared and observed that the CNT did not increase the thermal conductivity of both hybrid nanofluids. The thermal conductivity rather decreased because of either lack of interaction between gold-CNT nanomaterials or the addition of CNTs to copper nanoparticles reduces the dispersibility which results in agglomeration of nanoparticles [48].

The density and specific heat of hybrid water-based suspension of alumina nanoparticles and micro encapsulated phase change material were studied. They compared the measured values of density and specific heat with that of theoretical mixture theory formula [49] and validated the correctness of the formula.

Thermal conductivity is increased for Al_2O_3-Cu/water hybrid nanofluids from 0.1% to 2% volume fraction. The viscosity of hybrid nanofluids for various volume concentration was measured and showed that the effective viscosity increases with volume concentrations and the experimental results of viscosity were higher than those predicted by the theoretical models [49].

The thermal conductivity ratio of Al-5 wt% Zn dispersed in ethylene glycol as a function of concentration of nanoparticles was studied and showed that the thermal conductivity ratio non-linearly increases for an increase in particle concentration. They have also compared the experimental values with the theoretical Hamilton–Crosser model, and the data showed huge deviation. They also demonstrated that the thermal conductivity increased linearly with temperature [53].

The thermal conductivity of water-based silica-MWNT hybrid nanofluids was measured and the enhancement of thermal conductivity of hybrid nanofluids is in between the increase of MWNT

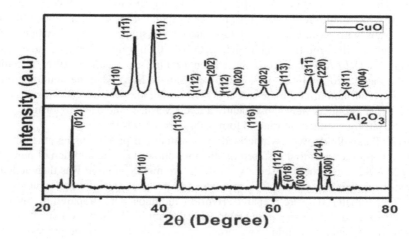

FIGURE 2.4   XRD characterization of hybrid nanofluids. (From Kannaiyan, S. et al., *J. Mol. Liq.*, 244, 469–477, 2017.)

nanofluids and silica nanofluids. They have shown CNT was the reason behind the increase in thermal conductivity and also revealed that addition of surfactant causes an adverse effect on thermal properties [50].

The thermal conductivity of water-based hybrid nanofluids (γ-Al$_2$O$_3$/MWCNT) depends on the synthesis method and the stability of hybrid nanofluids. Thermal conductivity enhancement of 14.5% could be achieved at 0.01% concentration [51].

The dependence of temperature on the effective thermal conductivity of hybrid nanofluids containing silver/MWNT was studied and showed that higher thermal conductivity could be achieved with hybrid nanofluids compared to nanofluids containing MWNT [54].

The thermal conductivity of graphene/MWNT as a function of temperature for various volume fractions in deionized water and ethylene glycol was investigated. Because of the high aspect ratio of graphene/MWNT composite, thermal transfer properties were high [55].

The thermal conductivity of Ag/TiO$_2$-water nanofluids for various weight concentrations at different temperatures was calculated. Hybrid nanofluids showed an increase in thermal conductivity with an increase in temperature, and thermal conductivity can be improved by introducing flattened Ag particles to the TiO$_2$-based solution [56].

Hybrid fillers added synergistic effect on heat conductive networks for Fe$_2$O$_3$-MWNT hybrid nanofluids was shown in the literature. The thermal conductivity of hybrid nanofluids was more when compared to mono nanofluids [57]. The thermal conductivity enhancement of water-based nanofluid containing 0.05 wt% MWNTs and 0.02 wt% Fe$_2$O$_3$ nanoparticles is 27.75%, which is higher than that of nanofluid containing 0.2 wt% single MWNTs or Fe$_2$O$_3$ nanoparticles.

Al$_2$O$_3$ and cupric oxide nanoparticles (made using a simple and scalable process) are used as co-dispersoids in an ethylene glycol-water dispersion medium. The hybrid nanofluids showed an enhancement of nearly 45% in their thermal conductivity. However, thermal conductivity and viscosity trends suggest a definite need to account for interfacial chemical interactions in reported-nanofluids [58].

2.5 Theoretical Models Involved in Calculation of Thermophysical Properties

2.5.1 Models for Thermal Conductivity

The different factors which potentially influence the heat transfer enhancement of nanofluids are Brownian motion of nanoparticles, clustering of nanoparticles, nanolayering of the liquid at the liquid/nanoparticle interface, ballistic transport and non-local effect, thermophoretic effect, and near-field radiation. Most of the studies till date have been done on the Brownian motion of the nanoparticles, molecular-level layering of the liquid at the liquid/particle interface (nanolayer), nanoparticle clustering, and a combination of these factors together with other conditional parameters as temperature, nanoparticles size, and volume fraction.

For hybrid nanofluids, these thermophysical property models can be extended by using the mixture theory formula to calculate the particle thermophysical property [59].

There are various models involved in the calculation of thermal conductivity and the models are listed below:

2.5.1.1 Maxwell Model

Maxwell model is based on effective medium theory, and this model can be used for randomly dispersed and uniformly sized spherical particles [58].

$$\frac{k_{eff}}{k_f} = \frac{k_p + 2k_f + 2\phi\left(k_p - k_f\right)}{k_p + 2k_f - \phi\left(k_p - k_f\right)} \tag{2.1}$$

where, k_{eff} is the effective thermal conductivity of nanofluid, k_f is the thermal conductivity of base fluid, k_p is thermal conductivity of nanoparticle, and ϕ is the volume fraction of nanoparticle.

2.5.1.2 Hamilton–Crosser Model

Hamilton–Crosser model is applicable for spherical and cylindrical particles, and it was developed by using shape factor [61].

$$\frac{k_{eff}}{k_f} = \frac{k_p + (n-1)k_f - (n-1)\phi(k_f - k_p)}{k_p + (n-1)k_f + \phi(k_f - k_p)} \tag{2.2}$$

where n is the shape factor and $n = 3$ for spheres and $n = 6$ for cylinder.

Figure 2.5 shows the comparison of experimental and calculated thermal conductivity values using Hamilton–Crosser model for hybrid nanofluids.

2.5.1.3 Bruggeman Model

Bruggeman model is applicable for binary mixtures of homogeneous randomly dispersed and spherical nanoparticles [7].

$$\frac{k_{eff}}{k_f} = \frac{1}{4}\left[(3\phi - 1)\frac{k_p}{L_f} + (2 - 3\phi) + \frac{k_f}{4}\sqrt{\Delta} \right] \tag{2.3}$$

where, Δ is the thermal conductivity of suspension regardless of particle motion and vibration.

2.5.1.4 Xuan and Li Model

This model is not applicable for spherical particles and shape factor is unity [62].

$$\frac{k_{eff}}{k_f} = \frac{k_p + 2k_f - 2\phi(k_f - k_p)}{k_p + 2k_f + \phi(k_f - k_p)} \tag{2.4}$$

where, k_{eff} is the effective thermal conductivity of nanofluid, k_f is the thermal conductivity of base fluid, k_p is the thermal conductivity of nanoparticle, and ϕ is the volume fraction of nanoparticle.

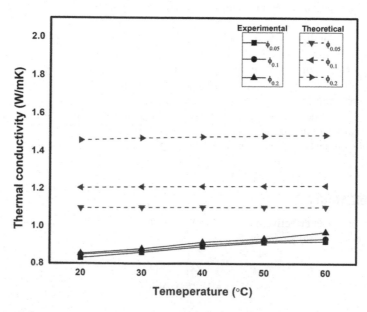

FIGURE 2.5 Comparison of experimental and theoretical thermal conductivity for hybrid nanofluid. (From Kannaiyan, S. et al., *J. Mol. Liq.*, 244, 469–477, 2017.)

2.5.1.5 Maxwell and Garnett's Model

This model is based on electromagnetic theory and particles interaction is not taken into account [60].

$$\frac{k_{eff}}{k_f} = \frac{(1-\phi)(k_p + 2k_f) + 3\phi k_p}{(1-\phi)(k_p + 2k_f) + 3\phi k_f} \qquad (2.5)$$

where, k_{eff} is the effective thermal conductivity of nanofluid, k_f is the thermal conductivity of base fluid, k_p is the thermal conductivity of nanoparticle, and ϕ is the volume fraction of nanoparticle.

In all the above models, particle thermal conductivity was calculated using the mixture theory formula for hybrid nanofluids.

2.5.2 Classical Models for Viscosity

2.5.2.1 Einstein Model

The nanofluid viscosity formula was first developed by Einstein in 1906, and the model was based on the assumption that the viscous fluid contains spherical particles at a very low volume fraction ($\phi < 0.02$). The Einstein model shows a linear increase in viscosity with particle volume concentration, and it has the limitation, since it does not the structure and particle concentration [63]. Further, this model cannot be applied to high particle concentration. The Einstein model is given below:

$$\frac{\mu_{nf}}{\mu_f} = 1 + 2.5\phi \qquad (2.6)$$

where, μ_{nf} is the viscosity of the nanofluid, μ_f is the viscosity of the base fluid, and ϕ is the volume fraction of the particle in suspension.

Figure 2.6 represents the comparison of experimental and theoretical viscosity (calculated using Einstein model) for hybrid nanofluids.

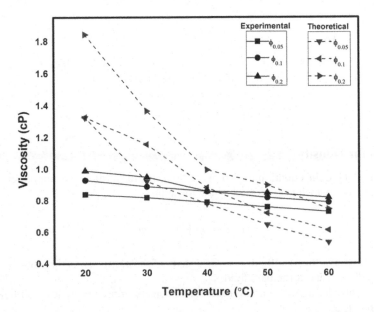

FIGURE 2.6 Comparison of experimental and theoretical viscosity for hybrid nanofluid. (From Kannaiyan, S. et al., *J. Mol. Liq.*, 244, 469–477, 2017.)

2.5.2.2 Mooney Model

For higher concentrations of spherical suspension, Mooney proposed a model in 1951, and the model is given by the below expression [64].

$$\frac{\mu_{nf}}{\mu_f} = e^{\left(\frac{\xi\phi}{1-k\phi}\right)}$$

(2.7)

where k is a constant, called self-crowding factor ($1.35 < k < 1.91$) and ξ is called the fitting parameter whose value is 2.5.

2.5.2.3 Krieger and Dougherty Model

For randomly dispersed spherical particles, a semi empirical model was proposed by Krieger and Dougherty in 1959 [65]. The model is stated below:

$$\frac{\mu_{nf}}{\mu_f} = \left[1 - \frac{\phi}{\phi_m}\right]^{-\eta\phi_m}$$

(2.8)

where ϕ_m is the maximum particle packing fraction, which varies from 0.495 to 0.54 and is approximately 0.605 at higher shear rates, and η is the intrinsic viscosity whose value is 2.5 for monodispersed suspensions of hard spheres.

2.5.2.4 Nielsen Model

For particle volume fraction of more than 0.02, Nielsen proposed the power law model in 1970, and the mathematical expression is as follows [66]:

$$\mu_{nf} = \left(1 + 1.5\phi\right)e^{\frac{\phi}{(1-\phi_m)\mu_f}}$$

(2.9)

2.5.2.5 Batchelor Model

Batchelor modified Einstein's model by considering Brownian motion effect, isotropic suspension of rigid, and spherical nanoparticles in 1972 [67]. The expression is given below:

$$\mu_{nf} = \left(1 + 2.5\phi + 6.5\phi^2\right)\mu_f$$

(2.10)

2.5.3 Models for Density

Pak and Cho [9] adopted the equation of two phase mixtures for particles of micrometer size for nanometer size particles too, which is expressed by the formula:

$$\rho_{nf} = \rho_p\varphi + \rho_{bf}\left(1-\varphi\right)$$

(2.11)

where, ρ_{nf} is the density of the nanofluid, ρ_p is the density of the particle, ρ is the particle volume concentration, and ρ_{bf} is the density of the base fluid.

Another model considering the nanolayer thickness, density of nanoparticles, and fluid and nanoparticle size is given below.

$$\rho_{nf-new} = \frac{\rho_{nf}}{(1-\varphi) + \varphi(r_p + t_v)^3 \Big/ r_p^3} \tag{2.12}$$

where, r_p is the radius of the particle and t_v is the equivalent thickness of void.

2.5.4 Models for Specific Heat

Two models are widely used in the literature for calculating the specific heat, and it is simple to calculate specific heat. Model I is similar to mixing theory for ideal gas mixtures, which is a straight average relating specific heat of nanofluid, $C_{p,nf}$, to the specific heat of base fluid, $C_{p,f}$, specific heat of nanoparticles, $C_{p,n}$, and volume fraction, ϕ [66]. The nanofluid specific heat using Model I is given as follows:

$$C_{p,nf} = \varphi C_{p,n} + (1-\varphi)C_{p,f} \tag{2.13}$$

Model II was proposed based on the assumption of thermal equilibrium between the particles and the surrounding fluid. Model II considers the effect of particle and fluid densities (ρ_n and ρ_f, respectively) on specific heat and the Model II is given below:

$$C_{p,nf} = \frac{\varphi(\rho C_p)_n + (1-\varphi)(\rho C_p)_f}{\varphi \rho_n + (1-\varphi)\rho_f} \tag{2.14}$$

2.6 Effect of Various Parameters on Thermophysical Properties of Hybrid Nanofluids

In heat transfer applications, the efficiency of nanofluid is assessed by evaluating its thermophysical properties like thermal conductivity, viscosity, specific heat capacity, density, and the flow regime of the fluid. There are several factors like particle composition, particle crystal structure, particle loading, particle morphology, fabrication method, temperature, the surface modifier, stability, and base liquid, which are influencing these thermophysical properties since nanofluids are the complex suspension.

2.6.1 Thermal Conductivity of Nanofluids

One of the most significant thermophysical properties to study the capability of nanofluids in heat transfer area is its thermal conductivity. For effective nanofluids, higher thermal conductivity is needed for heat transfer applications. Thermal conductivity influences Nusselt and Prandtl numbers, which are required for heat transfer characteristics of the fluid [7].

There are various parameters, which can control thermal conductivity of nanofluids, and they are discussed below:

2.6.1.1 Effect of pH and Surface Modifier on Thermal Conductivity

The thermal conductivity enhancement decreases the pH of the suspension closer to its iso electric point [67].

A proper surface modifier should be chosen to avoid agglomeration of nanoparticles in the liquid medium. The addition of these surface modifiers affects the thermophysical properties of nanofluids. The addition of surfactants not only decreases the thermal conductivity of nanofluids, but also modifies the surface of the nanoparticle [70].

2.6.1.2 Effect of Temperature and Base Liquid

From the literature, it has been reported that thermal conductivity increases with an increase in temperature. The effect of the base liquid on thermal conductivity is not well understood [71,72].

2.6.1.3 Effect of Nanoparticle Size, Shape, and Concentration on Thermal Conductivity

The size and shape of the nanoparticles also influence the thermal conductivity of nanofluids. Some of the investigations [27] show smaller size nanoparticles increase the thermal conductivity and few others shows larger size nanoparticles increase the thermal conductivity depending on the agglomerates in the base liquid.

The role of thermal conductivity on two different shapes (spherical and cylindrical) was studied in the literature. The thermal conductivity is more for spherical shape when compared to that of cylindrical-shaped particles [73].

Nanoparticle concentration also plays a vital role in thermal conductivity enhancement. Thermal conductivity is increased by increasing the concentration of the nanoparticles for the same particle composition.

2.6.1.4 Effect of Fabrication Method on Thermal Conductivity

The dispersion methods used for synthesizing nanofluids affect the stability as well as the thermal conductivity of nanofluids. Two-step method yields a higher thermal conductivity and better stability using bath ultrasonication method.

2.6.2 Viscosity of Nanofluids

For cooling applications, it is not only sufficient to achieve high thermal conductivity, but also viscosity should be assessed. Viscosity is defined as the internal resistance to flow, and it is a valuable property in all the thermal applications. Viscosity increases by adding nanoparticles to the base fluid, and this often causes an adverse effect on pumping power and heat transfer coefficient. Viscosity is influenced by several factors, and they will be discussed in the following sections.

2.6.2.1 Surface Modifiers

The viscosity of nanofluids could be affected by the type of surface modifier and its concentration. Increase in concentration of the surfactant, increases nanofluids viscosity [74].

2.6.2.2 Effect of Nanoparticles Size, Shape, and Concentration on Viscosity

Nanofluids with smaller particle size increase its viscosity because of its large solid/liquid interface area. The nanoparticle shape too influences the viscosity of nanofluids. Non-spherical nanoparticles have higher viscosity compared to the spherical nanoparticles [75]. If the nanoparticle concentration increases, the viscosity of nanofluids increases either non-linearly or linearly.

2.6.2.3 Effect of Temperature, pH, and Preparation Method on Viscosity

Based on the previous studies, when the temperature increases, the viscosity of nanofluids decreases [73]. The viscosity of nanofluids is also affected by its pH. From the literature [75], the viscosity fluctuates within the pH range between 5 and 7 when the nanoparticle size is less than 20 nm. For particle size greater than 20 nm, viscosity is not a function of pH.

The viscosity of nanofluids will be more in two-step method when compared to the single-step method since the formation of agglomerates is unavoidable [76].

2.6.3 Density of Nanofluids

The pressure drop and pumping power in engineering applications also depend upon the density of nanofluids. The density of nanofluids depends on several factors mentioned below:

2.6.3.1 Effect of Temperature, Concentration, and Surface Modifiers on Density

An increase in the concentration of nanofluids increases the density, whereas an increase in temperature decreases the density values [77].

The addition of surfactants enhance the density of nanofluids since the mass of the substance is more in a given volume [78].

2.6.4 Specific Heat of Nanofluids

One of the major thermal properties in energy systems is specific heat, and it is one of the fundamental thermophysical properties which characterize any thermal fluid. The influences of various parameters on the specific heat of nanofluids are as given below:

2.6.4.1 Effect of Temperature and Volume Fraction on Specific Heat

Increase in temperature showed a marginal increase in specific heat and an increase in volume fraction decrease in the specific heat of nanofluids [77].

2.6.4.2 Effect of Nanoparticle Size and Shape on Specific Heat

From the literature studies, an increase in particle size reduces the specific heat of the nanofluids [77].

The specific heat increases with temperature irrespective of the shape of the particle. The specific heat of spherical particles is more when compared to the rod-shaped particles [77].

2.6.4.3 Influence of Surfactant on Specific Heat

Small amount of surfactant does not cause a change in specific heat, whereas the surfactant added in more quantities increases the specific heat of nanofluid [78].

2.7 Applications of Nanofluids

Nanofluids usage was increased in heat transfer applications since its origination a decade ago. Some of the applications of nanofluids are discussed here.

2.7.1 Electronic Application

Heat dissipation will become more difficult while designing the electronic components with more compactness because of its higher density of chips. Thermal management of advanced electronic devices will be challenging since these devices produce high heat energy and also available surface area should be reduced, which will increase the cost [79]. For the smooth operation of advanced electronic devices, an alternate approach should be used. One approach is to find the optimum geometry for cooling devices, and another approach is to increase the heat transfer capacity. Based on the recent research [80], use of nanofluids was found to increase the heat transfer rate.

2.7.2 Transportation

The universally used automotive coolant, mixture of ethylene glycol and water, is a poor heat transfer fluid compared to water alone. Engine oils work even worse as a heat transfer medium. If the nanoparticles are added to the engine coolant, engine cooling rate would be increased with a reduced size in the cooling system which would lead to increased fuel economy. The low pressure operation of an ethylene glycol nanofluid and its high boiling point, which is desirable for maintaining single-phase coolant, will make nanofluid-based coolant more suitable in the engine cooling system [81].

2.7.3 Industrial Cooling

Significant energy savings and reduction in emissions can be achieved through the use of nanofluids in industrial cooling. The replacement of water with nanofluids in American industries can save up to 1 trillion Btu of energy [82].

2.7.4 Heating Buildings and Reducing Pollution

Nanofluids can be used in building heating systems in the cold regions. Usage of nanofluids needs smaller heating systems, which can deliver the same thermal energy as that of larger heating systems with low cost. Except for the nanofluids cost, the initial equipment cost is quite low and also reduces the environmental pollutants since smaller heating units use less power and less liquid and material waste [83]. The experimental setup for calculating the heat transfer coefficient in conventional finned tube exchanger used in cold countries was shown in Figure 2.7.

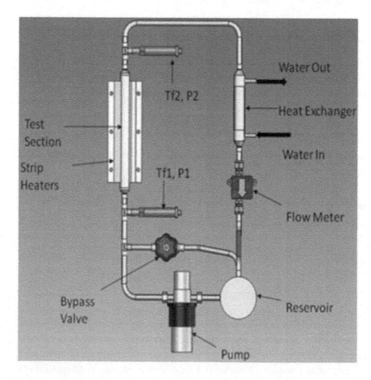

FIGURE 2.7 Experimental set up for the application of heating building and reducing pollution. (From Kulkarni, D.P. et al., *Appl. Energy*, 86, 2566–2573, 2009.)

2.7.5 Space and Defense

Nanofluids have been used in space and defense, in military systems, which include military vehicles, submarines, and high-power laser diodes. Transformer cooling is necessary for navy and power generation industry with the goal of reducing transformer size and weight [84]. Critical heat flux in pool boiling was studied by conducting an experiment, and the results showed an enormous enhancement of critical heat flux for nanofluid compared to water, and the experimental setup was shown in Figure 2.8.

2.7.6 Other Applications

Nanofluids as a phase change material can be used as an efficient thermal energy storage system. These nanofluids have high thermal conductivity compared with the base material. The efficiency of solar thermal collectors can be increased by using nanofluids as the absorption mechanism. The nanofluid-based solar collector has longer payback period for domestic hot water systems [85].

Nanofluids can also be used in surgical areas, which enhance patient's chance of survival and reduce the risk of organ damage by producing cooling efficiently. Nanofluids can be used to provide higher temperature around tumors, which kills cancerous cells without harming healthy cells. Iron-based nanofluids are used as vehicles for drug or for application of radiation in cancer patients. Magnetic nanofluids are used to carry the particles from the blood stream to a tumor with magnets which help doctors to deliver high local doses to the infectious tissue without affecting the healthy tissues [86].

Nanofluids are also used in hydraulic braking systems because of their high boiling point, viscosity, and conductivity. The usage of nanofluids reduces the occurrence of vapor lock and increases safety while driving. Nanofluids-based solar collector is an emerging technology which leads to higher sunlight absorption [82].

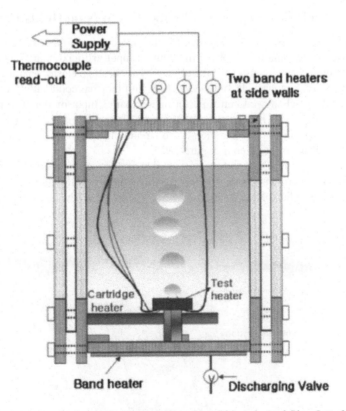

FIGURE 2.8 Experimental setup for the pool boiling test. (From You, S.M. et al., *Appl. Phys. Lett.*, 83, 3374–3376, 2003.)

Nanofluids are the best source to enhance the heat transfer rate without compromising quality, quantity, and cost of a product or process. Nanofluids save energy in heating, ventilating, and air-conditioning systems and reduce environmental pollutants. These nano coolants can be used in process industries like chemical, materials, oil and gas, paper and printing, and textiles.

Hybrid nanofluids are expected to be used for similar applications like single-phase nanofluids with better performance. There is a limited number of research on application-oriented hybrid nanofluids. Silica/MWCNT hybrid nanofluids are used as drilling fluid [50]. Copper titania hybrid nanofluid is used in tube type counter flow heat exchanger for better heat transfer effectiveness.

2.8 Challenges in the Synthesis and Application of Hybrid Nanofluid

The stability of nanocomposite particles in the base fluid is a major challenge in the synthesis of hybrid nanofluids. While for single-phase nanofluids, stability of the particles is achieved with well-established techniques, for nanocomposite-based nanofluids, the dispersion of two different materials in the base fluid poses a considerable problem due to the surface charge (positive or negative), which varies from one particle to another one. Therefore, when preparing hybrid nanofluids, it should be given to the following aspects: (i) selection of appropriate materials, (ii) synthesis of the nanocomposite materials, (iii) bonding between the materials involved in the composite, and (iv) use of adequate surfactant. An unfortunate drawback associated with the hybrid nanofluids is their increase in viscosity, when compared to that of the base fluid. The nanofluids, when used in internal flows, lead to an increase of pressure drop with the consequent increase in pumping power.

2.9 Optimization of Thermophysical Properties of Nanofluids Using ANN

The thermophysical properties of nanofluids can be optimized using ANN in MATLAB. The thermal conductivity of oxide-water nanofluids for a 3 input and 4 input artificial neural network model can be investigated. The 4 input model considered the effect of nanoparticle cluster average size in addition to the effect of temperature, nanoparticle volume fraction, and thermal conductivity of nanoparticle. The result showed 4 input model showed better performance than 3 input model. The effect of Brownian motion of the nanoparticles and nanoparticles agglomeration on thermal conductivity was highlighted by ANN model [86]. A pictorial representation for the modeling of density using multilayer perceptron feed forward network structure is given in Figure 2.9.

The thermal conductivity of alumina water nanofluids at different temperature and volume fraction could be modelled and a correlation was developed using experimental data with the modelled data in terms of temperature and solid volume fraction. The training data for ANN was the thermal conductivity

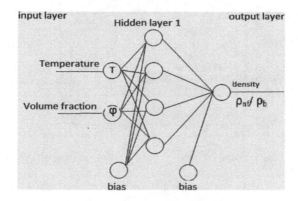

FIGURE 2.9   ANN structure for the modeling of density. (From Kannaiyan, S. et al., *Chin. J. Chem. Eng.*, 2018.)

of nanofluids at different temperatures. The results indicated good ability of the proposed correlation for predicting the thermal conductivity of nanofluids, and the model showed good agreement with the experimental data [88].

The performance of nanofluids by applying central composite rotatable design (CCRD) and ANN can be optimized. The evaluation of the performance was measured by the stability and thermal conductivity ratio of nanofluids which was based on independent variables like temperature, particle volume fraction, and pH of the solution. They constructed second order polynomial equation for the target output by accomplishing 20 experiments. The results indicated that the predicted values were in good agreement with the experimental data and proved that ANN was a promising method compared to CCRD because of the complexity involved in CCRD [89].

The thermal conductivity enhancement based on volume fraction, particle size, and temperature using ANN for magnesium oxide (MgO)/ethylene glycol nanofluids may be modelled. The properties of nanofluids were measured experimentally and an empirical correlation was developed using modelled data and experimental data. At low concentrations, changing particle size was not taken into account, whereas at high concentrations and when the particle size was changed from 60 to 20 nm, the enhancement of thermal conductivity was about 8% and 10%. ANN modelling was done using a 3 input model, and the results obtained were in reasonable agreement with the experimental data [90].

The thermal conductivity ratio of various nanofluids like alumina in water, alumina in ethylene glycol, copper in water, copper in ethylene glycol etc., using 5 input ANN model can be investigated. A total of 776 experimental data points were collected from 21 sources of data available in the literature and a network was designed and trained. Input variables of the network were average diameter, volume fraction, thermal conductivity of nanoparticles, and temperature, and the target was the thermal conductivity ratio. The results showed that experimental data were in good agreement with the developed optimal ANN model [91].

The thermal conductivity of non-Newtonian nanofluids as a function of temperature, concentration, and thermal conductivity of nanofluids by proposing a neural network model can be predicted [92].

2.10 Conclusions

The present work is focused on the synthesis of hybrid nanofluids, its thermophysical properties, and various parameters which affect the thermophysical properties. Further, various models used in the calculation of nanofluids and the optimization of thermophysical properties were also studied. Due to the addition of nanoparticles, thermal conductivity increases with an increase in volume fraction. An increase in temperature substantially increases the thermal conductivity because of the molecular vibrations. Increase in volume fraction enhances the density of nanofluids, and an increase in temperature decreases the density values. An increase in temperature decreases the viscosity, whereas the increase in volume fraction significantly increase the viscosity of nanofluids. With the addition of particles, the specific heat decreases with an increase in volume fraction while the specific heat increases with increase in temperature. Dilute nanofluids system enhance the heat transfer rate and, hence, dilute hybrid nanofluids could be a better alternative to conventional coolants to improve the heat transfer efficiency.

ACRONYMS AND ABBREVIATIONS

CVFEM	Control volume based finite element method
Al_2O_3	Alumina
Fe_3O_4	Ferric Oxide
LBM	Lattice Boltzmann method
Cu	Copper
TiO_2	Titania
ANN	Artificial Neural Network
CNT	Carbon Nanotube
MWNT	Multi Walled Nanotube

Al	Aluminium
Zn	Zinc
MgO	Magnesium Oxide
k_{eff}	Effective thermal conductivity of nanofluid
k_f	Thermal conductivity of base fluid
k_p	Thermal conductivity of nanoparticle
ϕ	Volume fraction of nanoparticle
n	Shape factor
Δ	Thermal conductivity of suspension regardless of particle motion and vibration
μ_{nf}	viscosity of the nanofluid
μ_f	viscosity of the base fluid
k	self-crowding factor
ξ	Fitting parameter
ϕ_m	maximum particle packing fraction
η	intrinsic viscosity
ρ_{nf}	density of the nanofluid
ρ_p	density of the particle
ρ_{bf}	density of the base fluid
$C_{p,nf}$	specific heat of nanofluid
$C_{p,f}$	specific heat of base fluid
$C_{p,n}$	specific heat of nanoparticles
ANN	Artificial Neural Network
CCRD	Central composite rotatable design (CCRD)

REFERENCES

1. Ahuja, AS (1975). Augmentation of heat transport in laminar flow of polystyrene suspensions. I. Experiments and results. *Journal of Applied Physics* 46, 3408–3416.
2. Bergles, AE (1973). Recent development in convective heat transfer augmentation. *Applied Mechanical Reviews* 26, 675–682.
3. Eastman, JA, Choi, SUS, Li, S, Yu, W, & Thompson, LJ (2001). Anomalously increased effective thermal conductivities of ethylene glycol-based nanofluids containing copper nanoparticles. *Applied Physics Letters* 78(6), 718–720.
4. Choi, SUS & Eastman JA (1995). Enhancing thermal conductivity of fluids with nanoparticles. *ASME International Mechanical Engineering Congress and Exposition*, San Francisco, CA, pp. 1–7.
5. Zeinali Heris, S, Nasr Esfahany, M, & Etemad, G (2007). Numerical investigation of nanofluid laminar convective heat transfer through a circular tube. *Numerical Heat Transfer, Part A: Applications: An International Journal of Computation and Methodology* 52(11), 1043–1058.
6. Siddique, M, Khaled, ARA, Abdulhafiz, NI, & Boukhary, AY (2010). Recent advances in heat transfer enhancements: A review report. *International Journal of Chemical Engineering* 2010, 1–28.
7. Wang, XQ & Mujumdar, AS (2008). A review on nanofluids: Part II—Experiments and applications. *Brazilian Journal of Chemical Engineering* 25(4), 631–348.
8. Sheikholeslami, M, Gorji-Bandpay, M, & Ganji DD (2012). Magnetic field effects on natural convection around a horizontal circular cylinder inside a square enclosure filled with nanofluid. *International Communications in Heat and Mass* 39, 978–986.
9. Sheikholeslami, M, Mustafa, MT, & Ganji, D (2015). Nanofluid flow and heat transfer over a stretching porous cylinder considering thermal radiation. *Iranian Journal of Science and Technololy* 39, 433–440.
10. Pak, BC & Cho, YI (1998). Hydrodynamic and heat transfer study of dispersed fluids with submicron metallic oxide particles. *Experimental Heat Transfer* 11, 151–170.
11. Brinkman, H (1952). The viscosity of concentrated suspensions and solutions. *Journal of Chemical Physics* 20, 571.
12. Patel, HE, Sundarrajan, T, Pradeep, T, Dasgupta, A, Dasgupta, N, & Das, SK (2005). A micro convection model for thermal conductivity of nanofluid. *Pramana: Journal of Physics* 65, 863–869.

13. Sheikholeslami, M, Hayat, T, & Alsaedi, A (2017). Numerical study for external magnetic source influence on water based nanofluid convective heat transfer. *International Journal of Heat and Mass Transfer* 106, 745–755.

14. Sundar, LS & Singh, MK (2013). Convective heat transfer and friction factor correlations of nanofluid in a tube and with inserts: A review. *Renewable and Sustainable Energy Reviews* 20, 23–35. doi:10.1016/j. rser.2012.11.041.

15. Eastman, JA, Choi, US, Li, S, Thompson, LJ, & Lee, S (1996). Enhanced thermal conductivity through the development of nanofluids. MRS Online Proceedings Library Archive 457.

16. Sheikholeslami, M & Rokni, HB (2017). Simulation of nanofluid heat transfer in presence of magnetic field: A review. *International Journal of Heat and Mass Transfer* 115, 1203–1233.

17. Sheikholeslami, M (2017). Magnetohydrodynamic nanofluid forced convection in a porous lid driven cubic cavity using Lattice Boltzmann method. *Molliq.* doi:10.1016/j.molliq.2017.02.020.

18. Sheikholeslami, M (2016). Influence of Coulomb forces on Fe_3O_4-H_2O nanofluid thermal improvement. *International Journal of Hydrogen Energy.* doi:10.1016/j.ijhydene.2016.09.185.

19. Sheikholeslami, M & Shamlooei, M (2017). Fe_3O_4-H_2O nanofluid natural convection in presence of thermal radiation. *International Journal of Hydrogen Energy.* doi:10.1016/j.ijhydene.2017.02.031.

20. Sheikholeslami, M & Rokni, HB (2017). Free convection of CuO-H_2O nanofluid in a curved porous enclosure using mesoscopic approach. *International Journal of Hydrogen Energy.* doi:10.1016/j. ijhydene.2017.04.217.

21. Sheikholeslami, M & Shehzad, SA (2017). Thermal radiation of ferrofluid in existence of Lorentz forces considering variable viscosity. *International Journal of Heat and Mass Transfer* 109, 82–92.

22. Madhesh, D, Parameshwaran, R, & Kalaiselvam, S (2014). Experimental investigation on convective heat transfer and rheological characteristics of Cu–TiO_2 hybrid nanofluids. *Experimental Thermal and Fluid Science* 52, 104–115.

23. Sundar, LS, Singh, MK, & Sousa, ACM (2014). Enhanced heat transfer and friction factor of MWCNT–Fe3O4/water hybrid nanofluids. *International Communications in Heat and Mass Transfer* 52, 73–83.

24. Harandi, SS, Karimipour, A, Afrand, M, Akbari, M, & D'Orazio, A (2016). An experimental study on thermal conductivity of F-MWCNTs–Fe_3O_4/EG hybrid nanofluid: Effects of temperature and concentration. *International Communications in Heat and Mass Transfer* 76, 171–177.

25. Suresh, S, Venkitaraj, KP, Selvakumar, P, & Chandrasekar, M (2012). Effect of Al_2O_3–Cu/water hybrid nanofluid in heat transfer. *Experimental Thermal and Fluid Science* 38, 54–60.

26. Moghadassi, A, Ghomi, E, & Parvizian, F (2015). A numerical study of water based Al_2O_3 and Al_2O_3–Cu hybrid nanofluid effect on forced convective heat transfer. *International Journal of Thermal Science* 92, 50–57.

27. Kumar, K (2017). Study on single step method for synthesis of nanofluids. *International Journal for Research in Applied Science and Engineering Technology* 5(2), 586–594.

28. Yu, W & Xie, H (2012). A review on nanofluids: Preparation, stability mechanisms, and applications. *Journal of Nanomaterials* 2012, 1–17.

29. Allahyari, S, Behzadmehr, A, & Sarvari, SMH (2011). Conjugate heat transfer of laminar mixed convection of a nanofluid through an inclined tube with circumferentially non-uniform heating. *Nanoscale Research Letters* 6(1), 360.

30. Dilip Kumar, K & Uma Maheswar Gowd, B (2013). State of art to reduce the effect of clustering in graphene nano fluids. *International Journal of NanoScience and Nanotechnology* 4(1), 41–52.

31. Jama, M, Singh, T, Mahmoud Gamaleldin, S, Koc, M, Samara, A, Rima, J, & Muataz, A (2016). Critical review on nanofluids: Preparation, characterization, and applications. *Journal of Nanomaterials* 2016, 1–22.

32. Duncan, MA & Rouvray, DH (1989). Microclusters. *Scientific American* 261, 110–115.

33. Kimoto, K, Kamilaya, Y, Nonoyama, M, & Uyeda, R (1963). An electron microscope study on fine metal particles prepared by evaporation in argon gas at low pressure. *Japanese Journal of Applied Physics* 2, 702–713.

34. Granqvist, CG & Buhrman, RA (1976). Ultrafine metal particles. *Journal of Applied Physics* 47(5), 2200–2219.

35. Gleiter, H (1989). Nanocrystalline materials. *Progress in Material Science* 33, 223–315.

36. Zhu, H, Lin, Y, & Yin, Y (2004). A novel one-step chemical method for preparation of copper nanofluids. *Journal of Colloid and Interface Science* 227, 100–103.

37. Saidur, R (2011). A review on applications and challenges of nanofluids. *Renewable and Sustainable Energy Reviews* 15(3), 1646–1668.
38. Choi, US, Zhang, ZG, Yu, W, Lockwood, FE, & Grulke, EA (2001). Anomalous thermal conductivity enhancement in nanotube suspensions. *Applied Physics Letters* 79, 2252–2254.
39. Das, SK, Choi, SUS, Yu, W, & Pradeep, T, editors. (2007). *Nanofluids: Science and Technology.* John Wiley & Sons, New York.
40. Mukherjee, S & Paria, S (2013). Preparation and stability of nanofluids: A review. *IOSR Journal of Mechanical and Civil Engineering* 9(2), 63–69.
41. Ghadimi, A, Saidur, R, & Metselaar, HSC (2011). A review of nanofluid stability properties and characterization in stationary conditions. *International Journal of Heat and Mass Transfer* 54, 4051–4068.
42. Kamiya, H & Iijima, M (2010). Surface modification and characterization for dispersion stability of inorganic nanometer-scaled particles in liquid media. *Science and Technology of Advanced Materials* 11(4), 1–7.
43. Yang, L & Hu, Y (2017). Toward TiO_2 nanofluids: Part 1—Preparation and properties. *Nanoscale Research Letters* 12, 417–437.
44. Sheikholeslami, M & Zeeshan, A. (2017). Analysis of flow and heat transfer in water based nanofluid due to magnetic field in a porous enclosure with constant heat flux using CVFEM. *Computer Methods in Applied Mechanics and Engineering* 320, 68–81.
45. Buongiorno, J (2006). Convective transport in nanofluids. *ASME Journal of Heat Transfer* 128, 240–250.
46. Mehrali, M, Sadeghinezhad, E, Tahan Latibari, S, Mehrali, M, Togun, H, Zubir, MNM, Kazi, SN, & Metselaar, HSC (2014). Preparation, characterization, viscosity, and thermal conductivity of nitrogen-doped graphene aqueous nanofluids. *Journal of Materials Science* 49(20), 7156–7171.
47. Khaleduzzamana, SS, Sohela, MR, Saidura, R, & Selvaraj, J (2015). Stability of Al_2O_3-water nanofluid for electronics cooling system. *6th BSME International Conference on Thermal Engineering, Procedia Engineering* 105, 406–411.
48. Jana, S, Salehi-Khojin, A, & Zhong, WH (2007). Enhancement of fluid thermal conductivity by the addition of single and hybrid nano-additives. *Thermochimica Acta* 462(1–2), 45–55.
49. Suresh, S, Venkitaraj, KP, Selvakumar, P, & Chandrasekar, M (2011). Synthesis of Al_2O_3-Cu/water hybrid nanofluids using two step method and its thermo physical properties. *Colloids and Surfaces A: Physicochemical and Engineering Aspects* 388, 41–48.
50. Baghbanzadeh, M, Rashidi, A, Rashtchian, D, Lotfi, R, & Amrollahi, A (2012). Synthesis of spherical silica/multiwall carbon nanotubes hybrid nanostructures and investigation of thermal conductivity of related nanofluids. *Thermochimica Acta* 549, 87–94.
51. Abbasi, SM, Rashidi, A, Nemati, A, & Arzani, K (2013). The effect of functionalisation method on the stability and the thermal conductivity of nanofluid hybrids of carbon nano-tubes/gamma alumina. *Ceramics International* 39(4), 3885–3891.
52. Nine, MJ, Batmunkh, M, Kim, JH, Chung, HS, & Jeong, HM (2012). Investigation of Al_2O_3-MWCNTs hybrid dispersion in water and their thermal characterization. *Journal of Nanoscience & Nanotechnology* 12(6), 4553–4559.
53. Paul, GJ, Raj, PB, Das, PK, & Manna, I (2011). Synthesis, characterization, and thermal property measurement of nano-Al95Zn05 dispersed nanofluid prepared by a two-step process. *International Journal of Heat and Mass Transfer* 54(15–16), 3783–3788.
54. Munkhbayar, B, Tanshen, MR, Jeoun, J, Chung, H, & Jeong, H (2013). Surfactant-free dispersion of silver nanoparticles into MWCNT-aqueous nanofluids prepared by one-step technique and their thermal characteristics. *Ceramics International* 39(6), 6415–6425.
55. Aravind, SJ & Ramaprabhu, S (2013). Graphene–multi walled carbon nanotube-based nanofluids for improved heat dissipation. *Royal Society of Chemistry Advances* 3(13), 4199–4206.
56. Batmunkh, M, Tanshen, MR, Nine, MJ, Myekhlai, M, Choi, H, & Chung, H (2014). Thermal conductivity of TiO_2 nanoparticles based aqueous nanofluids with an addition of a modified silver particle. *Industrial and Engineering Chemistry Research* 53(20), 8445–8451.
57. Chen, LF, Cheng, M, Yang, DJ, & Yang, L (2014). Enhanced thermal conductivity of nanofluid by synergistic effect of multi-walled carbon nanotubes and Fe_2O_3 nanoparticles. *Applied Mechanics and Materials* 548–549, 118–123.
58. Kannaiyan, S, Boobalan, C, Umasankaran, A, Ravirajan, A, Sathyan, S, & Thomas, T (2017). Comparison of experimental and calculated thermophysical properties of alumina/cupric oxide hybrid nanofluids. *Journal of Molecular Liquids* 244, 469–477.

59. Ho CJ, Huang JB, Tsai PS, & Yang YM (2010). Preparation and properties of hybrid water-based suspension of Al_2O_3 nanoparticles and MEPCM particles as functional forced convection fluid. *International Communications in Heat & Mass Transfer* 37(5), 490–494.

60. Maxwell, JC, editor. (1873). *A Treatise on Electricity and Magnetism.* Clarendon, Oxford, UK.

61. Hamilton, RL & Crosser, OK (1962). Thermal conductivity of heterogeneous two-component systems. *IEC Fundamental* 1(3), 187–191.

62. Xuan, Y & Li, Q (2000). Heat transfer Enhancement of nanofluids. *International Journal of Heat and Fluids Flow* 21, 58–64.

63. Einstein, A (1996). Eineneuebestimmung der moleküldimensionen. *Annals of Physics* 324(2), 289–306.

64. Mooney, M (1951). The viscosity of a concentrated suspension of spherical particles. *Journal of Colloid Science* 6(2), 162–170.

65. Krieger, IM & Thomas, JD (1957). A mechanism for non-Newtonian flow in suspensions of rigid spheres. *Transactions of the Society of Rheology* 3(1), 137–152.

66. Nielsen, LE (1970). Generalized equation for the elastic moduli of composite materials. *Journal of Applied Physics* 41(11), 4626–4627.

67. Batchelor, GK (1977). The effect of Brownian motion on the bulk stress in a suspension of spherical particles. *Journal of Fluid Mechanics* 83(1), 97–117.

68. Zhou, S & Ni, R (2008). Measurement of the specific heat capacity of water based Al_2O_3 nanofluid. *Applied Physics Letters* 92, 093123.

69. Lee, S, Choi, SUS, Li, S, & Eastman, JA (1999). Measuring thermal conductivity of fluids containing oxide nanoparticles. *Journal of Heat Transfer* 121, 280–289.

70. Sridhara, V & Satapathy, LN (2015). Effect of nanoparticles on thermal properties enhancement in different oils: A review. *Critical Reviews in Solid State and Materials Sciences* 40(6), 399–424.

71. Moosavi, M, Goharshadi, EK, & Youssefi, A (2010). Fabrication, characterization, and measurement of some physicochemical properties of ZnO nanofluids. *International Journal of Heat and Fluid Flow* 31(4), 599–605.

72. Xie, HQ, Wang, JC, Xi, TG, Liu, Y, & Ai, F (2002). Dependence of the thermal conductivity of nanoparticle-fluid mixture on the base fluid. *Journal of Material Science Letters* 21, 1469–1471.

73. Pirahmadian, MH & Ebrahimi, A (2012). Theoretical investigation heat transfer mechanisms in nanofluids and the effects of clustering on thermal conductivity. *International Journal of Bioscience, Biochemistry and Bioinformatics* 2(2), 90–94.

74. Jarahnejad, M, Haghighi, EB, Saleemi, M, Nikkam, N, Khodabandeh, Björn Palm, R, Toprak, MS, & Muhammed, M (2015). Experimental investigation on viscosity of water-based Al_2O_3 and TiO_2 nanofluids. *Rheologica Acta* 54(5), 411–422.

75. Ravisankar, B & Tara Chand, V (2013). Influence of nanoparticle volume fraction, particle size and temperature on thermal conductivity and viscosity of nanofluids: A review. *International Journal of Automotive and Mechanical Engineering* 8, 1316–1338.

76. Kong, L, Sun, J, & Bao, Y (2017). Preparation, characterization and tribological mechanism of nanofluids. *Royal Society of Chemistry Advances* 7, 12599–12609.

77. Vajjha, RS & Das, DK (2008). Measurements of specific heat and density of Al_2O_3 nanofluid. *AIP Conference Proceedings*, pp. 361–370.

78. Madhu, P & Rajasekhar, GP (2017). Measurement of density and specific heat capacity of different nanofluids. *International Journal of Advance Research, Ideas and Innovations in Technology* 3(1), 165–170.

79. Tsai, CY, Chien, HT, Ding, PP, Chan, B, Luh, TY, & Chen, PH (2004). Effect of structural character of gold nanoparticles in nanofluid on heat pipe thermal Performance. *Materials Letters* 58, 1461–1465.

80. Kleinstreuer, C & Xu, Z (2016). Mathematical modeling and computer simulations of nanofluid flow with applications to cooling and lubrication. *Fluids* 1(16), 1–33.

81. Zhang, Z & Que, Q (1997). Synthesis, structure and lubricating properties of dialkyldithiophosphate-modified Mo-S compound nanoclusters. *Wear* 209, 8–12.

82. Wong, KV & Leon, OD (2010). Review article: Applications of nanofluids—Current and future. *Advances in Mechanical Engineering* 2010, 1–11.

83. Kulkarni, DP, Das, DK, & Vajjha, RS (2009). Application of nanofluids in heating buildings and reducing pollution. *Applied Energy* 86(12), 2566–2573.

84. You, SM, Kim, JH, & Kim, KH (2003). Effect of nanoparticles on critical heat flux of water in pool boiling heat transfer. *Applied Physics Letters* 83, 3374–3376.
85. Ravikumar, J & Vinod Kumar Goud, P (2014). Nanofluids: A promising future. *Journal of Chemical and Pharmaceutical Sciences*, 2014, 57–61.
86. Longo, GA, Zilio, C, Ceseracciu, E, & Reggiani, M (2012). Application of artificial neural network (ANN) for the prediction of thermal conductivity of oxide-water nanofluids. *Nano Energy* 1, 290–296.
87. Kannaiyan, S, Boobalan, C, Nagarajan, FC, & Sivaraman, S (2018). Modeling of thermal conductivity and density of alumina/silica in water hybrid nanocolloid by the application of artificial neural networks. *Chinese Journal of Chemical Engineering*. doi:10.1016/j.cjche.2018.07.018.
88. Esfe, MH, Afrand, M, Yan, WM, & Akbari, M (2015). Applicability of artificial neural network and nonlinear regression to predict thermal conductivity modeling of Al_2O_3: Water nanofluids using experimental data. *International Communications in Heat and Mass Transfer* 66, 246–249.
89. Beydokhti, AK, Namaghi, HA, Asgarkhani, MAH, & Heris, SZ (2015). Prediction of stability and thermal conductivity of SnO_2 nanofluid via statistical method and an artificial neural network. *Brazilian Journal of Chemical Engineering* 32(4), 903–917.
90. Esfe, MH, Saedodin, S, Bahiraei, M, Toghraie, D, Mahian, O, & Wongwises, S (2014). Thermal conductivity modeling of MgO/EG nanofluids using experimental data and artificial neural network. *Journal of Thermal Analysis and Calorimetry* 118, 287–294.
91. Ahmadloo, E & Azizi, S (2016). Prediction of thermal conductivity of various nanofluids using artificial neural network. *International Communications in Heat and Mass Transfer* 74, 69–75.
92. Hojjat, M, Etemad, SG, Bagheri, R, & Thibault, J (2011). Thermal conductivity of non-Newtonian nanofluids: Experimental data and modeling using neural network. *International Journal of Heat and Mass Transfer* 54, 1017–1023.

3

Hybrid Nanoparticles Enriched Cutting Fluids in Machining Processes

Anuj Kumar Sharma, Arun Kumar Tiwari, Amit Rai Dixit, and Rabesh Kumar Singh

CONTENTS

3.1 Introduction

During dry machining of steels, due to high degree of heat generation at the machining zone, the cutting velocity of the tool gets restricted. Moreover, the heat affects hardness and sharpness of the cutting tools and results in their premature breakage. Therefore, a suitable cutting fluid becomes necessary in order to overcome these difficulties in high speed machining. The cutting fluid plays a vital role by cooling and lubricating the cutting tool-workpiece interface and washing away the chips from machining zone [1]. This conventional way of cooling, however, serves the purpose up to an extent. The excessive use of the conventional cutting fluids pollutes the environment and may even be hazardous for human beings. Moreover, the cutting fluid used in machining occupies 16%–20% of the cost of production in the manufacturing industry, hence, the extravagant use of conventional cutting fluids should be restricted [2].

To have restriction over the extravagant consumption of cutting fluids in manufacturing industry, various techniques can be attempted. One way may be the dry machining [3]. However, in most machining situations, dry machining cannot be preferred with high depth of cut as it causes short tool life [4–5]. The minimum quantity lubrication (MQL) may be an alternative to dry machining in which the cutting fluid is sprayed with pressurized air over tool-workpiece interface optimally [6]. The use of MQL can effectively reduce the machining forces, especially at low cutting speeds [7–9]. Li and Chou [10] noticed that by using the MQL technique in milling, the tool flank wear length in the cutting tool could be reduced up to 60% compared to dry machining. A few researchers like Kishawy et al. [11] and Li and Lin [12] concluded that the use of the MQL technique improves the surface finish, tool life, and reduces impact of the cutting forces. Furthermore, Bhowmick et al. [13] observed that the use of MQL with fatty acid fluids provides a performance equal to that of flooded tapping with high quality threads. The MQL affects the cutting temperature significantly [14] over a wide range of speeds and yields a lower cutting tool wear rate as compared to completely dry machining [15]. Dhar et al. [16] noticed that MQL jet provided reduced tool wear, improved tool life, and better surface finish as compared to dry and wet turning of steel. This may be attributed due to the reduction in the cutting zone temperature and favourable changes in the chip-tool and work-tool interaction. Khan et al. [17] used MQL in turning of American Iron and Steel Institute (AISI) 9310 alloy steel and observed that MQL systems enabled reduction in average chip-tool interface temperature up to 10% as compared to wet machining with a substantial reduction in tool wear which in turn enhanced the tool life and surface finish. In their opinion, it can be a viable alternative to wet machining because the MQL technique can minimize both the manufacturing cost and the environmental hazards. However, the application of coolants with lower thermal conductivity, even with the help of MQL system cannot fulfil the need of green machining. There is another technique which seems more appropriate, i.e., spraying the cutting fluid of high thermal conductivity using MQL.

The conventional fluids may possess good lubrication properties, but poor thermal properties exhibited by them restrict their use as a cutting fluid for commercial purpose. Therefore, to overcome this problem, nanometre-sized particles have been added into conventional fluids, leading to the synthesis of a new generation of fluids, which are called "nanofluids." An improvement of up to 22.4% in thermal conductivity of conventional fluid at room temperature could be achieved by adding 6% of Al_2O_3 in the base fluid [18]. Yang [19] and Choi et al. [20] noticed a massive increment of approximately 200% and 150%, respectively, in thermal conductivity when multi-walled carbon nanotube was added to the base fluid. Tansen et al. [21] observed that a little inclusion of multi-walled carbon nanotube nanoparticle into water-based alumina solution makes it a potential heat transfer fluid to transport heat efficiently. Zhang et al. [22] used MoS_2-CNT hybrid nanofluid in grinding. They observed that for the same mass fraction, MoS_2-CNTs hybrid nanofluid achieved lower G ratio and surface roughness ($R_a = 0.328$ μm) than MoS_2 and CNTs. This may be attributed to the physical collaboration of the mixed nanoparticles.

In present investigation, the alumina-molybdenum disulphide (Al-MoS_2) hybrid nano-cutting fluid is developed by mixing alumina (Al_2O_3)-based nanofluid with molybdenum disulphide (MoS_2) nanoparticles in a fixed volumetric proportion of 90:10 at different nanoparticle fractions (0.25 vol%, 0.75 vol%, and 1.25 vol%). Esfe et al. [23] observed that the fixed volumetric proportion of 90:10 has shown significant improvement. Therefore, the authors have decided to perform investigation for fixed volumetric proportion (90:10) in the present work. The tribology and spreadability of all the nanofluid samples is investigated on pin-on-disc tribology test rig and goniometer, respectively [24]. Later, the machining performances of alumina and its hybrid nanofluid as a cutting fluid during turning of AISI 304 steel is evaluated and compared regarding tool flank wear (VB) under MQL technique.

3.2 Experimentation

The base fluid was prepared by mixing 5 vol% vegetable oil in distilled water. The commercially available colloidal suspension containing 25% of Al_2O_3 nanoparticles (45 nm in diameter) in water was procured from Alfa Aesar® and a water-based nanofluid having MoS_2 (30 nm in size) was purchased from Sigma-Aldrich. The hybrid nanofluid was prepared by mixing Al_2O_3-based nanofluid with MoS_2-based nanofluid in volumetric ratio of 90:10 with the same base fluid in three volumetric concentrations (0.25% vol., 0.75% vol., and 1.25% vol.). The transmission electron microscopy (TEM) images of the three types of nanofluids (alumina, MoS_2 and Al-MoS_2 hybrid nanofluids) are shown in Figure 3.1. The turning of AISI 304 steel was carried out on HMT (model NH 22/1500) lathe machine under mist of alumina and its hybrid nanofluid using MQL technique as explained in [24]. The schematic of experimental setup is illustrated in Figure 3.2 and the machining variables are tabulated in Table 3.1.

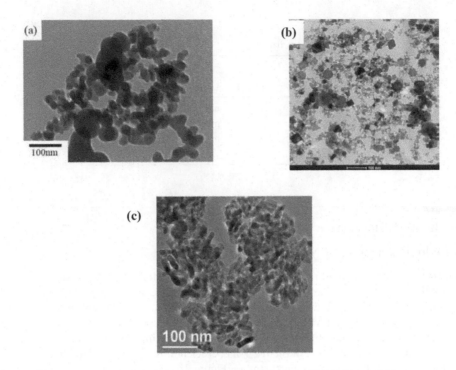

FIGURE 3.1 TEM images of (a) MoS$_2$, (b) Alumina nanofluid, and (c) Al-MoS$_2$ hybrid nanofluid.

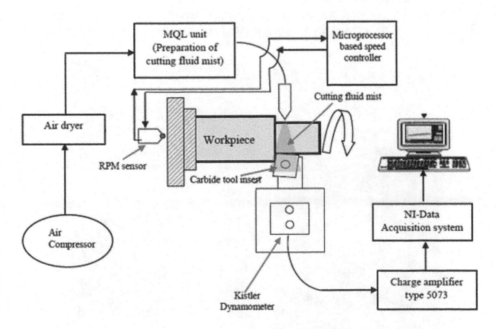

FIGURE 3.2 Schematic diagram of experimental setup. (From Sharma, A.K. et al., *J. Manuf. Process*, 30, 467–482, 2017.)

TABLE 3.1

Machining Variables and Their Levels

Machining Variables	Symbol	Units	Levels		
			1	2	3
Cutting speed	V	m/min	60	90	120
Feed rate	f	mm/rev	0.08	0.12	0.16
Depth of cut	d	mm	0.6	0.9	1.2
Nanoparticle volumetric fraction	ϕ	vol%	0.25	0.75	1.25

3.3 Results and Discussion

3.3.1 Tribological Testing of Nanofluids

All the cutting fluid samples (base fluid, alumina, and its hybrid nanofluid) were investigated on pin-on-disc tribotester for their tribological behaviour as explained in [24] (Figure 3.3). Figure 3.4 shows that Al-MoS$_2$ hybrid nanofluid (1.25 vol%) reported the lowest coefficient of friction. It is further observed that coefficient of friction showed increasing behaviour with the reduction of nanoparticle concentration. This may be attributed to the enhanced ball bearing effect of nanoparticles due to increase in concentration of nanoparticles in cutting fluid. This lower value of the friction coefficient might be helpful in reducing the friction force and, hence, the machining forces. Furthermore, Figure 3.5 illustrates that the pin wear got reduced with the increase of nanoparticle concentration. For hybrid nanofluid samples, the wear became almost uniform after certain time interval. This may be attributed due to the formation of nano-film between the sliding surfaces due to its superior spreadability over mono type nanofluid (Figure 3.7). Moreover, it establishes hybrid nanofluid to be the superior lubricant over alumina nanofluid and the base fluid. The presence of alumina nanoparticles in cutting fluid generates ball bearing effect between the sliding surfaces [7]. However, the blending of MoS$_2$ with alumina nanoparticles has further reduced the friction coefficient (Figure 3.4) between sliding surfaces due to synergic effect of hybrid nanoparticles as explained in [9].

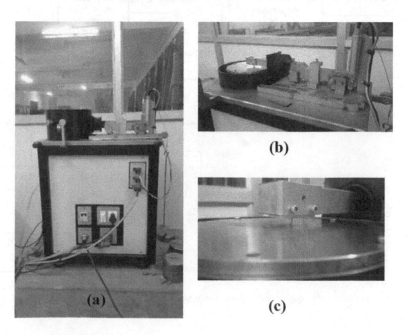

FIGURE 3.3   (a) Pin-on-disc tribotester, (b) pin and disc machine, and (c) closed view of sliding pin on rotating disc.

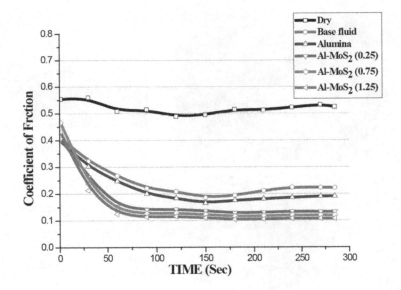

FIGURE 3.4 Coefficient of friction variation with time.

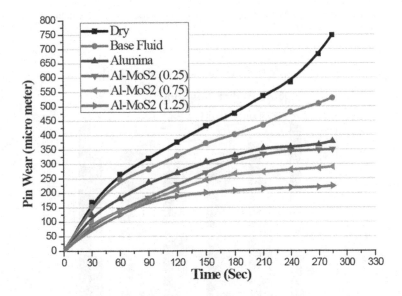

FIGURE 3.5 Variation of pin wear with time on pin on disc tribometer.

The spreadability of all the cutting fluid samples were investigated on goniometer (Figure 3.6) as explained in [24]. The wettability characteristics of any cutting fluid can be determined by the measurement of the contact angle between the solid surface and the droplet. The determination of contact angle is often based on Young's [25] contact angle equation (Eq. 3.1). The contact angle is given by:

$$\cos\theta = \frac{\sigma_{sv} - \sigma_{sl}}{\sigma_{lv}} \tag{3.1}$$

where θ is equilibrium contact angle, σ_{sl}, σ_{sv}, and σ_{lv} are solid-liquid, solid-vapour, and liquid-vapour interfacial tensions, respectively. The smallest contact angle for Al-MoS$_2$ and alumina nanofluids could be recorded as 38.6 (at 0.25 vol%) and 41.9 (at 1.0 vol%), respectively, (Figure 3.7) thus, gave maximum wetting area per unit liquid volume.

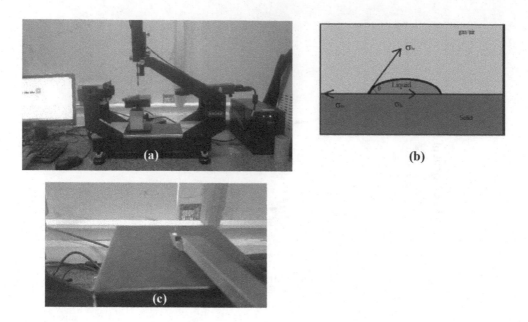

FIGURE 3.6 (a) Contact angle measurement setup, (b) schematic diagram showing a liquid droplet on solid surface, (c) closed view of dropper and carbide tool.

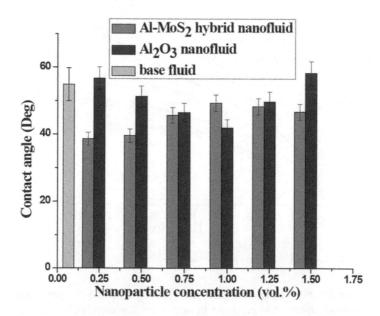

FIGURE 3.7 Contact angle variation with nanoparticle concentration of alumina and Al/MoS₂ hybrid nanofluids.

3.3.2 Machining with Nanofluids

Table 3.2 shows the results of ANOVA for VB. This analysis was carried out at a significance level of 5% (i.e., 95% confidence level). The left column of the table shows the influence of variation in input variables (significant or non-significant) on response (VB). It is found that nanoparticle volumetric fraction (ϕ) and its interaction with cutting speed has a statistical significance on VB.

The regression model of VB with a coefficient of determination (R^2) and adjusted R^2 equal to 96.5 and 92.41, respectively, is given below in Eq. (3.2).

TABLE 3.2

ANOVA table of VB for Al-MoS$_2$ Hybrid Nanofluid

Source	Sum of Squares	DF	Mean Square	F-value	Prob.	Remarks
Model	35346.87	14	2524.78	23.61	<0.0001	Significant
A–V	540.16	1	540.16	5.05	0.0442	Significant
B–f	8148.44	1	8148.44	76.19	<0.0001	Significant
C–d	20092.54	1	20092.54	187.87	<0.0001	Significant
D–ϕ	604.07	1	604.07	5.65	0.0350	Significant
AB	242.42	1	242.42	2.27	0.1580	
AC	545.92	1	545.92	5.10	0.0433	Significant
AD	2602.02	1	2602.02	24.33	0.0003	Significant
BC	9.77	1	9.77	0.091	0.7677	
BD	0.99	1	0.99	9.257E-003	0.9249	
CD	83.81	1	83.81	0.78	0.3934	
A^2	338.53	1	338.53	3.17	0.1005	
B^2	522.63	1	522.63	4.89	0.0472	Significant
C^2	533.11	1	533.11	4.98	0.0454	Significant
D^2	276.74	1	276.74	2.59	0.1337	
Residual	1283.41	12	106.95			
Lack of Fit	1238.27	10	123.83	5.49	0.1639	Not significant
Pure Error	45.14	2	22.57			
Cor Total	36630.28	26				

$$VB = 155.695 - 4.59178 \times V + 1650.99 \times f + 212.268 \times d - 240.89 \times \phi + 6.4875 \times V \times f$$

$$+ 1.29806 \times V \times d + 1.70033 \times V \times \phi - 130.208 \times f \times d + 24.875 \times f \times \phi \qquad (3.2)$$

$$+ 30.5167 \times d \times \phi + 0.00885231 \times V^2 - 6186.98 \times f^2 - 111.088 \times d^2 + 28.8133 \times \phi^2$$

To investigate the influence of nanoparticle concentration on various response variables, the response surfaces are drawn in Figure 3.8. The lowest tool wear was observed at a combination of highest $n_p\%$ and lowest cutting speed (Figure 3.8a) and at highest $n_p\%$ and lowest feed rate (Figure 3.8b).

The application of Al-MoS$_2$ hybrid nanofluid shows the lowest tool wear among all other nanofluids. This reduction in the tool wear could be due to the better diffusion and penetration of the nanofluid mist into the machining zone, better spreadability, and reduced friction force (Table 3.3).

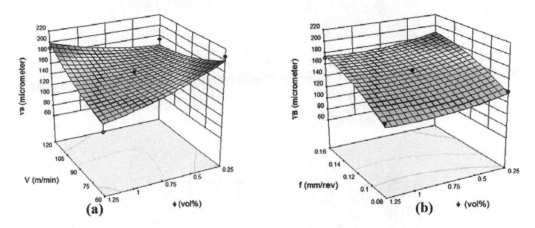

FIGURE 3.8 Estimated response surface plots for Al-MoS$_2$ nanoparticle volumetric fraction (ϕ) (a) versus cutting velocity (V) and (b) versus feed (f).

TABLE 3.3

Performance Comparison of Alumina and Its Hybrid (Al-MoS$_2$) Nanofluid

Nano-cutting Fluid	Trial No.	Tool Wear (VB) μm	Average Tool Wear (VB) μm
Dry Machining	1	193.22	
	2	183.57	194.56
	3	206.89	
Base Fluid	1	98.76	
	2	92.19	93.43
	3	89.34	
Al$_2$O$_3$	1	71.88	
	2	65.43	68.39
	3	67.86	
Al-MoS$_2$	1	59.96	
	2	58.69	60.16
	3	61.83	

In the case of Al-MoS$_2$ hybrid nano-cutting fluid, due to its lubricity (Figure 3.4) and better wettability (Figure 3.7) compared to the conventional fluid, the tool retains its original hardness for longer machining time. Another explanation could be the ball bearing effect of nanoparticles present in the cutting fluid. Lower friction force reduces the cutting force and nodal temperature as well. Because of its superior tribological properties and spreadability (Figures 3.4, 3.5, and 3.7) over tool surface, Al-MoS$_2$ hybrid nanofluid maintained the cutting tool's temperature at lower values, as a result of which, the tool sustained its hardness and sharpness of cutting edge and resulted in lower tool wear (Figure 3.9).

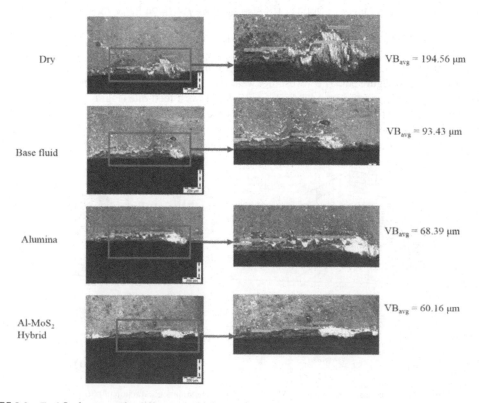

FIGURE 3.9   Tool flank wear under different machining environments.

3.4 Conclusions

- The hybridization of MoS_2 nanoparticles with alumina nanofluid has shown significant improvement in terms of tribological and machining performances.
- The nanoparticle concentration has shown significant effect on coefficient of friction. The coefficient of friction is reduced with an increase of nanoparticle concentration. The minimum value of coefficient of friction is recorded by using 1.25 vol% Al-MoS_2 hybrid nanofluid (~0.1) lower than monotype alumina-based nanofluid and much lower than base fluid.
- Both the nanofluids have demonstrated a significant variation in contact angle with increase of nanoparticle concentration. Al-MoS_2 hybrid exhibits minimum contact angle at 0.25 vol% and alumina-based nanofluid yields minimum value at 1.0 vol%.
- A significant reduction in tool flank wear is recorded using hybrid nanofluid. The hybrid nanofluid yielded a reduction of ~69%, ~36%, and ~12% as compared to dry machining, base fluid, and monotype alumina nanofluid, respectively.

The mixing of alumina and MoS_2 in a fixed volumetric ratio (90:10) enhanced its tribological and machining performances. However, the optimization of mixing proportion may further enhance its performances. Moreover, the further investigation can be attempted to optimize the nanoparticle volume fraction, their shape, and size.

REFERENCES

1. Baradie, M.A.E. (1996). Cutting fluids: Part I. Characterisation. *Journal of Materials Processing Technology* 56, 786–797.
2. Sreejith, P.S., Ngoi, B.K.A. (2000). Dry machining: Machining of the future. *Journal of Materials Processing Technology* 101, 287–291.
3. Granger, C. (1994). Dry machining's double benefit. *Machinery and Production Engineering* 152 (3873), 14–15, 17, 19–20.
4. Diniz, A.E., Micaroni, R. (2002). Cutting conditions for finish turning process aiming the use of dry cutting. *International Journal of Machine Tools and Manufacture* 42, 899–904.
5. Diniz, A.E., Oliveira, A.J. (2004). Optimizing the use of dry cutting in rough turning steel operations. *International Journal of Machine Tools and Manufacture* 44, 1061–1067.
6. Braga, D.U., Diniz, A.E., Miranda, G.W.A., Coppini, N.L. (2002). Using a minimum quantity of lubricant (MQL) and a diamond coated tool in the drilling of aluminum-silicon alloys. *Journal of Materials Processing Technology* 122, 127–138.
7. Sharma, A.K., Tiwari, A.K., Dixit, A.R. (2015). Progress of nanofluid application in machining: A review. *Materials and Manufacturing Processes* 30(7), 813–828.
8. Sharma, A.K., Tiwari, A.K., Singh, R.K., Dixit, A.R. (2016). Tribological investigation of TiO_2 nanoparticle based cutting fluid in machining under minimum quantity lubrication (MQL). *Materials Today: Proceedings* 3, 2155–2162.
9. Sharma, A.K., Tiwari, A.K., Dixit, A.R., Singh, R.K., Singh. M. (2018). Novel uses of alumina/graphene hybrid nanoparticle additives for improved tribological properties of lubricant in turning operation. *Tribology International* 119, 99–111.
10. Li, K.M., Chou, S.Y. (2010). Experimental evaluation of minimum quantity lubrication in near micromilling. *Journal of Materials Processing Technology* 210, 2163–2170.
11. Kishawy, H.A., Dumitrescu, M., Ng, E.G., Elbestawi, M.A. (2005). Effect of coolant strategy on tool performance, chip morphology and surface quality during high speed machining of A356 aluminum alloy. *International Journal of Machine Tools and Manufacture* 45, 219–227.
12. Li, K.M., Lin, C.P. (2012). Study on minimum quantity lubrication in micro-grinding. *The International Journal of Advanced Manufacturing Technology* 62, 99–105.
13. Bhowmick, S., Lukitsch, M.J., Alpas, A.T. (2010). Tapping of Al-Si alloys with diamond-like carbon coated tools and minimum quantity lubrication. *Journal of Materials Processing Technology* 210, 2142–2153.

14. Sharma, A.K., Tiwari, A.K., Dixit, A.R. (2018). Prediction of temperature distribution over cutting tool with alumina-MWCNT hybrid nanofluid using computational fluid dynamics (CFD) analysis. *The International Journal of Advanced Manufacturing Technology* 97(1–4), 427–439.

15. Li, K., Liang, S.Y. (2007). Performance profiling of minimum quantity lubrication in machining. *The International Journal of Advanced Manufacturing Technology* 35, 226–233.

16. Dhar, N.R., Kamruzzaman, M., Mahiuddin, A. (2006). Effect of minimum quantity lubrication (MQL) on tool wear and surface roughness in turning AISI-4340 steel. *Journal of Materials Processing Technology* 172, 299–304.

17. Khan, M.M.A., Mithu, M.A.H., Dhar, N.R. (2009). Effects of minimum quantity lubrication on turning AISI 9310 alloy steel using vegetable oil-based cutting fluid. *Journal of Materials Processing Technology* 209, 5573–5583.

18. Vajjha, R.S., Das, D.K. (2012). A review and analysis on influence of temperature and concentration of nanofluids on thermophysical properties, heat transfer and pumping power. *International Journal of Heat and Mass Transfer* 55, 4063–4078.

19. Yang, Y. (2006). Carbon nanofluids for lubricant application. PhD thesis. Lexington, KY: University of Kentucky.

20. Choi, S.U.S., Zhang, Z.G., Yu, W., Lockwood, F.E., Grulke, E.A. (2001). Anomalous thermal conductivity enhancement in nanotube suspensions. *Applied Physics Letters* 79(14), 2252–2254.

21. Tanshen, M.R., Lee, S., Kim, J., Kang, D., Noh, J., Chung, H., Jeong, H., Huh, S. (2014). Pressure distribution inside oscillating heat pipe charged with aqueous Al_2O_3 nanoparticles, MWCNTs and their hybrid. *Journal of Central South University of Technology* 21, 2341–2348.

22. Zhang, Y., Li, C., Jia, D., Li, B., Wang, Y., Yang, M., Hou, Y., Zhang, X. (2016). Experimental study on the effect of nanoparticle concentration on the lubricating property of nanofluids for MQL grinding of Ni-based alloy. *Journal of Materials Processing Technology* 232, 100–115.

23. Esfe, M.H., Afrand, M., Rostamian, S.H., Toghraie, D. (2017). Examination of rheological behavior of MWCNTs/ZnO-SAE40 hybrid nano-lubricants under various temperatures and solid volume fractions. *Experimental and Thermal Fluid Sciences* 80, 384–390.

24. Sharma, A.K., Tiwari, A.K., Dixit, A.R., Singh, R.K. (2017). Novel uses of alumina-MoS_2 hybrid nanoparticle enriched cutting fluid in hard turning of AISI 304 steel. *Journal of Manufacturing Processes* 30, 467–482.

25. Young, T. (1805). An essay on the cohesion of fluids. *Philosophical Transactions of the Royal Society* 95, 65–87.

Section II

Theoretical Perspectives

4

Nanofluid-Based Single-Phase Natural Circulation Loops

Nur Çobanoğlu, Ziya Haktan Karadeniz, and Alpaslan Turgut

CONTENTS

4.1 Introduction

Heat transfer problems arise in all scales from macro to micro. Although the heat transfer density has been increasing due to high technology devices and forced convection systems are much more visible, passive cooling solutions are still attracting attention. Natural circulation loops (NCLs) are one of the passive systems which can transfer heat from a hot medium to a cold one by natural convection. They use buoyancy force due to the density gradients instead of mechanical forces for heat transfer. Density gradient can be caused by temperature gradient in single-phase NCLs (SPNCLs) and phase change in two-phase NCLs. SPNCLs have advantages of simplicity, easy control, enhanced safety, and reliability in comparison with two-phase NCLs.

SPNCLs have been used in widespread engineering applications such as cooling of nuclear reactors, solar water heaters, and electronic chip cooling. The increasing demand of heat transfer at all scales increases the need for efficient solutions for the engineering problems. The well-known approaches of heat transfer enhancement such as increasing the surface roughness and fluid velocity have been pushed to their limits. Another option to increase heat transfer is changing the working fluid. Water is the most common working fluid used in the SPNCLs for low temperature applications, while molten salts and liquid metals are the alternatives for high temperature cases. Nanofluids, which have been under discussion for the last two decades, are one of the candidates of working fluid for SPNCL applications.

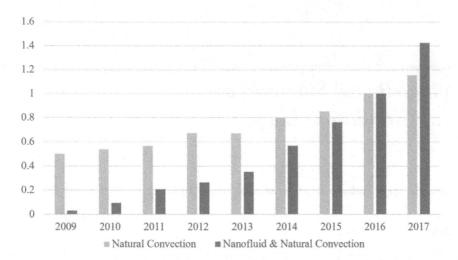

FIGURE 4.1 Normalized numbers of publications in fields of nanofluid and nanofluid & natural convection. Normalization is carried out with values of 2016. (Data taken at July 13, 2018 from ISI Web of Knowledge.)

Addition of the nanoparticles into liquid causes increase in thermophysical properties such as density, viscosity, and thermal conductivity. A promising increase in thermal conductivity with a slight increase in viscosity is reported for some nanofluids. This makes the nanofluids a highly efficient heat transfer fluid for all heat transfer applications. For some nanofluids, the increase in the thermal conductivity and viscosity is in the same order of magnitude. Although the higher thermal conductivity results in a higher heat transfer coefficient, higher viscosity increases the pumping power, which limits nanofluids' usage in the forced convection applications. However, such nanofluids are still useful for natural convection heat transfer systems due to increased flow stability and heat transfer and shifted boiling temperature. Comparison of the normalized numbers of publications in the fields of "natural convection" and "natural convection & nanofluid" is presented in Figure 4.1. 2016 values which are 476 for the keyword "nanofluid & natural convection" and 1678 for "natural convection" are used for normalization. Exponential increase of publications for nanofluid & natural convection is obvious.

Today, numerical and experimental studies on nanofluid-based SPNCLs have increased attraction in pursuance of engineering applications. This chapter firstly presents a general description of SPNCLs considering their geometry, working fluids used, efficiency, and stability issues. Then the details of nanofluid-based SPNCLs and miniaturized SPNCLs (SPNCmL) are given.

4.2 Conventional Single-Phase Natural Circulation Loops

In this part, SPNCLs are introduced by considering their geometry, working fluids used, efficiency, and stability issues. Geometry is the primary factor affecting the efficiency and stability of a SPNCL. Selection of the working fluid is another important issue due to operation conditions such as temperature range, capacity, and environmental impact. As a result of the design and operating conditions, performance of the SPNCL should be determined objectively to understand and compare efficiency of different systems. Finally, stability of the SPNCL is critical for continuous and reliable operation of the system.

4.2.1 Loop Geometry

SPNCLs has been built in different geometries for interested applications. Rectangular or toroidal loops are widely studied, while only a few studies were found on oval shaped point sourced loops. Rectangular loops comprise of four elbows, two straight pipes as vertical legs, and heater and cooler sections having

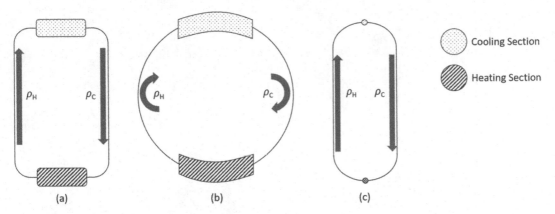

FIGURE 4.2 Loop geometries: (a) rectangular, (b) toroidal, and (c) oval.

an inner straight pipe in which the working fluid flows. There are either outer shells on the heater and cooler parts in which a secondary heat transfer fluid passes through or they are open to a heat transfer source or sink medium. Toroidal loops have circular arcs as heating and cooling portions [1–9]. Oval loops have two straight legs connected by two semi-circular pipes having point source heater and cooler sections at their tips [10–15]. Figure 4.2 represents schematically different geometries of the loops.

Heater and cooler orientations in rectangular loops are also an important parameter for both natural circulation performance and control. In order to increase the efficiency and stabilize the flow, studies on heater-cooler orientations have been carried out. In rectangular loops, there are four different configurations for the heater and cooler locations: horizontal-heater-horizontal-cooler (HHHC), vertical-heater-vertical-cooler, horizontal-heater-vertical-cooler, and vertical-heater-horizontal-cooler. Figure 4.3 represents the different orientations of rectangular loops. Many studies use HHHC orientation, but efficiency of different orientations is still studied [16].

In addition to different heater and cooler orientations, stability and heat transfer enhancement can be also achieved by changing the tilt angle in both lateral and forward/backward directions. Figure 4.4 shows different tilt angle conditions.

4.2.2 Working Fluids

Water is the most common working fluid for low temperature SPNCL applications. Water-based SPNCLs reported in the literature have a heat transfer capacity for a single SPNCL up to 3 kW. Although the

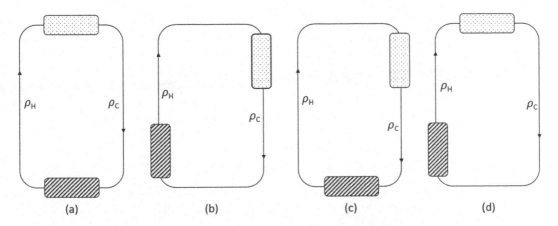

FIGURE 4.3 Orientations of rectangular loops: (a) vertical-heater-vertical-cooler, (b) HHHC, (c) vertical-heater-horizontal-cooler, and (d) horizontal-heater-vertical-cooler.

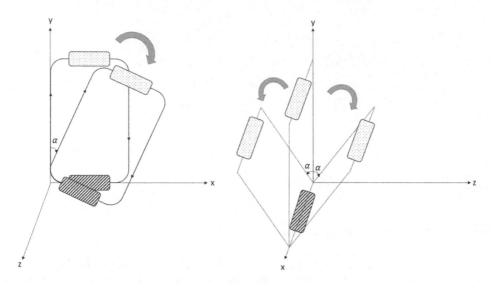

FIGURE 4.4   Different tilt angle conditions.

system size can be increased to increase the heat transfer capacity, stability problems arise as the system gets bigger. This limits the water usage in SPNCLs.

Molten salts (nitrates of potassium and sodium, and fluorides, etc.) are already utilized as thermal storage and heat transport media in solar thermal power systems and high temperature nuclear reactors [17–25]. Designing high temperature systems at low pressures is possible with molten salts due to their low melting point and high boiling point [26,27]. Moreover, their high volumetric capacity, low vapor pressure, and chemical stability can be counted as advantages [28]. The heat transfer capacity is also increased by using molten salts in SPNCLs.

Liquid metals such as (sodium, sodium-potassium alloy, and lead bismuth eutectics, etc.) can also be used as a working fluid in SPNCLs [29].

4.2.3 Performance

Maximum temperature in the loop, temperature difference between the hot and cold legs, and mass flow rate are the important indicators of the thermo-hydraulic performance. In order to investigate the performance of conventional SPNCLs, measurements of temperature were done from the legs of the loop (Figure 4.5).

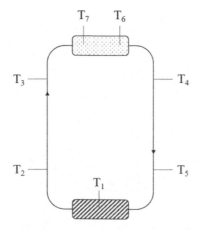

Thermocouple Number	Position
T_1	Heater
T_2	Heater side outlet
T_3	Cooler side inlet
T_4	Cooler side outlet
T_5	Heater side inlet
T_6	Cooler inlet
T_7	Cooler outlet

FIGURE 4.5   General locations for temperature measurements.

Performance of the NCL is also affected by hydrodynamic behavior of the working fluid. Therefore, mass flow rate is a significant parameter to characterize the loop performance, but it cannot be directly measured due to the experimental difficulties. Empirical correlations are mostly used to determine mass flow rate. Numerical study results are also very useful to obtain mass flow rate for different operating conditions.

Temperature vs. heat transfer rate and temperature difference (ΔT_{heater}) vs. heat transfer rate charts include valuable information for comparing the performance of different systems. However, non-dimensional parameters are more accurate, when the compared systems and operating conditions are different. Effectiveness factor (which is a well-known performance indicator for heat exchangers) is used for comparison of performance analysis because of the measurement difficulties of mass flow rate due to the smaller size of the loop [30]. It is defined as the ratio of the actual heat transfer to maximum possible heat transfer.

$$\varepsilon = \frac{T_2 - T_5}{T_2 - T_6} \tag{4.1}$$

In order to describe the flow regime of natural convection flows, Grashof number must be determined. Grashof number is defined as the ratio of gravitational forces to shear stresses and is an indicator of the natural convection intensity of the working fluid [31]. For a stable flow, the modified Grashof number (Gr_m) for a SPNCL is determined by:

$$Gr_m = \frac{D^3 \cdot \rho^2 \cdot Q_h \cdot \Delta z}{A \cdot \mu^3 \cdot c_p} \cdot \beta_T \cdot g \tag{4.2}$$

In Eq. (4.2), D is the inner diameter of the loop (m), Q_h is total heat input rate (W), and Δz is the height of the loop (m). g is gravitational acceleration (m/s²). β_T is thermal expansion coefficient (K⁻¹). c_p is the specific heat (J/kg·K), and A is the flow area (m²). μ is the dynamic viscosity (Pa·s), and ρ is the fluid density (kg/m³).

As mentioned earlier, determination of the mass flow rate is crucial to reveal the performance of the loop. A theoretical approach to calculate the mass flow rate in a natural circulation loop which uses an incompressible Boussinesq fluid, has uniform diameter, and negligible local pressure losses is possible under steady-state flow assumptions. In this simplified case, retarding frictional force balances the driving buoyancy force which is stated as:

$$\rho_0 \cdot \beta_T \cdot g \cdot \oint T dz = \left(\frac{f \cdot L_{total}}{D} \right) \left(\frac{\dot{m}^2}{2 \cdot \rho_0 \cdot A^2} \right) \tag{4.3}$$

by Vijayan et al. [32]. They introduced the mass flow rate equation by the integration of Eq. (4.3) over the closed loops and substitutions of:

$$\Delta T_{heater} = \frac{Q_h}{\dot{m} \cdot c_p} \text{ and } f = \frac{p}{Re_{ss}^b} \tag{4.4}$$

Here, p and b are a constant and an exponent in the friction factor equation, respectively. For laminar flow p is 64 and b equals to 1. For turbulent flow, p is 0.316 and b is 0.25. Steady-state Reynolds (Re_{ss}) number can be written as:

$$Re_{ss} = \frac{\dot{m} \cdot D}{\mu A} \tag{4.5}$$

The corresponding mass flow rate in the loop is than calculated by:

$$\dot{m} = \left[\frac{2 \cdot g \cdot D^b \cdot \rho^2 \cdot \beta_T \cdot A^{2-b} \cdot Q_h \cdot \Delta z}{p \cdot c_p \cdot \mu^b \cdot N_G} \right]^{1/(3-b)}$$

(4.6)

N_G in Eq. (4.6) is effective loss coefficient and introduced as:

$$N_G = \frac{L_{total}}{D} \sum_{i=1}^{N} \left(\frac{l_e}{d^{1+b} \cdot a^{2-b}} \right)_i$$

(4.7)

where a, d, and l are defined as non-dimensional numbers, $a_i = A_i/A_r$ 1, $a_i = L_i/L_{total}$, $d_i = D_i/D_r$. l_e is the effective length and equals to summation of L_{total} and L_{et} which are defined as total loop length (m) and the total equivalent length accounting for the sum of the individual local loss coefficients of the entire loop (m), respectively [33].

This theoretical approach to calculate the mass flow rate results in an opportunity to correlate steady-state Reynolds number (Eq. 4.2) and modified Grashof number (Eq. 4.2) for the laminar and turbulent flows as:

$$Re_{ss} = B \left(\frac{Gr_m}{N_G} \right)^C$$

(4.8)

where B and C are arbitrary constants, and their value depends on the loop geometrical parameters and working fluid's thermophysical properties. Representative characteristic and performance curves for SPNCLs are given (with respect to input power) on Figure 4.6. Increase in the input power increases the temperature difference between inlet and outlet of the heater and mass flow rate. As mass flow rate increases, the pass-through duration in the cooler decreases. Therefore, higher temperature in the cooler outlet results in decrease in effectiveness.

Stanton number is another performance indicator that is used to characterize heat transfer in forced convection flows. which is a dimensionless parameter relating heat transfer coefficient to heat capacity of the fluid stream per unit cross-sectional area per unit time. The modified Stanton number (St_m) is also used to determine the performance of SPNCLs, defined as:

$$St_m = \frac{4 \cdot Nu_m}{Re_{ss} \cdot Pr}$$

(4.9)

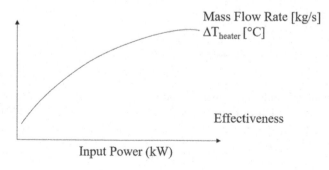

FIGURE 4.6 Change in performance indicators with the input power.

where Pr is the Prandtl number and evaluated at the average fluid temperature, and Nu_m is the modified Nusselt number calculated as:

$$Nu_m = \frac{U \cdot L_{total}}{k} \qquad (4.10)$$

where U is the overall heat transfer coefficient (W/m²·K), and k is the thermal conductivity of the fluid (W/m·K) [34]. Nusselt number is also used for the explanation of the heat transfer efficiency of the SPNCLs. Rayleigh (Ra) number is another non-dimensional number which can be considered as performance indicator due to being useful for investigation of buoyancy effect and thermophysical effects.

$$Ra = \frac{D^4 \cdot \rho^2 \beta_T \cdot g \cdot Q_h \cdot c_p}{\mu \cdot k^2} \qquad (4.11)$$

4.2.4 Stability

SPNCLs are suffering from instabilities because of weak driving force and non-linear nature. Instabilities depend upon the buoyancy and friction forces. When they are out-of phase, instabilities occur. In this circumstance, flow is affected by any disturbance in the driving force and oscillation is generated where the steady-state behavior is expected [35]. Though, these instabilities do not directly affect the average heat transport capability, they result in large amplitude flow and power oscillations as well as mechanical vibrations, premature occurrence of critical heat flux, difficulties in operation, and control of SPNCL system [36].

Because of the mass flow rate or velocity measurement difficulties, temperature measurement is the sole way of observing instabilities. Oscillations in temperature indicate that the natural circulation system exhibits unstable behavior. As shown in Figure 4.7, the time required to reach the stable temperature at lower powers is longer when compared to higher powers. Moreover, maximum temperature and temperature difference between heater inlet and outlet increases as the power increases.

Because of the instabilities, different regimes can be observed in the flow. Steady unidirectional pulsing flow, periodic unidirectional pulsing flow, periodic bidirectional pulsing flow, and oscillatory flow with chaotic switching between unidirectional and bidirectional pulsing flow are different types of regimes [36]. Combination of periodic high flow pulses and following stagnant flow is named as periodic unidirectional pulsing flow. Periodic bidirectional pulsing is comprised of alternation of clockwise and anticlockwise flow pulses. Figure 4.8 shows the relation between power difference and flow regime for the power raising and set-back process. It is shown that high powers influence the stability of the system negatively by generating chaotic flow regime. However, instabilities in the flow can be sustained by reduction of power.

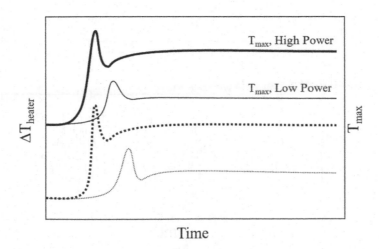

FIGURE 4.7 Comparison of power effects in terms of temperature.

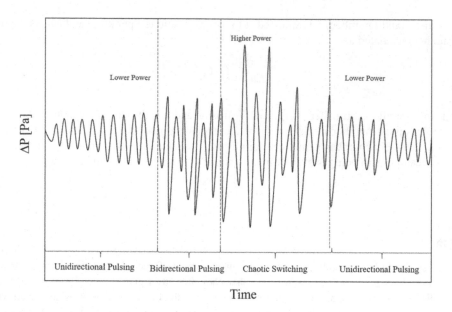

FIGURE 4.8 Pressure oscillations in a SPNCL and corresponding instability types.

Mass flow rate, temperature difference at the heater, and pressure drop are affected by the flow conditions. Temporal changes in these performance parameters can also indicate the stability of the system. Although, in case of stable flow, the parameters exhibit oscillations at the initial condition (Figure 4.9a), after a while, the oscillations are suppressed, and the parameters reach the stable value. But if these oscillations cannot be suppressed with time (Figure 4.9b), the system is considered as unstable.

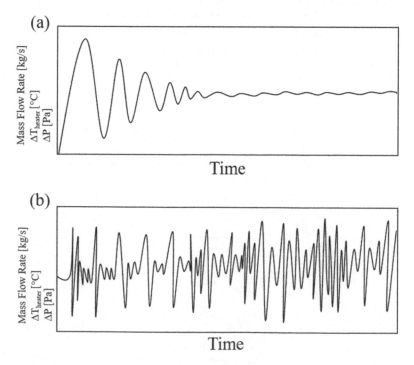

FIGURE 4.9 Time change of performance indicators when the system is (a) stable or (b) unstable.

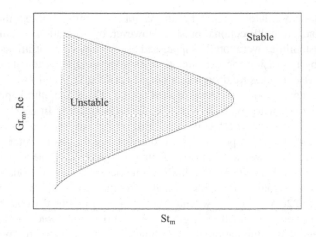

FIGURE 4.10   Stability map in the natural circulation loops.

There are several ways to retain stability. A fluidic way to suppress instabilities is to generate a local pressure drop. In order to do that, two sharp-edged orifices can be located in vertical legs of a rectangular loop. Although these orifices provide localized pressure drop, it increases the maximum temperature of the fluid because of the reduced mass flow rate. If the temperature increases to transition temperature, fluid transits from single-phase to two-phase and the system fails. A geometric way that has been used to overcome instabilities is miniaturization of the system. Tilt angle change also effect the stability in a manner that increased tilt angle improves stability of the SPNCL. Instabilities in flow can also be overcome by adding nanoparticles into base fluid. Thus, perturbing forces in flow can be dampened by enhancement in steady-state flow rate.

Stability is influenced by Gr_m, St_m, flow regime, length to diameter ratio, orientation, flow direction, and length scales [37]. Stability of an SPNCL can be represented by stability maps generated by the linear stability method [38]. In this method, a characteristic equation is obtained by the linearization of the non-linear conservation equations. The system is considered as unstable when the characteristic equation does not have any root with a positive real part. A representative stability map for the Re_{ss} or Gr_m, as a function of modified Stanton number is given in Figure 4.10.

4.3 Nanofluid-Based Single-Phase Natural Circulation Loops

As mentioned in the previous section, SPNCLs have many research issues to improve their performance. Nanofluids, as promising heat transfer fluids, have been one of the solutions for these issues by damping instabilities and also increasing the thermal performance of the system. Base fluids of nanofluids are mostly water, oils, and ethylene-glycol. Recent studies show that molten salts [39] and ionic-liquids [40,41] can also be used as base fluids. However, existing literature shows that water-based nanofluids are mostly studied in SPNCLs. This section summarizes the literature in two parts as: nanofluid-based SPNCLs and SPNCmLs.

4.3.1 Single-Phase Natural Circulation Loops

In the literature, the first nanofluid-based SPNCL was studied experimentally by Nayak et al. [33]. They reported the flow instabilities in a rectangular SPNCL could be suppressed by using alumina-water nanofluid (0.3–2 wt% and particle size 40–80 nm) without decreasing flow rate. When water was used as a working fluid, flow instability in SPNCL was observed for power raising and set-back processes. Unidirectional pulsing was observed when the power raised from 0 W to 300 W, and then it switched to bidirectional pulsing as increasing power up to 600 W. In set-back process, power was reduced from

600 W to 400 W and the oscillatory regime became chaotic switching. For less than 400 W in set-back process, it switched back to unidirectional pulsing. However, by utilization of alumina-water nanofluid as a working fluid, instabilities were totally suppressed even at low concentrations (0.3 wt%). Flow was getting more stable by increasing the concentration. In addition, the steady-state flow rate increased 20%–35% and also a rise in pressure drop which ranges between 28% and 44% was observed with the increasing particle concentration. This study opened up the field of nanofluid applications of SPNCLs by promising increased efficiency without worrying about instabilities. In 2009, Nayak et al. [36] presented a more detailed study on the effect of nanofluids on the instability of SPNCLs by using 1 wt% alumina-water nanofluid at various powers. When the working fluid was water, steady unidirectional flow was observed when the power was increased from 0 W to 150 W by 50 W power steps. For sudden increment in power from 0 W to 175 W, the flow became periodic unidirectional pulsing flow. Sudden addition of 300 W to the stagnant flow changes the flow regime to periodic bidirectional flow for water. And for water again at 220 W, chaotic switching between unidirectional to bidirectional pulsing was observed. And at higher powers which were greater than 500 W, subcooled boiling occurred, and the flow regime was observed as combination of single-phase-two-phase instability with bidirectional pulsing when water was used as working fluid. In parallel with their previous study, flow regime was stable for all conditions when nanofluid was used.

Thomas and Sobhan [42] studied transient performance analysis of CuO-water and alumina-water nanofluids in a rectangular SPNCL of aspect ratio 1 which has vertical-heater-vertical-cooler orientation. The transient performance analysis of the SPNCL was conducted by measuring time dependent temperature at various points. Heater was placed at the bottom of the left arm and heat inputs were provided as 90 W, 175 W, 300 W, and 475 W. For cooling, heat sink placed at upper of right arm and 10°C and 25°C constant temperatures are provided. Steady-state flow rates for the nanofluids were greater than the corresponding pure water cases. The enhancements in mass flow rates were reported as 12% and 14% for 0.075 vol% of alumina-water and 0.05 vol% of CuO-water nanofluid, respectively. 0.075 vol% concentration of alumina nanofluid was the most stable due to greatest zeta potential value (39.1 mV) and least particle size. Therefore, dispersion stability has a significant effect on choosing nanofluid as a working fluid. Considering the time dependent temperatures, the flow behavior was reported as stable. And also, the importance of heat sink temperature at low heat inputs was mentioned. SPNCL was starting earlier as increasing the heat input.

Kumar et al. [43] developed one dimensional homogenous model for steady-state analysis of SPNCL which used coaxial heat exchangers as a cooler and a heater. A code was developed for alumina-water nanofluid which has 0.04 vol% and 25 nm particle size. Increase in the loop height resulted in enhancement of the heat transfer rate. Also, increase in the length of the heat exchanger improved the heat transfer rate. Heat transfer rate increases as hot fluid inlet temperature increases and cold fluid inlet temperature decreases. Recently in 2018, Bejjam and Kumar [44] studied lateral tilt effect on nanofluid-based SPNCL numerically with three dimensional (3D) model. The analyses are conducted for 500–2500 W power inputs at the heating section for constant temperature of 293 K at the cooling section. 1 vol% and 2 vol% of alumina nanofluids were studied and result were compared with water. At 15° of lateral tilt, enhancement in effectiveness was determined as 4.31% and 9.29% for 1 vol% and 2 vol% of alumina nanofluids, respectively. Increase of heat input causes a reduction in effectiveness. In comparison with 0°, 60° inclination angle at 2000 W heat load enhances the effectiveness by 18.65%, 18.87%, and 19.83% for water, 1 vol% and 2 vol% of alumina nanofluids, respectively. It was reported that higher Ra values were indications of better thermophysical properties therefore better loop performance. Since increased loop tilt angle decreases the Rayleigh number and heat transfer coefficient, a tilt angle of up to 15 is suggested for SPNCLs in this study.

4.3.2 Single-Phase Natural Circulation Mini Loops

Misale et al. [45] proposed a single-phase natural circulation mini loop SPNCmL in order to improve flow stabilization. In 2012, alumina-water nanofluid was used as working fluid in this SPNCmL by Misale et al. [46]. Particle size of alumina nanoparticles was 45 nm. 0.5 vol% and 3 vol% were used. For high concentrations, operating conditions were limited. Maximum heat sink temperature was limited to 20°C, and the loop inclination varied between 0 and 60. The flow in mini loop was stable for both water and nanofluid at the concentration up to 3 vol%. Though at 75° loop inclination, utilization of nanofluid slightly enhanced the thermal performance, the thermal performance of nanofluid was comparable with the water. Experimental data were appropriate with Vijayan's correlation [38] which was developed by large scale loops.

Contrarily to these results, Turgut and Doganay [30] studied the thermal performance of SPNCmL from a point of effectiveness. Alumina-water nanofluid which has the average particle size of 10 nm and 30 nm was used as working fluid at concentrations of 1 vol%, 2 vol%, and 3 vol%. The experiments were conducted at constant temperature 20°C at heat sink and at 0° loop inclination. Power input ranged between 10 W and 50 W in steps of 10 W. In order to study the influence of concentration and particle size on thermal performance, they proposed effectiveness factor. Effectiveness was calculated higher when nanofluids were used in the system. Increment in power inputs decreases the effectiveness factor. However, increase in power input decreases the quiescent state time. Effectiveness increases as increasing of the concentration for both particle sizes at all power inputs. Although for all concentrations at 10 W, nanofluid with the 30-nm particles have higher effectiveness than the ones with 10-nm. At lower concentrations, small particles had higher effectiveness factor. However, it was reversed for higher concentration. There was not a significant effect of particle size on 2 vol%. Therefore, concentration of nanofluid had significant influence on the effectiveness for larger particle sizes.

Karadeniz et al. [47] proposed a numerical model for nanofluid-based SPNCLs. Alumina-distilled water nanofluids at 1 vol%, 2 vol%, and 3 vol% concentrations were used as working fluids, but modelled as a filler which was single-phase. Numerical studies which were based on alumina-distilled water as a working fluid were considered as a reference. Numerical results were compared with the experimental results of Turgut and Doganay [30]. Heating and cooling sections were modelled as boundary conditions. Cooler part was set to 20°C which equaled to cooling fluid temperature. Constant heat fluxes (10–50 W) applied at heater section. All other boundaries were modelled as adiabatic walls. However, simplified models were developed to understand unsteady behaviors of SPNCLs, and the 3D numerical models cannot explain the nature of the flow with simplifications. With developed 3D model in this study, it can be predictable of flow rate, local fluid velocity, and pressure drops. Miniaturization of the system increases the stability of the flow. Therefore, in this study, steady flow results were given due to the better prediction of the steady behavior of the SPNCmLs. When the numerical results were compared with the experimental results, T_{max} values exhibit same tendency with up to 10% error for numerical results. The numerical values were over-predicted when compared with the experimental results. The reason behind this error could be the heat loss to surrounding in experimental setup. In numerical study, generated heat in the heater was given to the working fluid. For T_{max}, numerical model was in good agreement with the experimental results for 1 vol% and 2 vol% concentration, but for 3% filler concentration, T_{max} increases with the increase of heating power input. And also ΔT_{heater}, which represented how much heat removed from the system, exhibits same behavior with T_{max}. It was reported that the reason of this difference between results of experimental and numerical studies should be the differences in thermophysical properties of nanofluids. And also, for change of viscosity and thermal conductivity with the temperature, the results of 1 vol% and 2 vol% were comparable with the experimental results, whereas for 3 vol%, same behavior was observed. Furthermore, inapplicability of the numerical model for determination of heat transfer and flow of nanofluids was discussed. But the tendency at comparisons weakened this possibility. Effectiveness values which were calculated by numerical models were comparable with the experimental values for 10 W power input for all working fluids excepting 3% filler content. All numerical results have same tendency with the experimental results. However, effectiveness of 3% filler content was obtained closer to experimental results due to the error in determination thermophysical properties.

Ho et al. [48] concentrated on a nanofluid-based SPNCmL with a mini channel heat sink and source. Utilization of mini channel heat sinks have the potential of enhancement of heat transfer coefficient. The influence of the weight concentration of nanofluid and wall temperature difference between hot and cold ends of the NCL was studied. They compared the Nu number change with respect to Rayleigh number at the hot and cold ends to reveal the performance under different conditions. Average Nusselt number is increased by addition of alumina nanoparticles into water for both cooling and heating ends. However, increase of Nu number is more significant at the cold end. For higher concentrations, increments were observed in the heat transfer enhancement of both the hot and cold walls. With the rise in temperature at constant concentration of nanofluid, Nusselt number at cooling end increased and at the heating section it decreased. Reduction of the cold wall temperature caused the increase in Nusselt number of heating end and decrease in the cooling end for constant hot wall temperature. Increasing Rayleigh number causes reduction of the fin thermal resistance, the flow thermal resistance, and the total thermal resistance of heating end. But for cooling end, slight decrement was observed in the flow thermal resistance, while the fin thermal resistance and the total thermal resistance increased. Utilization of a nanofluid decreased the total thermal resistance in the heating end, while it had no significant effect in cooling end. It was observed that Reynolds number increased with the increase of Rayleigh number. For constant Rayleigh number, by the addition of nanoparticles causes insignificant reduction in the Reynolds number.

Doganay and Turgut [49] investigated effects of inclination angle (0°, 30°, 60°, and 75°), heating power (10 W, 30 W, and 50 W), and heat sink temperature (10°C and 20°C) on SPNCmL by using effectiveness factor, for the first time. 1 vol%, 2 vol%, and 3 vol% concentrations of alumina-water nanofluids (10 nm and 30 nm particle size) were used as working fluids. The experimental setup was same with their previous study [30]. There was no significant change in thermal stability of the system for 30°, 60°, and 75° in comparison with 0°. Quiescent state time increases by the increase of inclination angle. Higher ΔT_{heater} values were observed for lower heat sink temperatures, higher power inputs, and higher concentrations. At 75° for water, the temperature of the heater rapidly increased and maybe resulted in evaporation of the fluid. Therefore, water could not be tested at 75° inclination angle. However, it was reported that alumina-water nanofluids could be used as a working fluid at higher temperature required systems. Effectiveness of nanofluids in 30° and 60° were found greater than water. Effectiveness ratio between nanofluids and water which was varied from 0.6% to 7.5% increased with the increase in particle concentration. Contrarily to Misale et al. [46], the thermal performance should not only have been determined in terms of ΔT_{heater}. For higher power inputs (30 W and 50 W) and lower heat sing temperature (10°C), effectiveness was calculated more than conditions which were lower power input (10 W) and higher heat sink temperature (20°C). Moreover, increasing the inclination angle improved the effectiveness. For 2 vol% alumina-water nanofluid (10 nm diameter particle size) at 75° tilt angle and 50 W power input, effectiveness was enhanced 26% with respect to 0° inclination angle.

Karadeniz et al. [50] studied numerically the inclination angle and heat powers on the thermal performance of alumina nanofluid-based SPNCmL and compared the numerical results with the experimental results of Doganay and Turgut [49]. Developed model was based on the previous model of Karadeniz et al. [47] and adopted from Doganay and Turgut [49]. Mass flow rate results of numerical study compared to results obtained by Vijayan's correlation [38]. They observed good agreement between results of experimental study and numerical study for 30 and 60. But for 50 W heat input and 75 inclination angle, unexpected behavior observed. It was thought that phase change was occurred due to the boiling of the fluid in the heater part. Results of numerical study were coherent with the Vijayan's correlation. Moreover, temperature distribution was uniform at the inlets of heater and cooler sections. For 1 vol% alumina-water nanofluid, for the heater inlet temperature, distribution was uniform for all inclination angles and power inputs, whereas for cooler inlet, local non-uniformities were observed with increase of the power inputs and inclination angle. They reported that for 10 W power input, temperature distribution at the outlet of the cooler section was also uniform for all working fluids. Increase in heating power inputs causes the decrease in the effective thermal expansivity due to the increment in the volumetric concentration. High temperature was observed at the above parts of the cross-section, but it shrinked by the increasing of the volumetric concentration. However, for heater outlet, increase in heating power input and volumetric concentration increased the average temperature. Non-uniform temperature distribution was observed at the low temperature cells at the bottom half of the cross-section. Reduction of average temperature at cooler outlet due to the rise in inclination angle was reported.

Koca et al. [51] studied with Ag-water-based nanofluid (polyvinyl pyrrolidone [PVP] stabilized) in SPNCmL and compared the thermal performance with alumina-water nanofluids. They reported that thermal conductivity of the nanofluid decreases with the addition of nanoparticles. Rise in the reduction of thermal conductivity was observed in the range of between 3% and 11.5% with respect to increase in concentration from 0.25 wt% to 1 wt%. Considering the thermal stability, mini loop had same performance for both water and 0.25 wt% for all power inputs. Quiescent time was higher for nanofluids, and it increased with the reduction in the power and increase in inclination angle. At 30° and 60° inclination angle, thermal stability behavior had no significant difference from 0° inclination angle. Increase in the nanofluid concentration improved the effectiveness for all power inputs. Nevertheless, increase in the power at constant cooling bath temperature reduced the effectiveness. When the Ag-water system was compared to alumina-water system, the effectiveness showed same behavior for both nanofluids and increased with the increasing concentration. Considering various powers, alumina-water nanofluids had almost constant effectiveness ratio, whereas effectiveness ratio of Ag-water nanofluid was increased significantly, especially for higher concentrations. Moreover, they mentioned that PVP had a crucial role in thermal performance of nanofluid. PVP behaved as a barrier on Ag nanoparticles at ambient temperature, and this behavior was getting lost with the increase in the temperature due to the weak polymer bindings. For higher concentrations, remarkable variations on relative viscosity and thermal performance were observed due to this behavior of PVP, whereas at lower concentrations, relative viscosity and thermal performance was almost constant with increase in temperature and power input, respectively.

Different from the water-based nanofluids, Mohan et al. [52] presented the efficiency of hexagonal boron nitride and mineral oil-based nanofluid in single-phase natural circulation mini loop with the orientation of HHHC. The aspect ratio of the loop was set to 0.77. Pure mineral oil, 0.02 wt% and 0.08 wt% hexagonal boron nitride nanosheet dispersed oil-based nanofluids were used in experiments. The heat inputs were 10 W, 20 W, 30 W, and 40 W. The capacity of mini loop was 100 mL. In comparison with the water, the temperature difference was increased in the range of 2% and 7.4% as increasing of the power. At 40 W, 4% enhancement was obtained in flow rate with the addition of 0.08 wt% nanoparticles into mineral oil. Filler fraction was increased 5% at same conditions. Furthermore, the maximum enhancement in Nu was measured as 5% at 0.08 wt%.

4.4 Conclusion

Single-phase natural circulation loops are passive systems in which temperature gradient causes the density gradient which generates buoyancy force and drives the system without any external power source (pump, fan, etc.). With the advantages of easy control, simplicity, enhanced safety, and reliability, they are preferred in widespread engineering applications such as solar thermal energy systems, cooling of nuclear reactors, etc.

Stability and the performance of the system are influenced by the geometry and working fluid. Effects of orientation of heater and cooler and shape and inclination of natural circulation loop on stability and efficiency have been investigated by many researchers. Selection of the working fluid is another important issue due to operation conditions such as temperature range, capacity, and environmental impact. Water, molten salts, and liquid metals are widely used in conventional SPNCLs.

Maximum temperature, temperature difference between hot leg and cold leg, effectiveness, steady-state mass flow rate, and non-dimensional numbers which are determined by thermophysical properties of working fluid and loop geometry are important indicators for investigation of thermohydraulic performance of the SPNCLs. Although temperatures can be measured easier at legs of the loop, measurement of the mass flow rate or velocity is quite difficult. Non-dimensional parameters (effectiveness, Nu, GR_m, Re_{ss}, and St) are more accurate for comparisons of different systems and operating conditions.

Imbalance of the buoyancy and friction forces causes the instabilities in the loop, and different flow regimes are observed due to these instabilities. In order to suppress the instability, some fludic (orifice addition) and geometric (miniaturization and tilt angle change) methods are proposed. Moreover, recent studies reported that addition of the nanoparticles also increases the stability of the system since nanofluids have higher boiling temperatures compared to base fluids. Therefore, it is possible to use them at higher temperatures in SPNCLs by retarding the local boiling and phase transition. In addition, increase in particle concentration increases the viscosity of the fluid and so it prevents the possible eddy currents by decreasing the Re_{ss} number.

Utilization of nanofluids does not only suppress the instabilities, but also increases the thermal performance of the natural circulation loop. In addition, when nanofluids are used as a working fluid, rise in pressure drop is observed compared to water. Different flow regimes are observable when water is used as a working fluid, but for nanofluids flow regime has been reported as stable at all conditions. Moreover, it is mentioned that dispersion stability of nanofluids has crucial importance for performance of natural circulation loops. Improved stability of the nanofluid increases the performance and stability of the loop because of uniform thermophysical properties of the nanofluid. If the effects of nanofluid are investigated in terms of performance indicators, increase in concentration enhances the effectiveness in SPNCmLs for all power inputs and particle size. At higher concentrations, smaller particles have greater effectiveness factor. However, it was reversed for larger particles. Therefore, concentration of nanofluid had significant influence on the effectiveness for larger particle sizes. T_{max} and ΔT_{heater} increase with the increase in concentration. Moreover, mass flow rate increases with the utilization of nanofluids.

Several studies on nanofluid-based SPNCLs exist in the literature, they have mentioned that nanofluids propose the enhanced effectiveness and improved stability to the SPNCLs. However, there are still challenges and open fields in the research of nanofluid-based SPNCLs. It is clear that, as the nanofluids will have better thermophysical properties, so do the SPNCLs. It is recommended that molten salt-based nanofluids as a working fluid in SPNCLs must be investigated in detail due to their better stability, enhanced heat transfer properties, small or negligible increment in viscosity, and chemical compatibility. Moreover, due to their better stability, much more study should be performed on nanofluid-based toroidal loops which is recommended as a future work.

NOMENCLATURE

A	flow area (m²)
b	constant in the friction factor equation
B	constant in steady state Reynold's number equation
C	constant in steady state Reynold's number equation
c_p	specific heat (J/kg·K)
D	inner diameter of the loop (m)
f	friction factor
g	gravitational acceleration (m/s²)
Gr	Grashof number
k	thermal conductivity (W/m·K)
L_{et}	total equivalent length accounting for the sum of the individual local loss coefficient of the entire loop
L_{total}	total length (m)
$\dot{m}$	mass flow rate (kg/s)
N_G	effective loss coefficient
Nu	Nusselt number
p	exponent in the friction factor equation
Pr	Prandtl number
Q_h	total heat input rate (W)
Ra	Rayleigh number
Re	Reynold's number
T	temperature (°C)
U	overall heat transfer coefficient (W/m²·K)
z	center line elevation (m)

GREEK SYMBOLS

α tilt angle (°)
β_T thermal expansion coefficient (K^{-1})
ε effectiveness
ρ fluid density (kg/m^3)
μ dynamic viscosity (N·s/m^2)
Δ interval

SUBSCRIPTS

m modified
ss steady state
e effective
h heater

ABBREVIATIONS

DIW   deionized water
FCL   forced circulation loop
hBN   hexagonal boron nitride
HHHC   horizontal heater horizontal cooler
HHVC   horizontal heater vertical cooler
NCL   natural circulation loop
PVP   polyvinyl pyrrolidone
SPNCL   single phase natural circulation loop
SPNCmL   single phase natural circulation mini loop
TPNCL   two phase natural circulation loop
wt   weight
vol   volume
VHHC   vertical heater horizontal cooler
VHVC   vertical heater vertical cooler

REFERENCES

1. Ridouane E.H., Hitt D.L., Danforth C.M. (2011). A numerical investigation of 3-D flow regimes in a toroidal natural convection loop. *International Journal of Heat and Mass Transfer* 54(25), 5253–5261.
2. Basu D.N., Bhattacharyya S., Das P.K. (2012). Performance comparison of rectangular and toroidal natural circulation loops under steady and transient conditions. *International Journal of Thermal Sciences* 57, 142–151.
3. Stern C.H., Greif R., Humphrey J.A.C. (1988). An experimental study of natural convection in a toroidal loop. *Journal of Heat Transfer* 110(4a), 877–884.
4. Lavine A.S., Greif R., Humphrey J.A.C. (1987). A three-dimensional analysis of natural convection in a toroidal loop: The effect of Grashof number. *International Journal of Heat and Mass Transfer* 30(2), 251–262.
5. Merzari E., Fischer P., Ninokata H. (2011). Numerical simulation of the flow in a toroidal thermosiphon. In: *ASME-JSME-KSME 2011 Joint Fluids Engineering Conference American Society of Mechanical Engineers*; New York: American Society of Mechanical Engineers. pp. 1549–1560.
6. Jiang Y.Y., Shoji M., Naruse M. (2002). Boundary condition effects on the flow stability in a toroidal thermosyphon. *International Journal of Heat and Fluid Flow* 23(1), 81–91.
7. Creveling H.F., De Paz J.F., Baladi J.Y., Schoenhals R.J. (1975). Stability characteristics of a single-phase free convection loop. *Journal of Fluid Mechanics* 67(1), 65–84.

8. Desrayaud G., Fichera A. (2008). Numerical analysis of general trends in single-phase natural circulation in a 2D-annular loop. *Science and Technology of Nuclear Installations* 2008.

9. Pacheco-Vega A., Franco W., Chang H.-C., Sen M. (2002). Nonlinear analysis of tilted toroidal thermosyphon models. *International Journal of Heat and Mass Transfer* 45(7), 1379–1391.

10. Zvirin Y. and Greif R. (1979). Transient behavior of natural circulation loops: Two vertical branches with point heat source and sink. *International Journal of Heat and Mass Transfer* 22(4), 499–504.

11. Welander P. (1967). On the oscillatory instability of a differentially heated fluid loop. *Journal of Fluid Mechanics* 29(1), 17–30.

12. Zvirin Y. (1980). Effects of a throughflow on the steady state and stability of a natural circulation loop. In: *AIChE Symposium Series* 76.

13. Zvirin Y. (1985). Throughflow effects on the transient and stability characteristics of a thermosyphon–nEinflüsse der Zuströmung in einen Thermosyphon auf sein transientes Verhalten und seine Stabilität. *Wärme-und Stoffübertragung* 19(2), 113–120.

14. Zvirin Y. (1979). On the behavior of a natural circulation loop with a throughflow. In: *14th Intersociety Energy Conversion Engineering Conference*, Washington, D.C.: American Chemical Society, pp. 1974–1978.

15. Zvirin Y. (1979). The effect of dissipation on free convection loops. *International Journal of Heat and Mass Transfer* 22(11), 1539–1546.

16. Kudariyawar J.Y., Vaidya A.M., Maheshwari N.K., Satyamurthy P. (2016). Computational study of instabilities in a rectangular natural circulation loop using 3D CFD simulation. *International Journal of Thermal Sciences* 101, 193–206.

17. Kudariyawar J.Y., Srivastava A.K., Vaidya A.M., Maheshwari N.K., Satyamurthy P. (2016). Computational and experimental investigation of steady state and transient characteristics of molten salt natural circulation loop. *Applied Thermal Engineering* 99, 560–571.

18. Hughes J.T. (2017). *Experimental and Computational Investigations of Heat Transfer Systems in Fluoride Salt-Cooled High-Temperature Reactors*. Albuquerque, NM: The University of New Mexico.

19. Yang Z., Meng Z., Yan C., Chen K. (2017). Heat transfer and flow characteristics of a cooling thimble in a molten salt reactor residual heat removal system. *Nuclear Engineering and Technology* 49(8), 1617–1628.

20. Kudariyawar J.Y., Vaidya A.M., Maheshwari N.K., Satyamurthy P., Srivatsav A.K., Lingade B.M. (2017). Investigation on heat transfer behaviour of molten salt natural circulation loop using numerical simulations. *Journal of Nuclear Energy Science & Power Generation Technology* 5.

21. Srivastava A.K., Chouhan R., Borgohain A., Jana S.S., Maheshwari N.K., Pilkhwal D.S. et al. (2017). An experimental and numerical study to support development of molten salt breeder reactor. *Journal of Nuclear Engineering and Radiation Science* 3(3), 31007–31008.

22. Chen K., Yan C., Meng Z., Wu X., Song S., Yang Z. et al. (2016). Experimental analysis on passive residual heat removal in molten salt reactor using single cooling thimble test system. *Energy* 112, 1049–1059.

23. German P., Fehér S., Czifrus S., Yamaji B. (2016). Analysis of heat source distribution in internal circulating surface heat transfer molten salt reactor. *Progress in Nuclear Energy* 92, 155–163.

24. Cammi A., Cauzzi M., Luzzi L., Pini A. (2016). DYNASTY: An experimental loop for the study of natural circulation with internally heated fluids. In: *Conference: 12th International Conference on Heat Transfer, Fluid Mechanics and Thermodynamics* (*HEFAT 2016*), Malaga, Spain.

25. Wu X., Yan C., Meng Z., Chen K., Song S., Yang Z. et al. (2016). Numerical analysis of the passive heat removal system for molten salt reactor at steady state. *Applied Thermal Engineering* 102, 1337–1344.

26. Srivastava A.K., Kudariyawar J.Y., Borgohain A., Jana S.S., Maheshwari N.K., Vijayan P.K. (2016). Experimental and theoretical studies on the natural circulation behavior of molten salt loop. *Applied Thermal Engineering* 98, 513–521.

27. Kudariyawar J.Y., Vaidya A.M., Maheshwari N.K., Satyamurthy P., Srivastava A.K. (2015). Estimating steady state and transient characteristics of molten salt natural circulation loop using CFD. *Kerntechnik* 80(1), 20–31.

28. Jeong Y.S., Bin S. S., Bang I.C. (2018). Natural convection heat transfer characteristics of molten salt with internal heat generation. *International Journal of Thermal Sciences* 129, 181–192.

29. Basu D.N., Bhattacharyya S., Das P.K. (2014). A review of modern advances in analyses and applications of single-phase natural circulation loop in nuclear thermal hydraulics. *Nuclear Engineering and Design* 280, 326–348.
30. Turgut A., Doganay S. (2014). Thermal performance of a single phase natural circulation mini loop working with nanofluid. *High Temperatures: High Pressures* 43(4).
31. Cheng H., Lei H., Dai C. (2017). Heat transfer of a single-phase natural circulation loop with heating and cooling fluids. *Energy Procedia* 142, 3926–3931.
32. Vijayan P.K., Nayak A.K., Saha D., Gartia M.R. (2008). Effect of loop diameter on the steady state and stability behaviour of single-phase and two-phase natural circulation loops. *Science and Technology of Nuclear Installations* 2008.
33. Nayak A.K., Gartia M.R., Vijayan P.K. (2008). An experimental investigation of single-phase natural circulation behavior in a rectangular loop with Al$_2$O$_3$ nanofluids. *Experimental Thermal and Fluid Science* 33(1), 184–189.
34. Misale M., Garibaldi P., Tarozzi L., Barozzi G.S. (2011). Influence of thermal boundary conditions on the dynamic behaviour of a rectangular single-phase natural circulation loop. *International Journal of Heat and Fluid Flow* 32(2), 413–423.
35. Misale M. (2014). Overview on single-phase natural circulation loops. In: *International Conference on Advances in Mechanical & Automation Engineering—MAE*, Rome, Italy; p. 12.
36. Nayak A.K., Gartia M.R., Vijayan P.K. (2009). Thermal–hydraulic characteristics of a single-phase natural circulation loop with water and Al2O3 nanofluids. *Nuclear Engineering and Design* 239(3), 526–540.
37. Vijayan P.K., Sharma M., Saha D. (2007). Steady state and stability characteristics of single-phase natural circulation in a rectangular loop with different heater and cooler orientations. *Experimental Thermal and Fluid Science* 31(8), 925–945.
38. Vijayan P.K. (2002). Experimental observations on the general trends of the steady state and stability behaviour of single-phase natural circulation loops. *Nuclear Engineering and Design* 215(1–2), 139–152.
39. Muñoz-Sánchez B., Nieto-Maestre J., Iparraguirre-Torres I., García-Romero A., Sala-Lizarraga J.M. (2018). Molten salt-based nanofluids as efficient heat transfer and storage materials at high temperatures: An overview of the literature. *Renewable and Sustainable Energy Reviews* 82, 3924–3945.
40. Minea A.A., Murshed S.M.S. (2018). A review on development of ionic liquid based nanofluids and their heat transfer behavior. *Renewable and Sustainable Energy Reviews* 91, 584–599.
41. França J.M.P., Lourenço M.J.V., Murshed S.M.S., Pádua A.A.H., Nieto de Castro C.A. (2018). Thermal conductivity of ionic liquids and ionanofluids and their feasibility as heat transfer fluids. *Industrial & Engineering Chemistry Research* 57(18), 6516–6529.
42. Thomas S., Sobhan C.B. (2018). Stability and transient performance of vertical heater vertical cooler natural circulation loops with metal oxide nanoparticle suspensions. *Heat Transfer Engineering* 39(10), 861–873.
43. Kumar K., Bejjam R.B., Najan A. (2015). Numerical investigation of nanofluid based thermosyphon system. *World Academy of Science, Engineering and Technology, International Journal of Mechanical, Aerospace, Industrial, Mechatronic and Manufacturing Engineering* 8(12), 2091–2096.
44. Bejjam R.B., Kiran Kumar K. (2018). Numerical investigation to study the effect of loop inclination angle on thermal performance of nanofluid-based single-phase natural circulation loop. *International Journal of Ambient Energy*, 1–9.
45. Misale M., Garibaldi P., Passos J.C., De Bitencourt G.G. (2007). Experiments in a single-phase natural circulation mini-loop. *Experimental Thermal and Fluid Science* 31(8), 1111–1120.
46. Misale M., Devia F., Garibaldi P. (2012). Experiments with Al2O3 nanofluid in a single-phase natural circulation mini-loop: Preliminary results. *Applied Thermal Engineering* 40, 64–70.
47. Karadeniz Z.H., Doganay S., Turgut A. (2014). Numerical study on nanofluid based single phase natural circulation mini loops. In: *Proceedings of CONV-14: International Symposium on Convective Heat and Mass Transfer*, June 8–13, Kusadasi, Turkey, ICHMT Digital Library Online, Begel House Inc.
48. Ho C.J., Chung Y.N., Lai C.-M. (2014). Thermal performance of Al2O3/water nanofluid in a natural circulation loop with a mini-channel heat sink and heat source. *Energy Conversion and Management* 87, 848–858.

49. Doganay S. and Turgut A. (2015). Enhanced effectiveness of nanofluid based natural circulation mini loop. *Applied Thermal Engineering* 75, 669–676.
50. Karadeniz Z.H., Doganay S., Turgut A. (2016). Numerical study on nanofluid based single phase natural circulation mini loops: A steady 3D approach. *High Temperatures—High Pressures* 45(4).
51. Koca, H. D., Doganay, S., & Turgut, A. (2017). Thermal characteristics and performance of Ag-water nanofluid: Application to natural circulation loops. Energy Conversion and Management, 135, 9–20.
52. Mohan M., Thomas S., Taha-Tijerina J., Narayanan T.N., Sobhan C.B., Ajayan P.M. (2013). Heat transfer studies in thermally conducting and electrically insulating nano-oils in a natural circulation loop. In: *ASME 2013 International Mechanical Engineering Congress and Exposition American Society of Mechanical Engineers*; p. V06BT07A040, New York: American Society of Mechanical Engineers.

5

Entropy Generation Analysis of Hybrid Nanofluids Flow in Ducts with Various Shapes

Gabriela Huminic and Angel Huminic

CONTENTS

5.1 Introduction

Forced convection heat transfer is generally accompanied by thermodynamic irreversibilities such as the heat transfer through a finite temperature difference between the fluid and wall, the friction through viscous or turbulent losses of the fluid, and chemical reactions or mixing and which may be measured by entropy generation rates. First studies in entropy generation were performed by Bejan [1–2] which studied different physical phenomena contributing to the entropy production. Plus, in these studies were developed some equations necessary to compute the entropy generation rate. In order to provide up-to-date information, a survey of publications on entropy generation in circular and non-circular ducts with conventional fluids and hybrid nanofluids from 2000 onwards is given below.

5.1.1 Entropy Generation Analysis in Ducts with Conventional Fluids

5.1.1.1 *Entropy Generation Analysis in Circular Ducts*

Şahin [3] analytically studied the effect of temperature dependence of viscosity in the heating process on entropy generation and pumping power in smooth duct at constant wall temperature. The analysis was performed in fully developed turbulent fluid flow. Two working fluids, water and glycerol, were investigated. Based on the analytical results, for viscous fluids (glycerol), the entropy generation per unit heat flux is minimum along the duct length, and the pumping power ratio decreases due to the decrease in viscosity. Later, in 2002, Şahin [4] made an analytical investigation on entropy generation and pumping power in smooth duct at constant wall heat flux. Their results showed that the viscosity variation had a significant effect on both the entropy generation and the pumping power and that if the dependence of viscosity on the temperature is not taken into account, then considerable errors might occur in the evaluation of entropy generation and pumping power.

Mahmud and Fraser [5] analytically investigated the second law of thermodynamics inside a channel with different configurations: one fixed plate and one moving plate (Couette flow) with isothermal boundary condition, two fixed plates (Poiseuille flow) with differentially heated isothermal boundary condition, circular cross-section for Newtonian fluid (Poiseuille flow) with the constant heat flux boundary condition and fully developed laminar flow, circular annulus with isoflux boundary condition, axially moving concentric cylinder (Couette flow) with the isoflux boundary condition, rotating concentric cylinder, circular cross-section for non-Newtonian fluid, and non-Newtonian fluid through a channel with two parallel plates. For each considered case, the expressions for dimensionless entropy generation number and Bejan number were determined.

Khaliq [6] performed a study on entropy generation under laminar flow in a duct with circular cross-section using water, glycerol, and air. The results concerning the variation of entropy generation number and pumping power were analyzed for constant viscosity and temperature-dependent viscosity. The results revealed that the dimensionless entropy generation based on constant viscosity had higher values than those based on temperature-dependent viscosity. Also, for constant viscosity, the higher pumping power was obtained for glycerol and air, while for water, the pumping power had the same values for constant viscosity and temperature-dependent viscosity.

The entropy generation analysis in a circular duct with internal longitudinal fins was investigated by Dağtekin et al. [7]. Three types of fins were studied: thin, triangular, and V-shaped. The study was performed in laminar flow. The results revealed that both entropy generation and pumping power were significantly influenced by the number of fins and by dimensionless length of the fins in the case of the thin fins and triangular fins, and also by the fin angle in the case of the triangular and V-shaped fins. Furthermore, in this study, it was found that the dimensionless temperature difference influences also both entropy generation and pumping power.

Öztop [8] analytically studied the effects of cross-sectional area and wall heat flux on entropy generation in semicylindrical ducts at constant wall heat flux under laminar flow. Water was used as working fluid. The results indicated that the dimensionless entropy generation increases with the increase in wall heat flux and also that the dimensionless entropy generation decreases with cross-sectional area increases. Plus, through the decrease in wall heat flux, the pumping power ratio increases and also the cross-sectional area increase lead to increase required pumping power.

Second law analysis of forced convection in a circular duct under fully developed forced convection for non-Newtonian fluids subjected to constant wall heat flux was analytically studied by Mahmud and Fraser [9]. They obtained few expressions for dimensionless entropy generation number (N_S), irreversibility distribution ratio (F), and Bejan number (Be) as functions of dimensionless radius (R), Peclet number (Pe), modified Eckert number (Ec), Prandtl number (Pr), dimensionless temperature difference (Ω), and fluid index (m or n). The equations were numerically solved for seven selected fluid indices and concluded that the numerically obtained asymptotic Nusselt number showed excellent agreement with the analytical Nusselt number.

Jankowski [10] investigated the entropy generation in fully developed convective heat transfer in ducts with different cross-sectional shape. The main findings can be summarized as follows: (a) in adiabatic flow, the circular duct is most efficient section which leads to the minimization of the entropy generation, but in situations where the flow is with heat transfer, the circular cross-section may not always be ideal and (b) for low flow rates, large cross-sectional areas, and high heat transfer rates, ducts with a large wetted perimeter (i.e., rectangular) lead to the minimization of the overall entropy generation.

Another numerical analysis of the effect of non-uniform heating on entropy generation of the laminar flow through pipe was performed by Esfahani and Shahabi [11]. Analyses were performed for a fluid with high Pr number used as working fluid and also for the same amount of heat rate, but with different heat flux distribution. The results revealed that the entropy generation for cases with decreasing heat flux distribution is more than the cases with the same amount of heat flux at sections, but with increasing distribution. Moreover, the authors showed the entropy generation for the cases with decreasing heat flux distribution was more than with uniform heat flux.

You et al. [12] studied the effects of alignment method and geometrical parameters on entropy generation in the horizontal circular tubes with conical strip inserts under laminar flow. Two types of conical strips ($\beta = 60°$ and $\beta = 90°$) were analyzed, and also the performances of tubes with strips of staggered and non-staggered alignments in terms of Nusselt number ratio (Nu/Nu_0) and friction factor ratio (f/f_0) were compared. Based on the results, the non-staggered alignment generates larger Nu/Nu_0 and f/f_0 than those of staggered alignment, and also the irreversible losses (S_g^t and N_S^t) of enhanced tubes with non-staggered strips were smaller than those of staggered alignment.

Abdous et al. [13] investigated the influence of the geometrical parameters (the number of fins, fin height, bottom width, and tube outer diameter) and flow conditions (mass velocity, inlet vapor quality, saturation temperature) during flow boiling on entropy generation in the micro-fin tube. It is observed that the decrease of hydraulic diameter leads to an increase of entropy generation due to pressure drop, while the entropy generation due to heat transfer decreases, and also it is founded that the entropy generation decreases at the lower value of fin height and bottom width. Concerning the effect of flow conditions on entropy generation, the results indicated that the increase in the value of mass velocity and inlet vapor quality leads to the decrease of entropy generation due to pressure drop, while the entropy generation due to heat transfer increases, and also that the entropy generation decreases with the increase in the saturation temperature.

Keklikcioglu and Ozceyhan [14] performed an experimental study on entropy generation for a circular tube with equilateral triangle cross-sectioned coiled-wire inserts with constant wall heat flux. The experiments were carried out at Reynolds numbers in the range 2731–27,732 and authors founded that the entropy generation number increases with increase of Reynolds number and decreases with increasing pitch ratio.

5.1.1.2 Entropy Generation Analysis in Rectangular Ducts

Hooman et al. [15] investigated the heat transfer and the entropy generation inside a duct of rectangular cross-section saturated by a porous medium by applying Darcy-Brinkman flow model. In this study, three different types of thermal boundary conditions were studied namely: (a) all walls were uniformly heated; (b) one of the walls was assumed to be adiabatic, and three walls were kept at a uniform temperature; and (c) the two side walls were assumed to be adiabatic, and the upper and lower walls are kept at a uniform temperature. The results obtained in this study could be used as benchmark checks on numerical findings for flow in parallel-plate channels or ducts with rectangular cross-section filled with or without a porous matrix. In another study, Silva and Garcia [16] made a theoretical and numerical investigation of the entropy generation analysis in rectangular ducts with three-dimensional (3-D) heat transfer coupling under fully developed laminar flow. Air, water, and oil at 300 K were used as working fluids, while carbon-steel were used as wall material.

Öztop et al. [17] performed an analytical study on entropy generation for rectangular ducts with semi-circular ends under constant wall temperature and constant wall heat flux boundary conditions in laminar flow. The results revealed that, for both boundary conditions, the increase of aspect-ratio value leads to the increase in total entropy generation for a fixed Reynolds number, and also the increase of both aspect ratio and Reynolds number lead to the pumping power ratio increase.

Yang et al. [18] investigated the flow reversal and the entropy generation in the entrance region of a 3-D vertical rectangular duct under constant wall temperature. The results showed that the buoyancy induced flow reversal phenomenon significantly influenced the local entropy generation due to heat transfer and fluid friction contributions. Another numerical analysis of evaluation of the thermal performance of laminar fully developed flow in a bundle with different non-circular ducts (rectangular, isosceles triangular, elliptical, trapezoidal, and double trapezoidal [hexagonal]) subjected to constant wall temperature boundary condition and common constraint $D^* = 1$ (specified hydraulic diameter of the ducts) was performed by Petkov et al. [19]. They found that the tube optimal shape and geometrical details depend on the constraints and the objectives imposed. In another study, Petkov et al. [20] studied the thermal performances of the bundle with different non-circular ducts (rectangular, isosceles triangular, elliptical, trapezoidal, and double trapezoidal), but with common constraint $A^* = 1$ (the cross-sectional area). Results from this study were completely different from those obtained by Petkov et al. [19].

Yang and Wu [21] numerically investigated the entropy generation in a 3-D rectangular channel under the influence of opposed buoyancy with uniform wall temperature. The results indicated that opposed buoyancy force could increase the entropy generation due to fluid friction contribution in the channel, which suggests more losses of useful work were generated by friction.

5.1.1.3 Entropy Generation Analysis in Curved Ducts

An entropy generation analysis of laminar forced convection in a curved rectangular duct with longitudinal ribs was performed by Ko [22–23]. The results showed the ribs number and arranging them mounted on curved duct wall can lead to the minimization of entropy generation, thus, a single rib on the heated duct wall significantly reduce the entropy generation, while the addition of multiple ribs does not result in a reduction in entropy generation. In another study, Ko and Ting [24] numerically studied the effects of external heat flux, Dean number, and cross-sectional aspect ratio on the entropy generation in a curved rectangular duct with constant wall heat flux and found that in the flow fields with higher Dean numbers and lower heat fluxes, the entropy generation was dominated by the frictional irreversibility, while at lower Dean numbers and higher heat fluxes, the entropy generation was dominated by the heat transfer irreversibility. Later, Ko and Wu [25] also performed a numerical study on entropy generation of turbulent forced convection in a 3-D curved rectangular duct. Three aspect-ratio (γ) cases were studied: 0.25; 1; and 4, and the authors founded that for all cases the resultant entropy generations in the flow fields was dominated by the frictional irreversibilities, and also that $\gamma = 1$ represents optimal choice for aspect ratio from the point of view of the minimal entropy generation principle.

The effect of various coil and flow parameters on the total entropy generation of laminar viscous flow with constant wall temperature were analytically investigated by Shokouhmand and Salimpour [26], and their results revealed that the increase in coil pitch can lead to the reduction of entropy generation.

Amani and Nobari [27] numerically investigated the entropy generation in the entrance region of curved pipes with constant wall temperature. In this study, the entropy generation rates and the optimal Reynolds numbers in the entrance region were numerically obtained and compared to the analytical prediction in the fully developed region using two different groups of non-dimensional parameters (*first group*: Reynolds number, Re, non-dimensional curvature ratio, δ, Pr, non-dimensional heat transfer parameter, q^*, non-dimensional length of the pipe, λ, and β defined as $\beta = \mu^3 / 32 \rho^2 a^2 k T_w$, where μ is dynamic viscosity, Pa s, ρ is density, kg/m^3, a is pipe radius, m, k is thermal conductivity, W/m K, and T_w is wall temperature, K; *second group*: Re, δ, Pr, λ, Eckert number, Ec, and temperature ratio, T^{**}). The results revealed that for the first group of non-dimensional parameters, the total entropy generation rate based on the analytical study in the fully developed region differs from the numerical ones by 7% in the entrance region, while for the optimum Reynolds number, the deviations between analytical and numerical analyses were 4% deviation in average. Concerning the second group of non-dimensional parameters, the deviations between numerical and analytical computations at the entrance region, in this case the total entropy generation rate was 30%, while for the optimum Reynolds number these deviations were 6%.

The effects of length, heat flux, as well as the effect of temperature dependence of thermophysical properties on entropy generation of laminar forced convection and optimal operation of helical coils were numerically investigated by Ahadi and Abbassi [28]. They founded that the entropy generation increases with increasing combined length and heat flux characteristic as well with the inlet temperature, and also that in the region dominated by the heat transfer irreversibility, the optimal values of the Reynolds number increase with increasing inlet temperature and decrease with increasing combined length and heat flux characteristic. Also, the effect of different flow conditions (mass velocity, inlet vapor quality, saturation temperature, and heat flux) on entropy generation in a helically coiled tube in flow boiling condition with constant heat flux were experimentally studied by Abdous et al. [29]. It was found that by increasing in the values of mass velocity, inlet vapor quality and heat flux resulted in an increase in entropy generation, and also an increase of saturation temperature led to increase of entropy generation. Moreover, this study showed that there is a favorable region for the use of the helically coiled tube with respect to the straight one.

An entropy generation analysis of laminar and turbulent flows in tube-in-tube helical heat exchangers was performed by Farzaneh-Gord et al. [30]. The authors developed a dimensionless function for the entropy generation number as a function of Pr, Dean number (De), the ratio of helical pipe diameter to the tube diameter, δ, and the duty parameter of the heat exchanger. This expression was minimized to obtain the optimum values of δ, Dean number in laminar flow, and Reynolds number in turbulent flow. Sepehr et al. [31] numerically studied the heat transfer, pressure drop, and entropy generation in shell and helically coiled tube heat exchangers and also proposed the correlations for estimation of the Nusselt number, friction factor, and entropy generation as a function of number of transfer units.

5.1.1.4 Other

The entropy generation minimization analysis for the optimization of the configuration of the heat exchange surfaces in a solar roof tile was performed by Giangaspero and Sciubba [32]. In this study, the authors investigated different geometries (i.e., pins, fins, plate-fins, pin-fins) and found that the geometry with pin-fins ensures the best performance among the tested ones, and also that the optimal pin array shape parameters (pitch and span) can be obtained by a critical analysis of the local entropy maps and of the temperature contours.

Gheorghian et al. [33] proposed a performance evaluation method of the heat transfer surfaces (louvered fin, plain fin, and strip fin) from the point of view of entropy generation, implementing this method on a cross flow heat exchanger. The results showed that the classification of heat transfer surfaces from the point of view of entropy generation depends on Reynolds number, and also that the proposed method predicted with good accuracy the Reynolds number value for which a heat transfer surface type becomes more or less advantageous than the reference surface.

Amirahmadi et al. [34] carried out an entropy generation analysis in a trapezoidal duct with tabulators and two vortex generators (surface roughness and beveled corners) in laminar regime. The main conclusion of this study was that the installation of the vortex generators inside the duct led to the decrease in total entropy generation up to approximately 27%. Later, Akbarzadeh et al. [35] numerically investigated the effects of corrugation profiles and Reynolds number on entropy generation and thermo-hydraulic performance of a wavy channel in laminar flow. Three corrugation profiles, namely, sinusoidal, trapezoidal, and triangular shapes, were studied, and the results revealed that the minimum entropy generation due to heat transfer contribution was obtained for sinusoidal profile, followed by trapezoidal and triangular profiles, the reduction in entropy generation being 111%, 100%, and 116%, the authors recommending the use of a sinusoidal wall in a channel due to the high performance and low entropy generation. In another study, Rashidi et al. [36] also performed a numerical analysis which studied the effects of Reynolds number, wave amplitude, and wavelength of the corrugated wall on entropy generation of turbulent forced convection in corrugated channel. They found that there is an optimal Reynolds number value ($Re = 20,000$), for all considered values of wave amplitude and wavelength of the corrugated wall, where the total entropy generation is minimum. Also, their results indicated that the thermal entropy generation decreases with increasing wave amplitude of the corrugated wall, for Reynolds numbers higher than 20,000, while the frictional entropy generation increases with increasing wave amplitude of the corrugated wall, for all studied Reynolds numbers.

Kiyasatfar [37] analytically studied the effects of slip and viscous dissipation on entropy generation of non-Newtonian power-law fluid flows in parallel-plate and circular microchannels subject to wall heat flux constant. The results revealed that the effects of slip and viscous dissipation were significant in the entropy generation analysis, thus, the increase in slip coefficient and the decrease in both power-law index and Brinkman number leads to decreasing entropy generation rate. Also, in the case of dilatant fluids, the entropy generation due to fluid friction contribution was higher than the entropy generation dominated by the heat transfer, while for pseudoplastic fluids with very low Brinkman number the thermal entropy generation was dominating. Moreover, this study showed that the parallel-plate microchannels tend to produce more entropy than circular microchannels.

Rastogi and Mahulikar [38] performed an optimization of micro-heat sink with constant wall heat flux based on theory of entropy generation in laminar flow. The results indicated that there is an optimum tube diameter and a corresponding optimum natural number where the sum-total entropy generation rate was minimum.

Genić et al. [39] developed a methodology based on combining second law of thermodynamics with the number of transfer units method in the case of heat exchangers with the flow configuration in counter and co-currents and also proposed of analysis of the optimum heat exchanger in regard to minimum operational and investment costs. They proposed a relation between the minimum value of the number of entropy generation units and minimum of total annual costs.

5.2 Entropy Generation Analysis in Ducts with Hybrid Fluids

Ahammed et al. [40] experimentally studied the entropy generation in a multiport minichannel heat exchanger in laminar flow at constant wall heat flux. The working fluids were alumina and graphene nanofluids and graphene-alumina hybrid nanofluids with a 0.1% volume concentration. Their results showed a decrease in total entropy generation of 19.6% for graphene-alumina hybrid nanofluids with increasing Reynolds number at a heat flux of 25,000 W/m^2, while for alumina and graphene nanofluids, the decreases in total entropy generation were 6.15% and 31.86%, respectively, which indicated that the graphene-water nanofluids have better thermodynamic characteristics.

Hussain et al. [41] numerically investigated the entropy generation of Al_2O_3 + Cu/water hybrid nanofluid flow in a horizontal channel with an open cavity under the influence of magnetohydrodynamics (MHD) field. They studied three different vertical locations of the obstacle placed in channel and founded that by increase in the Hartmann number, the entropy generation due to thermal irreversibility decreases, while the entropy generation due to frictional irreversibility and magnetic force increases. Moreover, by the increase in Richardson and Reynolds numbers, the entropy generation increases in channel.

In another study, Kasaeipoor et al. [42] carried out a numerical and experimental study on entropy generation in a cavity with refrigerant rigid body using the Lattice Boltzmann method. In this study, the working fluid was multi-walled carbon nanotubes (MWCNTs) + MgO/water hybrid nanofluids with different volume concentrations of nanoparticles. They observed an increase in total entropy generation with increasing Rayleigh number and also a decrease in total entropy generation with increasing volume concentration of nanoparticles. The authors explained the increase in entropy generation as being due to increasing temperature difference between side walls and refrigerant rigid body.

Mehrali et al. [43] performed an experimental study on entropy generation in a pipe under constant wall heat flux, magnetic field, and laminar flow. They founded that application of graphene oxide + Fe_3O_4/water decreases the entropy generation up 41% compared to distilled water. Another entropy generation analysis is also proposed by Bahiraei et al. [44] in order to find the optimal conditions which the entropy generation is minimum. Analyses were performed for a double-pipe minichannel heat exchanger using Fe_3O_4/CNT non-Newtonian hybrid nanofluids, and the results showed that the minimum point of total entropy generation rate was obtained at low concentrations of magnetite. The same team of researchers, Shahsavar et al. [45], performed an optimization of double-pipe heat exchanger based on theory of entropy generation and founded that the influence of concentration of Fe_3O_4 nanoparticles on the increase of global frictional entropy generation is greater than the effect of concentration of CNT, and also that the global total entropy generation rate increases with increasing Reynolds number, concentration of CNT, and concentration of Fe_3O_4. In another study, Shahsavar et al. [46] studied the effects of concentration and radius ratio on entropy generation of Fe_3O_4/CNT non-Newtonian hybrid nanofluid in a concentric annulus. The results revealed that in a concentric annulus, the entropy generation due to frictional irreversibility was much more that the entropy generation due to thermal irreversibility, such that total entropy generation exhibits a similar variation to the variation of the frictional entropy generation, namely, the total entropy increases with increasing both of CNT and Fe_3O_4 nanoparticles concentration.

Mansour et al. [47] analytically examined the effects of heat source and sink on entropy generation and magneto-hydrodynamics natural convection flow and heat transfer in a square porous cavity, and the study focuses on Cu + Al$_2$O$_3$/water hybrid nanofluids. Their results indicated that the entropy generation decreases with increasing Hartmann number, and also that the entropy generation increases with increasing nanoparticle volume fraction.

Huminic and Huminic [48] studied the effects of volume concentration of hybrid nanoparticles, inlet temperature, and Reynolds number on entropy generation of MWCNT + Fe$_3$O$_4$/water and nanodiamond (ND) + Fe$_3$O$_4$/water hybrid nanofluids flowing through a flattened tube and founded that with the increase of volume concentration of hybrid nanofluid, the total entropy generation decreases. Moreover, they showed that the MWCNT + Fe$_3$O$_4$/water hybrid nanofluids exhibit a higher reduction in entropy generation than ND + Fe$_3$O$_4$/water hybrid nanofluids. Also, thermophysical properties and convective heat transfer of hybrid nanofluids in heat transfer applications were reviewed by Huminic and Huminic [49].

Because various cross-sectional ducts are used in thermal devices due to the size and volume constraints to enhance heat transfer, in first part of the current chapter several numerical and experimental research concerning at the reduction of entropy generation in circular and non-circular ducts is reviewed. Then, the computational development and application of hybrid nanofluid flow in circular and non-circular ducts are discussed to illustrate the impact of hybrid nanofluids on entropy generation and pressure drop. Thus, this analysis could be a powerful tool to decide which studied geometry is more efficient.

5.3 Entropy Generation Theory

The use of the hybrid nanofluids as working fluids in various heat exchangers often leads to not only an increase of the heat transfer coefficient, but also to an increase in pumping power because more energy is required to ensure the working conditions of the heat exchangers.

First study on entropy generation in heat exchangers was performed by Bejan [50]. In this work, a proposal is made of a heat exchanger design method based on the concept of thermodynamic irreversibility.

For a control volume of thickness dx along of duct, the entropy generation is [51–52]:

$$d\dot{S}_{gen} = \dot{m}\,ds - \delta\dot{Q}/T_w \tag{5.1}$$

For an incompressible fluid:

$$ds = c_p \frac{dT}{T} - \frac{dP}{\rho T} \tag{5.2}$$

and

$$\delta Q = \dot{m} c_p\, dT = q'' p\, dx \tag{5.3}$$

In Eq. (5.3) $\dot{m} = \rho \bar{U} A_c$ is the mass flow rate, p is the perimeter of the duct, and by integrating of Eq. (5.3) the bulk-temperature variation of the fluid and the total heat-transfer rate along the duct is obtained:

$$T = T_0 + \left(\frac{4q''}{\rho \bar{U} D_h c_p} \right) x \tag{5.4}$$

and

$$\dot{Q} = q'' p L \tag{5.5}$$

Also, in the Eq. (5.2), $\Delta P = -\left(f \rho \bar{U}^2 / 2 D_h \right) dx$ is the pressure drop [53], f is friction factor, ρ is the fluid density, w is fluid bulk velocity, d_h is hydraulic diameter, and ΔT is the increase of the fluid temperature $T(x = L) - T_0$.

A dimensionless total entropy generation may be written as [54]:

$$\psi = \frac{\dot{S}_{gen}}{\dot{m}c_p} = \frac{\dot{S}_{gen}}{\left(\dot{Q}/\Delta T\right)}$$

(5.6)

equation which is based on the flow stream heat capacity rate $\left(\dot{m}c_p\right)$ or [52]:

$$\psi = \ln\left[\frac{\left(Re+\tau\,\Pi_1\right)\left(1+\tau\right)}{Re+\tau\,Re+\tau\Pi_1}\right] + \Pi_2 Re^2\,\ln\left[\frac{Re+\tau\,\Pi_1}{Re}\right]$$

(5.7)

where $\Pi_1 = 4\frac{Nu\,\lambda}{Pr}$ and $\Pi_2 = \frac{\mu^3\left(f\,Re\right)}{8\rho^2\,D_h^3\,q''}$, Nu is Nusselt number, $\lambda = \frac{L}{D_h}$ is dimensionless length of the duct, μ is dynamic viscosity of fluid, ρ is density, q'' is the heat flux imposed on the wall of the duct, f is friction factor, Re is Reynolds number, and $\tau = \frac{T_w - T_{in}}{T_w}$ is dimensionless parameter, T_w is wall temperature of the duct, and T_{in} is inlet fluid temperature.

The modified dimensionless entropy generation based on the total heat-transfer rate can be written as [52]:

$$\phi = \frac{Re}{\Pi_1}\psi$$

(5.8)

In Eq. (5.6), the entropy generation rate per unit duct length is described by two terms: the entropy generation due to heat transfer and the entropy generation due to fluid friction [55]:

$$\dot{S}_{gen} = \left(\dot{S}_{gen}\right)_{\Delta T} + \left(\dot{S}_{gen}\right)_{\Delta P} = \frac{q'^2}{\pi\,T^2\,k\,Nu} + \frac{2\dot{m}^3\,f}{\rho^2\,T\,D_h\,A_c^2}$$

(5.9)

where $q' = \dot{Q}/L$ is the wall heat-transfer rate per unit length (W/m), Nu is Nusselt number, f is friction factor, and T, k, and ρ are temperature, thermal conductivity, and bulk density.

The Eq. (5.9) is valid independently from the shape of the cross section of the duct.

For circular ducts, the equation of rate of entropy generation is:

$$\dot{S}_{gen} = \frac{q''^2\,\pi\,D_h^2\,L}{T^2\,k\,Nu} + \frac{32\dot{m}^3\,f\,L}{\pi^2\,\rho^2\,T\,D_h^5}$$

(5.10)

and for non-circular duct (flat ducts), the entropy generation rate becomes [56]:

$$\dot{S}_{gen} = \frac{\dot{Q}^2}{Nu\,k\,T_{in}\,T_{out}\,L} + \frac{\rho\,U^3\,f\,p\,L}{2\left(T_{out} - T_{in}\right)}\ln\left(\frac{T_{out}}{T_{in}}\right)$$

(5.11)

TABLE 5.1

The Parameters of Ducts with Various Cross-Section Areas

Geometry	Perimeter (p)	Cross-Section Area (A_c)	Shape Factor $\chi = \dfrac{p}{D_h}$ or $\chi = \dfrac{p^2}{4A_c} = \dfrac{4A_c}{D_h^2}$
Circle	πD	$\dfrac{\pi D^2}{4}$	π
Rectangle	$2(a+b)$	ab	$2 + a/b + b/a$
Ellipse	$\pi\left[3\left(\dfrac{a}{2}+\dfrac{b}{2}\right) - \sqrt{\left(3\dfrac{a}{2}+\dfrac{b}{2}\right)\left(\dfrac{a}{2}+3\dfrac{b}{2}\right)}\right]$	$\pi ab/4$	$(\pi/8)\left[3(a/b+b/a)+2\right]$

where the friction factor was expressed as:

$$f = \frac{2\Delta P\, D_h}{\rho U^2 L} \tag{5.12}$$

The pressure drop ΔP was taken from the numerical simulations, and the velocities U were calculated from the Reynolds numbers.

The pumping power to heat-transfer ratio for laminar flow can be written as [52]:

$$P_r = \Pi_2\, Re^2 \tag{5.13}$$

The perimeter, the cross-section area, and shape factor for ducts considered in this paper for the circular, rectangular, and elliptical cross-sections are tabulated in Table 5.1.

5.4 Thermophysical Properties of Hybrid Nanofluids

For a better understanding of the effect of hybrid nanofluids on entropy generation, following will discuss some relevant models for the thermal conductivity and the viscosity of hybrid nanofluids. In Tables 5.2 and 5.3 are presented few of the available experimental models proposed for hybrid nanofluids. As one can notice, most of the models are dependent of both the volume concentration of the nanoparticles and temperature.

TABLE 5.2

Correlations for the Thermal Conductivity of Hybrid Nanofluids

References	Correlation	Hybrid Nanofluids	Relevant Information
Esfe et al. [57]	$\dfrac{k_{nf}}{k_{bf}} = \dfrac{0.1747 \cdot 10^5 + \phi_p}{0.1747 \cdot 10^5 - 0.1498 \cdot 10^6 \cdot \phi_p + 0.1117 \cdot 10^7 \cdot \phi_p^2 + 0.1997 \cdot 10^8 \cdot \phi_p^3}$	Ag-MgO/water	$0 < \phi \leq 0.02$
Esfe et al. [58]	$\dfrac{k_{nf}}{k_{bf}} = 1.07 + 0.000589 \times T + \dfrac{-0.000184}{T \times \phi} + 4.44 \times T \times \phi$ $\times \cos\left(6.11 + 0.00673 \times T + 4.41 \times T \times \phi - 0.0414 \sin(T)\right) - 32.5 \times \phi$	Cu-TiO$_2$/water-ethylene glycol (60:40)	$0 < \phi \leq 0.02$ $30°C \leq T \leq 60°C$
Harandi et al. [59]	$\dfrac{k_{nf}}{k_{bf}} = 1 + 0.0162\phi^{0.7038} T^{0.6009}$	MWCNTs—Fe$_3$O$_4$/ethylene glycol	$0 < \phi \leq 0.023$ $25°C \leq T \leq 50°C$
Sundar et al. [60]	$\dfrac{k_{nf}}{k_{bf}} = a + b\phi$ $T = 20°C$: $a = 1.0149; b = 0.2403$; $T = 30°C$: $a = 1.0188; b = 0.3751$; $T = 40°C$: $a = 1.0157; b = 0.4728$; $T = 50°C$: $a = 1.0168; b = 0.5697$; $T = 60°C$: $a = 1.0150; b = 0.6818$	(ND)—Fe$_3$O$_4$/water ND—Fe$_3$O$_4$/EG—water (20:80, 40:60, and 60:40)	$0 < \phi \leq 0.002$ $20°C \leq T \leq 60°C$
Sundar et al. [61]	$\dfrac{k_{nf}}{k_{bf}} = 0.9978(1 + \phi)^{0.6556}$	ND—Co$_3$O$_4$/water, ND—Co$_3$O$_4$/ethylene glycol ND—Co$_3$O$_4$/ethylene glycol (EG)—water (20:80, 40:60, and 60:40)	$0 < \phi \leq 0.0015$ $20°C \leq T \leq 60°C$
Vafaei et al. [62]	$\dfrac{k_{nf}}{k_{bf}} = 0.9787 + \exp\left(0.3081\,\phi^{0.3097} - 0.002\,T\right)$	MgO—MWCNTs/ethylene glycol	$0 < \phi \leq 0.006$ $25°C \leq T \leq 50°C$
Afrand [63]	$\dfrac{k_{nf}}{k_{bf}} = 0.8341 + 1.1\phi^{0.243} T^{-0.289}$	MgO—functionalized multi-walled carbon nanotubes (fMWCNT)/ethylene glycol	$0 < \phi \leq 0.006$ $25°C \leq T \leq 50°C$

(Continued)

TABLE 5.2 (Continued)

Correlations for the Thermal Conductivity of Hybrid Nanofluids

References	Hybrid Nanofluids	Correlation	Relevant Information
Esfe et al. [64]	single-walled carbon nanotubes (SWCNTs)—Al$_2$O$_3$/ethylene glycol	$\dfrac{k_{nf}}{k_{bf}} = 0.963 + 0.008379\left(\phi^{0.4439}T^{0.9246}\right)$	$0 < \phi \leq 0.02$ $30°C \leq T \leq 50°C$
Esfe et al. [65]	SiO$_2$—WCNTs (85:15)/ethylene glycol	$\dfrac{k_{nf}}{k_{bf}} = 0.905 + 0.002069T\phi + 0.04375\phi^{0.09265}T^{0.3305} - 0.0063\phi^3$	$0 < \phi \leq 0.0195$ $30°C \leq T \leq 50°C$
Esfe et al. [66]	ZnO—MWCNTs/water—EG (50:50)	$\dfrac{k_{nf}}{k_{bf}} = 1.024 + 0.5988\phi^{0.6029}\exp\left(\dfrac{\phi}{T}\right) - \dfrac{8.059\phi T^{0.2} + 2.24}{6.052\phi^{0.2} + T}$	$0 < \phi \leq 0.01$ $30°C \leq T \leq 50°C$
Esfe et al. [67]	SWCNT—ZnO (30:70)/water—EG (40:60)	$\dfrac{k_{nf}}{k_{bf}} = 0.8707 + 0.0008883T\phi + 0.004435\phi^{0.252}T + 0.179\phi^{0.179}\exp(0.09624\phi^2)$	$0 < \phi \leq 0.016$ $26°C \leq T \leq 50°C$
Esfe et al. [68]	SWCNT—MgO/EG	$\dfrac{k_{nf}}{k_{bf}} = 0.90844 - 0.06613\phi^{0.3}T^{0.7} + 0.01266\phi^{0.31}T$	$0 < \phi \leq 0.02$ $30°C \leq T \leq 50°C$
Rostamian et al. [69]	SWCNT—CuO/water—EG (60:40)	$\dfrac{k_{nf}}{k_{bf}} = 1.0 + (0.04056(T\phi)) - (0.003252(T\phi)^2) + (0.0001181(T\phi)^3) - (0.000001431(T\phi)^4)$	$0 < \phi \leq 0.0075$ $20°C \leq T \leq 50°C$
Esfe et al. [70]	MWCNTs—SiO$_2$ (30:70)/ethylene glycol	$\dfrac{k_{nf}}{k_{bf}} = 1.01 + 0.007685T\phi - 0.5136\phi^2 T^{-0.1578} + 11.5\phi^3 T^{-1.175}$	$0 < \phi \leq 0.0086$ $20°C \leq T \leq 60°C$
Hamid et al. [71]	TiO$_2$—SiO$_2$ (30:70)/water—EG (60:40)	$\dfrac{k_{nf}}{k_{bf}} = 1.17(1+R)^{-0.1151}\left(\dfrac{T}{80}\right)^{0.0437}$ (R is mixture ratio of TiO$_2$ and SiO$_2$)	$\phi = 0.01$ $30°C \leq T \leq 80°C$
Akilu et al. [72]	SiO$_2$—CuO/C (80:2)/glycerol and ethylene glycol (60:40)	$\dfrac{k_{nf}}{k_{bf}} = 0.1332\left(1+\dfrac{\phi}{100}\right)^{6.1752}\left(\dfrac{T_{nf}}{T_0}\right)^{2.64208}$, $(T_0 = 273.15K)$	$0 < \phi \leq 0.02$ $30°C \leq T \leq 80°C$

TABLE 5.3

Correlations for the Viscosity of Hybrid Nanofluids

References	Correlation	Hybrid Nanofluid	Relevant Information
Esfe et al. [57]	$\dfrac{\mu_{nf}}{\mu_{bf}} = 1 + 32.795\phi - 7214\phi^2 + 714600\phi^3 - 0.1941 \cdot 10^8 \, \phi^4$	Ag—MgO/water	$0 < \phi \leq 0.02$
Asadi and Asadi [73]	$\mu_{nf} = 796.8 + 76.26\phi + 12.88T + 0.7695\phi T + \dfrac{-196.9T - 16.53\phi T}{\sqrt{T}}$	MWCNTs—ZnO/engine oil	$0.0625 \leq \phi \leq 0.01$ $5°C \leq T \leq 55°C$
Afrand et al. [74]	$T = 25°C: \dfrac{\mu_{nf}}{\mu_{bf}} = 0.9566 + 0.9841\phi - 4.4687\phi^2 + 7.8779\phi^3 - 4.0731\phi^4$ $T = 30°C: \dfrac{\mu_{nf}}{\mu_{bf}} = 1.023 - 0.1613\phi + 1.6674\phi^2 - 2.6513\phi^3 + 1.3953\phi^4$; $T = 35°C: \dfrac{\mu_{nf}}{\mu_{bf}} = 0.9956 + 0.1194\phi + 0.7286\phi^2 - 1.5119\phi^3 + 0.9488 \; \phi^4$; $T = 40°C: \dfrac{\mu_{nf}}{\mu_{bf}} = 1.0133 - 0.1316\phi + 1.5444\phi^2 - 2.326\phi^3 + 1.2019 \; \phi^4$; $T = 45°C: \dfrac{\mu_{nf}}{\mu_{bf}} = 1.007 - 0.1187\phi + 1.6253\phi^2 - 2.5437\phi^3 + 1.3528\phi^4$; $T = 50°C: \dfrac{\mu_{nf}}{\mu_{bf}} = 1.0182 - 0.2347\phi + 1.593\phi^2 - 1.7341\phi^3 + 0.7139\phi^4$; $T = 55°C: \dfrac{\mu_{nf}}{\mu_{bf}} = 1.0224 - 0.471\phi + 3.3185\phi^2 - 5.047\phi^3 + 2.5453\phi^4$; $T = 60°C: \dfrac{\mu_{nf}}{\mu_{bf}} = 0.9952 + 0.0305\phi + 0.7167\phi^2 - 0.8848\phi^3 + 0.5161\phi^4$.	SiO$_2$—MWCNTs/society for automotive engineers (SAE) 40	$0.0625 \leq \phi \leq 0.01$ $25°C \leq T \leq 60°C$
Dardan et al. [75]	$\dfrac{\mu_{nf}}{\mu_{bf}} = 1.123 + 0.3251\phi - 0.08994T + 0.0025522T^2 - 0.00002386T^3 + 0.9695\left(\dfrac{T}{\phi}\right)^{0.01719}$	Al$_2$O$_3$—MWCNTs/SAE 40	$0.0625 \leq \phi \leq 0.01$ $25°C \leq T \leq 50°C$

(Continued)

TABLE 5.3 (Continued)

Correlations for the Viscosity of Hybrid Nanofluids

References	Hybrid Nanofluid	Correlation	Relevant Information
Esfe et al. [76]	MWCNTs—SiO$_2$/SAE 40	$T = 25°C: \dfrac{\mu_{nf}}{\mu_{bf}} = 1.0343 + 0.2336\phi - 0.2604\phi^2 + 0.2375\phi^3;$ $T = 30°C: \dfrac{\mu_{nf}}{\mu_{bf}} = 1.0435 + 0.4417\phi - 0.5087\phi^2 + 0.2985\phi^3;$ $T = 35°C: \dfrac{\mu_{nf}}{\mu_{bf}} = 1.0202 + 0.7748\phi - 1.2154\phi^2 + 0.7073\phi^3;$ $T = 40°C: \dfrac{\mu_{nf}}{\mu_{bf}} = 0.9903 + 1.0245\phi - 1.7095\phi^2 + 0.9978\phi^3;$ $T = 45°C: \dfrac{\mu_{nf}}{\mu_{bf}} = 1.0279 + 0.7283\phi - 1.0703\phi^2 + 0.6115\phi^3;$ $T = 50°C: \dfrac{\mu_{nf}}{\mu_{bf}} = 1.0347 + 0.6317\phi - 1.0729\phi^2 + 0.6942\phi^3.$	$0.0625 \leq \phi \leq 0.01$ $25°C \leq T \leq 50°C$
Soltani and Akbari [77]	MgO—MWCNTs/ethylene glycol	$\dfrac{\mu_{nf}}{\mu_{bf}} = 0.191\phi + 0.240\left(T^{-0.342}\phi^{-0.473}\right)exp\left(1.45T^{-0.120}\phi^{0.158}\right)$	$0.001 \leq \phi \leq 0.01$ $30°C \leq T \leq 60°C$
Esfe et al. [78]	MWCNTs—ZnO/SAE 40	$T = 25°C: \dfrac{\mu_{nf}}{\mu_{bf}} = 1.0087 + 0.1553\phi - 0.0334\phi^2 + 0.0631\phi^3;$ $T = 30°C: \dfrac{\mu_{nf}}{\mu_{bf}} = 1.0085 + 0.2499\phi - 0.2865\phi^2 + 0.2043\phi^3;$ $T = 35°C: \dfrac{\mu_{nf}}{\mu_{bf}} = 1.0223 + 0.5341\phi - 0.6313\phi^2 + 0.366\phi^3;$ $T = 40°C: \dfrac{\mu_{nf}}{\mu_{bf}} = 1.0382 + 0.5376\phi - 0.5013\phi^2 + 0.261\phi^3;$ $T = 45°C: \dfrac{\mu_{nf}}{\mu_{bf}} = 1.013 + 0.6448\phi - 0.9427\phi^2 + 0.5225\phi^3;$ $T = 50°C: \dfrac{\mu_{nf}}{\mu_{bf}} = 1.0132 + 0.6596\phi - 0.913\phi^2 + 0.4822\phi^3.$	$0.05 \leq \phi \leq 0.01$ $25°C \leq T \leq 60°C$

(Continued)

TABLE 5.3 *(Continued)*

Correlations for the Viscosity of Hybrid Nanofluids

References	Correlation	Hybrid Nanofluid	Relevant Information
Sundar et al. [60]	$\dfrac{\mu_{nf}}{\mu_{bf}} = a\,e^{b\phi}$ $T = 20°C: a = 1.444; b = 1.402;$ $T = 30°C: a = 1.368; b = 1.472;$ $T = 40°C: a = 1.277; b = 1.625;$ $T = 50°C: a = 1.288; b = 1.771;$ $T = 60°C: a = 1.338; b = 1.655$	(ND)—Fe$_3$O$_4$/water ND—Fe$_3$O$_4$/EG—water (20:80, 40:60, and 60:40)	$0 < \phi \leq 0.002$ $20°C \leq T \leq 60°C$
Asadi et al. [79]	$\mu_{nf} = 328201 T^{-2.053}\,\phi^{0.09359}$	MWCNTs—MgO/SAE 40	
Sundar et al. [61]	$\dfrac{\mu_{nf}}{\mu_{bf}} = 0.9595 (1+\phi)^{2.399}$	ND—Co$_3$O$_4$/water, ND—Co$_3$O$_4$/ ethylene glycol ND—Co$_3$O$_4$/ EG—water (20:80, 40:60, and 60:40)	$0.0005 \leq \phi \leq 0.0015$ $20°C \leq T \leq 60°C$
Hamid et al. [71]	$\dfrac{\mu_{nf}}{\mu_{bf}} = 1.42 (1+R)^{-0.1063} \left(\dfrac{T}{80}\right)^{0.2321}$ (R is mixture ratio of TiO$_2$ and SiO$_2$)	TiO$_2$—SiO$_2$ (30:70)/ water—EG (60:40)	$\phi = 0.01$ $30°C \leq T \leq 80°C$
Akilu et al. [72]	$\dfrac{\mu_{nf}}{\mu_{bf}} = 0.9894 \left(1+\dfrac{\phi}{100}\right)^{6.6301} \left(\dfrac{T_{nf}}{T_0}\right)^{0.064}$, $(T_0 = 273.15\,K)$	SiO$_2$—CuO/C (80:2)/glycerol and ethylene glycol (60:40)	$0 < \phi \leq 0.02$ $30°C \leq T \leq 80°C$

TABLE 5.4

The Thermal Conductivity a nd Viscosity of GO—Co$_3$O$_4$/Water Hybrid Nanofluids at $T = 293\,K$

Thermophysical Properties	Water (Base Fluid) $\phi = 0.0\%$	GO—Ce$_3$O$_4$/Water Nanofluid $\phi = 0.10\%$	f $= 0.20\%$
Thermal conductivity [W/m^{2}K]	0.602	0.625	0.648
Viscosity [mPa·s]	0.79	0.92	1.18
Density [kg/m^3]	996.16	1000	1003
Specific heat [J/kg K]	4184.4	4180.8	4177.3

In current study, the thermal conductivity and viscosity of water and graphene oxide (GO)—cobalt oxide (Co$_3$O$_4$)/water hybrid nanofluids were taken from Sundar et al. [80] and listed in Table 5.4.

The density and specific heat of GO—Co$_3$O$_4$/water hybrid nanofluids were computed on basis of mixture theory:

$$\rho_{hnf} = (1-\phi)\rho_{bf} + \phi\left(\rho_{GO+Co3O4}\right)_p \tag{5.14}$$

and

$$c_{p,hnf} = (1-\phi)c_{bf} + \phi\left(c_{p,GO+Co3O4}\right)_p \tag{5.15}$$

where $\rho_{GO} = 1919\,kg/m^3$ is the density of graphene oxide, $\rho_{Co3O4} = 6110\,kg/m^3$ is the density of Co$_3$O$_4$, and the density of $\rho_{GO+Co3O4}$ nanoparticles is obtained as 3296 kg/m^3. Also, $c_{p,GO} = 710$ J/kg K is the heat specific of graphene oxide, $c_{p,Co3O4} = 460$ J/kg K is the specific heat of Co$_3$O$_4$, and the specific heat of $c_{p,GO+Co3O4}$ nanoparticles is obtained as 627.5 J/kg K.

5.5 Numerical Analysis

Figure 5.1 shows the geometrical models considered in the current study. It consists of flattened, elliptical, and circular ducts. Dimensions geometrical of the ducts are tabulated in Table 5.5.

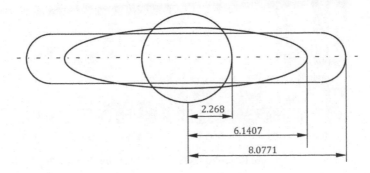

FIGURE 5.1 Schematic of the studied configurations.

TABLE 5.5

Geometrical Dimensions of Considered Configurations

Configuration	a [mm]	b [mm]	Hydraulic Diameter, D_h[mm]	Length, L [mm]
Flattened duct	16.1544	2.56	4.536	500
Elliptical duct	12.28	3.10		
Circular duct	4.536			

In this study, the single-phase approach was employed for the simulation of hybrid nanofluids flow inside ducts with various shapes. Simulations were performed in steady-state and laminar flow ($U_{in} = 0.043 - 0.347$ m/s). The continuity, momentum, and energy equations were solved using ANSYS-CFX. At the inlet of the elliptical tube, the uniform axial velocity and the temperature were imposed. At outlet, zero pressure is applied. Along the tube wall, a constant heat flux, $q'' = 2000 \frac{W}{m^2}$, was imposed. The conservative interface flux on the interference surfaces of the computational domain was considered. The gravitational effect was considered in all simulations, and also the viscous dissipation of kinetic energy of the flow and the radiation effects were negligible.

The residuals resulting from an iterative process were lower than 10^{-6}. Also, the high resolution advection scheme and the double-precision solver were selected.

5.6 Results and Discussions

Figures 5.2 through 5.4 show the variation of dimensionless entropy generation ψ with inlet velocity for the studied duct geometries. The hydraulic diameter and inlet temperature were kept constant. For all studied cases, it is noted that the dimensionless entropy generation decreases with the increases in inlet velocity and the increases with volume concentration of hybrid nanoparticles. The flattened duct geometry gives the lowest dimensionless entropy generation, while for the circular duct was obtained the highest value of dimensionless entropy generation at all inlet velocities and volume concentrations of nanoparticles.

Figure 5.5 shows the dimensionless entropy generation for water and MWCNT + Fe_3O_4/water hybrid nanofluids for different Reynolds numbers and inlet temperature in the flattened duct. It can be observed

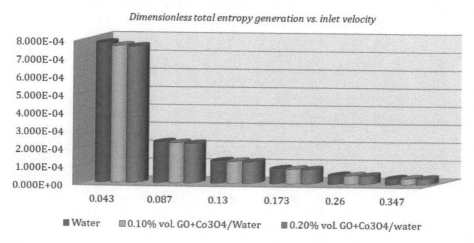

FIGURE 5.2 Dimensionless entropy generation versus inlet velocity for the flattened duct.

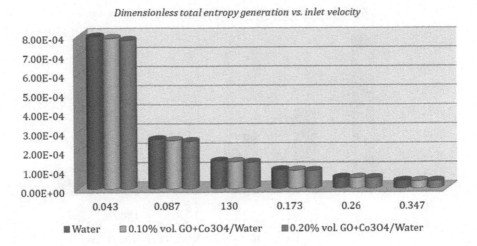

FIGURE 5.3 Dimensionless entropy generation versus inlet velocity for the elliptical duct.

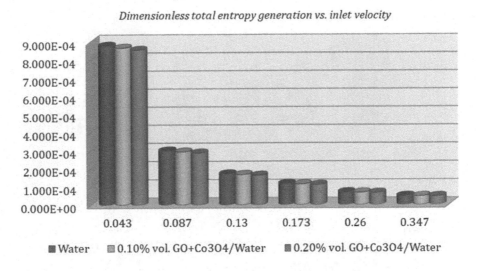

FIGURE 5.4 Dimensionless entropy generation versus inlet velocity for the circular duct.

that the dimensionless entropy generation for both fluids decreases with the increase in Reynolds number, the minimum entropy generation being in the case when the hybrid nanofluid is used. Also, the dimensionless entropy generation increases with increasing inlet temperature.

In Figure 5.6, it is shown the dimensionless entropy generation in the case when ND + Fe_3O_4/water hybrid nanofluid is used as working fluid in the flattened duct. Although from Figure 5.6 one can also see a decrease in dimensionless entropy generation for ND + Fe_3O_4/water hybrid nanofluid compared to water, however the MWCNT + Fe_3O_4/water hybrid nanofluids are more beneficial in the flattened ducts, because they produce lowest values of entropy generation compared to water.

Figures 5.7 through 5.9 show the modified dimensionless entropy generation for the studied geometry. Comparison of various geometries of ducts using the modified dimensionless entropy generation ϕ can be appropriate when the total heat-transfer rate is important. Figures 5.7 through 5.9 illustrated that modified dimensionless entropy generation decreases with increases in inlet velocity, whereas the volume concentration of nanoparticles increases. The lowest values of modified dimensionless entropy

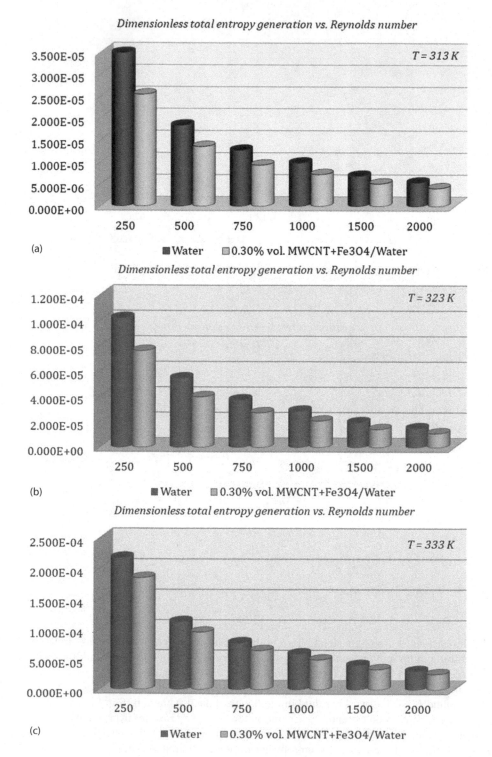

FIGURE 5.5   Dimensionless entropy generation versus Reynolds number and inlet temperature for water and MWCNT + Fe$_3$O$_4$/water hybrid nanofluids in the flattened duct for: a) T=313 K; b) T=323 K; c) T=333 K.

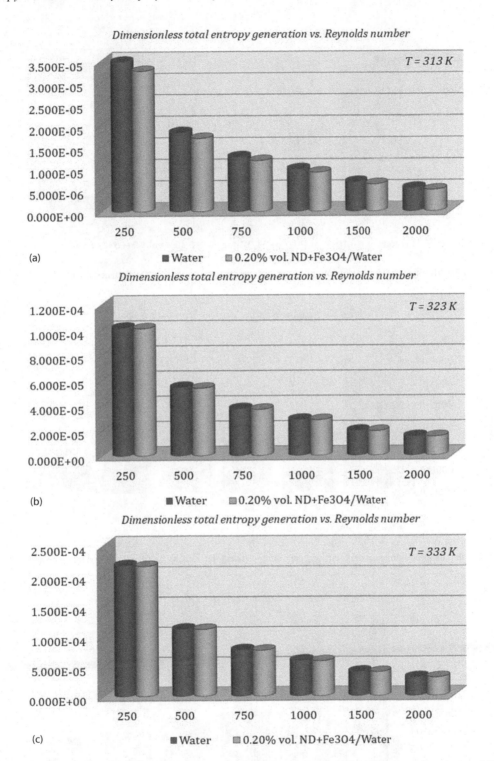

FIGURE 5.6 Dimensionless entropy generation versus Reynolds number and inlet temperature for water and ND + Fe$_3$O$_4$/water hybrid nanofluids in the flattened duct for: a) T=313 K; b) T=323 K; c) T=333 K.

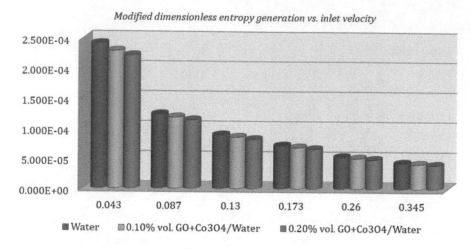

FIGURE 5.7   Modified dimensionless entropy generation versus inlet velocity for the flattened duct.

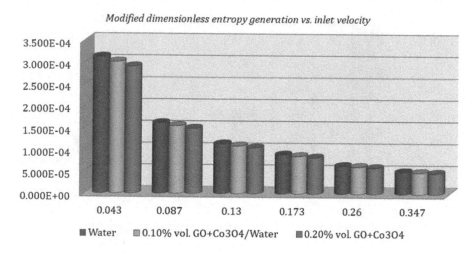

FIGURE 5.8   Modified dimensionless entropy generation versus inlet velocity for the elliptical duct.

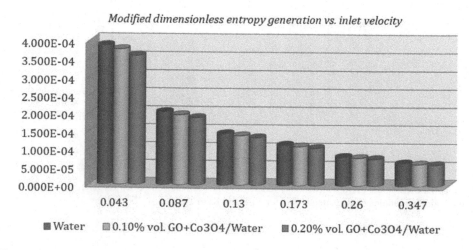

FIGURE 5.9   Modified dimensionless entropy generation versus inlet velocity for the circular duct.

generation were obtained for the flattened duct geometry, while for the circular duct were obtained the highest values of modified dimensionless entropy generation for all concentrations of nanoparticles and inlet velocities. From the performed analysis, it can be said that the minimum entropy generation is obtained for the flattened duct geometry, and that the hybrid nanofluids generate less entropy than the base fluid for all studied cases.

Figures 5.10 through 5.12 show variation of the pumping power to heat-transfer ratio, defined as the ratio of the pumping power to heat-transfer ratio of hybrid nanofluid ($P_{r,nf}$) to the pumping power to heat-transfer ratio of base fluid ($P_{r,bf}$), for different inlet velocities. From these figures, one can see that hybrid nanofluids have a higher pumping power to heat-transfer ratio than the base fluid for any given inlet velocity. Plus, one can notice that the pumping power to heat-transfer ratio is slightly higher for the flattened duct, followed by the elliptical duct and the circular duct.

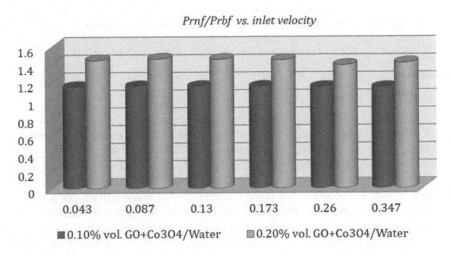

FIGURE 5.10 Pumping power to heat-transfer ratio versus inlet velocity for the flattened duct.

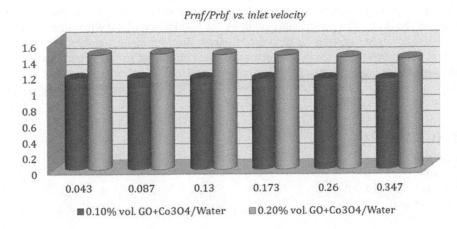

FIGURE 5.11 Pumping power to heat-transfer ratio versus inlet velocity for the elliptical duct.

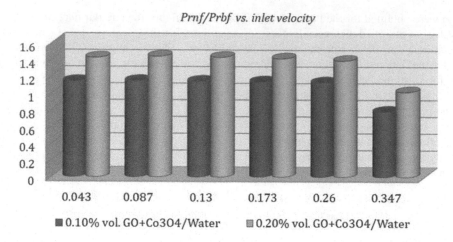

FIGURE 5.12   Pumping power to heat-transfer ratio versus inlet velocity for the circular duct.

5.7 Conclusions

In this paper, the entropy generation in ducts with various shapes subjected to constant wall heat flux using GO + Co_3O_4/water hybrid nanofluids with different volume concentrations of nanoparticle (0.10% vol. and 0.20% vol.) was studied numerically in laminar flow. The simulations were performed for three cross-sectional areas, namely, flattened, elliptical, and circular with the same hydraulic diameter. Results showed that the adding hybrid nanoparticles in the base fluid leads to a reduction in dimensionless entropy generation with the increase in volume concentration of nanoparticles due to their improved thermal transport mechanism. Also, the results indicated that the dimensionless entropy generation decreases with increasing inlet velocity. From the performed analysis, it can be noticed that the minimum entropy generation is obtained for the flattened duct geometry, and that the hybrid nanofluids generate less entropy than the base fluid for all studied cases. Also, an analysis of different hybrid nanofluids in flattened duct was performed. MWCNT + Fe_3O_4/water and ND + Fe_3O_4/water hybrid nanofluids at various Reynolds numbers and inlet temperatures were studied. Results showed that the MWCNT + Fe_3O_4/water hybrid nanofluids exhibited higher reduction in entropy generation compared to ND + Fe_3O_4/water hybrid nanofluids.

In addition, the pumping power to heat-transfer ratio was computed in order to evaluate the overall performance of the ducts. The performance analysis has highlighted that the hybrid nanofluids have a higher pumping power to heat-transfer ratio than the base fluid for any given inlet velocity due to the increase in pumping power, caused by the increase in the viscosity of hybrid nanofluids. Thus, further studies are also needed to find some hybrid nanofluids to which there are a reasonable ratio between thermal conductivity increase and viscosity increase.

NOMENCLATURE

a major axis (m)
a constant
A_c cross-sectional area (m)
b minor axis (m)
b constant
c_p specific heat at constant pressure (J/kg K)
D_h hydraulic diameter (m)
f friction factor

L length (m)
k thermal conductivity (W/m K)
$\dot{m}$ mass flow rate (kg/s)
Nu Nusselt number
ΔP pressure drop (Pa)
ΔT temperature difference (K)
p perimeter (m)
$\dot{Q}$ heat transfer rate (W)
q' wall heat transfer rate per unit length (W/m)
q'' heat flux (W/m^2)
Re Reynolds number
s specific entropy (J/kg K)
$\dot{S}_{gen}$ entropy generation (W/K)
T temperature (K)
U velocity (m/s)
$\dot{V}$ volumetric flow rate (m^3/s)

GREEK SYMBOL

λ dimensionless length of the duct
μ dynamic viscosity (Pa s)
ρ density (kg/m^3)
τ dimensionless parameter
ϕ modified dimensionless entropy generation
ϕ volume concentration of nanoparticles
ψ dimensionless entropy generation
χ shape factor

SUBSCRIPT

bf   base fluid
in   inlet
nf   nanofluid
out   outlet
p   particle
w   wall

REFERENCES

1. Bejan, A. (1982). Second-law analysis in heat transfer and thermal design. *Advances in Heat Transfer* 15, 1–58.
2. Bejan, A. (1996). *Entropy Generation Minimization*. Boca Raton, FL: CRC Press.
3. Şahin, A. (2000). Entropy generation in turbulent liquid flow through a smooth duct subjected to constant wall temperature. *International Journal of Heat and Mass Transfer* 43, 1469–1478.
4. Şahin, A. (2002). Entropy generation and pumping power in a turbulent fluid flow through a smooth pipe subjected to constant heat flux. *Exergy:An International Journal* 2, 314–321.
5. Mahmud, S. and R. A. Fraser (2003). The second law analysis in fundamental convective heat transfer problems. *International Journal of Thermal Sciences* 42, 177–186.
6. Khaliq, A. (2004). Thermodynamic optimization of laminar viscous flow under convective heat-transfer through an isothermal walled duct. *Applied Energy* 78, 289–304.
7. Dağtekin, I., H. F. Öztop and A. Z. Şahin (2005). An analysis of entropy generation through a circular duct with different shaped longitudinal fins for laminar flow. *International Journal of Heat and Mass Transfer* 48, 171–181.

8. Öztop, H. F. (2005). Effective parameters on second law analysis for semicircular ducts in laminar flow and constant wall heat flux. *International Communications in Heat and Mass Transfer* 32, 266–274.

9. Mahmud, S. and R. A. Fraser (2006). Second law analysis of forced convection in a circular duct for non-Newtonian fluids. *Energy* 31, 2226–2244.

10. Jankowski, T. A. (2009). Minimizing entropy generation in internal flows by adjusting the shape of the cross-section. *International Journal of Heat and Mass Transfer* 52, 3439–3445.

11. Esfahani, J. A. and P. B. Shahabi (2010). Effect of non-uniform heating on entropy generation for the laminar developing pipe flow of a high Prandtl number fluid. *Energy Conversion and Management* 51, 2087–2097.

12. You, Y., A. Fan, Y. Liang, S. Jin, W. Liu and F. Dai (2015). Entropy generation analysis for laminar thermal augmentation with conical strip inserts in horizontal circular tubes. *International Journal of Thermal Sciences* 88, 201–214.

13. Abdous, M. A., H. Saffari, H. B. Avval and M. Khoshzat (2016). The study of entropy generation during flow boiling in a micro-fin tube. *International Journal of Refrigeration* 68, 76–93.

14. Keklikcioglu, O. and V. Ozceyhan (2017). Entropy generation analysis for a circular tube with equilateral triangle cross sectioned coiled-wire inserts. *Energy* 139, 65–75.

15. Hooman, K., H. Gurgenci and A. A. Merrikh (2007). Heat transfer and entropy generation optimization of forced convection in porous-saturated ducts of rectangular cross-section. *International Journal of Heat and Mass Transfer* 50, 2051–2059.

16. Silva, R. L. and E. C. Garcia (2008). Temperature and entropy generation behavior in rectangular ducts with 3-D heat transfer coupling (conduction and convection). *International Communications in Heat and Mass Transfer* 35, 240–250.

17. Öztop, H. F., I. Dağtekin and A. Z. Şahin (2009). Second law analysis of fully developed laminar flow for rectangular ducts with semicircular ends. *International Communications in Heat and Mass Transfer* 36, 725–730.

18. Yang, G., J. Y. Wu and L. Yan (2013). Flow reversal and entropy generation due to buoyancy assisted mixed convection in the entrance region of a three dimensional vertical rectangular duct. *International Journal of Heat and Mass Transfer* 67, 741–751.

19. Petkov, V. M., V. D. Zimparov and A. E. Bergles (2014). Performance evaluation of ducts with non-circular shapes: Laminar fully developed flow and constant wall temperature. *International Journal of Thermal Sciences* 79, 220–228.

20. Petkov, V. M., V. D. Zimparov and A. E. Bergles (2012). Performance evaluation of ducts with non-circular shapes and laminar fully developed flow. *International Review of Chemical Engineering* 4, 379–391.

21. Yang, G. and J. Y. Wu (2015). Entropy generation in a rectangular channel of buoyancy opposed mixed convection. *International Journal of Heat and Mass Transfer* 86, 809–819.

22. Ko, T. H. (2006). A numerical study on entropy generation and optimization for laminar forced convection in a rectangular curved duct with longitudinal ribs. *International Journal of Thermal Sciences* 45, 1113–1125.

23. Ko, T. H. (2006). Numerical investigation on laminar forced convection and entropy generation in a curved rectangular duct with longitudinal ribs mounted on heated wall. *International Journal of Thermal Sciences* 45, 390–404.

24. Ko, T. H. and K. Ting (2006). Entropy generation and optimal analysis for laminar forced convection in curved rectangular ducts: A numerical study. *International Journal of Thermal Sciences* 45, 138–150.

25. Ko, T. H. and C. P. Wu (2009). A numerical study on entropy generation induced by turbulent forced convection in curved rectangular ducts with various aspect ratios. *International Communications in Heat and Mass Transfer* 36, 25–31.

26. Shokouhmand, H. and M. R. Salimpour (2007). Entropy generation analysis of fully developed laminar forced convection in a helical tube with uniform wall temperature. *Heat and Mass Transfer* 44, 213–220.

27. Amani, E. and M. R. H. Nobari (2011). A numerical investigation of entropy generation in the entrance region of curved pipes at constant wall temperature. *Energy* 36, 4909–4918.

28. Ahadi, M. and A. Abbassi (2015). Entropy generation analysis of laminar forced convection through uniformly heated helical coils considering effects of high length and heat flux and temperature dependence of thermophysical properties. *Energy* 82, 322–332.

29. Abdous, M. A., H. Saffari, H. B. Avval and M. Khoshzat (2015). Investigation of entropy generation in a helically coiled tube in flow boiling condition under a constant heat flux. *International Journal of Refrigeration* 60, 217–233.

30. Farzaneh-Gord, M., H. Ameri and A. Arabkoohsar (2016). Tube-in-tube helical heat exchangers performance optimization by entropy generation minimization approach. *Applied Thermal Engineering* 108, 1279–1287.

31. Sepehr, M., S. S. Hashemi, M. Rahjoo, V. Farhangmehr and A. Alimoradi (2018). Prediction of heat transfer, pressure drop and entropy generation in shell and helically coiled finned tube heat exchangers. *Chemical Engineering Research and Design* 134, 277–291.

32. Giangaspero, G. and E. Sciubba (2013). Application of the entropy generation minimization method to a solar heat exchanger: A pseudo-optimization design process based on the analysis of the local entropy generation maps. *Energy* 58, 52–65.

33. Gheorghian, A. T., A. Dobrovicescu, L. G. Popescu, M. Cruceru and B. M. Diaconu (2015). Entropy generation assessment criterion for compact heat transfer surfaces. *Applied Thermal Engineering* 87, 137–149.

34. Amirahmadi, S., S. Rashidi and J. A. Esfahani (2016). Minimization of exergy losses in a trapezoidal duct with turbulator, roughness and beveled corners. *Applied Thermal Engineering* 107, 533–543.

35. Akbarzadeh, M., S. Rashidi and J. A. Esfahani (2017). Influences of corrugation profiles on entropy generation, heat transfer, pressure drop, and performance in a wavy channel. *Applied Thermal Engineering* 116, 278–291.

36. Rashidi, S., M. Akbarzadeh, R. Masoodi and E. M. Languri (2017). Thermal-hydraulic and entropy generation analysis for turbulent flow inside a corrugated channel. *International Journal of Heat and Mass Transfer* 109, 812–823.

37. Kiyasatfar, M. (2018). Convective heat transfer and entropy generation analysis of non-Newtonian power-law fluid flows in parallel-plate and circular microchannels under slip boundary conditions. *International Journal of Thermal Sciences* 128, 15–27.

38. Rastogi, P. and S. P. Mahulikar (2018). Optimization of micro-heat sink based on theory of entropy generation in laminar forced convection. *International Journal of Thermal Sciences* 126, 96–104.

39. Genić, S., B. Jaćimović and A. Petrovic (2018). A novel method for combined entropy generation and economic optimization of counter-current and co-current heat exchangers. *Applied Thermal Engineering* 136, 327–334.

40. Ahammed, N., L. G. Asirvatham and S. Wongwises (2016). Entropy generation analysis of graphene–alumina hybrid nanofluid in multiport minichannel heat exchanger coupled with thermoelectric cooler. *International Journal of Heat and Mass Transfer* 103, 1084–1097.

41. Hussain, S., S. E. Ahmed and T. Akbar (2017). Entropy generation analysis in MHD mixed convection of hybrid nanofluid in an open cavity with a horizontal channel containing an adiabatic obstacle. *International Journal of Heat and Mass Transfer* 114, 1054–1066.

42. Kasaeipoor, A., E. H. Malekshah and L. Kolsi (2017). Free convection heat transfer and entropy generation analysis of MWCNT-MgO (15%–85%)/water nanofluid using Lattice Boltzmann method in cavity with refrigerant solid body-Experimental thermo-physical properties. *Powder Technology* 322, 9–23.

43. Mehrali, M., E. Sadeghinezhad, A. R. Akhiani, S. T. Latibari, H. S. C. Metselaar, A. Sh. Kherbeet and M. Mehrali (2017). Heat transfer and entropy generation analysis of hybrid graphene/Fe_3O_4 ferro-nanofluid flow under the influence of a magnetic field. *Powder Technology* 308, 149–157.

44. Bahiraei, M., M. Berahmand and A. Shahsavar (2017). Irreversibility analysis for flow of a non-Newtonian hybrid nanofluid containing coated CNT/Fe_3O_4 nanoparticles in a minichannel heat exchanger. *Applied Thermal Engineering* 125, 1083–1093.

45. Shahsavar, A., Z. Rahimi and M. Bahirae (2017). Optimization of irreversibility and thermal characteristics of a mini heat exchanger operated with a new hybrid nanofluid containing carbon nanotubes decorated with magnetic nanoparticles. *Energy Conversion and Management* 150, 37–47.

46. Shahsavar, A., M. Moradi and M. Bahirae (2018). Heat transfer and entropy generation optimization for flow of a non-Newtonian hybrid nanofluid containing coated CNT/Fe_3O_4 nanoparticles in a concentric annulus. *Journal of the Taiwan Institute of Chemical Engineers* 84, 28–40.

47. Mansour, M. A., S. Siddiqa, R. S. R. Gorla and A. M. Rashad (2018). Effects of heat source and sink on entropy generation and MHD natural convection of Al_2O_3-Cu/water hybrid nanofluid filled with square porous cavity. *Thermal Science and Engineering Progress* 6, 57–71.

48. Huminic, G. and A. Huminic (2018). The heat transfer performances and entropy generation analysis of hybrid nanofluids in a flattened tube. *International Journal of Heat and Mass Transfer* 119, 813–827.

49. Huminic, G. and A. Huminic (2018). Hybrid nanofluids for heat transfer applications—A state-of-the-art review. *International Journal of Heat and Mass Transfer* 125, 82–103.

50. Bejan, A. (1977). The concept of irreversibility in heat exchanger design: Counter flow heat exchangers for gas-to-gas applications. *Journal of Heat Transfer* 99, 374–380.

51. Bejan, A. (1988). *Advanced Engineering Thermodynamics*. New York: John Wiley and Sons.

52. Şahin, A. (1998). Irreversibilities in various duct geometries with constant wall heat flux and laminar flow. *Energy* 23, 465–473.

53. Krieth, F. and M. S. Bohn. (1993). *Principles of Heat Transfer*, 5th ed. New York: West Publication.

54. Jarungthammachote, S. (2010). Entropy generation analysis for fully developed laminar convection in hexagonal duct subjected to constant heat flux. *Energy* 35, 5374–5379.

55. Bejan, A. (1980). Second law analysis in heat transfer. *Energy* 5, 721–731.

56. Zhao, N., J. Yang, H. Li, Z. Zhang and S. Li (2016). Numerical investigations of laminar heat transfer and flow performance of Al_2O_3–water nanofluids in a flat tube. *International Journal of Heat and Mass Transfer* 92, 268–282.

57. Esfe, M. H., A. A. A. Arani and M. Rezaie (2015). Experimental determination of thermal conductivity and dynamic viscosity of Ag-MgO/water hybrid nanofluids. *International Communications in Heat and Mass Transfer* 66, 189–195.

58. Esfe, M. H., S. Wongwises, A. Naderi, A. Asadi, M. R. Safaei, H. Rastamian, M. Dahari and A. Karimipour (2015). Thermal conductivity of Cu/TiO_2-water/EG hybrid nanofluid: Experimental data and modeling using artificial neural network and correlation. *International Communications in Heat and Mass Transfer* 66, 100–104.

59. Harandi, S. S., A. Karimipour, M. Afrand, M. Akbari and A. D'Orazio (2016). An experimental study on thermal conductivity of f-MWCNTs–Fe_3O_4/EG hybrid nanofluid: Effects of temperature and concentration. *International Communications in Heat and Mass Transfer* 76, 171–177.

60. Sundar, L. S., E. Venkata Ramana, M. P. F. Graça, M. K. Singh and A. C. M. Sousa (2016) Nanodiamond-Fe_3O_4 nanofluids: Preparation and measurement of viscosity, electrical and thermal conductivities. *International Communications in Heat and Mass Transfer* 73, 62–74.

61. Sundar, L. S., G. O. Irurueta, E. V. Ramana, M. K. Singh and A. C. M. Sousa (2016). Thermal conductivity and viscosity of hybrid nanofluids prepared with magnetic nanodiamond-cobalt oxide ($ND\text{-}Co_3O_4$) nanocomposite. *Case Studies in Thermal Engineering* 7, 66–77.

62. Vafaei, M., M. Afrand, N. Sina, R. Kalbasi, F. Sourani, and H Teimouri (2017). Evaluation of thermal conductivity of MgO-MWCNTs/EG hybrid nanofluids based on experimental data by selecting optimal artificial neural networks. *Physica E: Low-Dimensional Systems and Nanostructures* 85, 90–96.

63. Afrand, M. (2017). Experimental study on thermal conductivity of ethylene glycol containing hybrid nano-additives and development of a new correlation. *Applied Thermal Engineering* 110, 1111–1119.

64. Esfe, M. H., M. Rejvani, R. Karimpour and A. A. A. Arani (2017). Estimation of thermal conductivity of ethylene glycol-based nanofluid with hybrid suspensions of SWCNT–Al_2O_3 nanoparticles by correlation and ANN methods using experimental data. *Journal of Thermal Analysis Calorimetry* 128, 1359–1371.

65. Esfe, M. H., P. M. Behbahani, A. A. A. Arani and M. R. Sarlak (2017). Thermal conductivity enhancement of SiO2–MWCNT (85:15%)–EG hybrid nanofluids: ANN designing, experimental investigation, cost performance and sensitivity analysis. *Journal of Thermal Analysis Calorimetry* 128, 249–258.

66. Esfe, M. H., S. Esfandeh, S. Saedodin and H. Rostamian (2017). Experimental evaluation, sensitivity analyzation and ANN modeling of thermal conductivity of ZnO-MWCNT/EG-water hybrid nanofluid for engineering applications. *Applied Thermal Engineering* 125, 673–685.

67. Esfe, M. H., A. A. A. Arani and M. Firouzi (2017). Empirical study and model development of thermal conductivity improvement and assessment of cost and sensitivity of EG-water based SWCNT-ZnO (30%:70%) hybrid nanofluid. *Journal of Molecular Liquids* 244, 252–261.

68. Esfe, M. H., A. Alirezaie and M. Rejvani (2017). An applicable study on the thermal conductivity of SWCNT-MgO hybrid nanofluid and price-performance analysis for energy management. *Applied Thermal Engineering* 111, 1202–1210.

69. Rostamian, S. H., M. Biglari, S. Saedodin and M. H. Esfe (2017). An inspection of thermal conductivity of CuO-SWCNTs hybrid nanofluid versus temperature and concentration using experimental data, ANN modeling and new correlation. *Journal of Molecular Liquids* 231, 364–369.

70. Esfe, M. H., S. Esfandeh and M. Rejvani (2018). Modeling of thermal conductivity of MWCNT-SiO2 (30:70%)/EG hybrid nanofluid, sensitivity analyzing and cost performance for industrial applications. *Journal of Thermal Analysis Calorimetry* 131, 1437–1447.

71. Hamid, K. A., W. H. Azmi, M. F. Nabil, R. Mamat and K. V. Sharma (2018). Experimental investigation of thermal conductivity and dynamic viscosity on nanoparticle mixture ratios of TiO_2-SiO_2 nanofluids. *International Journal of Heat and Mass Transfer* 116, 1143–1152.

72. Akilu, S., A. T. Baheta, M. A. M. Said, Λ. A. Minea and K. V. Sharma (2018). Properties of glycerol and ethylene glycol mixture based SiO_2-CuO/C hybrid nanofluid for enhanced solar energy transport. *Solar Energy Materials and Solar Cells* 179, 118–128.

73. Asadi, M. and A. Asadi (2016). Dynamic viscosity of MWCNT/ZnO-engine oil hybrid nanofluid: An experimental investigation and new correlation in different temperatures and solid concentrations. *International Communications in Heat and Mass Transfer* 76, 41–45.

74. Afrand, M., K. N. Kajafabadi and M. Akbari (2016). Effects of temperature and solid volume fraction on viscosity of SiO_2-MWCNTs/SAE40 hybrid nanofluids as a coolant and lubricant in heat engines. *Applied Thermal Engineering* 102, 45–54.

75. Dardan, E., M. Afrand and A. H. Meghdadi Isfahani (2016). Effect of suspending hybrid nano-additives on rheological behavior engine oil and pumping power. *Applied Thermal Engineering* 109, 524–534.

76. Esfe, M. H, M. Afrand, W. M. Yan, H. Yarmand, D. Toghraie and M. Dahari (2016). Effects of temperature and concentration on rheological behavior of MWCNT/SiO_2 (20:80)-SAE40 hybrid nano-lubricant. *International Communications in Heat and Mass Transfer* 76, 133–138.

77. Soltani, O. and M. Akbari (2016). Effects of temperature and particles concentration on the dynamic viscosity of MgO-MWCNT/ethylene glycol hybrid nanofluid: Experimental study. *Physica E* 84, 564–570.

78. Esfe, M. H., M. Afrand, S. H. Rostamian and D Toghraie (2017). Examination of rheological behavior of MWCNTs/ZnO-SAE40 hybrid nano-lubricants under various temperatures and solid volume fractions. *Experimental Thermal and Fluid Science* 80, 384–390.

79. Asadi, A., M. Asadi, M. Rezaei, M. Siahmargoi and F. Asadi (2016). The effect of temperature and solid concentration on dynamic viscosity of MWCNT/MgO (20:80) hybrid nano-lubricant and proposing a new correlation: An experimental study. *International Communications in Heat and Mass Transfer* 78, 48–53.

80. Sundar, L. S., M. K. Singh, M. C. Ferro and A. C. M. Sousa (2017). Experimental investigation of the thermal transport properties of graphene oxide/Co_3O_4 hybrid nanofluids. *International Communications in Heat and Mass Transfer* 84, 1–10.

6

Viscosity of Nanofluid Systems: A Critical Evaluation of Modeling Approaches

Amir Varamesh and Abdolhossein Hemmati-Sarapardeh

CONTENTS

6.1 Introduction

Attempts for enhancing the energy efficiency of industrial processes have led to increasing investigations on the technologies which can improve the effectiveness of heat transfer mechanism and devices. Hence, many efforts have been made to modify and improve thermophysical properties of working fluids. Since the conventional thermal fluids have relatively poor heat transfer properties, the idea of dispersing some sort of small solid particles was emerged [1]. Solid particles provide superior higher thermal conductivity in comparison to traditional fluids. Colloidal suspension of solid particles into a single phase fluid dates back to Maxwell's [2] study in 1873, when he observed enhanced thermal conductivity by dispersion of solids into the single phase fluids. However, the addition of disperse solid particles causes several problems including erosion on the flow passage walls, increasing the pressure drop across the installations, poor suspension, poor stability, sedimentation, and consequent channel clogging [3,4]. Later on, Masuda et al. [5] dispersed micro size Al_2O_3, SiO_2, and TiO_2 particles in base fluids and observed modified thermophysical properties, but also faced same problems as mentioned before. In 1995, Choi [6] from the Argonne National Laboratory, came up with an innovating idea based on Maxwell's [2] study and proposed a new method to improve thermal performance of traditional fluids by dispersing nanometer size particles (typically lower than 100 nm). This new generation of fluids is named as nanofluids. Nanofluids exhibit superior thermal conductivity and present higher specific surface area compared to traditional colloidal suspensions. Moreover, nanofluids are easily transportable in flow paths and have better long term stability. They also can efficiently convey energy with reduced amount of pressure drop in comparison to micro/millimeter size particles. Different types of nanoparticles can be used for nanofluid preparation such as metals, metal oxides, carbon nanotubes, graphene, graphene oxide, diamond, metal carbide, metal nitride, etc. [7,8]. Metal oxides are the preferred nanomaterials for nanofluid preparation [8]. Nanoparticles are suspended in various base fluids including water, oils, glycols, glycol-water mixture, gear oil, transformer oil, etc. [8,9]. Since water is the most commonly used coolant, majority of available investigations about nanofluids are on water-based nanofluids [8].

The incorporation of nanomaterial in a base fluid changes its thermophysical properties significantly. Different nanoparticles alter the thermophysical properties to different extents. Thermophysical properties are the essence factor, which influence the flow and heat transfer behaviors and have vital roles in the study of nanofluids. Viscosity, which describes the resistance of nanofluids to flow is one of the most important thermophysical properties of nanofluids [10]. The amount of heat transfer of a fluid relies more on the viscous state of the fluid rather than on its thermal conductivity [11]. Moreover, pumping power, pressure drop, and convective heat transfer coefficient are dependent on viscosity [3,12]. The factors that affect the viscosity of nanofluids are temperature, nanomaterial concentration, viscosity of base fluid, size and shape of the nanoparticles, aggregate size, surfactant used for formulation, interaction between nanoparticle and base-fluid molecules, dispersion pH, etc. [7,8]. In the last few decades, many experimental studies were conducted to investigate the viscosity of different types of nanofluids. Mahbubul et al. [10], Sundar et al. [13], Azmi et al. [14], Raja et al. [15], Bashrinezhad et al. [16], Akilu et al. [7], Gupta et al. [3], Murshed and Estellé [17], Suganthi and Rajan [8], Yang et al. [18], and Koca et al. [19] reviewed experimental developments on the viscosity of nanofluids.

Experimental determination of viscosity of nanofluids usually is a difficult, time consuming, and expensive task. Moreover, uncertainty is always unavoidable in the measurement of nanofluids properties [7]. To overcome these limitations, many models have been proposed for estimation of viscosity of nanofluids. Generally, these models can be categorized as theoretical models, empirical correlations, and computer-aided models. The present study critically reviews the existing models for estimation of viscosity of nanofluids. In the following sections, first the theoretical models are introduced, then the available empirical correlations are reviewed, and, finally, the proposed computer-aided models are presented.

6.2 Theoretical Models

Many theoretical models have been developed for prediction of viscosity of nanofluids. The first and also the most popular model was developed by Einstein [20]. Based on the phenomenological hydrodynamic equations, Einstein [20] calculated the effective viscosity of suspended small particles into a fluid. He modeled the energy dissipated by the fluid flow around an uncharged hard sphere by presumption of linearly viscous fluid which includes infinite suspensions of spherical particles and obtained the following equation by considering the required energy for displacing the spherical particles relatively to the surrounding fluid:

$$\mu_{nf} = \mu_{bf}(1 + 2.5\varphi). \tag{6.1}$$

In this equation, μ_{nf} denotes for the viscosity of suspension (nanofluid), and μ_{bf} is the viscosity of the base fluid. This linear equation is based on the assumption that there is not any interaction between suspended particles and applicable for very low volume concentration ranges ($\varphi < 2\%$).

After Einstein's [20] work, numerous theoretical models have been proposed for prediction of viscosity of nanofluids. Most of these models were developed by extending the Einstein's [20] pioneering work. Hatschek [21] proposed a linear equation for viscosity of two phase systems. The proposed equation is valid for solid particle volume fraction up to 40%. The Hatschek [21] model is as follows:

$$\mu_{nf} = \mu_{bf}(1 + 4.5\varphi). \tag{6.2}$$

Eilers [22] presented the following non-linear model for viscosity of systems containing suspension of spherical non-deformable particles:

$$\mu_{nf} = \mu_{bf}(1 + 2.5\varphi + 4.75\varphi^2 + \ldots). \tag{6.3}$$

De Bruijn [23] derived a formula for viscosity of suspended spherical particles with application up to high volume concentrations.

$$\mu_{nf} = \mu_{bf}\left(\frac{1}{1-2.5\varphi+1.552\varphi^2}\right). \tag{6.4}$$

Vand [24] considered the mutual interaction between the particles and their collisions and proposed the following theoretical equation for viscosity of rigid spheres:

$$\mu_{nf} = \mu_{bf}(1+2.5\varphi+7.349\varphi^2+\ldots). \tag{6.5}$$

Saito [25] proposed the following model for viscosity of suspensions:

$$\mu_{nf} = \mu_{bf}(1+2.5\varphi+14.1\varphi^2). \tag{6.6}$$

Einstein [20] viscosity model for an infinitely dilute suspension of spherical particles was extended by Mooney [26] for a suspension of finite concentration as below:

$$\mu_{nf} = \mu_{bf}\exp\left(\frac{2.5\varphi}{1-v\varphi}\right), \tag{6.7}$$

where v is the crowding factor and should be determined by empirical means. For this reason, Mooney equation can be considered as a semi-empirical model. This equation accounts for suspension of rigid spheres with wide spectrum of continuous size. Mooney model is accurate in the volume concentration range of 0.52–0.74 [13].

Brinkman [27] extended the Einstein [20] model for moderate volume fraction ranges (up to 4%) and proposed the following equation:

$$\mu_{nf} = \mu_{bf}\frac{1}{(1-\varphi)^{2.5}}. \tag{6.8}$$

Roscoe [28], by extending the Einstein model, proposed an equation for viscosity of rigid solid spheres with all concentrations (low to high). According to Roscoe [28], this equation is valid for spheres of equal size and for spheres with very wide sizes, instead of this equation, Brinkman [27] model should be used.

$$\mu_{nf} = \mu_{bf}\frac{1}{(1-1.35\varphi)^{2.5}}. \tag{6.9}$$

Simha [29] developed an equation for prediction of viscosity of nanofluids including spherical particles as follows:

$$\mu_{nf} = \mu_{bf}\left(1+2.5\varphi+\left(\frac{125}{64\varphi_{max}}\right)\varphi^2+\ldots\right), \tag{6.10}$$

where φ_{max} stands for the maximum particle volume fraction.

Krieger and Dougherty [30] proposed a semi-empirical power law-based model which covers the whole spectrum of nanoparticles as follows:

$$\mu_{nf} = \mu_{bf}\left(1-\frac{\varphi}{\varphi_{max}}\right)^{[\eta]\varphi_{max}}, \tag{6.11}$$

in the above equation, φ_{max} is the maximum particle volume concentration (maximum packing fraction) at which flow can still occur. φ_{max} varies from 0.495 to 0.54 under quiescent condition and is equal to 0.605 at high shear rates. $[\eta]$ is the intrinsic viscosity and equals to 2.5 for monodispersed suspensions of hard spheres, however, at actual flow conditions, particles have polydisperse nature, therefore, the Krieger and Dougherty [30] assumption is not valid for all particle suspensions. This model can be applied for both spherical and non-spherical particles.

Frankel and Acrivos [31] used asymptotic technique to develop an equation for viscosity of solid suspensions with considering the effect of maximum volume concentration as below:

$$\mu_{nf} = \frac{9}{8}\left(\frac{(\varphi/\varphi_{max})^{1/3}}{1-(\varphi/\varphi_{max})^{1/3}}\right)\mu_{bf}. \tag{6.12}$$

Nielsen [32] presented another power law-based equation which is applicable for low volume fractions. Originally, this model was developed for relative elastic moduli of composite materials and can be used for viscosity prediction. Nielsen [32] model is as follows:

$$\mu_{nf} = \mu_{bf}\left(1+1.5\varphi\right)\exp\left(\frac{\varphi}{1-\varphi_{max}}\right), \tag{6.13}$$

where φ_{max} approximately equals to 0.64 for randomly dispersed spheres.

Lundgren [33], by taking into account the Brownian motion of isotropic suspension of rigid spherical particles and the resulting bulk stress on particles, developed the following equation in the form of Taylor series in φ.

$$\mu_{nf} = \mu_{bf}\left(1+2.5\varphi+6.25\varphi^2+f\left(\varphi^3\right)\right). \tag{6.14}$$

Brenner and Condiff [34], by considering the shape impact of rod-like particles in dilute suspension, offered a model for viscosity. This model is developed for high shear rates and can be used for volume concentration up to $1/r^2$ (r is the aspect ratio of the particles). Brenner and Condiff model [34] is as below:

$$\mu_{nf} = \mu_{bf}\left(1+\eta\varphi\right), \tag{6.15}$$

where η is defined as:

$$\eta = \frac{0.312r}{\ln 2r-1.5}+2-\frac{0.5}{\ln 2r-1.5}-\frac{1.872}{r}. \tag{6.16}$$

Jeffrey and Acrivos [35] offered another model for the viscosity of extensional flow containing non-dilute suspension of rod-like particles. This model indicates that the aspect ratio of particles (r) can have substantial effect in increasing the viscosity of fluid. The Jeffrey and Acrivos [35] model is given by:

$$\mu_{nf} = \mu_{bf}\left(3+\frac{4}{3}\left(\frac{\varphi r^2}{\ln\left(\frac{\pi}{\varphi}\right)}\right)\right). \tag{6.17}$$

Batchelor [36] considered the effect of hydrodynamic interactions or Brownian motion on viscosity of suspension of isotropic spherical and rigid particles and presented the following equation:

$$\mu_{nf} = \mu_{bf}\left(1+2.5\varphi+6.5\varphi^2\right). \tag{6.18}$$

Graham [37] proposed a generalized viscosity model by considering Brownian motion, zero inertia and uniform spheres, electro-viscous, and van der Waals forces. The model was developed for solid spheres by assuming uniform diameter for the spheres. This model approaches to Einstein [20] model as the volume fraction tends to zero and approaches to the Frankel and Acrivos [31] model as the volume fraction tends to infinity. The Graham model is in the following formula:

$$\mu_{nf} = \left(1 + 2.5\varphi + 4.5 \times \left[\frac{1}{\left(h/d_P \right)\left(2 + \left(h/d_P \right) \right)\left(1 + \left(h/d_P \right) \right)^2} \right] \right) \mu_{bf},$$ (6.19)

where h and d_P are inter particle spacing and particle diameter, respectively.

A simple model was presented by Kitano et al. [38] based on the effective volume concentration of aggregates and maximum particle volume fraction for prediction of viscosity of nanofluids:

$$\mu_{nf} = \left(1 - \frac{\varphi_a}{\varphi_{max}} \right)^{-2} \mu_{bf},$$ (6.20)

where φ_a is effective volume concentration of aggregates.

Metzner [39] represented the following exponential formula for viscosity of particle-fluid mixtures with maximum particle volume fraction less than 0.68.

$$\mu_{nf} = \left(1 - \frac{\varphi}{\varphi_{max}} \right)^{-2} \mu_{bf}.$$ (6.21)

Leighton and Acrivos [40] recommended the following equation for prediction of viscosity of nanofluids containing spherical particles with maximum volume concentration less than 0.58.

$$\mu_{nf} = \left(1 - \frac{0.5\,\mu_{in}\varphi}{1 - \varphi/\varphi_{max}} \right)^{-2} \mu_{bf},$$ (6.22)

where $\varphi_{max} = 0.58$ and $\mu_{in} = 3.0$.

Barnes et al. [41] proposed the following equation for $0.63 \le \varphi_{max} \le 0.71$ and $2.71 \le \mu_{in} \le 3.13$.

$$\mu_{nf} = \left(1 - \frac{\varphi}{\varphi_{max}} \right)^{-\mu_{in}/\varphi_{max}} \mu_{bf}.$$ (6.23)

Thomas and Muthukumar [42] developed the following simple mathematical equation for viscosity of nanofluids. This equation is valid for spherical particles with volume fraction less than 25% [7].

$$\mu_{nf} = \mu_{bf}(1 + 2.5\varphi + 4.8292\varphi^2 + 6.4028\varphi^3 + \dots).$$ (6.24)

Neto [43] developed a theoretical equation, based on the mean free path of the nanoparticles, which is as follows:

$$\mu_{nf} = \frac{a}{\lambda^n} \mu_{bf},$$ (6.25)

where λ stands for the mean free path of the nanoparticles, and a and n are constants. λ can be calculated by the Fullman [44] model of the mean free path, which is expressed as:

$$\lambda = \frac{2}{3} d_P \left(\frac{1-\varphi}{\varphi} \right).$$

(6.26)

De Noni et al. [45] presented a model by extending Neto [43] model for viscosity of suspensions as a function of volume fraction of dispersed particles and a bidimensional balance of forces acting on a particle in suspension.

$$\mu_{nf} = \mu_{bf} \left(1 + b \left(\frac{\varphi}{1-\varphi_{max}} \right)^n \right),$$

(6.27)

where b and n are constants that should be calculated with regression of experimental data. De Noni et al. model is valid for volume concentration ranges between 2% and 24% [45].

An exponential formula was proposed by Cheng and Law [46] for modeling the viscosity of particle-fluid mixtures. The model is an extension of Einstein [20] model, and the effects of dynamic particle interactions and fluid turbulence were neglected in development of the model. This model is in the following form:

$$\mu_{nf} = \mu_{bf} \left(\begin{array}{l} 1 + \frac{5}{2}\varphi + \left(\frac{35}{8} + \frac{5}{4}\beta \right)\varphi^2 + \left(\frac{105}{16} + \frac{35}{8}\beta + \frac{5}{12}\beta^2 \right)\varphi^3 + \left(\frac{1155}{128} + \frac{935}{96}\beta + \frac{235}{96}\beta^2 + \frac{5}{48}\beta^3 \right)\varphi^4 \\ + \left(\frac{3003}{256} + \frac{1125}{64}\beta + \frac{1465}{192}\beta^2 + \frac{95}{96}\beta^3 + \frac{1}{48}\beta^4 \right)\varphi^5 + \ldots \end{array} \right).$$

(6.28)

Chen et al. [47] modified Krieger and Dougherty [30] model, which was based on packing fraction of monodispersed particles without agglomeration and developed a new model as presented in Eq. (6.29). Chen et al. [47] developed their model by considering fractal index of the agglomerates and maximum packing fraction of agglomerates.

$$\mu_{nf} = \mu_{bf} \left(1 - \frac{\varphi_a}{\varphi_{max}} \right)^{[\eta]\varphi_{max}},$$

(6.29)

in the above equation, $\varphi_a = \varphi/\varphi_{ma}$, where φ_{ma} is the packing fraction of aggregates. D is the fractal index, and $\varphi_a = \varphi(a_a/a)^{3-D}$, where a_a/a is the ratio of effective radii of aggregates and primary nanoparticles.

By taking into account the effect of Brownian motion and the relative velocity between nanoparticles and base fluid, Masoumi et al. [48] presented a model for viscosity of nanofluids. The model was developed as a function of five parameters including the mean nanoparticle diameter, temperature, the nanoparticle volume fraction, the nanoparticle density, and the base-fluid physical properties. For development of the model, a correction factor is calculated to consider the simplifications that were used on the boundary condition. Masoumi et al. [48] model is as follows:

$$\mu_{nf} = \mu_{bf} + \mu_{app},$$

(6.30)

where the μ_{app} is the apparent viscosity which accounts for the effects of nanoparticles on the viscosity of nanofluids and calculated as:

$$\mu_{app} = \frac{\rho_P V_B d_P^2}{72C\delta},$$ (6.31)

where ρ_P is the nanoparticle density, V_B stands for the Brownian velocity, d_P stands for the mean nanoparticle diameter, C demonstrates the correction factor, and δ is the distance between nanoparticles and calculated by Eq. (6.32).

$$\delta = \sqrt[3]{\frac{\pi}{6\varphi}} d_P.$$ (6.32)

A model was developed by Brouwers [49] for relative viscosity of concentrated suspensions of multi-modal and monosized particles, including of equally sized particles. Brouwers [49] applied geometrical considerations to propose a model for viscosity as below:

$$\mu_{nf} = \left[\left(\frac{1-\varphi}{1-\varphi/\varphi_{max}} \right)^{\left(\frac{b\varphi_{max}}{1-\varphi_{max}} \right)} \right] \mu_{bf}.$$ (6.33)

Hosseini et al. [50], by selecting a set of appropriate dimensionless groups, introduced a new model for estimation of viscosity of nanofluids. The dimensionless groups were chosen in a way to consider the impacts of hydrodynamic volume concentration of nanoparticles, diameter of nanoparticles, base-fluid viscosity, temperature changes, and thickness of capping layer on nanoparticles. These dimensionless groups were defined as Eqs. (6.34 through 6.37), and the nanofluid viscosity model presented as Eq. (6.38).

$$\pi_1 = \frac{\mu_{nf}}{\mu_{bf}}$$ (6.34)

$$\pi_2 = \varphi_h$$ (6.35)

$$\pi_3 = \frac{d}{1+r}$$ (6.36)

$$\pi_4 = \frac{T}{T_0}$$ (6.37)

$$\mu_{nf} = \exp\left[m + \alpha\left(\frac{T}{T_0}\right) + \beta(\varphi_h) + \gamma\left(\frac{d}{1+r}\right) \right] \mu_{bf},$$ (6.38)

where, π to π_4 are the dimensionless groups, φ_h is the hydrodynamic volume fraction of solid nanoparticles, d demonstrates the nanoparticle diameter, r stands for the thickness of the capping layer, T is the measured temperature of the nanofluids, and T_0 represents the reference temperature. In Eq. (6.38) α, β, and γ are constants which can be determined from experimental data, and m is a factor which is related to the properties of nanofluid system.

Abedian et al. [51] developed a theoretical model for prediction of viscosity of nanofluids. For development of this model, interactions between particles were neglected. The model was expressed as follows:

$$\mu_{nf} = \frac{1}{1 - 2.5\varphi}. \tag{6.39}$$

Based on Masoumi et al. [48] model, another model was developed for calculation of viscosity of nanofluids by Shahriari et al. [52]. The Brownian motion was considered as the key mechanism in development of this model. The model was expressed as a function of nanoparticle volume concentration, temperature, mean nanoparticle diameter, and thermophysical features of both of nanoparticles and base fluids. Moreover, to take care of simplification assumption for a free stream boundary condition outside the boundary layer, a correction factor was introduced. Shahriari et al. [52] model is expressed in the form of Eq. (6.40). Comparison of the results obtained by the developed model and available experimental data revealed that the model outputs are in good agreement with experimental viscosity data of various nanofluids [52]. It was demonstrated that the model is valid for volume concentration in the range of 0%–11%, nanoparticle size in the range of 13 nm–95 nm and temperature range of 290 K–350 K.

$$\mu_{nf} = \mu_{bf}\left[1 + \frac{A}{B - \varphi}\left(\frac{V_B d_P^2 \rho_B}{36\delta\sqrt{T}}\right)\right], \tag{6.40}$$

in the above equation, δ is covered distance where the kinetic energy of the Brownian velocity is dissipated due to friction. δ can be calculated by Eq. (6.41). Also, A and B are calculated by Eq. (6.42) and Eq. (6.43), respectively.

$$\delta = \sqrt[3]{\frac{4}{\varphi}} d_P \tag{6.41}$$

$$A = \frac{36\delta\sqrt{T}\varphi^{1/3}}{V_B d_P^2 \rho_P}\left(9.193 \times 10^{28} d_P^4 - 2.404 \times 10^{22} d_P^3 + 2.0250\right.$$
$$\left. \times 10^{15} d_P^2 - 5.8860 \times 10^7 d_P + \frac{1}{488774}\rho_P + \frac{2507}{4446}\right) \tag{6.42}$$

$$B = -8.344 \times 10^{36} d_P^5 + 2.212 \times 10^{30} d_P^4 - 2.16 \times 10^{23} d_P^3 + 9.492 \times 10^{15} d_P^2 - 1.804 \times 10^8 d_P$$
$$- \frac{2}{327327}\rho_P + \frac{3575}{291360}. \tag{6.43}$$

Based on extended irreversible thermodynamics, Lebon and Machrafi [53] proposed a new approach to calculate the viscosity of nanofluids. The model was introduced as an extension of Einstein [20] model by considering the flowing elements including effects of a layer surrounding nanoparticles and a thermodynamic description of the role of size impacts. Lebon and Machrafi [53] presented Eq. (6.44) by coupling of the effects of liquid layering, nanoparticle's size, and volume fraction. The obtained results by the model were in good agreement with molecular dynamic simulation and experimental data.

$$\mu_{nf} = \mu_{bf}\frac{1 + 2.5\varphi\left(1 + h/r\right)^3}{1 + 4\pi^2\varphi^2\left(1 + h/r\right)^2\left(l^2/r^2\right)}. \tag{6.44}$$

Available theoretical models are summarized in Table 6.1. Although many theoretical models were proposed over the past years, there is no widely accepted theoretical model yet. Each model has its application limitations and cannot be applied for a wide range of operating condition.

TABLE 6.1

Available Theoretical Models for Prediction of Viscosity of Nanofluids

Researchers	Year	Theoretical Model
Einstein [20]	1906	$\mu_{nf} = \mu_{bf}\left(1 + 2.5\varphi\right)$
Hatschek [21]	1913	$\mu_{nf} = \mu_{bf}\left(1 + 4.5\varphi\right)$
Eilers [22]	1941	$\mu_{nf} = \mu_{bf}\left(1 + 2.5\varphi + 4.75\varphi^2 + ...\right)$
de Brujin [23]	1942	$\mu_{nf} = \mu_{bf}\left(\dfrac{1}{1 - 2.5\varphi + 1.552\varphi^2}\right)$
Vand [24]	1948	$\mu_{nf} = \mu_{bf}\left(1 + 2.5\varphi + 7.349\varphi^2 + ...\right)$
Saito [25]	1950	$\mu_{nf} = \mu_{bf}\left(1 + 2.5\varphi + 14.1\varphi^2\right)$
Mooney [26]	1951	$\mu_{nf} = \mu_{bf}\exp\left(\dfrac{2.5\varphi}{1 - v\varphi}\right)$
Brinkman [27]	1952	$\mu_{nf} = \mu_{bf}\dfrac{1}{\left(1 - \varphi\right)^{2.5}}$
Roscoe [28]	1952	$\mu_{nf} = \mu_{bf}\dfrac{1}{\left(1 - 1.35\varphi\right)^{2.5}}$
Simha [29]	1952	$\mu_{nf} = \mu_{bf}\left(1 + 2.5\varphi + \left(\dfrac{125}{64\varphi_{max}}\right)\varphi^2 + ...\right)$
Krieger and Dougherty [30]	1959	$\mu_{nf} = \mu_{bf}\left(1 - \dfrac{\varphi}{\varphi_{max}}\right)^{\left[\eta\right]\varphi_{max}}$
Frankel and Acrivos [31]	1967	$\mu_{nf} = \dfrac{9}{8}\left(\dfrac{\left(\varphi/\varphi_{max}\right)^{1/3}}{1 - \left(\varphi/\varphi_{max}\right)^{1/3}}\right)\mu_{bf}$
Nielsen [32]	1970	$\mu_{nf} = \mu_{bf}\left(1 + 1.5\varphi\right)\exp\left(\dfrac{\varphi}{1 - \varphi_{max}}\right)$
Lundgren [33]	1972	$\mu_{nf} = \mu_{bf}\left(1 + 2.5\varphi + 6.25\varphi^2 + f\left(\varphi^3\right)\right)$
Brenner and Condiff [34]	1974	$\mu_{nf} = \mu_{bf}\left(1 + \eta\varphi\right)$ $\eta = \dfrac{0.312r}{\ln 2r - 1.5} + 2 - \dfrac{0.5}{\ln 2r - 1.5} - \dfrac{1.872}{r}$
Jeffrey and Acrivos [35]	1976	$\mu_{nf} = \mu_{bf}\left(3 + \dfrac{4}{3}\left(\dfrac{\varphi r^2}{\ln\left(\dfrac{\pi}{\varphi}\right)}\right)\right)$
Batchelor [36]	1977	$\mu_{nf} = \mu_{bf}\left(1 + 2.5\varphi + 6.5\varphi^2\right)$
Graham [37]	1981	$\mu_{nf} = \left(1 + 2.5\varphi + 4.5 \times \left[\dfrac{1}{\left(h/d_P\right)\left(2 + \left(h/d_P\right)\right)\left(1 + \left(h/d_P\right)\right)^2}\right]\right)\mu_{bf}$
Kitano et al. [38]	1981	$\mu_{nf} = \left(1 - \dfrac{\varphi_a}{\varphi_{max}}\right)^{-2}\mu_{bf}$

(Continued)

TABLE 6.1 (*Continued*)

Available Theoretical Models for Prediction of Viscosity of Nanofluids

Researchers	Year	Theoretical Model
Metzner [39]	1985	$\mu_{nf} = \left(1 - \dfrac{\varphi}{\varphi_{max}}\right)^{-2} \mu_{bf}$
Leighton and Acrivos [40]	1987	$\mu_{nf} = \left(1 - \dfrac{0.5\mu_{in}\varphi}{1 - \varphi/\varphi_{max}}\right)^{-2} \mu_{bf}$
Barnes et al. [41]	1989	$\mu_{nf} = \left(1 - \dfrac{\varphi}{\varphi_{max}}\right)^{-\mu_{in}/\varphi_{max}} \mu_{bf}$
Thomas and Muthukumar [42]	1991	$\mu_{nf} = \mu_{bf}\left(1 + 2.5\varphi + 4.8292\varphi^2 + 6.4028\varphi^3 + \ldots\right)$
Neto [43]	1999	$\mu_{nf} = \dfrac{a}{\lambda^n}\mu_{bf}$ $\lambda = \dfrac{2}{3}d_P\left(\dfrac{1-\varphi}{\varphi}\right)$
De Noni et al. [45]	2002	$\mu_{nf} = \mu_{bf}\left(1 + b\left(\dfrac{\varphi}{1-\varphi_{max}}\right)^n\right)$
by Cheng and Law [46]	2003	$\mu_{nf} = \mu_{bf}\left(\begin{array}{l} 1 + \dfrac{5}{2}\varphi + \left(\dfrac{35}{8} + \dfrac{5}{4}\beta\right)\varphi^2 + \left(\dfrac{105}{16} + \dfrac{35}{8}\beta + \dfrac{5}{12}\beta^2\right)\varphi^3 \\[2mm] + \left(\dfrac{1155}{128} + \dfrac{935}{96}\beta + \dfrac{235}{96}\beta^2 + \dfrac{5}{48}\beta^3\right)\varphi^4 \\[2mm] + \left(\dfrac{3003}{256} + \dfrac{1125}{64}\beta + \dfrac{1465}{192}\beta^2 + \dfrac{95}{96}\beta^3 + \dfrac{1}{48}\beta^4\right)\varphi^5 + .. \end{array} \right)$
Chen et al. [47]	2007	$\mu_{nf} = \mu_{bf}\left(1 - \dfrac{\varphi_a}{\varphi_{max}}\right)^{[\eta]\varphi_{max}}$
Masoumi et al. [48]	2009	$\mu_{nf} = \mu_{bf} + \mu_{app}$ $\mu_{app} = \dfrac{\rho_P V_B d_P^2}{72C\delta}$ $\delta = \sqrt[3]{\dfrac{\pi}{6\varphi}}d_P$
Brouwers [49]	2010	$\mu_{nf} = \left[\left(\dfrac{1-\varphi}{1 - \varphi/\varphi_{max}}\right)^{\left(\frac{b\varphi_{max}}{1-\varphi_{max}}\right)}\right]\mu_{bf}$
Hosseini et al. [50]	2010	$\mu_{nf} = \exp\left[m + \alpha\left(\dfrac{T}{T_0}\right) + \beta(\varphi_h) + \gamma\left(\dfrac{d}{1+r}\right)\right]\mu_{bf}$ $\pi_1 = \dfrac{\mu_{nf}}{\mu_{bf}} \quad \pi_2 = \varphi_h \quad \pi_3 = \dfrac{d}{1+r} \quad \pi_4 = \dfrac{T}{T_0}$
Abedian et al. [51]	2010	$\mu_{nf} = \dfrac{1}{1 - 2.5\varphi}$

(Continued)

TABLE 6.1 (Continued)

Available Theoretical Models for Prediction of Viscosity of Nanofluids

Researchers	Year	Theoretical Model
Shahriari et al. [52]	2017	$\mu_{nf} = \mu_{bf}\left[1 + \dfrac{A}{B-\varphi}\left(\dfrac{V_B d_P^2 \rho_B}{36\delta\sqrt{T}}\right)\right]$
		$\delta = \sqrt[3]{\dfrac{4}{\varphi}} d_P s$
		$A = \dfrac{36\delta\sqrt{T}\,\varphi^{1/3}}{V_B d_P^2 \rho_P}\left(9.193\times10^{28} d_P^4 - 2.404\times10^{22} d_P^3\right.$
		$\left. + 2.0250\times10^{15} d_P^2 - 5.8860\times10^7 d_P\right.$
		$\left. + \dfrac{1}{488774}\rho_P + \dfrac{2507}{4446}\right)$
		$B = -8.344\times10^{36} d_P^5 + 2.212\times10^{30}$
		$d_P^4 - 2.16\times10^{23} d_P^3 + 9.492\times10^{15}$
		$d_P^2 - 1.804\times10^8 d_P - \dfrac{2}{327327}\rho_P + \dfrac{3575}{291360}$
Lebon and Machrafi [53]	2018	$\mu_{nf} = \mu_{bf}\dfrac{1 + 2.5\varphi\left(1 + h/r\right)^3}{1 + 4\pi^2\varphi^2\left(1 + h/r\right)^2\left(l^2/r^2\right)}$

6.3 Empirical Correlations

Due to disability of the theoretical model in prediction of viscosity of nanofluids, researchers have developed many empirical correlations based on their own experimental data. In this section, some selected empirical correlations will be reviewed. All of these models were developed by fitting experimental viscosity data of particular type(s) of nanofluids.

Tseng and Lin [54] experimentally measured the viscosity of water-based TiO_2 nanoparticles with average nanoparticle size of 7–20 nm at constant temperature (25°C) over particle volume fraction range of 5%–12%. Based on the obtained experimental data, Tseng and Lin [54] proposed the following exponential formula:

$$\mu_{nf} = \left(13.47\exp(35.98\varphi)\right)\mu_{bf}. \tag{6.45}$$

By performing a least-square curve fitting of available experimental data, Maïga et al. [55] proposed two correlations for estimation of viscosity of Al_2O_3-water and Al_2O_3-ethylene glycol nanofluids:

$$\mu_{nf} = \left(1 + 7.1\varphi + 123\varphi^2\right)\mu_{bf} \quad \text{for } Al_2O_3\text{-water} \tag{6.46}$$

$$\mu_{nf} = \left(1 - 0.19\varphi + 306\varphi^2\right)\mu_{bf} \quad \text{for } Al_2O_3\text{-ethylene glycol} \tag{6.47}$$

Buongiorno [56], based on the obtained experimental viscosity data for Al_2O_3-water and TiO_2-water by Pak and Cho [57], proposed two different correlations as follows:

$$\mu_{nf} = \left(1 + 39.11\varphi + 533.9\varphi^2\right)\mu_{bf} \quad \text{for } Al_2O_3\text{-water} \tag{6.48}$$

$$\mu_{nf} = \left(1 + 5.45\varphi + 108.2\varphi^2\right)\mu_{bf} \quad \text{for } TiO_2\text{-water} \tag{6.49}$$

Kulkarni et al. [58] developed a generalized correlation for estimation of viscosity of nanofluids as a function of temperate and particle volume fraction. They [58] proposed their model based on experimental data of CuO nanoparticles with average size of 29 nm suspended in deionized water with volume concentration in the range of 5%–15%. These data were measured at temperature ranging from 278 to 323 K. The developed model is as follows:

$$\ln \mu_{nf} = A\left(\frac{1}{T}\right) - B, \tag{6.50}$$

where A and B, respectively, are given by:

$$A = 20587\varphi^2 + 15857\varphi + 1078.3 \tag{6.51}$$

$$B = -107.12\varphi^2 + 53.548\varphi + 2.8715. \tag{6.52}$$

Chen et al. [47] measured viscosity of TiO_2-ethylene glycol nanofluid over the temperature range of 293–333 K with volume concentration up to 8 wt% and proposed the following correlation based on the obtained data. It was expressed that the normalized shear viscosity with respect to the base-fluid viscosity is not dependent to temperature and can be correlated only as a function of volume concentration as follows:

$$\mu_{nf} = \left(1 + 10.6\varphi + \left(10.6\varphi\right)^2\right)\mu_{bf}. \tag{6.53}$$

Namburu et al. [59] developed an exponential viscosity correlation by suspending CuO nanoparticles in 60:40 (wt%) ethylene glycol and water mixture with volume fraction in the range of 0%–6.12% and in the temperature range of −35°C to 50°C, which is given as:

$$\mu_{nf} = A\exp\left(-BT\right), \tag{6.54}$$

where A and B are a function of particle volume fraction as:

$$A = 1.8375\varphi^2 - 29.643\varphi + 165.56 \tag{6.55}$$

$$B = 4 \times 10^{-6}\varphi^2 - 0.001\varphi + 0.0186. \tag{6.56}$$

Nguyen et al. [60] experimentally investigated the dynamic viscosity of Al_2O_3 nanofluids including two different particle sizes (36 and 47 nm). The experiments were conducted for volume concentration up to 13% and temperature up to 75°C. Based on the obtained experimental viscosity data, they proposed the following equations for 36 and 47 nm nanoparticles, respectively, as a function of volume fraction:

$$\mu_{nf} = \left(1 + 0.025\varphi + 0.015\varphi^2\right)\mu_{bf} \tag{6.57}$$

$$\mu_{nf} = \left(0.904\exp\left(0.1483\varphi\right)\right)\mu_{bf}. \tag{6.58}$$

Nguyen et al. [60] also proposed the following equations as a function of temperature for prediction of viscosity of Al_2O_3 nanofluids with 1% and 4% volume fractions, respectively:

$$\mu_{nf} = \left(1.125 - 0.0007T\right)\mu_{bf} \tag{6.59}$$

$$\mu_{nf} = \left(2.1275 - 0.0215T + 0.0002T^2\right)\mu_{bf} \tag{6.60}$$

By considering experimental data of Al_2O_3-water, TiO_2-water, Al_2O_3-propylene glycol, TiO_2-ethylene glycol, SiO_2-ethanol, and Cu-ethylene glycol nanofluids, Corcione [61] proposed an empirical correlation for viscosity of nanofluids as represented in Eq. (6.61). The model was developed based on large amounts of experimental data with volume fraction ranging from 0.01% to 7.1%, nanoparticle size in the range of 20–250 nm, and temperature in the range of 293–333 K.

$$\mu_{nf} = \left(\frac{1}{1 - 34.87\left(d_p / d_f\right)^{-0.3} \varphi^{1.03}}\right)\mu_{bf}, \tag{6.61}$$

in which d_f stands for the equivalent diameter of the base-fluid molecule which can be calculated by:

$$d_f = 0.1\left(\frac{6M}{N\pi\rho_{f0}}\right), \tag{6.62}$$

where M demonstrates the molecular weight of the base fluid, N is the Avogadro's number, and ρ_{f0} is the mass density of the base fluid at temperature of 293 K.

Based on experimental viscosity data of spherical CuO nanoparticles of 40 nm size dispersed in gear oil at temperature range of 10°C–80°C, Kole and Dey [62] proposed an empirical correlation. The correlation is a modified version of Krieger and Dougherty [30] model and was developed by considering the aggregation mechanism of nanoparticles. Kole and Dey [62] model is as follows:

$$\mu_{nf} = \left(1 - \frac{\varphi}{0.5}\left(\frac{a_a}{a}\right)^{1.3}\right)^{-1.25} \mu_{bf}, \tag{6.63}$$

where a_a and a are effective radii of aggregates and radius of nanoparticles, respectively.

Khanafer and Vafai [63] offered an equation for viscosity of Al_2O_3 water-based nanofluids. It was shown that the model is valid for $1\% \leq \varphi \leq 9\%$, $20°C \leq T \leq 70°C$ and $13\,nm \leq d_p \leq 131\,nm$. Khanafer and Vafai [63] model predicts viscosity changes with variation of temperature for some of the available experimental data with good precision. The model is as follows:

$$\mu_{nf} = -0.4491 + \frac{28.837}{T} + 0.574\varphi - 0.1634\varphi^2 + 25.053\frac{\varphi^2}{T^2} + 0.0132\varphi^3$$
$$- 2534.735\frac{\varphi}{T^3} + 23.498\frac{\varphi^2}{d_P^2} - 3.0185\frac{\varphi^3}{d_P^2}. \tag{6.64}$$

Based on 29 experimental data of ZnO-water nanofluid with volume fraction and temperature ranges of 0%–1.5% and 10°C–35°C, respectively, Suganthi and Rajan [64] developed the following equation:

$$\mu_{nf} = (1+11.97\varphi)\mu_{bf}. \tag{6.65}$$

According to Suganthi and Rajan [64], by replacing the volume fraction with agglomerate volume fraction (φ_a), Eq. (6.65) can be written in terms of Einstein [20] equation as follows:

$$\mu_{nf} = (1+2.5\varphi_a)\mu_{bf}, \tag{6.66}$$

where $\varphi_a = 4.788\varphi$.

Azmi et al. [65] proposed a viscosity correlation based on 233 experimental data points from 10 different sources in the literature. The data include water-based Al_2O_3, TiO_2, CuO, and SiC nanoparticles with particle size in the range of 20–170 nm and volume fraction less than 4%. By regression of the gathered data, Azmi et al. [65] proposed the following correlation which is a function of particle volume fraction, temperature, and particle size:

$$\mu_{nf} = \mu_{bf}\left(1+\frac{\varphi}{100}\right)^{11.3}\left(1+\frac{T}{70}\right)^{-0.038}\left(1+\frac{d_P}{170}\right)^{-0.061}. \tag{6.67}$$

Azmi et al. [65] conducted new experiments on viscosity of Al_2O_3 (50 nm), ZnO (100 nm), and TiO_2 (150 nm) water-based nanofluids at 21.5°C. They showed that the obtained model is valid for all of the obtained data, and the model can accurately describe the effect of different materials, particle sizes, and volume fractions on viscosity of nanofluids.

Vakili-Nezhad and Dorany [66] developed two empirical correlations based on experimental data of single wall carbon nanotube with about 2 nm diameter and length in the range of 10–15 μm dispersed in lube oil. They proposed Eq. (6.68) for prediction of the viscosity of these nanofluids as a function of volume fraction at 25°C and developed Eq. (6.69) for viscosity estimation as a function of temperate at volume fraction of 0.1 wt%.

$$\mu_{nf} = (1+1.59\varphi-16.36\varphi^2+50.4\varphi^3)\mu_{bf} \tag{6.68}$$

$$\mu_{nf} = (1048-30.3T+0.2T^2)\mu_{bf}. \tag{6.69}$$

Heyhat et al. [67] developed a correlation for estimation of viscosity of nanofluids based on experimental data of Al_2O_3 nanoparticles with 40 nm size dispersed in distilled water. The model is valid for volume concentration ranging from 0.1% to 2.0% and temperature ranging from 20°C to 60°C. The developed model has exponential form, given by:

$$\mu_{nf} = \exp\left(\frac{5.989\varphi}{0.278-\varphi}\right)\mu_{bf}. \tag{6.70}$$

Rashin and Hemalatha [68] developed two empirical correlations for prediction of viscosity of CuO-coconut oil nanofluids. One of the correlations is based on volume fraction (in weight percent), and the other one is based on temperature. The models were developed based on experimental data of 20 nm CuO-coconut oil with mass concentration range of 0–2.5 wt% and temperature range of 308–328 K. The models are as follows:

$$\mu_{nf} = (1+a\phi-b\phi^2)\mu_{bf} \tag{6.71}$$

$$\mu_{nf} = \left(c \exp(-0.03T) \right) \mu_{bf},$$ (6.72)

where ϕ is the mass fraction of nanoparticles, and a, b, and, c are fitting parameters and can be calculated by regression of experimental data.

Yiamsawas et al. [69] experimentally measured viscosity of Al_2O_3 (120 nm) and TiO_2 (21 nm) nanoparticles dispersed in water. The experiments were conducted over volume fraction range of 1%–8% and temperature range of 15°C–60°C. Based on the obtained data, Yiamsawas et al. [69] proposed an equation as a function of particle volume fraction, temperature, and base-fluid viscosity as follows:

$$\mu_{nf} = A\varphi^B T^C \mu_{bf}^D,$$ (6.73)

where A, B, C, and D are coefficients that can be calculated by regression analysis and are given by the author for Al_2O_3 and TiO_2 nanoparticles as shows in Table 6.2.

Based on experimental data of Fe_3O_4-water nanofluids with volume concentration ranging from 0% to 2% and at temperature range of 30°C–60°C, Sundar et al. [70] proposed the following correlation:

$$\mu_{nf} = \mu_{bf} \left(1 + \frac{\varphi}{2.5} \right)^{6.356}.$$ (6.74)

By regression analysis for the obtained experimental data, Hemmat Esfe and Saedodin [71] proposed a viscosity correlation as a function of temperature, volume fraction and base fluid viscosity [Eq. (6.75)]. The model was developed by considering experimental data of ZnO-ethylene glycol nanofluids including nanoparticles with average size of 18 nm with volume fraction ranging from 0.25% to 5%. The used data were in the temperature ranging from room temperature up to 50°C.

$$\mu_{nf} = \left[0.9118 \exp\left(5.49\varphi - 0.0001359T^2 \right) + 0.0303 \ln T \, \mu_{bf} \right].$$ (6.75)

Hemmat Esfe et al. [72], based on obtained experimental data by dispersing 18 nm sized ZnO nanoparticles in ethylene glycol, presented an equation as a function of temperature and volume fraction for viscosity of nanofluids [Eq. (6.76)]. The experimental data were obtained at the temperatures from room temperature up to 50°C and volume fraction ranging from 0.25% up to 5%.

$$\mu_{nf} = \left[0.9118 \exp\left(5.49\varphi - 0.00001359T^2 \right) + 0.0303 \ln T \right] \mu_{bf}.$$ (6.76)

An exponential equation, with particle volume fraction as input parameter, was developed by Sundar et al. [73] for estimation of viscosity of nanofluids [Eq. (6.77)]. This model was proposed based on 135 experimental data of Al_2O_3 (36 nm) nanoparticles dispersed in mixture of ethylene glycol and water. This model is valid for $20°C \leq T \leq 60°C$ and $0.3\% \leq \varphi \leq 1.5\%$.

$$\mu_{nf} = \left(A \exp(B\varphi) \right) \mu_{bf},$$ (6.77)

TABLE 6.2

Coefficients of the Yiamsaws [70] Model

Nanoparticle	Coefficient			
	A	B	C	D
Al_2O_3	0.891842	0739192	0.099205	0.9844
TiO_2	0.837931	0.188264	0.089069	1.100945

$A = 0.9396$ and $B = 24.16$ for EG-W nanofluid 20:80 vol%

$A = 0.9299$ and $B = 67.43$ for EG-W nanofluid 40:60 vol%

$A = 1.1216$ and $B = 77.56$ for EG-W nanofluid 60:40 vol%.

Sharifpur et al. [74] developed an empirical correlation for viscosity of nanofluids using dimensionless analysis. In development of their model, they used 132 obtained experimental data points of Al_2O_3-glycerol nanofluid including 19, 139, and 160 nm particles size with volume fraction ranging from 0% to 5% in the temperature range of 20°C–70°C. By regressing of the experimental data, they developed a model as a function of temperature, volume fraction, nanoparticle size, and capping layer thickness as introduced in Eq. (6.78).

$$\mu_{nf} = \left[1 + \Re[\eta] \left(\left(\frac{T}{T_0} \right)^{\alpha} \varphi^{\beta} \left(\frac{d}{h} \right)^{\gamma} \right) \right] \mu_{bf},$$
(6.78)

where $\Re$ stands for the system parameter, $[\eta]$ is the intrinsic viscosity, d is the nanoparticle diameter, h stands for capping layer thickness, α, β, and γ are correlation coefficients. For the used experimental data set, Sharifpur et al. [74] reported that the values of $\Re$, α, β, and γ are equal to 240.19, 0.807, 2.480, and −0.522, respectively.

A viscosity correlation for BioGlycol/water-based TiO_2 nanofluids was proposed by Abdolbaqi et al. [75]. They used 60 obtained data points with particle size of 50 nm, volume fraction between 0.5% and 2%, and temperature in the range of 30°C–80°C. Abdolbaqi et al. [75] stated that the viscosity of nanofluids increases exponentially with increasing volume fraction and decreases exponentially with increasing temperature and proposed the following equation:

$$\mu_{nf} = \left[0.918 \exp\left(14.696\varphi + 0.161\frac{T}{80} \right) \right] \mu_{bf}$$
(6.79)

They showed that for the used experimental data, the maximum deviation of the model is about 5%, average error is about 1.2%, and standard deviation is about 1.11%.

Abdolbaqi et al. [76] also developed another correlation for viscosity of BioGlycol/water-based SiO_2 (22 nm) nanofluids, based on 60 obtained experimental data points. The model is similar to Eq. (6.79) and given by:

$$\mu_{nf} = \left[0.906 \exp\left(10.975\varphi + 0.169\frac{T}{80} \right) \right] \mu_{bf}$$
(6.80)

Meybodi et al. [77] developed a correlation as a function of temperature, nanoparticle volumetric concentration, nanoparticle size, and water viscosity for prediction of viscosity of water-based nanofluids. They used 701 data points of Al_2O_3, TiO_2, CuO, and SiO_2 water-based nanofluids for development of their model (Eq. 6.81). It was demonstrated the average absolute relative error of the developed model is equal to 8.26%.

$$\frac{\mu_{nf}}{\mu_{bf}} =$$

$$\frac{\left(133.54064976 - 343.82413843 \times \exp\left(\varphi / S \right) + 290.11804759 \times \left(\exp\left(\varphi / S \right) \right)^2 - 78.993120761 \times \left(\exp\left(\varphi / S \right) \right)^3 \right)}{\left(0.91161630781 + 32.33014233\frac{Ln(S)}{T} - 11.732514460\frac{(Ln(S))^2}{T} \right)}.$$
(6.81)

Adio et al. [78], based on dimensionless analysis, developed a correlation for viscosity of nanofluids by considering temperature, volume fraction, nanoparticle size, and the density of nanofluid, the density of base fluid, viscosity of the base fluid, and capping layer thickness as inputs of the model. They used experimental data of MgO-ethylene glycol nanofluids including nanoparticles with 21, 105, and 125 nm sizes. The model is valid for temperature range of 20°C–70°C and volume concentration range of 0%–5%. Adio et al. [78] model is:

$$\mu_{nf} = \left[\begin{array}{c} 1 + 7.0764\varphi - 0.1246\left(\dfrac{T}{T_0}\right)\varphi - 0.0346\left(\dfrac{d_p}{h}\right)\varphi - 0.0024\left(\left(\dfrac{d_p}{h}\right)\varphi\right)^2 \\[4mm] -1.2357\left(\left(\dfrac{T}{T_0}\right)\varphi\right)^2 + 53.6946\varphi^2 + 0.0436\left(\dfrac{T}{T_0}\right)^2 \varphi^{1/3} \end{array} \right] \mu_{bf}. \qquad (6.82)$$

Based on experimental data of Al_2O_3 (13 nm) dispersed in ethylene glycol and water mixtures, a correlation was proposed by Chiam et al. [79]. The model was presented by taking into account the base ratio as a novel input parameter alongside volume fraction and temperature as the other inputs, which is given by:

$$\mu_{nf} = \left[\left(1 + \frac{\varphi}{100}\right)^{32} \left(\frac{T}{70}\right)^{-0.001} (1 + BR)^{0.08} \right] \mu_{bf}, \qquad (6.83)$$

where BR is the base ratio, shows the ratio of volume percentage of the ethylene glycol in the base-fluid mixture. Comparison by the experimental data revealed that the model is able to predict viscosity of nanofluids with average deviation of about 4.7%.

Based on experimental data of Fe_3O_4 magnetic nanoparticles dispersed in water at temperature range of 25°C–60°C and volume concentration range of 0%–4%, Bahiraei and Hangi [80] proposed an equation for estimation of viscosity of nanofluids as follows:

$$\mu_{nf} = 1.5474 \times 10^{10} (\varphi + 1)^{0.358} T^{-5.3892}. \qquad (6.84)$$

Aberoumand and Jafarimoghaddam [81] proposed a polynomial for estimation of viscosity of oil-based nanofluids. They used experimental data of Cu-engine oil nanofluids with average diameter of 50 nm and with weight concentration in the range of 0.2%–1.0% and temperature ranging from 40°C to 100°C.

$$\mu_{nf} = \left[1.15 + 1.061\varphi - 0.5442\varphi^2 + 0.1181\varphi^3 \right] \mu_{bf}. \qquad (6.85)$$

Elcioglu et al. [82] proposed two equations for estimation of the viscosity of nanofluids and the relative viscosity of nanofluids (the viscosity of nanofluids over the viscosity of base fluid) as functions of volume fraction, temperature, and nanoparticle diameter. The models were developed based on experimental data of Al_2O_3-water nanofluids and valid for particle volume fraction in the range of 1%–3%, temperature in the range of 20°C–50°C, and nanoparticle diameter in the interval of 10–30 nm. The proposed models are as follows:

$$\mu_{nf} = 1.0973 - 0.0148T + 34.2292\varphi + 0.0041d_P - 0.3458T\varphi \qquad (6.86)$$

$$\mu_r = 0.739716 - 0.000563T + 29.054\varphi + 0.006d_P \qquad (6.87)$$

In addition to the above mentioned models, many other empirical correlations have been proposed over the past years. These empirical correlations are summarized in Table 6.3. Most of these models are only valid for a particular type of nanofluid with limited range of temperature and particle volume fraction and size.

TABLE 6.3

Developed Empirical Correlations for Prediction of Viscosity of Nanofluids

Researchers	Year	Empirical Correlation	Nanofluid
Tseng and Lin [54]	2003	$\mu_{nf} = \left(13.47\exp\left(35.98\varphi\right)\right)\mu_{bf}$	$TiO_2 - W$
Tseng and Chen [100]	2003	$\mu_{nf} = \left(0.4513\exp\left(0.6965\varphi\right)\right)\mu_{bf}$	$Ni - terpineol$
Maiga et al. [55]	2004	$\mu_{nf} = \left(1 + 7.3\varphi + 123\varphi^2\right)\mu_{bf}$	$Al_2O_3 - W$
		$\mu_{nf} = \left(1 - 0.19\varphi + 306\varphi^2\right)\mu_{bf}$	$Al_2O_3 - EG$
Song et al. [101]	2005	$\mu_{nf} = \left(1 + 56.5\varphi\right)\mu_{bf}$	$SiO_2 - W$
Buongiorno [56]	2006	$\mu_{nf} = \left(1 + 39.11\varphi + 533.9\varphi^2\right)\mu_{bf}$	$Al_2O_3 - W$
		$\mu_{nf} = \left(1 + 5.45\varphi + 108.2\varphi^2\right)\mu_{bf}$	$TiO_2 - W$
Kulkarni et al. [58]	2006	$\ln\mu_{nf} = A\left(\dfrac{1}{T}\right) - B$ $A = 20587\varphi^2 + 15857\varphi + 1078.3$ $B = -107.12\varphi^2 + 53.548\varphi + 2.8715$	$CuO - W$
Prasher et al. [102]	2006	$\mu_{nf} = \left(1 + 10\varphi\right)\mu_{bf}$	$Al_2O_3 - PPG$
Chen et al. [59]	2007	$\mu_{nf} = \left(1 + 10.6\varphi + \left(10.6\varphi\right)^2\right)\mu_{bf}$	$TiO_2 - EG$
Namburu et al. [60]	2007	$\mu_{nf} = A\exp\left(-BT\right)$ $A = 1.8375\varphi^2 - 29.643\varphi + 165.56$ $B = 4\times10^{-6}\varphi^2 - 0.001\varphi + 0.0186$	$CuO - EG{:}W$ (60:40 wt%)
Kulkarani et al.			
Chevalier et al. [103]	2007	$\mu_{nf} = \left(1 + 8.3\varphi\right)\mu_{bf}$	$SiO_2 - ethanol$
Garg et al. [104]	2008	$\mu_{nf} = \left(1 + 11\varphi\right)\mu_{bf}$	$Cu - EG$
Williams et al. [105]	2008	$\mu_{nf} = \exp\left(\dfrac{4.91\varphi}{0.2092 - \varphi}\right)\mu_{bf}$	$Al_2O_3 - W$
		$\mu_{nf} = \exp\left(\dfrac{11.19\varphi}{0.1960 - \varphi}\right)\mu_{bf}$	$ZrO_2 - W$
Nguyen et al. [61]	2008	$\mu_{nf} = \left(1 + 0.025\varphi + 0.015\varphi^2\right)\mu_{bf}$	Al_2O_3 (36 nm) $- W$
		$\mu_{nf} = \left(0.904\exp\left(0.1483\varphi\right)\right)\mu_{bf}$	Al_2O_3 (47 nm) $- W$
		$\mu_{nf} = \left(1.125 - 0.0007T\right)\mu_{bf}$	Al_2O_3 (1 vol%) $- W$
		$\mu_{nf} = \left(2.1275 - 0.0215T + 0.0002T^2\right)\mu_{bf}$	Al_2O_3 (4 vol%) $- W$
Rea et al. [106]	2009	$\mu_{nf} = \left(1 + 46.80\varphi + 550.82\varphi^2\right)\mu_{bf}$	$Zirconia - W$
Duangthongsuand and Wongwises [107]	2009	$\mu_{nf} = \left(1.0226 + 0.0447\varphi - 0.0112\varphi^2\right)\mu_{bf}$	$TiO_2 - W$
Sahoo et al. [108]	2009	$\mu_{nf} = \left(A\exp\left(\dfrac{B}{T} + C\varphi\right)\right)\mu_{bf}$	$Al_2O_3 - EG{:}W$ (60:40 vol%)

(Continued)

TABLE 6.3 (*Continued*)

Developed Empirical Correlations for Prediction of Viscosity of Nanofluids

Researchers	Year	Empirical Correlation	Nanofluid
Chandrasekar et al. [109]	2010	$\mu_{nf} = \left[1 + A\left(\dfrac{\varphi}{1-\varphi}\right)^n\right]\mu_{bf}$	$Al_2O_3 - W$
Godson et al. [110]	2010	$\mu_{nf} = \left(1.005 + 0.497\varphi - 0.1149\varphi^2\right)\mu_{bf}$	$Ag - W$
Ho et al. [111]	2010	$\mu_{nf} = \left(14.93\varphi + 222.4\varphi^2\right)\mu_{bf}$	$Al_2O_3 - W$
Corcione [62]	2011	$\mu_{nf} = \left(\dfrac{1}{1 - 34.87\left(d_p/d_f\right)^{-0.3}\varphi^{1.03}}\right)\mu_{bf}$ $d_f = 0.1\left(\dfrac{6M}{N\pi\rho_{f0}}\right)$	$Al_2O_3 - W$, $TiO_2 - W$, $Al_2O_3 - PG$, $TiO_2 - EG$, $SiO_2 - ethanol$, $Cu - EG$
Kole and Dey [63]	2011	$\mu_{nf} = \left(1 - \dfrac{\varphi}{0.5}\left(\dfrac{a_a}{a}\right)^{1.3}\right)^{-1.25}\mu_{bf}$	$CuO - gear\ oil$
Khanafer and Vafai [64]	2011	$\mu_{nf} = -0.4491 + \dfrac{28.837}{T} + 0.574\varphi - 0.1634\varphi^2$ $+ 25.053\dfrac{\varphi^2}{T^2} + 0.0132\varphi^3 - 2534.735\dfrac{\varphi}{T^3}$ $+ 23.498\dfrac{\varphi^2}{d_P^2} - 3.0185\dfrac{\varphi^3}{d_P^2}$	$Al_2O_3 - W$
Yu et al. [112]	2011	$\mu_{nf} = \exp\left(\dfrac{1736.6}{T}\right)\mu_{bf}$	$SiC - W$
Naik and Sundar [113]	2011	$\mu_{nf} = \left(3.444\left(\dfrac{T_{max}}{T_{min}}\right)^{0.514}\varphi^{0.1829}\right)\mu_{bf}$	$CuO - PPG{:}W$ (30:70 vol%)
Suganthi and Rajan [65]	2012	$\mu_{nf} = \left(1 + 11.97\varphi\right)\mu_{bf}$	$ZnO - W$
Azmi et al. [66]	2012	$\mu_{nf} = \mu_{bf}\left(1 + \dfrac{\varphi}{100}\right)^{11.3}\left(1 + \dfrac{T}{70}\right)^{-0.038}\left(1 + \dfrac{d_P}{170}\right)^{-0.061}$	$Al_2O_3 - W$ $TiO_2 - W$ $CuO - W$ $SiC - W$
Vaikili-Nezhad and Dorany [67]	2012	$\mu_{nf} = \left(1 + 1.59\varphi - 16.36\varphi^2 + 50.4\varphi^3\right)\mu_{bf}$ $\mu_{nf} = \left(1048 - 30.3T + 0.2T^2\right)\mu_{bf}$	$SWCNT - lube\ oil$
Heyhat et al. [68]	2013	$\mu_{nf} = \exp\left(\dfrac{5.989\varphi}{0.278 - \varphi}\right)\mu_{bf}$	$Al_2O_3 - W$
Rashin and Hemalatha [69]	2013	$\mu_{nf} = \left(1 + a\phi - b\phi^2\right)\mu_{bf}$ $\mu_{nf} = \left(c\exp\left(-0.03T\right)\right)\mu_{bf}$	$CuO - coconut\ oil$
Yiamsawas et al. [70]	2013	$\mu_{nf} = A\varphi^B T^C \mu_{bf}{}^D$	$Al_2O_3 - W$ $TiO_2 - W$
Sundar et al. [71]	2013	$\mu_{nf} = \mu_{bf}\left(1 + \dfrac{\varphi}{2.5}\right)^{6.356}$	$Fe_3O_4 - W$
Hemmat Esfe and Saedodin	2014	$\mu_{nf} = \left[0.9118\exp\left(5.49\varphi - 0.0001359T^2\right) + 0.0303\ln T\,\mu_{bf}\right]$	$ZnO - EG$

(Continued)

TABLE 6.3 (*Continued*)

Developed Empirical Correlations for Prediction of Viscosity of Nanofluids

Researchers	Year	Empirical Correlation	Nanofluid
Sundar et al. [74]	2014	$\mu_{nf} = \left(A\exp\left(B\varphi\right)\right)\mu_{bf}$ $A = 0.9396$ and $B = 24.16$ $A = 0.9299$ and $B = 67.43$ $A = 1.1216$ and $B = 77.56$	$Al_2O_3 - EG:W$ (20:80 vol%) $Al_2O_3 - EG:W$ (40:60 vol%) $Al_2O_3 - EG:W$ (60:40 vol%)
Sekhar and Shrama [114]	2015	$\mu_{nf} = 0.935\mu_{nf}\left(1+\dfrac{T_{nf}}{70}\right)^{0.5602}\left(1+\dfrac{d_P}{80}\right)^{-0.05915}\left(1+\dfrac{\varphi}{100}\right)^{10.51}$	$Al_2O_3 - W$
Sharifpur et al. [75]	2015	$\mu_{nf} = \left[1+\Re[\eta]\left(\left(\dfrac{T}{T_0}\right)^{\alpha}\varphi^{\beta}\left(\dfrac{d}{h}\right)^{\gamma}\right)\right]\mu_{bf}$	Al_2O_3-glycerol
Abdolbaqi et al. [76]	2016	$\mu_{nf} = \left[0.918\exp\left(14.696\varphi + 0.161\dfrac{T}{80}\right)\right]\mu_{bf}$	$TiO_2 - BG/W$
Abdolbaqi et al. [77]	2016	$\mu_{nf} = \left[0.906\exp\left(10.975\varphi + 0.169\dfrac{T}{80}\right)\right]\mu_{bf}$	$SiO_2 - BG/W$
Meybodi et al. [78]	2016	$\dfrac{\mu_{nf}}{\mu_{bf}} = \dfrac{\left(\begin{array}{c}133.54064976 - 343.82413843\times\exp\left(\varphi/S\right)\\[6pt] +290.11804759\times\left(\exp\left(\varphi/S\right)\right)^2\\[6pt] -78.993120761\times\left(\exp\left(\varphi/S\right)\right)^3\end{array}\right)}{\left(\begin{array}{c}0.91161630781 + 32.33014233\dfrac{Ln(S)}{T}\\[6pt] -11.732514460\dfrac{\left(Ln(S)\right)^2}{T}\end{array}\right)}$	$Al_2O_3 - W$ $TiO_2 - W$ $CuO - W$ $SiO_2 - W$
Adio et al. [79]	2016	$\mu_{nf} = \left[\begin{array}{c}1 + 7.0764\varphi - 0.1246\left(\dfrac{T}{T_0}\right)\varphi - 0.0346\left(\dfrac{d_p}{h}\right)\varphi\\[6pt] -0.0024\left(\left(\dfrac{d_p}{h}\right)\varphi\right)^2 - 1.2357\left(\left(\dfrac{T}{T_0}\right)\varphi\right)^2\\[6pt] +53.6946\varphi^2 + 0.0436\left(\dfrac{T}{T_0}\right)^2\varphi^{1/3}\end{array}\right]\mu_{bf}$	$MgO - *EG$
Bahiraei and Hangi [81]	2016	$\mu_{nf} = 1.5474\times10^{10}\left(\varphi+1\right)^{0.358}T^{-5.3892}$	$Fe_3O_4 - W$
Aberoumand and Jafarimoghadam [82]	2017	$\mu_{nf} = \left[1.15 + 1.061\varphi - 0.5442\varphi^2 + 0.1181\varphi^3\right]\mu_{bf}$	$Cu - $ engine oil
Chiam et al. [80]	2017	$\mu_{nf} = \left[\left(1+\dfrac{\varphi}{100}\right)^{32}\left(\dfrac{T}{70}\right)^{-0.001}\left(1+BR\right)^{0.08}\right]\mu_{bf}$	
Elcioglu et al. [83]	2018	$\mu_{nf} = 1.0973 - 0.0148T + 34.2292\varphi + 0.0041d_P - 0.3458T\varphi$ $\mu_r = 0.739716 - 0.000563T + 29.054\varphi + 0.006d_P$	$Al_2O_3 - W$

(Continued)

TABLE 6.3 (*Continued*)

Developed Empirical Correlations for Prediction of Viscosity of Nanofluids

Researchers	Year	Empirical Correlation	Nanofluid
Zadeh and Toghraie [115]	2018	$\mu_{nf} = \left(0.9232 + 0.006307 T^{1.5281} \varphi^{0.97409} \exp\left(-0.0303 T\right)\right) \mu_{bf}$	Ag – EG
Saeedi et al. [116]	2018	$\mu_{nf} = \left(\begin{array}{c} 781.4 T^{-2.117} \varphi^{0.2722} + \dfrac{0.05776}{T^{-0.7189} \varphi^{-0.04009}} \\ + 0.511 \varphi^{2} - 0.1799 \varphi^{3} \end{array} \right) \mu_{bf}$	CeO2 – EG

* ethylene glycol

6.4 Computer-Aided Models

Modeling viscosity of nanofluids is a complicated task because of two main reasons. Firstly, viscosity of nanofluids is dependent on many variables, and, secondly, the relationships between these variables are very complex. These problems always influence the accuracy of conventional modeling approaches. However, artificial intelligent modeling approaches through their high computational ability are able to overcome these limitations and problems. In the recent years, many intelligent models were developed for estimation of viscosity of nanofluids. Almost all of these models will be reviewed in this section.

An artificial neural network based on genetic algorithm (GA) was proposed by Karimi et al. [83] to estimate viscosity of nanofluids. They used 381 experimental data from eight different nanofluids to develop the model. Detailed description of the used data set is shown in Table 6.4. 70% of this data set was used to train the model, and the rest 30% was applied for testing the model. The input variables of the model contain temperature, particle volume concentration, particle size, fluid base viscosity, and density ratio of fluid base to particle. Statistical analyses revealed that the outputs of the developed model are in good agreement with the experimental data with a mean average relative error equal to 2.48%.

Considering 563 experimental data of four different water-based nanofluids (TiO$_2$, Al$_2$O$_3$, SiO$_2$, and CuO), Mehrabi et al. [84] proposed a fuzzy C-means clustering-based adaptive neuro fuzzy system (FCM-ANFIS) model to estimate viscosity of nanofluids. In their study, the input variables were size

TABLE 6.4

Detailed Description of the Used Data Set

Nanofluids	Temperature [K]	Volume Fraction	Particle size [nm]	Base-fluid Viscosity [pa.s]	Nanofluids Viscosity [pa.s]
Al$_2$O$_3$–H$_2$O	293.15–343.15	0.005–0.094	30–47	0.00039–0.0010	0.00044–00022
Al$_2$O$_3$–PG	303.15–333.15	0.005–0.03	27–50	0.01304–0.05260	0.0136–0.0718
CuO–H$_2$O	278–339.25	0.050–0.09	29	0.00042–0.00151	0.00046–0288
CuO in 40% H$_2$O and 60% EG	238.25–323.15	0.01–0.06	29	0.133–0.656	0.152–0.875
TiO$_2$–H$_2$O	288.15–308.15	0.005–0.08	21–29	0.00069–0.65600	0.00077–0.67800
TiO–EG	293.15–333	0.005–0.08	85	0.00522–0.0420	0.00600–0.04405
SiO$_2$–H$_2$O	293–323	0.0045–0.04	12	0.00042–0.00101	0.000484–0.0040
SiO$_2$–EtOH	298	0.001–0.07	35–94	0.0013	0.00131–0.00254

Source: Karimi, H. et al., *Heat Mass Transf.*, 47, 1417–1425, 2011.

of nanoparticle, temperature, and nanoparticle volume fraction. Their analyses revealed that the prediction of the FCM-ANFIS well agreed with experimental data. Also, they showed that the results of the developed model are much closer to the experimental data than the most cited correlations of nanofluids viscosity.

Atashrouz et al. [85] proposed hybrid self-organizing polynomial neural networks (PNN) on the basis of group method of data handling (GMDH) to predict viscosity of nine different nanofluids including Al_2O_3-ethylene glycol, Al_2O_3-propylene glycol, Al_2O_3-Ethylene glycol and water mixture (60:40 vol%), SiO_2-water, TiO_2-ethylene glycol, CuO-water, CuO-Ethylene glycol and water mixture (60:40 vol%), TiO_2-water, and Al_2O_3-water. The GMDH polynomials were developed base on experimental data of 561 nanofluids from 10 literatures. In the proposed GMDH polynomials, relative viscosity of nanofluids were modeled as a function of temperature, volume fraction, and nanoparticle diameter. From their analysis and comparison, it was proved that all the presented GMDH-PNN correlations could effectively estimate the viscosity of nanofluids with an average absolute relative deviation of 2.14% and regression correlation of 0.9978. Moreover, the results demonstrated that the GMDH-PNN correlations are superior to five of the investigated theoretical models and empirical correlations.

Meybodi et al. [86] applied the least-square support vector machine (LSSVM) modeling approach to present a model for estimation of viscosity of water-based nanofluids of Al_2O_3, TiO_2, CuO, and SiO_2. A total of 801 experimental data were used for development of the model by setting particle size, volume fraction, temperature, base-fluid viscosity, and type of nanoparticle as input variables. Based on statistical analyses, it was demonstrated that LSSVM models have high capability in predicting viscosity of nanofluids with an average absolute relative error of 2.12% and coefficient of determination of 0.998. Besides, it was also revealed that LSSVM models could accurately predict the viscosity of nanofluids with variation of temperature and nanoparticle volume fraction. Furthermore, the developed LSSVM model showed much better performance in comparison to some of the proposed empirical and theoretical models.

Zhao et al. [87] designed two radial basis function neural networks (RBF-NN) to predict relative viscosity of water-based nanofluids. A total number of 721 experimental data points of Al_2O_3-water and CuO-water nanofluids were employed to develop and evaluate the models. The main difference of the two developed RBF-NNs was the number of their input parameters. The first model had five input parameters including temperature, volume concentration, nanoparticle density, nanoparticle diameter, and viscosity of base fluid. The second model was a model with four input parameters containing the same inputs as the first model without considering the temperature as input variable. Evaluation of the developed models revealed that the five input parameter RBF models predict the relative viscosity of Al_2O_3-water and CuO-water nanofluids with a mean absolute percentage error of 2.343% and 0.898%, respectively. For the four inputs RBF models, the values of mean absolute percentage error for Al_2O_3-water and CuO-water nanofluids were 3.169% and 3.298%. Therefore, they concluded that considering the effect of the temperature provides higher accuracy for RBF neural networks in predicting the viscosity of nanofluids. Moreover, they showed that the developed models give more accurate results compared to existing equations.

Adio et al. [78] applied genetic algorithm-polynomial neural network (GA-PNN) and FCM-ANFIS to develop models for estimation of viscosity of nanofluids. They used 198 experimental data of MgO-ethylene glycol nanofluids containing nanoparticles with 21, 105, and 125 nm sizes at temperature range of 20°C–70°C with volume fraction range of 0%–5%. Nanoparticle size, temperature, and nanoparticle volume fraction were considered as the input variables of the model. As reported by the authors, the developed GA-PNN and FCM-ANFIS models are able to predict the viscosity with average absolute relative errors equal to 2.67% and 1.32%, respectively.

In another research work, Atashrouz et al. [88] experimentally measured viscosity of Fe_3O_4 nanoparticles dispersed in a mixture of water and ethylene glycol (60:40 wt%) at temperature range of 288.15–343.15 K and at low volume fractions ranging from 0.22% to 0.55%. Based on the obtained experimental data, they developed a hybrid GMDH-type polynomial neural network, as a function of shear rate, temperature, and nanoparticle volume fraction for estimation of viscosity. As reported in their study, the developed model's estimates are in good agreement with their corresponding experimental data and a mean absolute relative deviation of 3.88% was obtained for the developed model.

As a function of five different parameters, including temperature, volume fraction of nanoparticle, size of nanoparticle, bulk density of nanofluid, and viscosity of base fluid, Barati-Harooni and Najafi-Marghmaleki [89] developed and validated a RBF-NN to estimate the relative viscosity of nanofluids. They developed their model by considering a data set of 1490 experimental data from various nanofluids collected from different published literatures. Statistical details about this data set are summarized in Table 6.5. According to their investigation, it was shown that the relative viscosities obtained by using the developed RBF-NN agreed well with the corresponding experimental data. The overall regression coefficient, average absolute relative error, and root mean square error of the developed model were reported to be 0.99996, 0.2, and 0.0089, respectively. Also, it was shown that the developed model outperforms the literature theoretical and empirical correlations.

By considering the same experimental data set as mentioned above, Heidari et al. [90] proposed a feed-forward back-propagation multilayer perceptron artificial neural network based on Levenberg-Marquardt training algorithm to predict relative viscosity of nanofluids. These authors stated that temperature, volume fraction, particle density, particle size, and base-fluid viscosity have the most influence on the viscosity of nanofluids. Therefore, they set these five parameters as model inputs. The optimum structure for the artificial neural network was determined to have two hidden layers with 25 neurons and 14 neurons. Their obtained results by the developed model showed that the model is capable to predict viscosity of nanofluids with excellent accuracy.

Longo et al. [91] presented an ANN model for predicting dynamic viscosity of oxide nanoparticles suspended in water and ethylene glycol. They used experimental data of 30 data points of Al_2O_3 water, 40 data points of TiO_2-water, and 36 data points of Al_2O_3-ethylene glycol and TiO_2-ethylene glycol nanofluids to train the model. The used experimental data include nanofluids with nanoparticle diameter ranging from 5 to 80 nm, cluster average size ranging from 100 nm to 300 nm, and particle volume fraction in the range of 1%–6% at temperature in the interval of 0°C–70°C. In this model, the relative viscosity of nanofluids was modeled as a function of temperature, particle volume fraction, nanoparticle diameter, cluster of nanoparticles average size, and type of base fluids. The optimum configuration of the ANN model was found to be 5-9-1. The performance of the develop model was evaluated by considering experimental data of TiO_2-ethylene glycol and A_2O_3-ethylene glycol from other literatures. The results showed that the model is able to fairly predict the viscosity of these nanofluids with a mean absolute percentage error equal to 4.15% with more than 94% of the data points within ±10%.

By considering temperature and particle volume fraction as input variables, Bahiraei and Hangi [80] developed three artificial neural networks trained with resilient back propagation, quasi-Newton, and Levenberg-Marquardt algorithms for estimation of viscosity of nanofluids. They used experimental data of Fe_3O_4-water nanofluids at the temperature range of 25°C–60°C with particle volume fraction ranging from 0% to 4% to propose these models. The outputs of the developed models were compared against each other, and it was found that the developed ANN model trained with Bayesian regularization-based Levenberg-Marquardt algorithm provides the best performance. Moreover, the results of this model were compared against Einstein theoretical model [20], and it was shown that the developed model outperforms the Einstein model [20].

TABLE 6.5

Statistical Details of Used Experimental Data

Parameter	Min.	Max.	Average	Standard Deviation
Temperature (K)	238.15	344.35	307.07	17.1
Size of nanoparticle (nm)	7	190	32.22	16.74
Volume fraction of nanoparticle (%)	0	9	3.13	3.03
Density of nanoparticle (g/cm³)	2.65	6.31	4.58	1.32
Viscosity of base fluid (cP)	0.394	452.599	6.01	25.29
Relative viscosity of nanofluid	0.56099	9.77636	1.82	1.41

Source: Barati-Harooni, A and Najaf-Marghmaleki A. *J. Mol. Liq.*, 224, 580–588, 2016.

TABLE 6.6

Statistical Details of Used Experimental Data

Parameter	Min.	Max.	Average	Standard Deviation
Temperature (K)	257.94	344.35	309.66	15.31
Size of nanoparticle (nm)	7.00	220	33.73	17.17
Volume fraction of nanoparticle (%)	0.01	9	2.86	3.01
Density of nanoparticle (g/cm^3)	0.2	6.31	4.40	1.44
Viscosity of base fluid (cP)	0.39	6.03	1.08	0.96
Viscosity of nano fluid (cP)	0.44	6.48	1.63	1.16

Source: Barati-Harooni, A. et al., *J. Mol. Liq.*, 241, 1079–1090, 2017.

By considering experimental data of 1140 different nanofluids, Barati-Harooni et al. [92] proposed four viscosity models based on multilayer perceptron neural networks (MLP-NNs), LSSVM, ANFIS, and gene expression programming (GEP) approaches. Statistical details about the experimental data, which were gathered from 15 available literatures, are summarized in Table 6.6. In these models, viscosity of nanofluids were modeled as a function of volume fraction of nanoparticle, temperature, size of nanoparticle, density of nanoparticle, and viscosity of base fluid. Accuracy evaluation of the models revealed that all of them are able to give acceptable prediction of viscosity of different nanofluids, while the best performance is for the developed LSSVM model with correlation factor (R^2) of 0.9989 and average absolute relative deviation of 1.44%. Figure 6.1 shows the cross plot of the developed models. This plot shows that all of the models provide accurate results, and the LSSVM models have best performance.

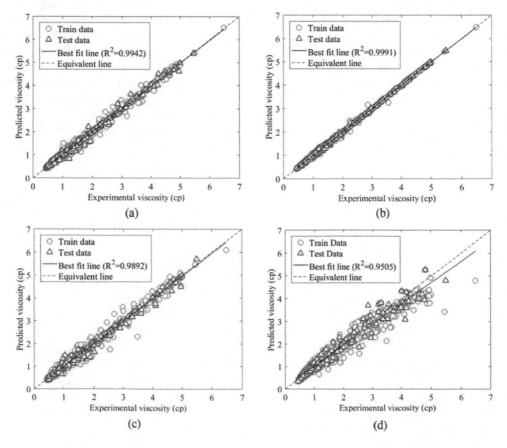

FIGURE 6.1 Cross plot of the developed models. (a) MLP-NN, (b) CSA-LSSVM, (c) Hybrid-Anfis, (d) GEP (From Barati-Harooni, A. et al., *J. Mol. Liq.*, 241, 1079–90, 2017.)

Aminian [93] developed a multilayer perceptron artificial neural network trained by Levenberg-Marquardt back propagation algorithm to model viscosity of nanofluids. They used a total number of 622 experimental data to develop their model. The model was developed by considering temperature, volume fraction, nanoparticle density, and diameter as input variables. The used data set covers temperature from 10°C to 90°C, volume fraction in the range of 0.01%–11.22%, and nanoparticle size ranging from 7 nm to 100 nm. It was found that the ANN network with one hidden layer which includes nine neurons provides the minimum error in predicting viscosity of nanofluids. The structure of the proposed model is shown in Figure 6.2. They used experimental data of eight different nanofluids including Al_2O_3-water, TiO_2-water, CuO-water, Fc_3O_4-water, SiC-water, MWCNT-water, Ni-water, and Ag-water to evaluate the accuracy of the developed model. The obtained results by the developed model were in good agreement with the experimental data with an average absolute relative error of 6.66% and R^2 value of 0.9482. Also, the proposed ANN model was more accurate in comparison to some of the available models.

A hybrid artificial neural network-genetic algorithm modeling approach was used by Vakili et al. [94] to propose a viscosity model for water-based graphene nanoplatelets nanofluid based on the obtained experimental data. This model considers the nanoparticle weight fraction and temperature as the input parameters of the model. The schematic of the proposed artificial neural network structure is depicted in Figure 6.3. In this hybrid model, genetic algorithm was applied to improve training of the artificial neural network. Proper values for weights and biases of the artificial neural network were obtained by genetic algorithm.

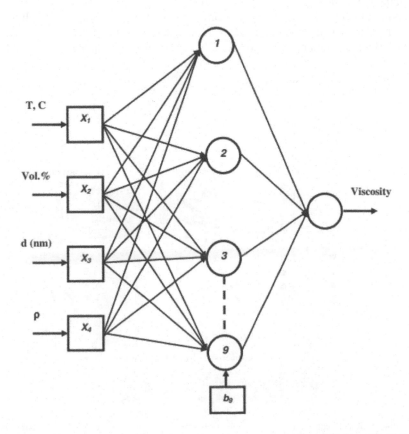

FIGURE 6.2 Structure of the proposed ANN model. (From Aminian, A. *J. Mol. Liq.*, 229, 300–308, 2017.)

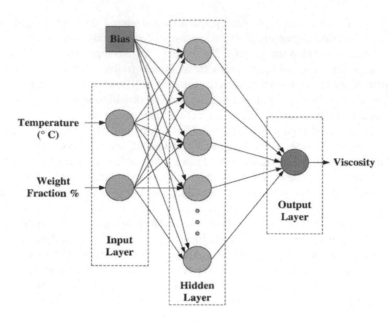

FIGURE 6.3   Schematic of the proposed ANN model by (From Vakili, M. et al., *Int. J. Heat Mass Transf.*, 82, 40–48, 2017.)

Figure 6.4 shows the combination of neural networks and genetic algorithms structure, and Figure 6.5 shows the computational flowchart for the proposed hybrid artificial neural network-genetic algorithm model. The optimum structure of the artificial neural network was found to be 2-8-1. It was shown that this model is able to predict viscosity of the graphene nanoplatelets nanofluid (at the temperature range of 20°C–60°C and weight fraction range of 0.025%–0.1%) with a mean absolute percentage error of 0.777%.

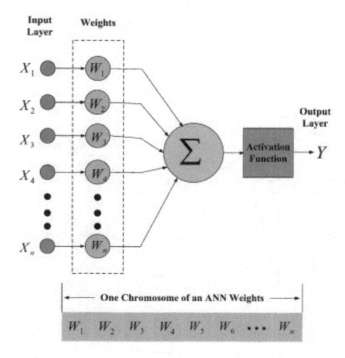

FIGURE 6.4   Combination of neural networks and genetic algorithms structure. (From Vakili, M. et al., *Int. J. Heat Mass Transf.*, 82, 40–48, 2017.)

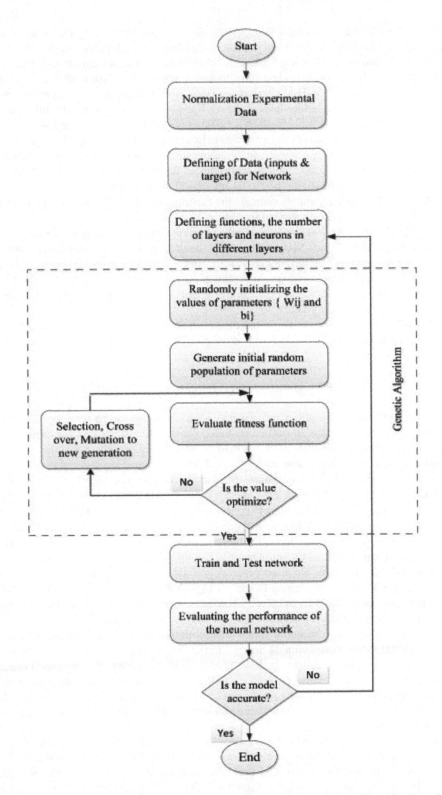

FIGURE 6.5 Computational flowchart of the proposed model. (From Vakili, M. et al., *Int. J. Heat Mass Transf.*, 82, 40–48, 2017.)

As a function of particle concentration, nanoparticle size, nanoparticle density, temperature, and shear rate values, Ansari et al. [95] proposed and validate a feed-forward back-propagation ANN to estimate relative viscosity of nanofluids. They used 1620 experimental data of various nanofluids to train, test, and validate the model. Different training algorithms including Levenberg-Marquardt (LM), scaled conjugate gradient (SCG), Bayesian regulation back propagation, and resilient back propagation were used to optimize the ANN model. In the hidden layer of the ANN model, six different transfer functions containing Hard-limit (hardlim), Radial basis (radbas), Log-sigmoid (logsig), Tan-sigmoid (tansig), Triangular basis (tribas), and Soft max (softmax) were evaluated to introduce the best choice which gives the highest accuracy. Based on the statistical accuracy analysis, the structure for the developed ANN was determined to be 5-23-1 with tan-sigmoid and purelin transfer functions in the hidden and output layers. Moreover, the LM was selected as the best training algorithm. The obtained results showed that the developed ANN model could accurately estimate the experimental data with mean square error (MSE) of 0.00901 and regression coefficient (R^2) of 0.9954.

Derakhshanfard and Mehralizadeh [96] conducted experiments to measure viscosity of crude oil-based nanofluids containing NiO, WO_3, TiO_2, ZnO, and FeO_3 nanoparticles with average size of 10–40 nm at temperature ranging from 40°C and 100°C and mass fractions in the range of 0.2%–2% wt. Based on the obtained experimental data, they developed RBF method of the ANN to estimate the viscosity. The results revealed that the outputs of the developed RBF ANN are in good agreement with the experimental data.

To predict the relative viscosity of CuO-liquid paraffin nanofluids, Karimpour et al. [97] proposed a correlation using hybrid GMDH-type neural network with temperature and mass fraction as input parameters which is given by:

$$\mu_r = 0.995246 - 0.000293119T + 29.054T\phi + 0.125761\phi. \tag{6.88}$$

This equation is valid for CuO-liquid paraffin nanofluids at the temperature range of 25°C–100°C and mass fraction range of 0.25–6 wt%. It was demonstrated that viscosity of most of the used experimental data can be predicted with an absolute error less than 5% with the proposed correlation.

Based on 674 experimental data gathered from nine different literatures, Gholami et al. [98] proposed several artificial intelligent models to estimate the viscosity of alumina nanofluids dispersed in various base fluids. They developed four different models based on MLP-NN, cascade feed-forward neural network (CFNN), RBF-NN, and least-square support vector machines approaches. To choose the independent variable of the models, correlation matrix analysis was performed. The results indicated that the diameter (D_P), and volume concentration (V_P) of nanoparticles, inverse of reduced temperature (T_C/T), inverse of critical pressure ($1/P_c$), and acentric factor of the base fluids are the most affecting independent variables on nanofluids viscosity. The used experimental data for development of the models include alumina nanofluids with diameters ranging from 8 nm to 50 nm suspended in seven different base fluids over the temperature range of 273 K–345 K with volume concentration up to 9.4%. The performance of the developed models was analyzed and compared with ten empirical correlations, and then the best model was introduced with the procedure which is illustrated in Figure 6.6. Based on statistical accuracy analysis, the MLP-NN model with 14 neurons in the hidden layer was selected as the best developed model. This model gave excellent results in predicting the viscosity of investigated experimental data with an average absolute relative error equal to 4.13%. Furthermore, it was demonstrated that the RBF-NN is the weakest model. Also, the results revealed that the accuracy of the ten empirical correlations are only better than the RBF-NN model and performance of the other proposed models is superior to the empirical correlations.

The most accurate and general model for estimation of viscosity of nanofluids was developed by Hemmati-Sarapardeh et al. [99] based on committee machine intelligent system (CMIS). They developed seven different intelligent models, and then combined all of them into a single accurate model by using CMIS approach as represented in Figure 6.7. The developed intelligent models include two RBF-NN models optimized with particle swarm optimization (PSO) and GA, four MLP models optimized with LM, resilient back propagation, SCG, and Bayesian regularization (BR), and one LSSVM model optimized with coupled simulated annealing (CSA) algorithm. For development of these models,

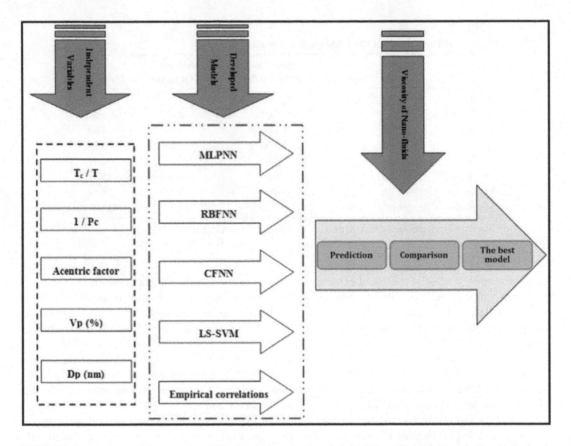

FIGURE 6.6   Schematic of the procedure of finding the best developed model. (From Gholami, E. et al., *Powder Technol.*, 323, 495–506, 2018.)

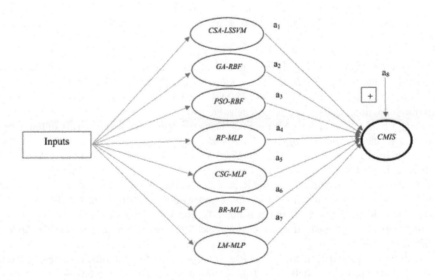

FIGURE 6.7   The proposed CMIS model. (From Hemmati-Sarapardeh, A. et al., *Renew. Sust. Energ. Rev.*, 81, 313–329, 2018.)

TABLE 6.7

Statistical Description of the Used Input Parameters

	T(°C)	Size (nm)	Particle Volume Fraction (%)	Density of Particle (g/cm³)	Viscosity of Fluid (cP)
Mean	34.89	34.37	2.21	4.26	5.24
Standard Error	0.31	0.49	0.04	0.02	1.07
Median	35.00	29.00	1.00	3.98	1.00
Mode	25.00	30.00	1.00	3.98	0.89
Kurtûsis	0.59	5.48	0.99	1,75	316.48
Skewness	−0.30	2.29	1.44	0.94	15.52
M ïiiimura	−35.03	4.60	0.00	2.22	0.24
Maximum	80.00	190.00	10.00	13.61	452.60

Source: Hemmati-Sarapardeh, A. et al., *Renew. Sust. Energ. Rev.*, 81, 313–29, 2018.

a comprehensive review of published experimental viscosity data was made and 3144 data points were collected. This data bank contains 42 different types of nanofluids and covers wide range of operational conditions. The CMIS model was developed by considering five input variables including temperature, nanoparticle volume fraction, nanoparticle size, nanoparticle density, and base-fluid viscosity. The statistical description of these input parameters is shown in Table 6.7. Statistical analysis revealed that the developed CMIS model provides excellent accuracy in predicting viscosity of all of the used experimental data points with an average absolute relative error of 3.95%. The performance of the developed CMIS was compared against five theoretical models and four empirical correlations. It was demonstrated that the CMIS model outperforms all of the other available models and unlike to the CMIS model, none of the existing models were able to give accurate prediction of viscosity of nanofluids over broad range of operating conditions. Hemmati-Sarapardeh et al. [99] also evaluated the quality of the gathered 3144 data points and concluded that the experimental data have very good reliability except small percent of the data, which are out of the applicability domain of the model.

REFERENCES

1. Salimi-Yasar H, Zeinali Heris S, Shanbedi M, Amiri A, Kameli A. Experimental investigation of thermal properties of cutting fluid using soluble oil-based TiO₂ nanofluid. *Powder Technology* 2017;310: 213–220.
2. Maxwell J. *A Treatise Electricity and Magnetism.* Clarendon Press, UK. 1873.
3. Gupta M, Singh V, Kumar R, Said Z. A review on thermophysical properties of nanofluids and heat transfer applications. *Renewable and Sustainable Energy Reviews* 2017;74: 638–670.
4. Sundar LS, Sharma KV, Singh MK, Sousa ACM. Hybrid nanofluids preparation, thermal properties, heat transfer and friction factor—A review. *Renewable and Sustainable Energy Reviews* 2017;68: 185–198.
5. Masuda H, Ebata A, Teramae K. Alteration of thermal conductivity and viscosity of liquid by dispersing ultra-fine particles. *Dispersion of Al₂O₃, SiO₂ and TiO₂ Ultra-fine Particles.* 1993; 227–233.
6. Chol S, Eastman J. Enhancing thermal conductivity of fluids with nanoparticles. *ASME-Publications-Fed* 1995;231: 99–106.
7. Akilu S, Sharma KV, Baheta AT, Mamat R. A review of thermophysical properties of water based composite nanofluids. *Renewable and Sustainable Energy Reviews* 2016;66: 654–678.
8. Suganthi KS, Rajan KS. Metal oxide nanofluids: Review of formulation, thermo-physical properties, mechanisms, and heat transfer performance. *Renewable and Sustainable Energy Reviews* 2017;76: 226–255.
9. Meyer JP, Adio SA, Sharifpur M, Nwosu PN. The viscosity of nanofluids: A review of the theoretical, empirical, and numerical models. *Heat Transfer Engineering* 2016;37(5): 387–421.
10. Mahbubul IM, Saidur R, Amalina MA. Latest developments on the viscosity of nanofluids. *International Journal of Heat and Mass Transfer* 2012;55(4): 874–885.

11. Timofeeva EV. Nanofluids for heat transfer–potential and engineering strategies. *Two Phase Flow, Phase Change and Numerical Modeling.* InTech; 2011.

12. Zhao N, Li S, Yang J. A review on nanofluids: Data-driven modeling of thermalphysical properties and the application in automotive radiator. *Renewable and Sustainable Energy Reviews* 2016;66: 596–616.

13. Sundar LS, Sharma KV, Naik MT, Singh MK. Empirical and theoretical correlations on viscosity of nanofluids: A review. *Renewable and Sustainable Energy Reviews* 2013;25: 670–686.

14. Azmi WH, Sharma KV, Mamat R, Najafi G, Mohamad MS. The enhancement of effective thermal conductivity and effective dynamic viscosity of nanofluids—A review. *Renewable and Sustainable Energy Reviews* 2016;53: 1046–1058.

15. Raja M, Vijayan R, Dineshkumar P, Venkatesan M. Review on nanofluids characterization, heat transfer characteristics and applications. *Renewable and Sustainable Energy Reviews* 2016;64: 163–173.

16. Bashirnezhad K, Bazri S, Safaei MR, Goodarzi M, Dahari M, Mahian O. et al. Viscosity of nanofluids: A review of recent experimental studies. *International Communications in Heat and Mass Transfer* 2016;73: 114–123.

17. Murshed SMS, Estellé P. A state of the art review on viscosity of nanofluids. *Renewable and Sustainable Energy Reviews* 2017;76: 1134–1152.

18. Yang L, Xu J, Du K, Zhang X. Recent developments on viscosity and thermal conductivity of nanofluids. *Powder Technology* 2017;317: 348–369.

19. Koca HD, Doganay S, Turgut A, Tavman IH, Saidur R, Mahbubul IM. Effect of particle size on the viscosity of nanofluids: A review. *Renewable and Sustainable Energy Reviews* 2018;82: 1664–1674.

20. Einstein A. A new determination of molecular dimensions. *Annals Physics* 1906;19: 289–306.

21. Hatschek E. The general theory of viscosity of two-phase systems. *Transactions of the Faraday Society* 1913;9: 80–92.

22. Eilers VH. Die viskosität von emulsionen hochviskoser stoffe als funktion der konzentration. *Kolloid-Zeitschrift* 1941;97(3): 313–321.

23. De Bruijn H. The viscosity of suspensions of spherical particles.(The fundamental η-c and φ relations). *Recueil des Travaux Chimiques des Pays-Bas* 1942;61(12): 863–874.

24. Vand V. Viscosity of solutions and suspensions. I. Theory. *The Journal of Physical Chemistry* 1948;52(2): 277–299.

25. Saitô N. Concentration dependence of the viscosity of high polymer solutions. I. *Journal of the Physical Society of Japan* 1950;5(1): 4–8.

26. Mooney M. The viscosity of a concentrated suspension of spherical particles. *Journal of Colloid Science* 1951;6(2): 162–170.

27. Brinkman H. The viscosity of concentrated suspensions and solutions. *The Journal of Chemical Physics* 1952;20(4): 571.

28. Roscoe R. The viscosity of suspensions of rigid spheres. *British Journal of Applied Physics* 1952;3(8): 267.

29. Simha R. A treatment of the viscosity of concentrated suspensions. *Journal of Applied physics* 1952;23(9): 1020–1024.

30. Krieger IM, Dougherty TJ. A mechanism for non-Newtonian flow in suspensions of rigid spheres. *Transactions of the Society of Rheology* 1959;3(1): 137–152.

31. Frankel N, Acrivos A. On the viscosity of a concentrated suspension of solid spheres. *Chemical Engineering Science* 1967;22(6): 847–853.

32. Nielsen LE. Generalized equation for the elastic moduli of composite materials. *Journal of Applied Physics* 1970;41(11): 4626–4627.

33. Lundgren TS. Slow flow through stationary random beds and suspensions of spheres. *Journal of Fluid Mechanics* 1972;51(2): 273–299.

34. Brenner H, Condiff DW. Transport mechanics in systems of orientable particles. IV. Convective transport. *Journal of Colloid and Interface Science* 1974;47(1): 199–264.

35. Jeffrey DJ, Acrivos A. The rheological properties of suspensions of rigid particles. *AIChE Journal* 1976;22(3): 417–432.

36. Batchelor G. The effect of Brownian motion on the bulk stress in a suspension of spherical particles. *Journal of fluid mechanics* 1977;83(1): 97–117.

37. Graham AL. On the viscosity of suspensions of solid spheres. *Applied Scientific Research* 1981;37(3–4): 275–286.

38. Kitano T, Kataoka T, Shirota T. An empirical equation of the relative viscosity of polymer melts filled with various inorganic fillers. *Rheologica Acta* 1981;20(2): 207–209.
39. Metzner A. Rheology of suspensions in polymeric liquids. *Journal of Rheology* 1985;29(6): 739–775.
40. Leighton D, Acrivos A. The shear-induced migration of particles in concentrated suspensions. *Journal of Fluid Mechanics* 1987;181: 415–439.
41. Barnes HA, Hutton JF, Walters K. *An Introduction to Rheology.* Amsterdam, the Netherlands, Elsevier; 1989.
42. Thomas CU, Muthukumar M. Three-body hydrodynamic effects on viscosity of suspensions of spheres. *The Journal of Chemical Physics* 1991;94(7): 5180–5189.
43. Neto JR. Deffloculation mechanisms of colloidal clay suspensions. *Materials Science and Engineering* (in Portuguese), UFSC, Florianopolis 1999, Ph.D. thesis.
44. Fullman R. Measurement of particle sizes in opaque bodies. *Journal of The Minerals* 1953;5(3): 447–452.
45. De Noni Jr A, Garcia DE, Hotza D. A modified model for the viscosity of ceramic suspensions. *Ceramics International* 2002;28(7): 731–735.
46. Cheng N-S, Law AW-K. Exponential formula for computing effective viscosity. *Powder Technology* 2003;129(1–3): 156–160.
47. Chen H, Ding Y, He Y, Tan C. Rheological behaviour of ethylene glycol based titania nanofluids. *Chemical Physics Letters* 2007;444(4–6): 333–337.
48. Masoumi N, Sohrabi N, Behzadmehr A. A new model for calculating the effective viscosity of nanofluids. *Journal of Physics D: Applied Physics* 2009;42(5): 055501.
49. Brouwers H. Viscosity of a concentrated suspension of rigid monosized particles. *Physical Review E* 2010;81(5): 051402.
50. Hosseini SM, Moghadassi A, Henneke DE. A new dimensionless group model for determining the viscosity of nanofluids. *Journal of Thermal Analysis and Calorimetry* 2010;100(3): 873–877.
51. Abedian B, Kachanov M. On the effective viscosity of suspensions. *International Journal of Engineering Science* 2010;48(11): 962–965.
52. Shahriari A, Jahantigh N, Rakani F. Providing an analytical model in determining nanofluids. *International Journal of Engineering-Transactions C: Aspects* 2017;30(12): 1919–1924.
53. Lebon G, Machrafi H. A thermodynamic model of nanofluid viscosity based on a generalized Maxwell-type constitutive equation. *Journal of Non-Newtonian Fluid Mechanics* 2018;253: 1–6.
54. Tseng WJ, Lin K-C. Rheology and colloidal structure of aqueous TiO_2 nanoparticle suspensions. *Materials Science and Engineering: A* 2003;355(1): 186–192.
55. Maïga SEB, Nguyen CT, Galanis N, Roy G. Heat transfer behaviours of nanofluids in a uniformly heated tube. *Superlattices and Microstructures* 2004;35(3): 543–557.
56. Buongiorno J. Convective transport in nanofluids. *Journal of Heat Transfer* 2006;128(3): 240–250.
57. Pak BC, Cho YI. Hydrodynamic and heat transfer study of dispersed fluids with submicron metallic oxide particles. *Experimental Heat Transfer an International Journal* 1998;11(2): 151–170.
58. Kulkarni DP, Das DK, Chukwu GA. Temperature dependent rheological property of copper oxide nanoparticles suspension (nanofluid). *Journal of Nanoscience and Nanotechnology* 2006;6(4): 1150–1154.
59. Namburu PK, Kulkarni DP, Misra D, Das DK. Viscosity of copper oxide nanoparticles dispersed in ethylene glycol and water mixture. *Experimental Thermal and Fluid Science* 2007;32(2): 397–402.
60. Nguyen C, Desgranges F, Galanis N, Roy G, Maré T, Boucher S et al. Viscosity data for Al_2O_3–water nanofluid—hysteresis: Is heat transfer enhancement using nanofluids reliable? *International Journal of Thermal Sciences* 2008;47(2): 103–111.
61. Corcione M. Empirical correlating equations for predicting the effective thermal conductivity and dynamic viscosity of nanofluids. *Energy Conversion and Management* 2011;52(1): 789–793.
62. Kole M, Dey T. Effect of aggregation on the viscosity of copper oxide–gear oil nanofluids. *International Journal of Thermal Sciences* 2011;50(9): 1741–1747.
63. Khanafer K, Vafai K. A critical synthesis of thermophysical characteristics of nanofluids. *International Journal of Heat and Mass Transfer* 2011;54(19–20): 4410–4428.
64. Suganthi K, Rajan K. Temperature induced changes in ZnO–water nanofluid: Zeta potential, size distribution and viscosity profiles. *International Journal of Heat and Mass Transfer* 2012;55(25–26): 7969–7980.
65. Azmi W, Sharma K, Mamat R, Alias A, Misnon II. Correlations for thermal conductivity and viscosity of water based nanofluids. *IOP Conference Series: Materials Science and Engineering.* 36. IOP Publishing; 2012: 012029.

66. Vakili-Nezhaad G, Dorany A. Effect of single-walled carbon nanotube on the viscosity of lubricants. *Energy Procedia* 2012;14: 512–517.

67. Heyhat M, Kowsary F, Rashidi A, Momenpour M, Amrollahi A. Experimental investigation of laminar convective heat transfer and pressure drop of water-based Al_2O_3 nanofluids in fully developed flow regime. *Experimental Thermal and Fluid Science* 2013;44: 483–489.

68. Rashin MN, Hemalatha J. Viscosity studies on novel copper oxide–coconut oil nanofluid. *Experimental Thermal and Fluid Science* 2013;48: 67–72.

69. Yiamsawas T, Dalkilic AS, Mahian O, Wongwises S. Measurement and correlation of the viscosity of water-based Al_2O_3 and TiO_2 nanofluids in high temperatures and comparisons with literature reports. *Journal of Dispersion Science and Technology* 2013;34(12): 1697–1703.

70. Syam Sundar L, Singh MK, Sousa ACM. Investigation of thermal conductivity and viscosity of Fe_3O_4 nanofluid for heat transfer applications. *International Communications in Heat and Mass Transfer* 2013;44: 7–14.

71. Esfe MH, Saedodin S. An experimental investigation and new correlation of viscosity of ZnO–EG nanofluid at various temperatures and different solid volume fractions. *Experimental Thermal and Fluid Science* 2014;55: 1–5.

72. Hemmat Esfe M, Saedodin S. An experimental investigation and new correlation of viscosity of ZnO–EG nanofluid at various temperatures and different solid volume fractions. *Experimental Thermal and Fluid Science* 2014;55: 1–5.

73. Syam Sundar L, Venkata Ramana E, Singh MK, Sousa ACM. Thermal conductivity and viscosity of stabilized ethylene glycol and water mixture Al_2O_3 nanofluids for heat transfer applications: An experimental study. *International Communications in Heat and Mass Transfer* 2014;56: 86–95.

74. Sharifpur M, Adio SA, Meyer JP. Experimental investigation and model development for effective viscosity of Al_2O_3–glycerol nanofluids by using dimensional analysis and GMDH-NN methods. *International Communications in Heat and Mass Transfer* 2015;68: 208–219.

75. Abdolbaqi MK, Sidik NAC, Aziz A, Mamat R, Azmi WH, Yazid MNAWM et al. An experimental determination of thermal conductivity and viscosity of BioGlycol/water based TiO_2 nanofluids. *International Communications in Heat and Mass Transfer* 2016;77: 22–32.

76. Abdolbaqi MK, Sidik NAC, Rahim MFA, Mamat R, Azmi WH, Yazid MNAWM et al. Experimental investigation and development of new correlation for thermal conductivity and viscosity of BioGlycol/water based SiO_2 nanofluids. *International Communications in Heat and Mass Transfer* 2016;77: 54–63.

77. Meybodi MK, Daryasafar A, Koochi MM, Moghadasi J, Meybodi RB, Ghahfarokhi AK. A novel correlation approach for viscosity prediction of water based nanofluids of Al_2O_3, TiO_2, SiO_2 and CuO. *Journal of the Taiwan Institute of Chemical Engineers* 2016;58: 19–27.

78. Adio SA, Mehrabi M, Sharifpur M, Meyer JP. Experimental investigation and model development for effective viscosity of MgO–ethylene glycol nanofluids by using dimensional analysis, FCM-ANFIS and GA-PNN techniques. *International Communications in Heat and Mass Transfer* 2016;72: 71–83.

79. Chiam HW, Azmi WH, Usri NA, Mamat R, Adam NM. Thermal conductivity and viscosity of Al_2O_3 nanofluids for different based ratio of water and ethylene glycol mixture. *Experimental Thermal and Fluid Science* 2017;81: 420–429.

80. Bahiraei M, Hangi M. An empirical study to develop temperature-dependent models for thermal conductivity and viscosity of water-Fe_3O_4 magnetic nanofluid. *Materials Chemistry and Physics* 2016;181: 333–343.

81. Aberoumand S, Jafarimoghaddam A. Experimental study on synthesis, stability, thermal conductivity and viscosity of Cu–engine oil nanofluid. *Journal of the Taiwan Institute of Chemical Engineers* 2017;71: 315–322.

82. Elcioglu EB, Guvenc Yazicioglu A, Turgut A, Anagun AS. Experimental study and Taguchi Analysis on alumina-water nanofluid viscosity. *Applied Thermal Engineering* 2018;128: 973–981.

83. Karimi H, Yousefi F, Rahimi MR. Correlation of viscosity in nanofluids using genetic algorithm-neural network (GA-NN). *Heat and Mass Transfer* 2011;47(11): 1417–1425.

84. Mehrabi M, Sharifpur M, Meyer JP. Viscosity of nanofluids based on an artificial intelligence model. *International Communications in Heat and Mass Transfer* 2013;43: 16–21.

85. Atashrouz S, Pazuki G, Alimoradi Y. Estimation of the viscosity of nine nanofluids using a hybrid GMDH-type neural network system. *Fluid Phase Equilibria* 2014;372: 43–48.

86. Meybodi MK, Naseri S, Shokrollahi A, Daryasafar A. Prediction of viscosity of water-based Al_2O_3, TiO_2, SiO_2, and CuO nanofluids using a reliable approach. *Chemometrics and Intelligent Laboratory Systems* 2015;149: 60–69.

87. Zhao N, Wen X, Yang J, Li S, Wang Z. Modeling and prediction of viscosity of water-based nanofluids by radial basis function neural networks. *Powder Technology* 2015;281: 173–183.

88. Atashrouz S, Mozaffarian M, Pazuki G. Viscosity and rheological properties of ethylene glycol+water+Fe3O4 nanofluids at various temperatures: Experimental and thermodynamics modeling. *Korean Journal of Chemical Engineering* 2016;33(9): 2522–2529.

89. Barati-Harooni A, Najafi-Marghmaleki A. An accurate RBF-NN model for estimation of viscosity of nanofluids. *Journal of Molecular Liquids* 2016;224: 580–588.

90. Heidari E, Sobati MA, Movahedirad S. Accurate prediction of nanofluid viscosity using a multilayer perceptron artificial neural network (MLP-ANN). *Chemometrics and Intelligent Laboratory Systems* 2016;155: 73–85.

91. Longo GA, Zilio C, Ortombina L, Zigliotto M. Application of artificial neural network (ANN) for modeling oxide-based nanofluids dynamic viscosity. *International Communications in Heat and Mass Transfer* 2017;83: 8–14.

92. Barati-Harooni A, Najafi-Marghmaleki A, Mohebbi A, Mohammadi AH. On the estimation of viscosities of Newtonian nanofluids. *Journal of Molecular Liquids* 2017;241: 1079–1090.

93. Aminian A. Predicting the effective viscosity of nanofluids for the augmentation of heat transfer in the process industries. *Journal of Molecular Liquids* 2017;229: 300–308.

94. Vakili M, Khosrojerdi S, Aghajannezhad P, Yahyaei M. A hybrid artificial neural network-genetic algorithm modeling approach for viscosity estimation of graphene nanoplatelets nanofluid using experimental data. *International Communications in Heat and Mass Transfer* 2017;82: 40–48.

95. Ansari HR, Zarei MJ, Sabbaghi S, Keshavarz P. A new comprehensive model for relative viscosity of various nanofluids using feed-forward back-propagation MLP neural networks. *International Communications in Heat and Mass Transfer* 2018;91: 158–164.

96. Derakhshanfard F, Mehralizadeh A. Application of artificial neural networks for viscosity of crude oil-based nanofluids containing oxides nanoparticles. *Journal of Petroleum Science and Engineering* 2018;168: 263–272.

97. Karimipour A, Ghasemi S, Darvanjooghi MHK, Abdollahi A. A new correlation for estimating the thermal conductivity and dynamic viscosity of CuO/liquid paraffin nanofluid using neural network method. *International Communications in Heat and Mass Transfer* 2018;92: 90–99.

98. Gholami E, Vaferi B, Ariana MA. Prediction of viscosity of several alumina-based nanofluids using various artificial intelligence paradigms-Comparison with experimental data and empirical correlations. *Powder Technology* 2018;323: 495–506.

99. Hemmati-Sarapardeh A, Varamesh A, Husein MM, Karan K. On the evaluation of the viscosity of nanofluid systems: Modeling and data assessment. *Renewable and Sustainable Energy Reviews* 2018;81: 313–329.

100. Tseng WJ, Chen C-N. Effect of polymeric dispersant on rheological behavior of nickel–terpineol suspensions. *Materials Science and Engineering: A* 2003;347(1–2): 145–153.

101. Song S, Peng C, Gonzalez-Olivares M, Lopez-Valdivieso A, Fort T. Study on hydration layers near nanoscale silica dispersed in aqueous solutions through viscosity measurement. *Journal of Colloid and Interface Science* 2005;287(1): 114–120.

102. Prasher R, Song D, Wang J, Phelan P. Measurements of nanofluid viscosity and its implications for thermal applications. *Applied Physics Letters* 2006;89(13): 133108.

103. Chevalier J, Tillement O, Ayela F. Rheological properties of nanofluids flowing through microchannels. *Applied Physics Letters* 2007;91(23): 233103.

104. Garg J, Poudel B, Chiesa M, Gordon J, Ma J, Wang J et al. Enhanced thermal conductivity and viscosity of copper nanoparticles in ethylene glycol nanofluid. *Journal of Applied Physics* 2008;103(7):074301.

105. Williams W, Buongiorno J, Hu L-W. Experimental investigation of turbulent convective heat transfer and pressure loss of alumina/water and zirconia/water nanoparticle colloids (nanofluids) in horizontal tubes. *Journal of Heat Transfer* 2008;130(4): 042412–042417.

106. Rea U, McKrell T, Hu L-w, Buongiorno J. Laminar convective heat transfer and viscous pressure loss of alumina–water and zirconia–water nanofluids. *International Journal of Heat and Mass Transfer* 2009;52(7–8): 2042–2048.

107. Duangthongsuk W, Wongwises S. Measurement of temperature-dependent thermal conductivity and viscosity of TiO$_2$-water nanofluids. *Experimental Thermal and Fluid Science* 2009;33(4): 706–714.
108. Sahoo BC, Vajjha RS, Ganguli R, Chukwu GA, Das DK. Determination of rheological behavior of aluminum oxide nanofluid and development of new viscosity correlations. *Petroleum Science and Technology* 2009;27(15): 1757–1770.
109. Chandrasekar M, Suresh S, Bose AC. Experimental investigations and theoretical determination of thermal conductivity and viscosity of Al$_2$O$_3$/water nanofluid. *Experimental Thermal and Fluid Science* 2010;34(2): 210–216.
110. Godson L, Raja B, Lal DM, Wongwises S. Experimental investigation on the thermal conductivity and viscosity of silver-deionized water nanofluid. *Experimental Heat Transfer* 2010;23(4): 317–332.
111. Ho C, Liu W, Chang Y, Lin C. Natural convection heat transfer of alumina-water nanofluid in vertical square enclosures: An experimental study. *International Journal of Thermal Sciences* 2010;49(8): 1345–1353.
112. Yu W, Xie H, Li Y, Chen L. Experimental investigation on thermal conductivity and viscosity of aluminum nitride nanofluid. *Particuology* 2011;9(2): 187–191.
113. Naik M, Sundar LS. Investigation into thermophysical properties of glycol based CuO nanofluid for heat transfer applications. *World Academy of Science, Engineering and Technology* 2011;59: 440–446.
114. Sekhar YR, Sharma K. Study of viscosity and specific heat capacity characteristics of water-based Al$_2$O$_3$ nanofluids at low particle concentrations. *Journal of Experimental Nanoscience* 2015;10(2): 86–102.
115. Zadeh AD, Toghraie D. Experimental investigation for developing a new model for the dynamic viscosity of silver/ethylene glycol nanofluid at different temperatures and solid volume fractions. *Journal of Thermal Analysis and Calorimetry* 2018;131(2): 1449–1461.
116. Saeedi AH, Akbari M, Toghraie D. An experimental study on rheological behavior of a nanofluid containing oxide nanoparticle and proposing a new correlation. *Physica E: Low-Dimensional Systems and Nanostructures* 2018;99: 285–293.

Section III

Nanofluids in Thermal Applications

7

Application of Nanofluids in Heat Transfer Enhancement of Refrigeration Systems

Khalil Khanafer and Kambiz Vafai

CONTENTS

7.1 Introduction

Several studies were conducted in the literature related to the application of nanoparticles in refrigeration, heating, ventilation, and air conditioning (HVAC), heat pumps, and other thermal systems to enhance the efficiency and reliability of these systems due to the superior thermophysical characteristics of nanoparticles [1–13]. For example, Khanafer and Vafai [1] presented a critical review on various results of thermophysical properties of nanofluids reported in the literature. They demonstrated in their review that several experimental results for the effective viscosity and thermal conductivity of nanofluids reported by many researchers in the literature were in disagreement.

Recently, this concept of nanofluids has been applied to a new class of refrigerant called "nanorefrigerant" [14–16]. The nanorefrigerant research is divided into two categories. The first one is associated with adding nanoparticles directly to the base refrigerant (i.e., nanorefrigerant), while the second category is referred to the addition of nanoparticles to the lubricant (i.e., nanolubricant), which is eventually circulating along with the refrigerant. The application of nanorefrigerants in refrigeration, HVAC, and other systems will lead eventually to a smaller refrigeration systems and consequently less compressor power consumption [17–23].

Haque et al. [20] analyzed experimentally the characteristics of a domestic refrigerator using different types and sizes of nanoparticles. Al_2O_3 and TiO_2 nanoparticles were added to the polyolester (POE) oil using different volume fractions of nanoparticles (0.05% and 0.1%). Their experimental results indicated that their system consumed 27.73% and 14.19% less energy when adding 0.1% volume fraction of Al_2O_3 and TiO_2 nanoparticles to the POE oil, respectively. Bi et al. [21] investigated experimentally the performance of a domestic refrigerator using mineral oil with TiO_2 nanoparticles mixture as a lubricant instead of polyolester oil in the R134a refrigerator. Their results indicated 26.1% less energy consumption using 0.1% mass fraction of TiO_2 nanoparticles compared to the R134a and POE oil system. Jwo et al. [22] conducted an experimental study to analyze the effect of replacing R134a refrigerant and polyolester lubricant oil with hydrocarbon refrigerant and mineral oil containing Al_2O_3 nanoparticles on the power consumption and coefficient of performance of the refrigeration system. Their results indicated that the use of 60% R134a refrigerant with 0.1% mass fraction of Al_2O_3 reduced the power consumption by 2.4% and increased the coefficient of performance by 4.4% compared with R134a and POE oil lubricant. Moreover, the authors indicated in their study that replacing R134a refrigerant with hydrocarbon refrigerant and Al_2O_3 nanoparticles added to the lubricant oil decreased effectively the power consumption and increased the coefficient of performance of the refrigeration system.

Boiling and two-phase flow phenomena, which are major areas of research, were used in various industrial applications such as refrigeration, HVAC, heat exchanger, energy conversion systems, and heat pumps. The study of boiling heat transfer phenomena is complex in nature and the addition of nanoparticles to the refrigerant will add more complexities. The published studies in the literature on nanorefrigerants reported contradictory results for boiling heat transfer. Therefore, more studies should be conducted in this area because the study of thermophysical properties, boiling heat transfer, and two-phase flow phenomena of nanorefrigerants are still in their early stages.

7.2 Thermophysical and Rheological Properties of Nanorefrigerants

Several studies were conducted in the literature associated with measuring both experimentally and theoretically the thermophysical properties of single-phase nanofluids [1], and less attention was given to the thermophysical properties of the refrigerant-based nanofluids. Jiang et al. [24] measured experimentally the thermal conductivity of carbon nanotube (CNT)-R113 nanorefrigerant and found a significant enhancement in thermal conductivity. Figure 7.1 illustrates the thermal conductivity ratio of CNT-R113 nanorefrigerant (k_{nf}/k_f) compared with pure refrigerant for various volume fractions of CNT nanoparticles of different diameters and aspect ratios. Figure 7.1 clearly shows that the thermal conductivity of CNT-R113 nanorefrigerant increased substantially with an increase in the CNT volume fraction. Further, the experimental results for 1% volume fraction of CNT indicated that the thermal conductivity of CNT-R113 nanorefrigerant increased by 82%, 104%, 43%, and 50% using four types of CNT (1-CNT, 2-CNT, 3-CNT, 4-CNT), respectively.

Rashidi and Nikou [25] developed a model to predict the thermal conductivity of CNT-R113 nanorefrigerant as a function of thermal conductivity of the base fluid, CNTs, and dimensions of carbon nanotubes. Their results indicated that the thermal conductivity of CNT-R113 nanorefrigerant increased nonlinearly with the concentrations of CNTs. Jiang et al. [26] investigated experimentally thermal

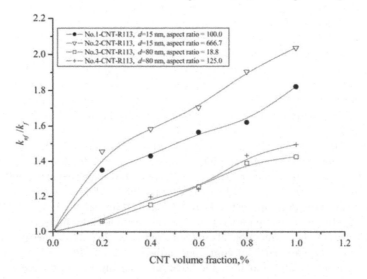

FIGURE 7.1 Effect of varying the volume fraction of CNT on k_{nf}/k_f of four kinds of CNT-R113 nanorefrigerants. (Reprinted from *Int. J. Therm Sci.*, 48, Jiang, W. et al., Measurement and model on thermal conductivities of carbon nanotube nanorefrigerants, 1108–1115, Copyright 2009, with permission from Elsevier.)

conductivity of nanorefrigerant using R113 as a base refrigerant and five nanoparticles including copper, nickel, aluminum, aluminum oxide, and copper oxide. Their results indicated that the thermal conductivity of nanorefrigerant increased sharply with an increase in the volume fraction of nanoparticles (Figure 7.2). Further, the thermal conductivities of nanorefrigerants using five different nanoparticles were close to each other for the same nanoparticles volume fraction. The authors developed a model in that study, and the deviation of the model prediction from the experimental data for the thermal conductivities of nanorefrigerants was estimated −5% to 5% and −11% to 8% for other nanofluids (Figure 7.3).

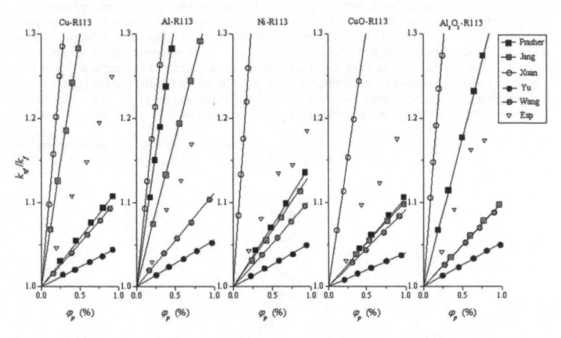

FIGURE 7.2 Experimental and calculated k_{nf}/k_f of nanorefrigerants for various volume fractions of nanoparticles. (With kind permission from Taylor & Francis Group: *HVAC&R Res.*, Experimental and model research on nanorefrigerant thermal conductivity, 15, 651–669, Jiang, W. et al., 2009.)

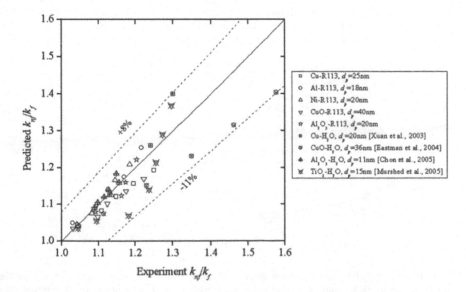

FIGURE 7.3 Experimental k_{nf}/k_f versus predicted k_{nf}/k_f by the new model. (With kind permission from Taylor & Francis Group: *HVAC&R Res.*, Experimental and model research on nanorefrigerant thermal conductivity, 15, 651–669, Jiang, W. et al., 2009.)

Mahbubul et al. [27,28] conducted an experimental study to measure the thermal conductivity of R141b-Al$_2$O$_3$ nanorefrigerant. Their results showed that the thermal conductivity of the nanorefrigerant increased with an increase in the volume fractions of Al$_2$O$_3$ nanoparticles and temperature (Figure 7.4).

Very few studies were conducted in the literature to determine the viscosity of nanorefrigerants. Generally, the effective viscosity of nanofluids is larger than their base fluids. Khanafer and Vafai [1] showed in their review that the effective viscosity of nanofluids depends strongly on the temperature, volume fractions of nanoparticles, and nanoparticle's diameter. Kedzierski [29] conducted an experimental study to measure the kinematic viscosity of polyolester-based aluminum oxide (nanolubricant) using two diameters of Al$_2$O$_3$ nanoparticles (10 nm and 60 nm) under atmospheric pressure and for a wide range of temperature (288 K–318 K) (Figure 7.5). A new model was developed for the kinematic viscosity of the nanolubricant as a function of temperature, mass fraction of nanoparticles, mass fraction of surfactant, and nanoparticle's diameter (Figure 7.6). Figure 7.6 illustrates a comparison of the nanolubricant kinematic viscosity between the experimental results and the predicted results obtained by their model. The author reported that the developed model predicted the experimental results within 15% of the measurements.

Mahbubul et al. [30] measured experimentally the viscosity of TiO$_2$-R123 nanorefrigerant for various volume fractions of TiO$_2$ (0.5%–2%) and different ranges of temperatures from 5°C to 20°C.

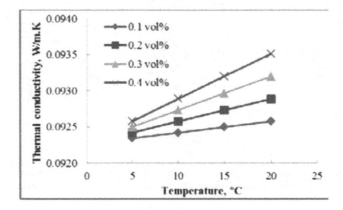

FIGURE 7.4 Effect of varying the volume fraction of Al$_2$O$_3$ nanoparticles and temperature on the thermal conductivity of nanorefrigerant. (Reprinted from *Procedia Eng.*, 56, Mahbubula, I.M. et al., Thermal conductivity, viscosity and density of R141b refrigerant based nanofluid, 310–315, Copyright 2013, with permission from Elsevier.)

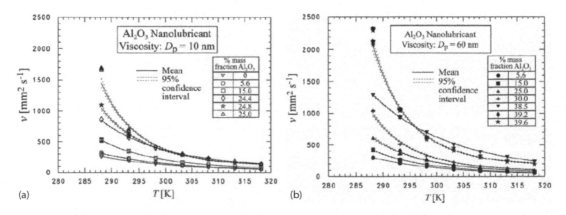

FIGURE 7.5 Effect of varying the temperature and nanoparticle's diameter on the experimental results of the kinematic viscosity of the Al$_2$O$_3$ nanolubricant under atmospheric pressure condition. (a) $D_p = 10$ nm and (b) $D_p = 60$ nm. (Reprinted from *Int. J. Refrig.*, 36, Kedzierski, M.A., Viscosity and density of aluminum oxide nanolubricant, 1333–1340, Copyright 2013, with permission from Elsevier.)

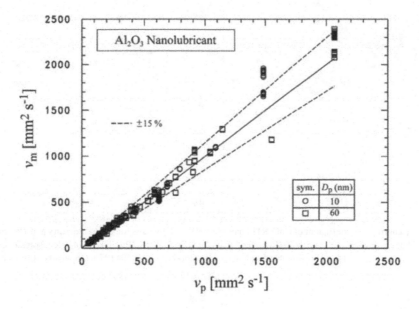

FIGURE 7.6 Comparison of experimental results of the kinematic viscosity of Al_2O_3 nanolubricant to the model predictions. (Reprinted from *Int. J. Refrig.*, 36, Kedzierski, M.A., Viscosity and density of aluminum oxide nanolubricant, 1333–1340, Copyright 2013, with permission from Elsevier.)

Their experimental results revealed that the viscosity of the nanorefrigerant increased with an increase in the volume fraction of nanoparticles and decreased with an increase in the temperature. Since the pumping power and pressure drop depend on the viscosity of the nanorefrigerant, the results in that study showed that the pressure drop increased substantially with an increase in the volume fraction of nanoparticles and vapor quality. The same authors [27] measured the viscosity of Al_2O_3-R141b nanorefrigerant and found that the viscosity increased with an increase in the volume fraction of nanoparticles and decreased with an increase in the temperature.

One can notice from the above citation that the research on the thermal conductivity and viscosity of nanorefrigerants is very limited in the literature and therefore more experimental and theoretical studies should be conducted. The lack of knowledge on accurate prediction of thermal conductivity and viscosity is a challenge to study accurately boiling heat transfer and two-phase phenomena of refrigerant-based nanofluids.

7.3 Boiling Behavior of Refrigerant-Based Nanofluids

The boiling behavior of the refrigerant-based nanofluids, both pool boiling and two-phase flow boiling plays a major role in affecting the performance of the refrigeration, HVAC, and heat pumps systems.

7.3.1 Flow Boiling Heat Transfer in Refrigerant-Based Nanofluids

The addition of nanoparticles into phase change fluids may enhance the boiling heat transfer performance of many thermal systems. Peng et al. [23] investigated experimentally the effect of adding CuO nanoparticles to R113 refrigerant on the flow boiling heat transfer performance of the nanorefrigerant (CuO + R113 mixture) inside a smooth pipe. Their experimental results indicated that the heat transfer coefficient of the nanorefrigerant was larger than the pure refrigerant and the maximum enhancement was 29.7% (Figure 7.7).

Henderson et al. [31] investigated experimentally the effect of adding nanoparticles (SiO_2 and CuO) on the flow boiling of R134a and R134a/polyolester mixtures. The experimental results showed that the addition of CuO nanoparticles to the mixture of R134a refrigerant and polyolester oil increased the

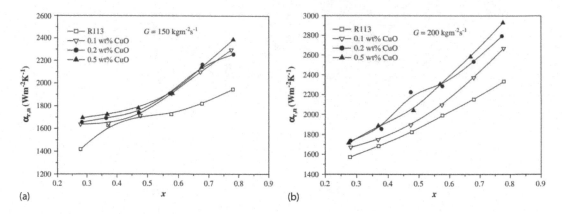

(a)

(b)

FIGURE 7.7 Heat transfer coefficient of CuO/R113 nanofluid (W/m²K) versus local vapor quality at different mass fluxes. (a) G = 150 kg/m²s and (b) G = 200 kg/m²s. (Reprinted from *Int. J. Refrig.*, 32, Peng, H. et al., Heat transfer characteristics of refrigerant-based nanofluid flow boiling inside a horizontal smooth tube, 1259–1270, Copyright 2009, with permission from Elsevier.)

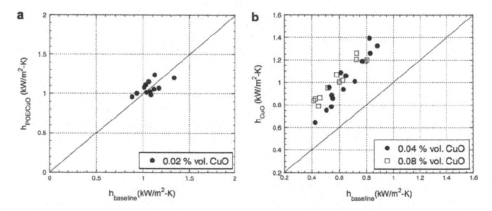

FIGURE 7.8 Effect of varying the volume fraction of CuO nanoparticles on the flow boiling heat transfer coefficient of R134a/POE mixtures. (a) 0.02% vol. CuO and (b) 0.04% vol. & 0.08% vol. CuO. (Reprinted from *Int. J. Heat Mass Transf.*, 53, Henderson, K. et al., Flow boiling heat transfer of R134a-based nanofluids in a horizontal tube, 944–951, Copyright 2010, with permission from Elsevier.)

heat transfer coefficient more than 100% compared with R134a/POE mixture (Figure 7.8). However, the addition SiO_2 nanoparticles to R134a refrigerant decreased the flow boiling heat transfer coefficient of R134a-based nanofluids by 55% compared with pure R134a refrigerant (Figure 7.9). The authors referred this suppression in heat transfer coefficient to the sedimentation of SiO_2 along the walls of the tubing and consequently increasing the thermal resistance to heat transfer.

Bartelt et al. [32] studied experimentally the effect of adding CuO nanoparticles on the flow-boiling heat transfer coefficient of R134a/polyolester mixture in a horizontal tube. The results presented in that study revealed that the addition of 0.5% mass fraction of nanolubricant in the R134a refrigerant had an insignificant effect on the heat transfer coefficient. However, for 1% mass fraction of nanolubricant, 42%–82% enhancement in heat transfer coefficient was observed. Further increase in the mass fraction of nanolubricant to 2%, significant enhancement in the heat transfer coefficient was observed between 50% and 101%. The results of that investigation also revealed that the use of nanoparticles had an insignificant effect on the pressure drop of the system. Sun and Yang [33] analyzed experimentally the performance of four nanorefrigerants flowing in an internal thread copper pipe (Cu-R141b, Al-R141b, Al_2O_3-R141b, and CuO-R141b) of various nanoparticles mass fraction. The results of that investigation illustrated that the heat transfer coefficients of four types of nanorefrigerants enhanced by 17%–25% and the maximum heat transfer coefficient of Cu-R141b nanorefrigerant enhanced by 25% (Figure 7.10).

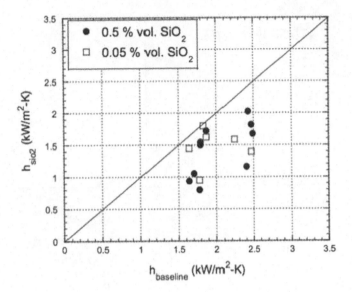

FIGURE 7.9 Effect of varying the volume fraction of SiO₂ nanoparticles on the flow boiling heat transfer coefficient of R134a. (Reprinted from *Int. J. Heat Mass Transf.*, 53, Henderson, K. et al., Flow boiling heat transfer of R134a-based nanofluids in a horizontal tube, 944–951, Copyright 2010, with permission from Elsevier.)

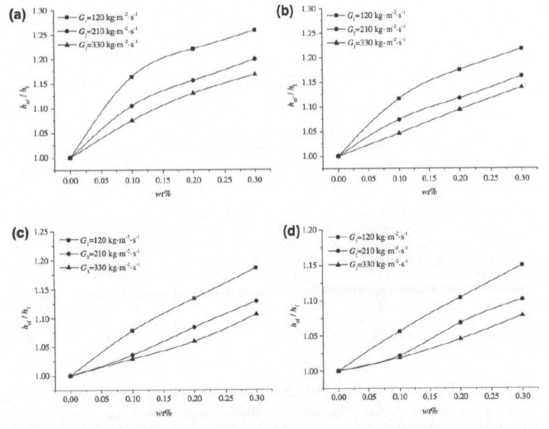

FIGURE 7.10 Effect of varying the mass fraction of nanoparticles on the heat transfer coefficients ratio (k_{nf}/k_f) for the four nanoparticles: (a) Cu, (b) Al, (c) Al₂O₃, and (d) CuO. (Reprinted from *Int. J. Heat Mass Transf.*, 64, Sun, B. and Yang, D., Experimental study on the heat transfer characteristics of nanorefrigerants in an internal thread copper tube, 559–566, Copyright 2013, with permission from Elsevier.)

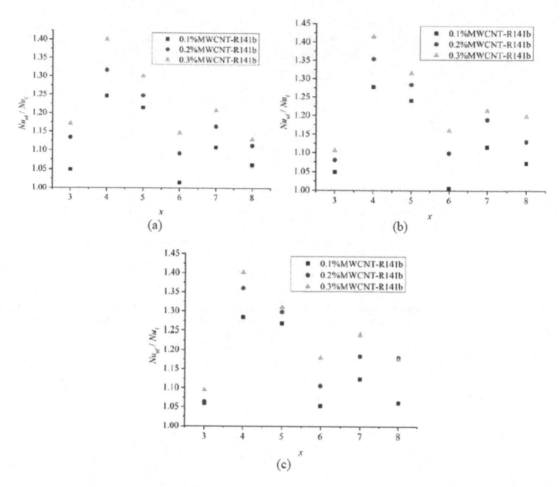

FIGURE 7.11 Effects of vapor quality (*x*) and concentration of nanoparticles on the average Nusselt number for various flow rates (a) $G = 9.4 \times 10^{-3}$ kg/s, (b) $G = 16.5 \times 10^{-3}$ kg/s, and (c) $G = 25.9 \times 10^{-3}$ kg/s. (Reprinted from *Int. J. Refrig.* 56, Yang, D. et al., Experimental study on the heat transfer and flow characteristics of nanorefrigerants inside a corrugated tube, 213–223, Copyright 2015, with permission from Elsevier.)

Yang et al. [34] investigated experimentally flow and heat transfer characteristics of multi-walled carbon nanotubes (MWCNTs)-R141b nanorefrigerant in a corrugated tube at various mass fractions. The results of that investigation illustrated that the addition of 0.3% by weight of MWCNT–R141b nanorefrigerant achieved an optimum heat transfer enhancement compared with pure refrigerants. Moreover, it was found that the average Nusselt number enhanced by 40% (Figure 7.11).

Zhang et al. [35] analyzed experimentally flow boiling heat transfer characteristics and pressure drop of MWCNT–R123-based nanofluids flowing through a horizontal pipe. The results of that study showed that the flow boiling heat transfer coefficient and frictional pressure drop increased with an increase of the concentration of nanoparticles, mass flux, and vapor quality as depicted in Figures 7.12 and 7.13.

7.3.2 Pool Boiling Heat Transfer in Refrigerant-Based Nanofluids

Several experimental studies were conducted in the literature associated with the effect of adding nanoparticles to refrigerants on the nucleate pool boiling heat transfer. Literature reported conflicting experimental results on nucleate pool boiling heat transfer with the addition of nanoparticles. Park and Jung [36,37] investigated experimentally the influence of adding 1% volume fraction of CNTs on nucleate boiling heat transfer in R22 refrigerant and water. Their experimental results indicated that

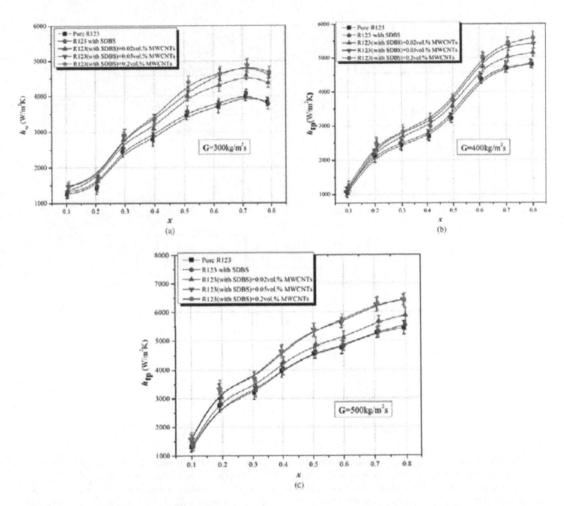

FIGURE 7.12 The effect of varying the volume fraction of MWCNTs and vapor quality on the flow boiling heat transfer coefficients of MWCNTs–R123 nanorefrigerant for different mass fluxes (a) G = 300 kg/m²s, (b) G = 400 kg/m²s, and (c) G = 500 kg/m²s. (With kind permission from Taylor & Francis Group: *Nanosc. Microsc. Therm.*, Characteristics of flow boiling heat transfer and pressure drop of MWCNT–R123 nanorefrigerant: Experimental investigations and correlations, 20, 97–120, Zhang, S.Y. et al., 2016.)

the addition of CNTs increased boiling heat transfer coefficients and maximum enhancement of 28.7% was achieved for heat fluxes below 30 kW/m² (Figure 7.14). The authors indicated in their study that the enhancement in the boiling heat transfer coefficient was decreased due to the development of bubbles in the test section.

Liu and Yang [38] investigated experimentally the effect of adding Au nanoparticles to R141b refrigerant on nucleate pool boiling heat transfer in a horizontal tube. Their results showed that the boiling heat transfer coefficient increased with an increase in the concentration of Au nanoparticles. Their results for pure refrigerant agreed very well with those obtained by the correlation of Cooper [39]. Moreover, the authors showed that the low concentration of Au nanoparticles (0.09%) had insignificant effect on the pool boiling heat transfer coefficient. Further, 1% concentration of Au nanoparticles resulted in boiling heat transfer coefficient more than twice larger than pure refrigerants. The experimental results of that study agreed with those reported by Wen and Ding [40] and in disagreement with the results of both Das et al. [9] and Bang and Chang [41], who noticed a decrease in the boiling heat transfer coefficients when adding nanoparticles. For confirming their experimental results, Liu and Yang [38] repeated the same test four times with intervals of every 5 days and found that pool boiling heat transfer coefficient decreased for each test and finally close to those of pure R141b refrigerant. Liu and Yang [38] conducted a

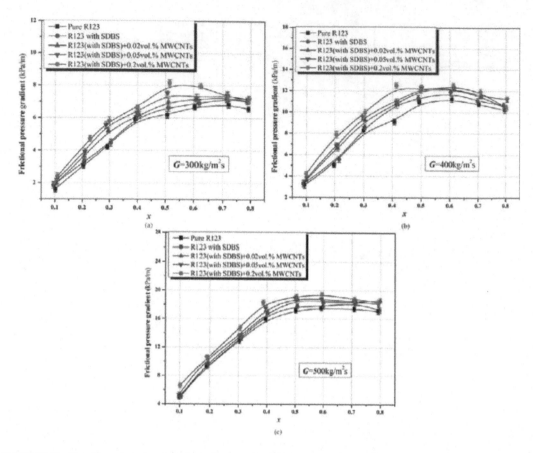

FIGURE 7.13 The effect of varying the volume fraction of MWCNTs and vapor quality on the unit frictional pressure drop of MWCNTs–R123 nanorefrigerant for different mass fluxes (a) G = 300 kg/m²s, (b) G = 400 kg/m²s, and (c) G = 500 kg/m²s. (With kind permission from Taylor & Francis Group: *Nanosc. Microsc. Therm.*, Characteristics of flow boiling heat transfer and pressure drop of MWCNT–R123 nanorefrigerant: Experimental investigations and correlations, 20, 97–120, Zhang, S.Y. et al., 2016.)

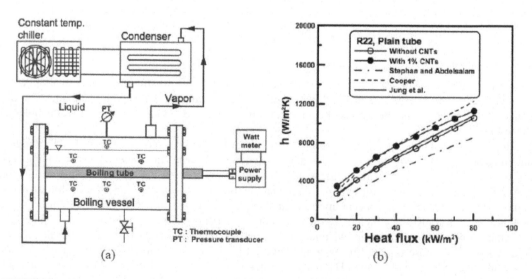

FIGURE 7.14 (a) Schematic diagram of the test facility (b) nucleate boiling heat transfer coefficients with 1.0 vol% CNTs for R22. (Reprinted from *Energ. Buildings*, 39, Park, K.J. and Jung, D., Boiling heat transfer enhancement with carbon nanotubes for refrigerants used in building air-conditioning, 1061–1064, Copyright 2007, with permission from Elsevier.)

scanning probe analysis and attributed the reduction in the boiling heat transfer coefficient to the trapped nanoparticles on the surface and reduced number of active nucleation locations. Further, Liu and Yang [38] reported that the roughness of the tube surface reduced and the aggregation of the nanoparticles size increased after the test which both reduced boiling heat transfer.

Trisaksri and Wongwises [42] analyzed experimentally the effect of adding various volume fractions of TiO_2 (0.01%, 0.03%, and 0.05%) nanoparticles on the nucleate pool boiling heat transfer coefficient of a R141b-TiO_2 nanorefrigerant in a cylindrical copper tube. Their results indicated that the nucleate pool boiling heat transfer coefficient reduced with increase in the volume fraction of TiO_2, particularly at high fluxes. Figure 7.15 shows that a small volume fraction of nanoparticles (0.01%) exhibited insignificant effect on the heat transfer coefficient compared with pure R141b refrigerant. This result is consistent with the results of Liu and Yang [38] for small volume fraction of nanoparticles at 0.09%. Further, increase in the volume fractions of TiO_2 nanoparticles (0.03% and 0.05%) deteriorated the nucleate pool heat transfer coefficient and shifted the curves to the right (Figure 7.15), which contradicted the results of Liu and Yang [38]. Figure 7.16a illustrates that the excess temperatures of R141b-TiO_2 nanorefrigerant were higher than the pure refrigerant for different values of pressure at 0.05% volume fraction. Moreover, Figure 7.16b shows that the pool boiling heat transfer coefficients of R141b-TiO_2 nanorefrigerant were lower than the pure refrigerant at all pressures.

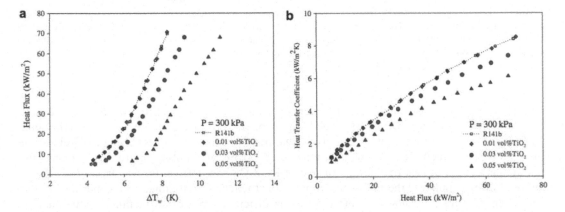

FIGURE 7.15 Effect of varying the volume fraction of TiO_2 nanoparticles on the nucleate pool boiling heat transfer of R141b-TiO_2 nanorefrigerant at 300 kPa. (a) Heat Flux (kW/m²) and (b) Heat Transfer Coefficient (kW/m²K). (Reprinted from *Int. J. Heat Mass Transf.*, 52, Trisaksri, V. and Wongwises, S., Nucleate pool boiling heat transfer of TiO_2–R141b nanofluids, 1582–1588, Copyright 2009, with permission from Elsevier.)

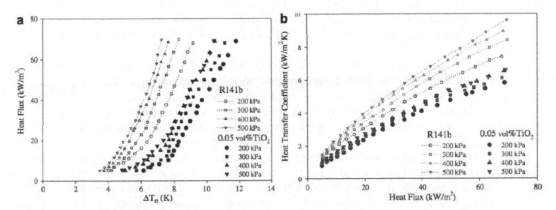

FIGURE 7.16 Variation of pool boiling heat transfer coefficients of R141b-TiO_2 nanorefrigerant with pressure for 0.05% volume fraction of TiO_2 nanoparticles. (a) Heat Flux (kW/m²) and (b) Heat Transfer Coefficient (kW/m²K). (Reprinted from *Int. J. Heat Mass Transf.*, 52, Trisaksri, V. and Wongwises, S., Nucleate pool boiling heat transfer of TiO_2–R141b nanofluids, 1582–1588, Copyright 2009, with permission from Elsevier.)

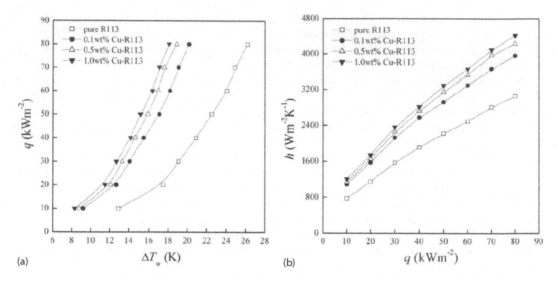

(a)

(b)

FIGURE 7.17   Effect of varying the mass fraction of nanoparticles on nucleate pool boiling heat transfer coefficient and excess temperature of pure R113 and R113-Cu nanofluids. (a) Heat Flux (kW/m²) and (b) Heat Transfer Coefficient (kW/m²K). (Reprinted from *Int. J. Refrig.*, 34, Peng, H. et al., Influences of refrigerant-based nanofluid composition and heating condition on the migration of nanoparticles during pool boiling. Part I: Experimental measurement, 1823–1832, Copyright 2011, with permission from Elsevier.)

Peng et al. [43] investigated experimentally the influence of using three types of surfactants (sodium dodecyl sulfate (SDS), cetyltrimethyl ammonium bromide, (CTAB) and sorbitan mono-oleate (Span-80)) on the nucleate pool boiling heat transfer coefficient of R113-Cu nanorefrigerant for various mass fractions of Cu nanoparticles (0.1%, 0.5%, and 1.0%). Figure 7.17 demonstrates that the nucleate pool heat transfer coefficient of R113-Cu nanorefrigerant was larger than that of pure R113 refrigerant and the maximum enhancement reached 55.4% under the conditions of that experiment. Figure 7.18 shows the effect of using different types of surfactants on the nucleate pool boiling heat transfer coefficient of R113-Cu nanorefrigerant for various mass fractions of Cu nanoparticles. The results presented in that study illustrated that the application of surfactants generally enhanced the nucleate pool boiling heat transfer of R113-Cu nanofluid compared with pure R113 refrigerant. However, the nucleate pool boiling heat transfer coefficient was found to decrease at high surfactant concentrations (Figure 7.18). The authors in that study reported that the ratios of nucleate pool boiling heat transfer coefficient of nanorefrigerant with surfactant to nanorefrigerant without surfactant were in the ranges of 1.12–1.67 for SDS, 0.94–1.39 for CTAB, and 0.85–1.29 for Span-80, respectively.

Kedzierski [44] investigated the effect of adding Al_2O_3 nanoparticles on the pool boiling heat transfer coefficient of R134a/polyolester mixture on a roughed horizontal surface for different mass fractions of the lubricant. The experimental results presented in that investigation showed that the addition of 0.5% mass fraction of nanolubricant (Al_2O_3 nanoparticles + polyolester oil) caused the pool boiling heat transfer coefficient to increase by 400% compared with pure R134a/polyolester mixture for the lowest heat flux (Figure 7.19).

Kedzierski [45] analyzed the effect of adding Al_2O_3 nanoparticles on the pool boiling characteristics of R134a/polyolester mixtures on a Turbo-BII-HP surface. Various volume fractions of Al_2O_3 nanoparticles (1.6%, 2.3%, and 5.1%) in the polyolester lubricant were mixed with R134a refrigerant at two different mass fractions (0.5% and 1%). Figure 7.20 summarizes the effect of adding Al_2O_3 nanoparticles on the pool boiling heat transfer of R134a/polyolester mixture. This figure plots the ratio of heat flux with nanolubricant to the flux with no nanoparticles versus the heat flux with no nanoparticles. The results presented in that investigation indicated that R134a/Al_2O_3 nanolubricant with volume fractions of 1.6%

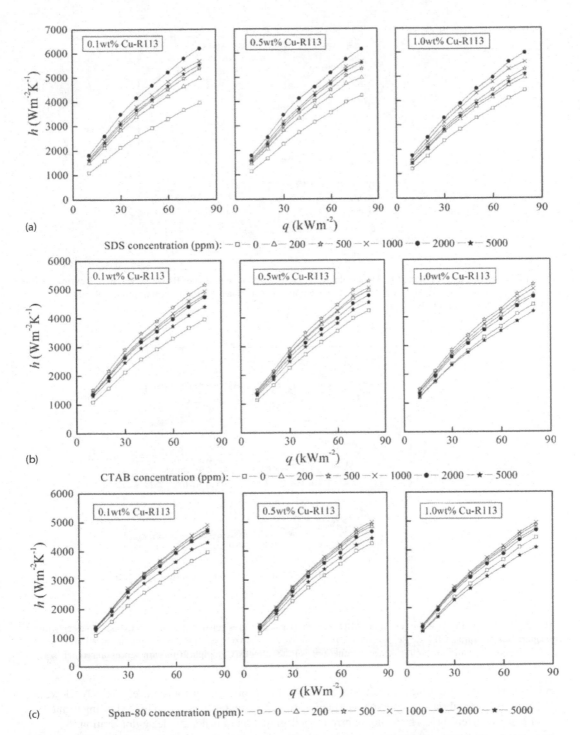

FIGURE 7.18 Effect of varying the type and concentration of surfactants on the nucleate pool boiling heat transfer of R113-Cu nanorefrigerant for different mass fractions of Cu nanoparticles. (a) SDS, (b) CTAB, and (c) Span-80. (Reprinted from *Int. J. Refrig.*, 34, Peng, H. et al., Influences of refrigerant-based nanofluid composition and heating condition on the migration of nanoparticles during pool boiling. Part I: Experimental measurement, 1823–1832, Copyright 2011, with permission from Elsevier.)

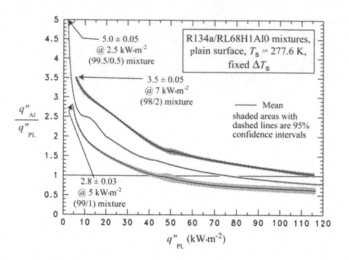

FIGURE 7.19   Boiling heat flux of R134a/nanolubricant mixtures relative to that of R134a/polyolester. (Reprinted from *Int. J. Refrig.*, 34, Kedzierski, M.A., Effect of Al$_2$O$_3$ nanolubricant on R134a pool boiling heat transfer, 498–508, Copyright 2011, with permission from Elsevier.)

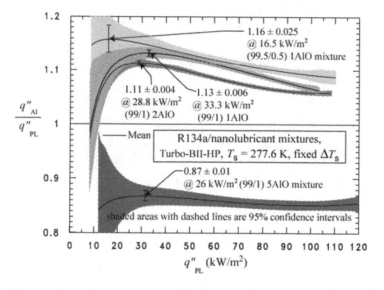

FIGURE 7.20   Boiling heat flux of R134a/Al$_2$O$_3$ nanolubricant mixtures relative to that of R134a/polyolester without nanoparticles for Turbo-BII-HP. (Reprinted from *Int. J. Refrig.*, 49, Kedzierski, M.A., Effect of concentration on R134a/Al$_2$O$_3$ nanolubricant mixture boiling on a re-entrant cavity surface, 36–48, Copyright 2015, with permission from Elsevier.)

and 2.3% showed an enhancement of 13% and 10% in the boiling heat transfer, respectively. However, R134a/Al$_2$O$_3$ nanolubricant with a volume fraction of 5.1% exhibited a reduction in the boiling heat transfer of approximately 14%. The author referred this degradation to nanoparticle agglomeration.

Diao et al. [46] analyzed experimentally pool boiling heat transfer and critical heat flux of Cu-R141b-sodium dodecyl benzene sulfonate (SDBS) nanorefrigerant at different volume fractions (0.008%, 0.015%, and 0.05%) on a flat surface under ambient pressure. SDBS was utilized in that study as a surfactant to stabilize the suspensions of nanoparticles. Their experimental results revealed that the dispersion of nanoparticles enhanced the pool boiling heat transfer compared with pure R141b refrigerant (Figure 7.21). Furthermore, their results also showed that the critical heat flux of the nanorefrigerant was lower than that of pure R141b refrigerant.

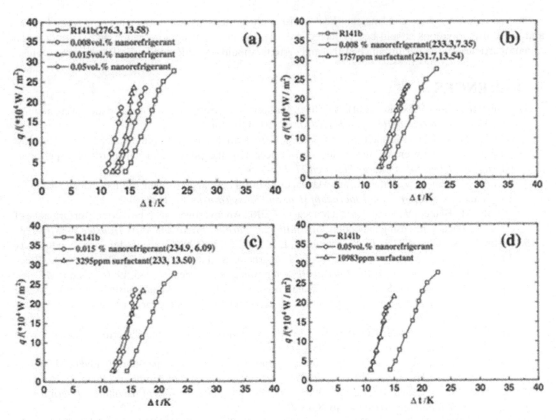

FIGURE 7.21 Boiling heat transfer of Cu-R141b nanorefrigerant with sodium dodecyl benzene sulfonate. (a) Cu-R141b, (b) Cu-R141b-SDBS (0.008%), (c) Cu-R141b-SDBS (0.015%), and (d) Cu-R141b-SDBS (0.015%). (Reprinted from *Int. J. Heat Mass Transf.*, 89, Diao, Y.H. et al., Experimental investigation on the pool boiling characteristics and critical heat flux of Cu-R141b nanorefrigerant under atmospheric pressure, 110–115, Copyright 2015, with permission from Elsevier.)

7.4 Concluding Remarks and Future Studies

Nanorefrigerants have the potential to increase the heat transfer performance, lower the power consumption, and increase the coefficient of performance (COP) of many thermal systems such as refrigeration and HVAC systems due to the superior thermophysical properties of nanoparticles. For example, Mehbubul et al. [47] found that the use of Al_2O_3/R134a nanorefrigerant increased COP by 15%. Sabareesh et al. [48] reported 17% enhancement in the COP of a vapor compression system using TiO_2 nanoparticles in mineral oil with R12 base refrigerant. The cited studies in this review indicated that the thermal conductivity and viscosity of nanorefrigerants increased with an increase in the volume fraction of nanoparticles. Moreover, the cited studies in the literature also indicated that the thermal conductivity of nanorefrigerants increased with increasing the operating temperature, while the viscosity decreased with increasing the temperature. This review illustrated that the flow boiling heat transfer characteristics of nanorefrigerants increased when adding nanoparticles. However, conflicting experimental results were reported in the literature associated with the effect of adding nanoparticles on the nucleate pool boiling heat transfer of nanorefrigerants.

There are many challenges and issues that need to be resolved before using nanorefrigerants in various industrial applications. Studies on the thermophysical properties of nanorefrigerants such as thermal conductivity, viscosity, specific heat, density, surface tension, etc. are limited and therefore, more experimental and theoretical studies should be conducted for various pertinent parameters to accurately predict the effect of nanoparticles on the nucleate pool boiling heat transfer. Thus, the mechanisms of increasing or decreasing nucleate pool boiling heat transfer should be further discussed to resolve the conflicts between various studies in the literature.

Studies regarding flow condensation heat transfer phenomena in nanorefrigerants are very limited and more investigations should be conducted. Finally, the effect of agglomeration and sedimentation of nanoparticles on the performance of thermal systems should be evaluated in detail in future studies.

REFERENCES

1. Khanafer, K. and Vafai, K. (2001). A critical synthesis of thermophysical characteristics of nanofluids. *International Journal of Heat and Mass Transfer* 54, 4410–4428.
2. Ali, A., Vafai, K., and Khaled, A.-R.A. (2003). Comparative study between parallel and counter flow configurations between air and falling film desiccant in the presence of nanoparticle suspensions. *International Journal of Energy Research* 27, 725–745.
3. Shafahi, M., Bianco, V., Vafai, K., and Manca, O. (2010). Thermal performance of flat-shaped heat pipes using nanofluids. *International Journal of Heat and Mass Transfer* 53, 1438–1445.
4. Shafahi, M., Bianco, V., Vafai, K., and Manca, O. (2010). An investigation of the thermal performance of cylindrical heat pipes using nanofluids. *International Journal of Heat and Mass Transfer* 53, 376–383.
5. Eastman, J.A., Choi, S.U.S., Li, S., Thompson, L.J., and Lee, S. (1996). Enhanced thermal conductivity through the development of nanofluids. In: Komarneni, S., Parker, J.C., Wollenberger, H.J. (Eds.), *Nanophase and Nanocomposite Materials II: Symposium*, held December 2–5, 1996, Boston, MA.
6. Eastman, J.A., Choi, S.U.S., Li, S., Yu, W., and Thompson, L.J. (2001). Anomalously increased effective thermal conductivities of ethylene glycol-based nanofluids containing copper nanoparticles. *Applied Physics Letters* 8, 718–720.
7. Jang, S.P. and Choi, S.U.S. (2004). Role of Brownian motion in the enhanced thermal conductivity of nanofluids. *Applied Physics Letters* 84, 4316–4318.
8. Lee, S. and Choi, S.U.S. (1996). Application of metallic nanoparticle suspensions in advanced cooling systems. In: *1996 International Mechanical Engineering Congress and Exhibition*, Atlanta, GA.
9. Das, S.K., Putra, N., and. Roetzel, W. (2003). Pool boiling characteristics of nanofluids. *International Journal of Heat and Mass Transfer* 46, 851–862.
10. Das, S.K., Putra, N., Thiesen, P., and Roetzel, W. (2003). Temperature dependence of thermal conductivity enhancement for nanofluids. *Journal of Heat Transfer* 125, 567–574.
11. Putra, N., Roetzel, W., and Das, S.K. (2003). Natural convection of nanofluids. *Heat and Mass Transfer* 39, 775–784.
12. Wang, B.X., Zhou, L.P., and Peng, X.F. (2003). A fractal model for predicting the effective thermal conductivity of liquid with suspension of nanoparticles. *International Journal of Heat and Mass Transfer* 46, 2665–2672.
13. Das, S.K., Putra, N., and Roetzel, W. (2003). Pool boiling of nanofluids on horizontal narrow tubes. *International Journal of Multiphase Flow* 29, 1237–1247.
14. Nair, V., Tailor, P.R., and Parekh, A.D. (2016). Nanorefrigerants: A comprehensive review on its past, present and future. *International Journal of Refrigeration* 67, 290–307.
15. Cheng, L. and Liu, L. (2013). Boiling and two-phase flow phenomena of refrigerant-based nanofluids: Fundamentals, applications and challenges. *International Journal of Refrigeration* 36, 421–446.
16. Sanukrishna, S.S., Murukanb, M., and Jose, P. (2018). An overview of experimental studies on nanorefrigerants: Recent research, development and applications. *International Journal of Refrigeration* 88, 552–577.
17. Saidur, R, Kazi, S.N., Hossain, M.S., Rahman, M.M., and Mohammed, H.A. (2011). A review on the performance of nanoparticles suspended with refrigerants and lubricating oils in refrigeration systems. *Renewable and Sustainable Energy Reviews* 15, 310–323.
18. Yusof, T.M., Arshad, A.M, Suziyana, M.D., Chui, L.G., and Basrawi, M.F. (2015). Experimental study of a domestic refrigerator with POE-Al$_2$O$_3$ nanolubricant. *International Journal of Automotive and Mechanical Engineering* 11, 2243–2252.
19. Fadhillahanafi, N.M., Leong, K.Y., and Risby, M.S. (2018). Stability and thermal conductivity characteristics of carbon nanotube based nanofluids. *International Journal of Automotive and Mechanical Engineering* 8, 1376–1384.
20. Haque, M.E., Bakar, R.A., Kadirgama, K., Noor, M.M., and Shakaib, M. (2016). Performance of a domestic refrigerator using nanoparticles-based polyolester oil lubricant. *Journal of Mechanical Engineering and Sciences* 10, 1778–1791.

21. Bi, S., Shi, L., and Zhang, L. (2008). Application of nanoparticles in domestic refrigerators. *Applied Thermal Engineering* 28, 1834–1843.
22. Jwo, C-S., Jeng, L-Y., Teng, T-P., and Chang, H. (2009). Effects of nanolubricant on performance of hydrocarbon refrigerant system. *Journal of Vacuum Science and Technology B: Microelectronics and Nanometer Structures* 27, 1473–1477.
23. Peng, H., Ding, G., Jiang, W., Hu, H., and Gao, Y. (2009). Heat transfer characteristics of refrigerant-based nanofluid flow boiling inside a horizontal smooth tube. *International Journal of Refrigeration* 32, 1259–1270.
24. Jiang, W., Ding, G., and Peng, H. (2009). Measurement and model on thermal conductivities of carbon nanotube nanorefrigerants. *International Journal of Thermal Sciences* 48, 1108–1115.
25. Rashidi, H. and Nikou, M. (2012). Modeling of thermal conductivity of carbon nanotubes refrigerants fluids. *Journal of American Science* 8, 1–5.
26. Jiang, W., Ding, G., Peng, H., Gao, Y., and Wang, K. (2009). Experimental and model research on nanorefrigerant thermal conductivity. *HVAC&R Research* 15, 651–669.
27. Mahbubula, I.M., Saidura, R., and Amalina, M.A. (2013). Thermal conductivity, viscosity and density of R141b refrigerant based nanofluid. *Procedia Engineering* 56, 310–315.
28. Mahbubula, I.M., Saidura, R., and Amalina, M.A. (2013). Influence of particle concentration and temperature on thermal conductivity and viscosity of Al_2O_3/R141b nanorefrigerant. *International Communications in Heat and Mass Transfer* 43, 100–104.
29. Kedzierski, M.A. (2013). Viscosity and density of aluminum oxide nanolubricant. *International Journal of Refrigeration* 36, 1333–1340.
30. Mahbubul, I.M., Saidur, R., and Amalina, M. (2012). Investigation of viscosity of R123-TIO_2 nanorefrigerant. *International Journal of Mechanical and Materials Engineering* 7, 146–151.
31. Henderson, K., Park, Y.G., Liu, L., and Jacobi, A.M. (2010). Flow boiling heat transfer of R-134a-based nanofluids in a horizontal tube. *International Journal of Heat and Mass Transfer* 53, 944–951.
32. Bartelt, K., Park, Y., Liu, L., and Jacobi, A. (2008). Flow boiling of R-134a/POE/CuO nanofluids in a horizontal tube. In: *2008 International Refrigeration and Air Conditioning Conference at Purdue*, July 14–17, 2008.
33. Sun, B. and Yang, D. (2013). Experimental study on the heat transfer characteristics of nanorefrigerants in an internal thread copper tube. *International Journal of Heat and Mass Transfer* 64, 559–566.
34. Yang, D., Sun, B., Li, H., and Fan, X. (2015). Experimental study on the heat transfer and flow characteristics of nanorefrigerants inside a corrugated tube. *International Journal of Refrigeration* 56, 213–223.
35. Zhang, S.Y., Ge, Z., Wang, H.T., and Wang, H. (2016). Characteristics of flow boiling heat transfer and pressure drop of MWCNT–R123 nanorefrigerant: Experimental investigations and correlations. *Nanoscale and Microscale Thermophysical Engineering* 20, 97–120.
36. Park, K.J., and Jung, D. (2007). Boiling heat transfer enhancement with carbon nanotubes for refrigerants used in building air-conditioning. *Energy and Buildings* 39, 1061–1064.
37. Park, K.J. and Jung, D. (2007). Enhancement of nucleate boiling heat transfer using carbon nanotubes. *International Journal of Heat and Mass Transfer* 50, 4499–4502.
38. Liu, D.W. and Yang, C.-Y. (2007). Effect of nano-particles on pool boiling heat transfer of refrigerant 141b. In: *Proceedings of the Fifth International Conference on Nanochannels, Microchannels and Minichannels*, pp. 789–793.
39. Cooper, M.G. (1984). Saturation nucleate pool boiling: A simple correlation. *International Chemical Engineering Symposium Series* 86, 785–792.
40. Wen, D. and Ding, Y. (2005). Experimental investigation into the pool boiling heat transfer of aqueous based γ-alumina nanofluids. *Journal of Nanoparticle Research* 7, 265–274.
41. Bang, I.C. and Chang, S.H. (2005). Boiling heat transfer performance and phenomena of Al_2O_3-water nano-fluids from a plain surface in a pool. *International Journal of Heat and Mass Transfer* 48, 2407–2419.
42. Trisaksri, V. and Wongwises, S. (2009). Nucleate pool boiling heat transfer of TiO_2–R141b nanofluids. *International Journal of Heat and Mass Transfer* 52, 1582–1588.
43. Peng, H., Ding, G., and Hu, H. (2011). Influences of refrigerant-based nanofluid composition and heating condition on the migration of nanoparticles during pool boiling. Part I: Experimental measurement. *International Journal of Refrigeration* 34, 1823–1832.

44. Kedzierski, M.A. (2011). Effect of Al$_2$O$_3$ nanolubricant on R134a pool boiling heat transfer. *International Journal of Refrigeration* 34, 498–508.
45. Kedzierski, M.A. (2015). Effect of concentration on R134a/Al$_2$O$_3$ nanolubricant mixture boiling on a reentrant cavity surface. *International Journal of Refrigeration* 49, 36–48.
46. Diao, Y.H., Li, C.Z., Zhao, Y.H., Liu, Y., and Wang, S. (2015). Experimental investigation on the pool boiling characteristics and critical heat flux of Cu-R141b nanorefrigerant under atmospheric pressure. *International Journal of Heat and Mass Transfer* 89, 110–115.
47. Mahbubul, I.M., Saadah, A., Saidur, R., Khairul, M.A., and Kamyar, A. (2015). Thermal performance analysis of Al$_2$O$_3$/R-134a nanorefrigerant. *International Journal of Heat and Mass Transfer* 85, 1034–1040.
48. Sabareesh, R.K., Gobinath, N., Sajith, V., Das, S., and Sobhan, C.B. (2012). Application of TiO$_2$ nanoparticles as a lubricant-additive for vapor compression refrigeration systems-An experimental investigation. *International Journal of Refrigeration* 35, 1989–1996.

8

Graphene-Based Hybrid Nanofluids and Its Application in Heat Exchangers

Hooman Yarmand, Nurin Wahidah Binti Mohd Zulkifli, Mahidzal Dahari, and S.N. Kazi

CONTENTS

8.1 Introduction

Engine oil, ethylene glycol, and water are most common medium working fluids for transfer of heat in a lot of thermal industrial set-up such as cooling devices, solar collectors, and heat exchangers. Minor upgrading in efficiency of heat exchanger apparatus could lead to enormous saving in costs (initial and operational) as a large number of equipment in operation along with rapid growth of industries in the last decade. One way to reach this goal is to enhance thermal conductivity of working fluids, which transfer the heat. Since most of liquids do not have high thermal conductivity, many researchers are trying to add small solid particles in conventional fluids to enhance thermal conductivity property. Researchers earlier, who faced many obstacles such as, increasing in pressure drop, sedimentation of particles and erosion of equipment, tried dispersion of micrometre or millimetre sizes particles in working fluid. For first time in 1995, Choi and his co-worker have established a new type of fluids with suspension of nanomaterial, they called it 'nanofluid'. Later exploration confirms that nanoparticles have capability to suspend in the base fluid and increase thermal conductivity in various thermal industrial applications [1].

In the past two decades, scientists observed many varieties of nanomaterials to make nanofluids, some results were good and some were unsatisfactory. Aluminium oxide, copper oxide, and titanium dioxide have been examined more than other varieties of metal oxide nanomaterials [2], and recently materials related to carbon such as, graphene oxide, carbon nanotube, and graphene nanoplatelet were experimentally examined by researchers [3–6]. Nanofluids are interesting topic for investigators since the nanomaterials have unique chemical and physical performance, particularly in terms of thermal conductivity development of heat exchanging fluids. Experiments have shown that adding nanoparticles has not only improve thermal conductivity, but also enlarges convective heat transfer ability than that of traditional heat exchanging fluids.

In the last decade, important research has been directed towards the use of carbon-based nanomaterials such as, single-wall carbon nanotube, multi-wall carbon nanotube, graphene oxide, and graphene platelet to prepare nanofluids [7–10]. New research points out those graphene nanofluids could deliver higher thermal conductivity enrichment in comparison to other examined nanofluids. Graphene particles show better thermal conductivity and holds good mechanical strength, high electrical conductivity, etc. Supreme thermophysical properties of graphene has made it as one of the most promising nanofluids [11]. Besides, synthesis of graphene is not complicated, and it costs less than other nanofluids. The slight difference of thermal properties of graphene has been reported due to dissimilar methods used to manufacture graphene such as, oxide layer, exfoliation, deposition with chemical vapour and mechanical cleavage, etc. [11–14]. Most of research has confirmed that the thermal and heat transfer performances of one layer graphene are higher than carbon nanotube. Two-dimensional honeycomb lattice graphene with more than ten layers are named graphene nanoplatelets (GNPs), the GNPs not only provide properties advantage over single layer graphene such as, good mechanical strengths, high electrical-thermal conductivity, and excellent surface area, but also have characteristics advantage over carbon with highly ordered graphite, acceptable stability, with enough sources which are not expensive. Homogeneous dispersion of graphene with worthy stability is one of the most important issues that must be resolved. So by using proper ultrasonic, functionalisation method (amino function and acid treat), and adding surfactant could be able to make steady dispersed graphene nanofluids [15,16].

Majority of earlier examination regarding thermo-fluid ability of nanofluid was done on single-phase nanoparticles, based on them, graphene-based nanofluids had the heat transfer performance. Preparation of nanofluid based on nanocomposite is very innovative and exciting topic for scientist. Al_2O_3-Cu hybrid nanofluid has been experimentally examined by Suresh et al. [17]. They stated that about 14% improvement in Nusselt number for laminar flow was reached in comparison with water. Sunder et al. [18] blended multi-walled carbon nanotube (MWCNT)-Fe_3O_4 nanocomposite and water, it's hybrid nanofluid had shown 31% improvement in Nusselt number at 0.3% volume concentration at 22000 Reynolds number. Tessy et al. [19] synthesised and prepared CuO-hydrogen exfoliated graphene (HEG) hybrid nanofluid and found 28% enhancement in thermal conductivity for 0.05% volume concentration of functionalised graphene without any surfactant.

Munkhbayar et al. [20] prepared Ag/MWNT water-based hybrid nanofluid. MWNT single nanofluids were filled into an exploding glass, which was afterward fitted in the main section of the pulsed-wire evaporation apparatus. After that, silver nanoparticles were synthesised by the pulsed-wire evaporation technique and made direct contact with the base fluid inside the chamber wall. A water-based Ag/ MWCNT nanofluid without any surface stain was finally achieved.

Abbasi et al. [21] synthesised Al_2O_3/MWNT hybrid nanofluid. Alumina nano powders were dissolved in ethanol, after that the functionalised MWNT was added to the suspension and kept in an ultrasonic bath. Ammonia was added to the suspension dropwise, and then the suspension was placed in a Teflon-lined stainless steel autoclave chamber, after which solvothermal synthesis was completed. Finally, gum arabic was added to the deionised water and the solution was put in bath ultrasonic, with the hybrid nanocomposite added to the solution and dispersed with bath ultrasonication.

Baby and Ramaprabhu and Baby and Sundara [22,23] prepared MWNT-HEG nanocomposite by catalytic chemical vapor deposition (CCVD). Most of functional groups of graphene removed during the exfoliation process and the nanocomposite became hydrophobic. Therefore, the hybrid nanocomposite were functionalised with H2SO4 and HNO3 acid medium. A specified amount of silver nitrate solution was added to the above mentioned solution with stirring. After some time, nanofluids were synthesised by dispersing a specific quantity of hybrid nanoparticles in deionised water with the assistance of ultrasonication.

Suresh et al. [24,25] prepared alumina-copper hybrid nanoparticles by a thermophysical synthesis technique. They prepared water suspension of soluble nitrates of aluminum and copper. Then they spray dried these solutions to achieve the precursor powder, and after that the sample was heated in air atmosphere to get the powder mixture form of stable Al_2O_3 and copper oxide. The mixture powder was then put in an alumina boat and placed in a horizontal alumina tube of the furnace. Alumina was kept in its original forms without any change, but CuO was reduced in hydrogen and changed to metallic copper. Then the Al_2O_3-Cu hybrid nanocomposite was ball milled to get a homogeneous powder. Finally, the prepared sample was dispersed in water by assistance of sodium lauryl sulfate (SLS) as a surfactant and by ultrasonication.

Han et al. [26] made a CNT/sphere hybrid nanofluid by using two-step method. They prepared the spherical nanoparticles with spray pyrolysis, then grew CNT through catalytic procedure. Finally, hybrid nanocomposite has been dispersed in oil. Botha et al. [27] also prepared a hybrid oil-based nanofluid, they prepared a silver-silica nanocomposite with one-step method.

In this book chapter, preparation, thermophysical property, and heat transfer performance of the functionalised graphene and two various kinds of its hybrid nanofluids have been discussed and compared. Turbulent carbon-based hybrid nanofluids under constant heat flux boundary condition in a schematic heat exchanging apparatus are discussed in the current chapter.

8.2 Material Synthesis

GNPs with purity ~99.5%, maximum particle diameter of 2 μm and specific surface area 500 m^2/g were purchased from, XG Sciences, Lansing, MI, United States. The chemicals such as silver nitrate (AgNO$_3$), potassium tetrachloroplatinate II (K$_2$PtCl$_4$), sodium borohydride (NaBH$_4$), sodium hydroxide (NaOH), nitric acid (HNO$_3$), and sulphuric acid (H$_2$SO$_4$) were purchased from Sigma-Aldrich Co., Selangor, Malaysia.

Graphene nanoplatelet is naturally hydrophobic, and it cannot be dispersed in any solvent which is polar, like distilled water. Functionalisation by acid treatment is a suitable way to make GNPs hydrophilic. This functionalisation process helps to introduce functional groups such as carboxyl and hydroxyl groups on the surface of GNPs. Acid treatment process was performed by dispersing GNPs in a solution of HNO$_3$ and H$_2$SO$_4$ at 1:3 ratio (strong acid medium), for 3 hours under bath ultrasonication. After 3 hours, GNPs were washed several times with deionized (DI) water, and then the prepared rich sample was used in the next step to make nanofluids at different concentrations with the addition of the specific amounts of distilled water. The resulting nanofluids were stable and no sedimentation of particles were found for a long time. Figure 8.1 shows the schematic of functionalisation procedure and nanofluid preparation.

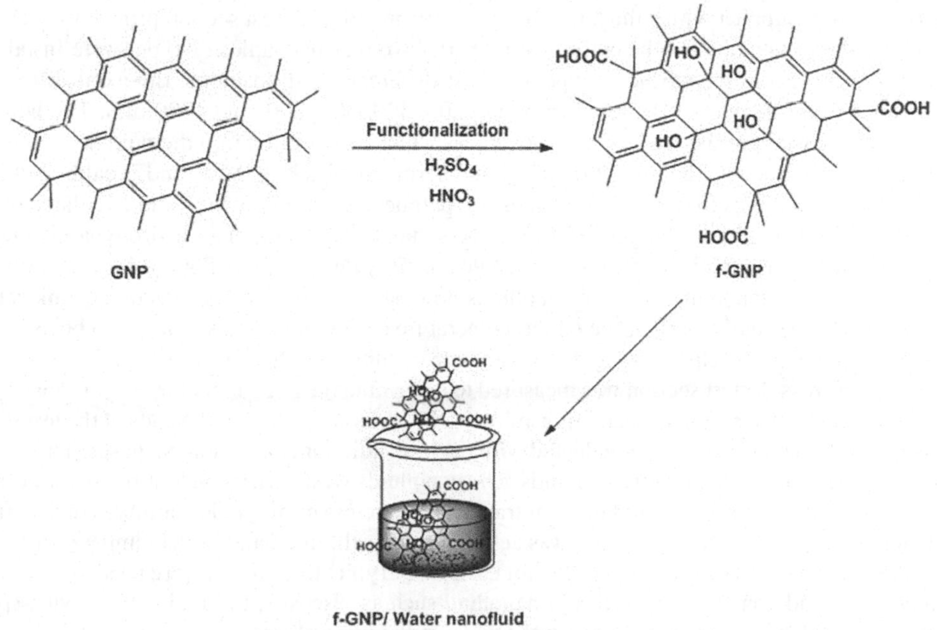

FIGURE 8.1 Schematic of functionalisation process and making of nanofluid. (Reprinted from *Energy Convers. Manag.*, Yarmand, H. et al., Experimental investigation of thermophysical properties, convective heat transfer and pressure drop of functionalized graphene nanoplatelets aqueous nanofluid in a square heated pipe, 114, 38–49. Copyright 2016, with permission from Elsevier (license number 4422350279402).)

Then the functionalised GNPs were decorated with platinum (*Pt*) and silver (*Ag*) by a chemical reaction method. The brief procedure of synthesis is stated for reference. The solution of ammonia-silver was prepared by adding dropwise ammonia (1 wt %) to 0.01 L silver nitrate solution (0.05 M) until fully reacted and silver colour disappeared. The Ag $(NH_3)_2OH$ solution (0.04 M) was mixed with 120 mL functionalised GNP (1 mg/mL) solution, at a weight ratio of 1:6. The irradiation of final solution was done under vigorous stirring for 4 hours. After that, GNP-Ag nanocomposites were collected after centrifuge at 11000 rpm for 40 min. The obtained composite was washed well with distilled water several times to remove reactants. The prepared rich sample was used in the next step to make nanofluids at different concentrations by adding specific amounts of distilled water. The resulting nanofluids were stable and no sedimentation of particles was found for up to 60 days. For preparation of GNP-Pt nanocomposite, 30 mg of functionalised GNP was dispersed into 10 mL of distilled water. This process was continued by the addition of 0.035 M K_2PtCl_4 to the dispersed functionalised GNP suspension with continuous stirring for 2 hours at room temperature, and then 2 mL of sodium borohydride (0.1 M) was added to the solution dropwisely at 60°C. The irradiation of final solution was done under vigorous stirring for 4 hours. Then GNP-Pt nanocomposites were collected after centrifuge at 11000 rpm for 40 minutes. The obtained composite was washed well with distilled water several times to remove reactants. The prepared rich sample was used in the next step to prepare nanofluids at different concentrations by adding specific amounts of distilled water. The resulting nanofluids were stable and the sedimentation of GNP-Pt hybrid nanofluid was less than 5.7% after 22 days. Finally, specific amount of functionalised GNPs and their hybrid dispersed in the distilled water with assistant of proper ultrasonication.

8.3 Experimental Set-Up

Figure 8.2 shows the schematic and photograph of the experimental set-up, which includes the flow loop, heater, chiller (for cooling), and a unit for measurement and control. The flow loop contains a pump, a tank, a test section, flow meter system, chiller, and a storage tank. A square stainless steel pipe with 1.4 m length, 10 mm inner width, and 12.8 mm outer width was installed as a test section. A direct current (DC) power supplier with a thick isolator layer surrounding the test section provides the constant heat flux boundary condition for the entire testing part. Five thermocouples (T-type) were installed on the test section to record the surface temperatures at different axial positions, the axial distances of thermocouples from the inlet of the test section are 200, 400, 600, 800, and 1000 mm. The bulk temperatures of the flow are measured by 2 thermocouples which are inserted into the fluid flow at the inlet and outlet of the test section, the resolution of all the thermocouples is ± 0.2°C and is calibrated before installation. The nanofluids from the storage tank are pumped into the flow loop and circulated through the test section and finally discharged back into the same storage tank. The hydrodynamic entrance length, $l = 140d$ was maintained to ensure that the flow is fully developed. A flow meter was installed at the discharge of the pump to measure the nanofluids flow rate. The stainless steel storage tank capacity was 14 L. With the aid of the chiller, the outlet temperature of nanofluids was reduced to become equal to the inlet temperature to help reaching the steady state condition in shorter time.

Pressure drop cross the test section was measured to determine the friction factor of hybrid nanofluids. For this purpose, the differential pressure transmitter (DPT) was connected to the ends of the test section. First sets of experimental runs were conducted with the base fluid for calibration purposes. Then a series of tests were performed for different nanofluids. The nanofluids with various weight fractions were used for investigating the friction factor and the heat transfer parameters for Reynolds numbers in the range of 5000–17500. When steady state condition was achieved, the wall, inlet, and outlet temperatures as well as mass flow rate of nanofluids were recorded for each case. Typically, 3 hours were needed for reaching the steady state condition. The properties of nanofluid such as viscosity, thermal conductivity, specific heat capacity, and density were considered at the mean bulk temperature.

The most important property of nanofluid which must be measured is thermal conductivity. Thermal conductivities of nanofluids were measured by the KD-2 pro device (Decagon, United States), where KS-1 probe sensors were used having 6 cm and 1.3 mm length and diameter, respectively. The accuracy of the measured thermal conductivity is 5%. To ensure the equilibrium of nanofluids,

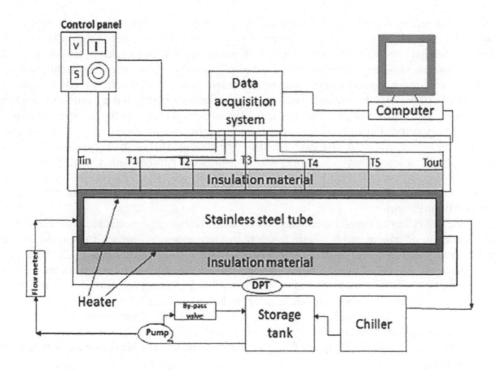

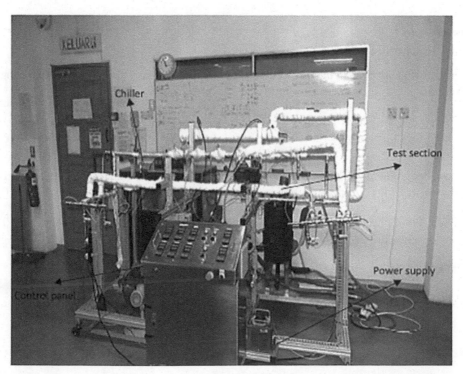

FIGURE 8.2 Schematic and photograph of the heat exchanger. (Reprinted from *Energy Convers. Manag.*, Yarmand, H. et al., Experimental investigation of thermophysical properties, convective heat transfer and pressure drop of functionalized graphene nanoplatelets aqueous nanofluid in a square heated pipe, 114, 38–49. Copyright 2016 with permission from Elsevier; Reprinted from *Energy Convers. Manag.*, Yarmand, H. et al., Graphene nanoplatelets–silver hybrid nanofluids for enhanced heat transfer. 100, 419–428 copyright 2015, with permission from Elsevier, (license numbers 4422261444316 and 4422350279402).)

an average of 16 measurements were recorded during 4 hours for each temperature and weight concentration. Calibration of instrument with DI water was performed before starting of the measurements of nanofluids. Thermal conductivity of DI water at 30°C was measured and a value of 0.61 W/mK found, which is in agreement with the previous investigations. Viscosity is also other important property of fluid flow. Pressure drop, pumping power, and heat transfer capability of fluid is directly dependent on the viscosity. Furthermore, the dynamic viscosity of nanofluid is investigated at various shear rates to find the rheological behavior of suspension, whether the suspension has Newtonian or non-Newtonian behavior. The viscosity of distilled water and different weight fractions of nanofluids were measured by rheometer Physica, MCR, Anton Paar, Austria. The rotational rheometer consists of a moving cylindrical plate and a stationary cylindrical surface which are parallel with a small gap. Viscosity of distilled water and nanofluid samples at various concentrations and temperatures in the range of 20°C–40°C at a shear rate of 500/s had been investigated. The measured viscosity of distilled water at 20°C is 1.10 (mPa sec).

Volumetric behaviour of nanofluid samples was obtained from experimental measurements of density as a function of nanoparticles weight fraction and temperature. The densities of the prepared nanofluids were measured experimentally by Mettler Toledo DE40 density meter. The accuracy of density measurement is 10–4 g/cm³. For each temperature and sample, the measurements have been recorded three times. Differential scanning calorimetry (DSC) is an appropriate device to investigate the specific heat capacity of nanofluids. The change in the amount of heat flow needed for heating up a sample reference pan and sample pan are measured as a function of temperature. In this study, a differential scanning calorimeter (DSC 8000, Perkin Elmer, United States), which is calibrated using indium (99.999%) with an accuracy of ± 1.0%, is used to find the specific heat capacity of all nanofluid samples directly.

8.4 Data Reduction

The amount of heat supplied to the test section and the amount of heat absorbed by the flowing fluid, respectively, are:

$$P = V \times I \tag{8.1}$$

$$Q = \dot{m} \times C_P \times (T_o - T_i) \tag{8.2}$$

A maximum deviation of ± 3% was observed between the measured amount of energy supplies and energy absorbed, which confirms negligible amount of heat transfer to the surrounding of the test section.

The Newton law for heat transfer is used to estimate the experimental heat transfer coefficient. That is:

$$h = \frac{Q}{A(T_w - T_b)}, \tag{8.3}$$

where $T_w = \frac{\sum T}{5}$ (T_w is average temperature of five thermocouples), $T_b = \frac{T_0 + T_i}{2}$, and $A = \pi DL$.
The Nusselt number is defined as:

$$Nu = \frac{h \times D_h}{k} \tag{8.4}$$

where $D_h = \frac{4A_c}{P}$, A_c is the cross-sectional area, and P is the perimeter of the square pipe.
The available Nusselt number correlations for single-phase fluids are listed in this section.
Dittus and Boelter [30] presented equation for evaluation of Nusselt number for water:

$$Nu = 0.023 Re^{0.8} Pr^{0.4}. \tag{8.5}$$

Eq. (8.5) is applicable for the range of Reynolds number $(Re) > 10^4$, $0.6 < $ Prandtl number $(Pr) < 200$. Petukhov [31] presented equation for evaluation of Nusselt number for water:

$$Nu = \frac{\left(\frac{f}{8}\right)Re\,Pr}{1.07 + 12.7\left(\frac{f}{8}\right)^{0.5}\left(Pr^{\frac{2}{3}} - 1\right)}. \tag{8.6}$$

Eq. (8.6) is applicable for the range of $0.5 < Pr < 2000$ and $3000 < Re < 5 \times 10^6$. Gnielinski [32] presented equation for evaluation of Nusselt number for water:

$$Nu = \frac{\left(\frac{f}{8}\right)(Re - 1000)Pr}{1 + 12.7\left(\frac{f}{8}\right)^{0.5}\left(Pr^{\frac{2}{3}} - 1\right)}, \tag{8.7}$$

where $f = (0.79\,ln\,Re - 1.64)^{-2}$. Eq. (8.7) is for the range $2300 < Re < 10^6$, $0.5 < Pr < 2000$.

Because the single-phase fluid Nusselt number correlations underestimate the heat transfer of nanofluids, the researchers have developed new Nusselt number correlations for nanofluids. Some of the available Nusselt number correlations are outlined here:

Maïga et al. [33] presented equation for evaluation of Nusselt number for Al_2O_3/water nanofluids as a function of Re and Pr:

$$Nu = 0.085\,Re^{0.71}Pr^{0.35} \tag{8.8}$$

$10^4 < Re < 5 \times 10^5$, $6.6 < Pr < 13.9$ and $0 < \varphi < 10$.

Pak and Cho [34] presented equation for evaluation of Nusselt number for Al_2O_3/water and TiO_2/water nanofluids as a function of Re and Pr:

$$Nu = 0.021\,Re^{0.8}Pr^{0.5} \tag{8.9}$$

$10^4 < Re < 10^5$, $6.54 < Pr < 12.33$, $0 < \varphi < 3.0\%$.

Sundar et al. [18] presented equation for evaluation of Nusselt number for MWCNT-Fe_3O_4/water hybrid nanofluids as a function of Re, Pr, and volume fraction:

$$Nu = 0.0215\,Re^{0.8}Pr^{0.5}(1 + \varphi)^{0.78} \tag{8.10}$$

$3000 < Re < 22000$, $0 < \varphi < 0.3\%$, $4.50 < Pr < 6.13$.

The friction factor of nanofluids was measured experimentally from the recorded pressure loss data across the test section. That is:

$$f = \frac{\Delta P}{\left(\frac{L}{D}\right)\left(\frac{\rho V^2}{2}\right)}. \tag{8.11}$$

The available friction factor expression for water and nanofluids are represented by Eqs. 8.12 through 8.14.

Blasius [35] presented equation for evaluation of friction factor for water flow:

$$f = 0.3164\,Re^{-0.25}, \tag{8.12}$$

where, $3000 < Re < 10^5$.

Petukhov [31] presented equation for evaluation of friction factor for water flow:

$$f = (0.790\ln(Re) - 1.64)^{-2}, \tag{8.13}$$

where, $2300 < Re < 5 \times 10^6$.

Sundar et al. [18] presented equation for evaluation of friction factor for nanofluid flow:

$$f = 0.3108\,Re^{-0.245}(1.0 + \varphi)^{0.42}, \tag{8.14}$$

where, $3000 < Re < 22000,\ 0 < \varphi < 0.6\%$.

Applying base-fluids including nanoparticles can be imposed on heat transfer equipment with an enhancement in both of the pressure drop (negative effect) and heat transfer coefficient (positive effect). To investigate this issue exactly, performance index (ε) can be selected as an appropriate parameter to clarify the range of temperature and velocity that can be used by synthesised coolant:

$$\varepsilon = \frac{h_{nf}/h_{bf}}{\Delta P_{nf}/\Delta P_{bf}} = \frac{R_h}{R_{\Delta P}}. \tag{8.15}$$

R_h is the ratio of the heat transfer enhancement in the presence of the nanoparticles to the base-fluid, and $R_{\Delta P}$ is the ratio of pressure drop of synthesised coolant to the base-fluid. In the turbulent region, the pumping power can be measured by Eq. 8.16 to study energy saving [36].

$$\frac{W_{nf}}{W_{bf}} = \left(\frac{\mu_{nf}}{\mu_{bf}}\right)^{0.25}\left(\frac{\rho_{bf}}{\rho_{nf}}\right)^2. \tag{8.16}$$

8.5 Thermophysical Property

Thermal conductivity of three different graphene-based nanofluids (functionalised GNP and its hybrid) with various weight percentages (0.02 wt%, 0.06 wt%, and 0.1 wt%) are recorded in the range of 20°C–40°C. Tiny amounts of weight percentages are chosen to avoid sedimentation and increase of effective viscosity. Figure 8.3 presents the thermal conductivity of water, GNP, GNP/Ag, and GNP/Pt nanofluids at different temperatures for various volume concentrations. It is shown that the thermal conductivity of nanofluids rises with the increase of weight fractions of nanoparticles and fluid temperature. For 0.1% weight fraction of f-GNP/Ag, the enhancement of thermal conductivity is 16.95% at 20°C and nearly 22.22% at 40°C. Enhancement in the effective thermal conductivity is due to the high thermal conductivity of GNP nanoparticles. With the increasing of nanoparticles weight fractions, the particles distance (free path) decreases. This fact is because of the percolation effect. More particles are in contact with each other, which increases the frequency of lattice vibration. It is obviously confirmed that thermal conductivity of hybrid nanofluids is more than single-phase

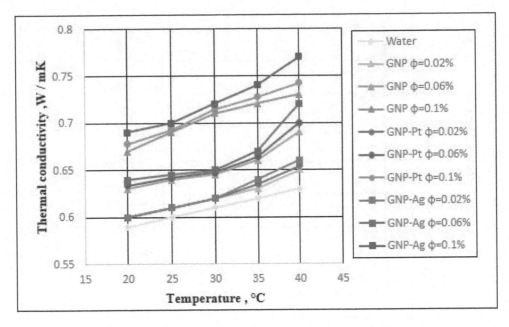

FIGURE 8.3 Thermal conductivity of functionalised GNP and its hybrid nanofluids as a function of temperature for different weight fractions.

nanofluids due to higher ability of nanocomposite as well as synergetic effect. Among the samples, GNP/Ag nanofluids showed the highest improvement due to larger thermal conductivity of guest nanomaterial. To elucidate the reasons for the strange growth of the thermal conductivity in nanofluids, some potential mechanisms have been studied earlier such as Brownian motion, the nature of heat transport in the nanoparticles, and the effects of nanoparticle clustering, and also several theoretical models have been investigated by choosing the carbon and carbon-based hybrid nanofluids. The thermal enhancement has been attributed to several reasons where more investigations are needed for further clarifications.

Figure 8.4 indicates the viscosity of water and functionalised GNP and its hybrid nanofluids at temperatures in the range of 20°C–40°C at different weight fractions for a shear rate of 500/s. It is found that with the increase of nanofluids weight fractions, the viscosity of nanofluids rises, as the increase in concentration would have a direct influence on the fluid internal shear stress. The viscosity decreases with the increment of temperatures, for the weakening of inter-molecular and inter-particle adhesion forces. Viscosity rise of about 33% is occurred at 0.1% weight percentage of GNP-Pt nanofluid compared to the viscosity of the water at 40°C.

Density of functionalised GNP and its hybrid nanofluids has been examined from experimental measurements. Table 8.1 shows the density of distilled water, graphene, and its hybrid nanofluids samples as a function of weight fraction and temperature. The data with same weight concentration and temperature were reported. The instrument was calibrated with distilled water and found as, 0.997 kg/m³ which is very close to the standard value (0.998 kg/m³). The density of GNP and its water-based hybrid nanofluids have increases with the rise of nanoparticle concentrations and decreases when the temperature increases. Some other experimental results with various base fluid and nanoparticles have shown the same tendency [37,38]. The highest amount of density enhancement is 0.11% for GNP-Pt nanofluid at 0.1 wt% and 40°C. This slight density upgrading might be attributed to the interface effects on the bulk fluid properties produced by the solid nanoparticle surface, and to the interactions between the nanomaterials themselves, which is typically considered negligible.

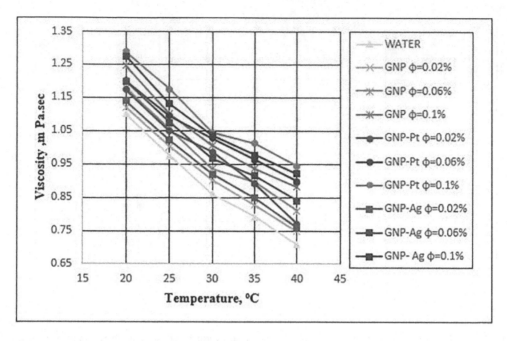

FIGURE 8.4   Viscosity of functionalised GNP and its hybrid nanofluids as a function of temperature for different weight fractions.

TABLE 8.1

Density of Water, Functionalised GNP, and Its Hybrid Water-Based Nanofluids at Different Weight Concentration and Temperature

Temperature	Water	GNP/Water			GNP-Pt/Water			GNP-Ag/Water		
	0.00%	0.02%	0.06%	0.10%	0.02%	0.06%	0.10%	0.02%	0.06%	0.10%
20	997.9	998	998.4	998.6	998.4	998.7	999	998.1	998.5	998.9
25	996.7	996.8	997.2	997.4	997.3	997.6	997.9	996.9	997.4	997.8
30	995.3	995.4	995.8	996	995.9	996.2	996.5	995.5	996	996.3
35	993.7	993.9	994.2	994.4	994.3	994.6	994.9	993.9	994.4	994.7
40	992	992.1	992.3	992.6	992.5	992.8	993.1	992.1	992.5	992.9

8.6 Heat Transfer and Friction Factor

Nanofluids at various weight fractions of functionalised GNP and its hybrid were experimentally investigated. The Nusselt numbers for the nanofluids are calculated by Eq. 8.4, and the outcomes are plotted in Figure 8.5. It is recognised that the Nusselt number enhances with the increase of Reynolds number and also with the increase of nanoparticles weight fractions. This is because the nanofluid contains suspended nanoparticles, which have higher conductivity compared to the base fluid. The Nusselt number enhancement for functionalised GNP and its hybrid nanofluid is also attributed to the thermophysical properties of the nanoparticles as well as particle Brownian motion [18]. The highest improvement of Nusselt numbers among single- and two phase-examined nanofluid is found for GNP-Ag nanofluids, with particle volume fraction of 0.1% are 21.89% and 32.70% corresponding to the Reynolds numbers of 5000 and 17500, respectively.

Eq. 8.11 is useful for calculating the friction factor of functionalised GNP and its hybrid nanofluids at different concentrations, and the outcome is illustrated in Figure 8.6. The highest increase of friction factor for 0.1% weight concentration of GNP-Pt nanofluid occurred with value of 10.98% at

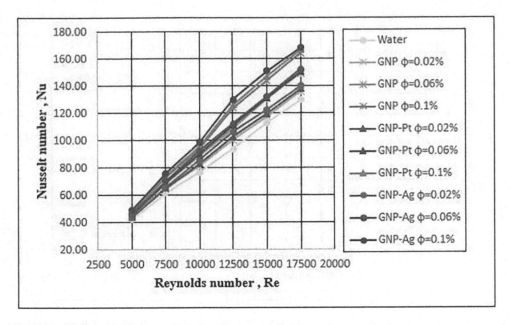

FIGURE 8.5 Nusselt number of functionalised GNP and its hybrid nanofluids as a function of temperature for different weight fractions.

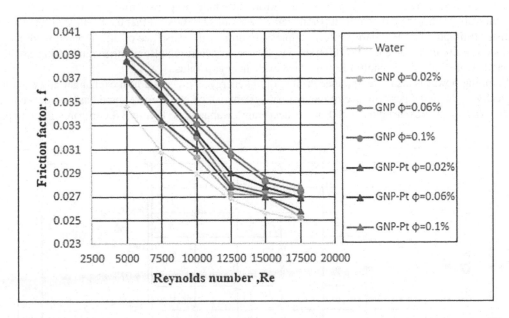

FIGURE 8.6 Friction factor of functionalised GNP and its hybrid nanofluids as a function of temperature for different weight fractions.

the Reynolds number of 17500. The enrichment of friction factor due to the suspension of nanoparticles or nanocomposites in the base fluid is insignificant in comparison to the heat transfer development.

Figure 8.7 shows the performance index of functionalised GNP and GNP-Pt nanofluids for different weight concentrations and Re numbers. It is noteworthy that the performance index of functionalised GNP and its hybrid nanofluids for all weight concentrations is more than one, except for Re number of 5000, confirming that the prepared nanofluids can be used as an appropriate alternative coolant for various

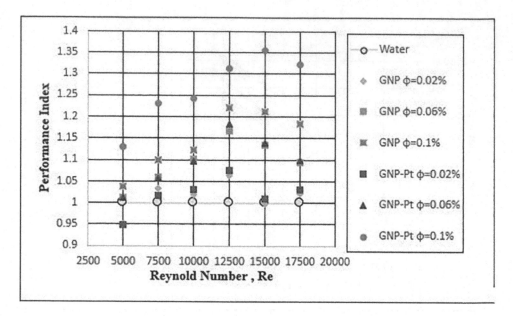

FIGURE 8.7   Performance index of functionalised GNP and its hybrid nanofluids at different weight concentrations.

types of heat exchangers at different boundary conditions. In addition, as the weight concentration of GNP and its hybrid in base fluid increases, the performance index improves, implying advanced influence of heat transfer factors to rheological parameters. The results established that the positive effects of heat transfer triumph over the negative effects of pressure drop in the existence of the prepared GNP and its nanocomposite in the base fluid at different weight concentrations and Re numbers, indicating excellent capability of synthesised nanofluids in the heat exchanger apparatus.

Pumping power is also a key parameter in defining energy savings of a heat exchanger and can rheologically evaluate the performance of the thermal system, which can prepare some basic evidence about operability of heat exchanging liquid in the heat transfer apparatus. Figure 8.8 compares the pumping

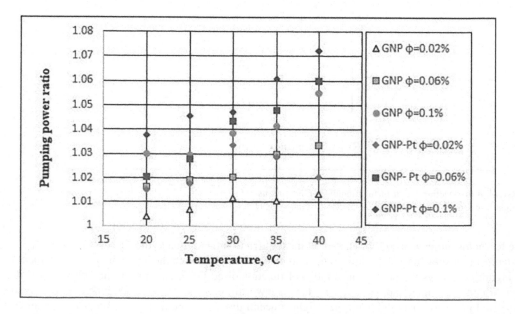

FIGURE 8.8   Pumping power ratio of functionalised GNP and its hybrid nanofluids at different weight concentrations.

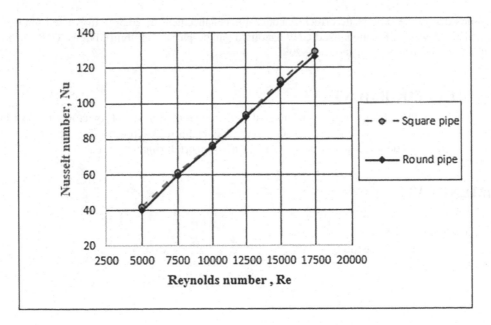

FIGURE 8.9   Effect of cross section on the Nusselt number.

power of the GNP and its hybrid nanofluids samples at various weight concentrations. Outcomes proposed are that there is no important alteration in pumping power in the attendance of nanomaterial and nanocomposite at different temperatures and concentrations. A trivial rise in the pumping power (highest increase is about 7.2%) with GNP-Pt loading can be noticed in Figure 8.8, which considers as a positive factor as compared with other conventional nanofluids.

Figure 8.9 shows the effect of cross section of heat exchanger on the Nusselt number. Two different cross sections have been examined with water at a same boundary and flow condition. Both square and round pipe have been chosen with same hydraulic diameter of 10 mm. The experimental results confirm that at same condition and hydraulic diameter change of cross section do not have significant influence on the Nusselt number and heat transfer performance of the convective heat transfer loop. An average deviation of 2.26% was found which might be due to the uncertainty of the device.

8.7 Conclusion

Recently many research groups have focused on using nanofluid as a heat exchanger working fluid. Using nanofluids is a promising way to enhance the efficiency of the thermal systems. Many research has confirmed the enhancement of the heat transfer performance of the thermal system by using nanofluids. The aims of the present book chapter are to study the synthesis, preparation, thermos-physical properties, friction loss, and heat transfer of the functionalised carbon-based and its hybrid nanofluids in a close conduit flow. Since graphene is naturally hydrophobic, so covalent functionalisation has a key effect on the stability of graphene hybrid-based nanofluids. Thus, selection of a proper functionalisation method can make its homogeneous stable suspension.

Nanocomposite weight concentration and temperature have great influence on the thermal conductivity and viscosity of the as prepared nanofluids. Presence of a small quantity (≤ 0.1 wt.%) of nanocomposite in water resulted in a significant thermal conductivity enhancement, but the influence of adding nanocomposite on the density and specific heat capacity of the nanofluids is not substantial. Experimental results of Nusselt numbers of all the nanofluids have revealed the remarkable enhancement of heat transfer performance in comparison to the base fluids, but it is important to highlight that the improvement for hybrid nanofluids is higher than the case of single-phase functionalised carbon-based nanofluids due to the synergistic effect. A noticeable enhancement of 26.50% in Nusselt number was obtained at 0.1 wt% of functionalised GNP nanofluid (single-phase nanofluid), but for GNP-Ag hybrid nanofluid, it was further escalated to 32.70%.

ACKNOWLEDGEMENT

The authors gratefully acknowledge Fundamental Research Grant Scheme (Project no.: FRGS-FP063-2015A) University of Malaya, Malaysia for support to conduct this research work. The authors would like to thank the IPPP section, University of Malaya for their support during this research.

NOMENCLATURE

A Area, m²
Cp   Specific heat, J/kg K
D Inner diameter of the tube, m
f Friction factor
GNP   Graphene nanoplatelet
h Heat transfer coefficient, W/m² K
k Thermal conductivity, W/m K
l Length of the tube, m
m Mass flow rate, kg/s
Nu   Nusselt number,
P Power, Watts
Pr   Prandtl number, $\mu C/k$
Q Heat flow, Watts
q Heat flux, W/m²
Re   Reynolds number,
T Temperature, °C
v Velocity, m/s

GREEK SYMBOLS

φ Weight concentration of nanoparticles, %
μ Viscosity, kg/m² s
ρ Density, kg/m
ε Performance index

REFERENCES

1. Chol, S.U.S. and J.A. Estman (1995). Enhancing thermal conductivity of fluids with nanoparticles. *ASME-Publications-Fed.* 231, 99–106.
2. Yarmand, H., S. Gharehkhani, S.N. Kazi, E. Sadeghinezhad, and M.R. Safaei (2014). Numerical investigation of heat transfer enhancement in a rectangular heated pipe for turbulent nanofluid. *The Scientific World Journal.* 2014, 9.
3. Choi, S., Z. Zhang, W. Yu, F. Lockwood, and E. Grulke (2001). Anomalous thermal conductivity enhancement in nanotube suspensions. *Applied Physics Letters.*79(14), 2252–2254.
4. Turgut, A., I. Tavman, M. Chirtoc, H. Schuchmann, C. Sauter, and S. Tavman (2009). Thermal conductivity and viscosity measurements of water-based TiO2 nanofluids. *International Journal of Thermophysics.* 30(4), 1213–1226.

5. Karthikeyan, N., J. Philip, and B. Raj (2008). Effect of clustering on the thermal conductivity of nano-fluids. *Materials Chemistry and Physics.* 109(1), 50–55.

6. Das, S.K., N. Putra, P. Thiesen, and W. Roetzel (2003).Temperature dependence of thermal conductivity enhancement for nanofluids. *Journal of Heat Transfer.* 125(4), 567–574.

7. Yu, W., H. Xie, and D. Bao (2010). Enhanced thermal conductivities of nanofluids containing graphene oxide nanosheets. *Nanotechnology.* 21(5), 055705.

8. Zheng, R., J. Gao, J. Wang, S.-P. Feng, H. Ohtani, J. Wang, and G. Chen (2011). Thermal percolation in stable graphite suspensions. *Nano letters.* 12(1), 188–192.

9. Ding, Y., H. Alias, D. Wen, and R.A. Williams (2006). Heat transfer of aqueous suspensions of carbon nanotubes (CNT nanofluids). *International Journal of Heat and Mass Transfer.* 49(1), 240–250.

10. Yu, W., H. Xie, X. Wang, and X. Wang (2011). Significant thermal conductivity enhancement for nano-fluids containing graphene nanosheets. *Physics Letters A.* 375(10), 1323–1328.

11. Novoselov, K.S., A.K. Geim, S. Morozov, D. Jiang, Y. Zhang, S. Dubonos, I. Grigorieva, and A. Firsov (2004). Electric field effect in atomically thin carbon films. *Science.* 306(5696), 666–669.

12. Land, T., T. Michely, R. Behm, J. Hemminger, and G. Comsa (1992). STM investigation of single layer graphite structures produced on Pt (111) by hydrocarbon decomposition. *Surface Science.* 264(3), 261–270.

13. Tung, V.C., M.J. Allen, Y. Yang, and R.B. Kaner (2008). High-throughput solution processing of large-scale graphene. *Nature Nanotechnology.* 4(1), 25–29.

14. Kaniyoor, A., T.T. Baby, and S. Ramaprabhu (2010). Graphene synthesis via hydrogen induced low temperature exfoliation of graphite oxide. *Journal of Materials Chemistry.* 20(39), 8467–8469.

15. Sridhar, V., J. H. Jeon, and I. K. Oh (2010). Synthesis of graphene nano-sheets using eco-friendly chemicals and microwave radiation. *Carbon.* 48(10), 2953–2957.

16. Zhang, W., W. He, and X. Jing (2010). Preparation of a stable graphene dispersion with high concentration by ultrasound. *The Journal of Physical Chemistry B.* 114(32), 10368–10373.

17. Suresh, S., K. Venkitaraj, P. Selvakumar, and M. Chandrasekar (2012). Effect of Al 2O3–Cu/water hybrid nanofluid in heat transfer. *Experimental Thermal and Fluid Science.* 38, 54–60.

18. Sundar, L.S., M.K. Singh, and A. Sousa (2014). Enhanced heat transfer and friction factor of MWCNT–Fe_3O_4 water hybrid nanofluids. *International Communications in Heat and Mass Transfer.* 52, 73–83.

19. Baby, T.T. and R. Sundara (2011). Synthesis and transport properties of metal oxide decorated graphene dispersed nanofluids. *The Journal of Physical Chemistry C.* 115(17), 8527–8533.

20. Munkhbayar, B., M.R. Tanshen, J. Jeoun, H. Chung, and H. Jeong (2013). Surfactant-free dispersion of silver nanoparticles into MWCNT-aqueous nanofluids prepared by one-step technique and their thermal characteristics. *Ceramics International.* 39(6), 6415–6425.

21. Abbasi, S.M., A. Rashidi, A. Nemati, and K. Arzani (2013). The effect of functionalisation method on the stability and the thermal conductivity of nanofluid hybrids of carbon nanotubes/gamma alumina. *Ceramics International.* 39(4), 3885–3891.

22. Baby, T.T. and S. Ramaprabhu (2011). Experimental investigation of the thermal transport properties of a carbon nanohybrid dispersed nanofluid. *Nanoscale.* 3(5), 2208–2214.

23. Baby, T.T. and R. Sundara (2013). Synthesis of silver nanoparticle decorated multiwalled carbon nanotubes-graphene mixture and its heat transfer studies in nanofluid. *AIP Advances.* 3(1), 012111.

24. Suresh, S., K. Venkitaraj, P. Selvakumar, and M. Chandrasekar (2011). Synthesis of Al 2 O 3–Cu/water hybrid nanofluids using two step method and its thermo physical properties. *Colloids and Surfaces A: Physicochemical and Engineering Aspects.* 388(1), 41–48.

25. Suresh, S., K. Venkitaraj, and P. Selvakumar (2011). Synthesis, characterisation of Al2O3-Cu nano composite powder and water based nanofluids. *In Advanced Materials Research.* 328, 1560–1567.

26. Han, Z., B. Yang, S. Kim, and M. Zachariah (2007). Application of hybrid sphere/carbon nanotube particles in nanofluids. *Nanotechnology.* 18(10), 105701.

27. Botha, S.S., P. Ndungu, and B.J. Bladergroen (2011). Physicochemical properties of oil-based nanofluids containing hybrid structures of silver nanoparticles supported on silica. *Industrial & Engineering Chemistry Research.* 50(6), 3071–3077.

28. Yarmand, H., S. Gharehkhani, S.F.S. Shirazi, A. Amiri, M.S. Alehashem, M. Dahari, and S. Kazi (2016). Experimental investigation of thermo-physical properties, convective heat transfer and pressure drop of functionalized graphene nanoplatelets aqueous nanofluid in a square heated pipe. *Energy Conversion and Management.* 114, 38–49.

29. Yarmand, H., S. Gharehkhani, G. Ahmadi, S.F.S. Shirazi, S. Baradaran, E. Montazer, M.N.M. Zubir, M.S. Alehashem, S. Kazi, and M. Dahari (2015). Graphene nanoplatelets–silver hybrid nanofluids for enhanced heat transfer. *Energy Conversion and Management*. 100, 419–428.

30. Dittus, F.W. and L.M.K. Boelter (1930). *Heat Transfer for Automobile Radiators of the Tubular Type*. Berkeley, CA, University of California Publications in Engineering. 2.

31. Petukhov, B. (1970). Heat transfer and friction in turbulent pipe flow with variable physical properties. *Advances in Heat Transfer*. 6, 503–565.

32. Gnielinski, V. (1975). New equations for heat and mass transfer in the turbulent flow in pipes and channels. *NASA STI/Recon Technical Report A*. 75, 22028.

33. Maïga, S.E.B., C.T. Nguyen, N. Galanis, G. Roy, T. Maré, and M. Coqueux (2006). Heat transfer enhancement in turbulent tube flow using Al_2O_3 nanoparticle suspension. *International Journal of Numerical Methods for Heat & Fluid Flow*. 16(3), 275–292.

34. Pak, B.C. and Y.I. Cho (1998). Hydrodynamic and heat transfer study of dispersed fluids with submicron metallic oxide particles. *Experimental Heat Transfer an International Journal*. 11(2), 151–170.

35. Blasius, H. (1980). Grenzschichten in Flussigkeiten mit kleiner Reibung (German). Zeitschrift für Angewandte Mathematik und Physik. 56, 1–37.

36. Mansour, R.B., N. Galanis, and C.T. Nguyen (2007). Effect of uncertainties in physical properties on forced convection heat transfer with nanofluids. *Applied Thermal Engineering*. 27(1), 240–249.

37. Mariano, A., M.J. Pastoriza-Gallego, L. Lugo, L. Mussari, and M.M. Piñeiro (2015). Co_3O_4 ethylene glycol-based nanofluids: Thermal conductivity, viscosity and high pressure density. *International Journal of Heat and Mass Transfer*. 85, 54–60.

38. Kumaresan, V. and R. Velraj (2012). Experimental investigation of the thermo-physical properties of water–ethylene glycol mixture based CNT nanofluids. *Thermochimica Acta*. 545, 180–186.

9

Lattice Boltzmann Modeling on Convective Heat Transfer of Nanofluids through Highly Conductive Metal Foams

H.J. Xu, Z.B. Xing, X. Fang, W. Zhang, and Z.Z. Zhou

CONTENTS

9.1 Literature Survey

With the continuous advancement of scientific research and production processes, the heat transfer performance of heat exchangers needs to be correspondingly improved. From the energy and economy point of view, people always hope to continuously enhance heat transfer technologies. Improving the structure and size of the heat exchanger leads to the good heat transfer enhancement effect. However, in some special conditions, the traditional heat transfer working medium can no longer meet the requirements, and the low thermal conductivity of the heat transfer medium is greatly limited. To enhance heat transfer, nanofluids will become a new generation of heat transfer fluids with better thermal conductivity. Researchers in the field of heat transfer are always looking for new solutions to optimize the heat transfer performance of heat exchange equipment. In various methods for enhancing heat transfer, dispersing the nanoscale solid particles into a common heat-transfer working medium (such as water, oil, etc.) can effectively improve the heat conduction capability of fluid, which are called nanofluids [1]. Nanofluids have higher thermal conductivity than common liquids, so they have great potential for improving the heat transfer rate in engineering systems, especially in heat exchanging devices and the cooling of electronics. There have been various studies on nanofluids in recent decades [2,3]. On the other hand, the use of highly conductive porous metals in heat exchangers is another effective technique for improving the heat transfer rate by providing a very high specific surface area, and high thermal conductivity. Porous metals have a rigid and open saturated cell structure, which means that the pores are fully connected and the fluid can fully fill and flow through the pores.

Flow and heat transfer characteristics of nanofluids have always been the research focus of scholars. Research on convective heat transfer of nanofluids is essential for practical applications. Buongiorno [4] analyzed seven factors that may affect the thermal properties of nanofluids. The study shows that Brownian diffusion and thermophoretic diffusion are the two most important factors and a mass, momentum, and energy conservation equation was established. The equations for the conservation equations of nanoparticle and flux are used to describe the flow and heat transfer of nanofluids. Kim et al. [5] studied the boiling process of water-based nanofluids containing Al_2O_3, ZrO_2, and SiO_2, and found that nanoparticles deposit on the wall in the boiling process of nanofluids, forming many porous structures with irregular surfaces. Shafah et al. [6] gave an empirical correlation for calculating the nanofluid viscosity based on temperature and nanoparticle volume fraction. Wu et al. [7] added nanoparticles to paraffin to make a new phase change material (PCM) by using five dispersants to stably suspend the copper nanoparticles in the molten paraffin and found that copper has the best heat transfer performance. Zeng et al. [8] studied the enhancement of thermal conductivity of organic PCMs with silver nanowires. Corcione [9] established a correlation for calculating the thermal conductivity of nanofluids through regression analysis. Corcione [10] gave an empirical correlation for calculating the Nusselt number for different channel aspect ratios, Prandtl numbers, and Rayleigh numbers by summarizing experimental data. Barber et al. [11] systematically summarized the research on boiling heat transfer enhancement using nanofluids. Guo [12] found that the use of Al_2O_3/paraffin PCMs as room ceilings can effectively reduce the required room cooling capacity in the summer, and the research is focused on the thermal conductivity and heat storage capacity of Al_2O_3/paraffin PCMs. Khanafer et al. [13] proposed a correlation for the calculation of the thermal conductivity of nanofluids at room temperature based on experimental data by taking into account the effects of the thermal conductivity of the base fluid and the nanoparticles, the diameter and volume fraction of the nanoparticles, and other factors. Lee et al. [14] made a comprehensive summary and analysis of the thermal conductivity, heat transfer mechanism, and mathematical model of nanofluids. Wu et al. [15] studied the influence of copper nanoparticles on the thermal conductivity and phase change heat transfer of copper/paraffin composite PCMs and found that the addition of nanomaterials improves the thermal conductivity and phase change heat transfer. Yang et al. [16] used Buongiorno's model [4] to study convection heat transfer between Al_2O_3/H_2O nanofluids and TiO_2/H_2O nanofluids in concentric ducts and discussed the effects of Brown diffusion, thermophoretic diffusion, boundary conditions, particle volume fraction, and ratio of inner to outer diameters on flow and heat transfer. Hadadian et al. [17] conducted a comparative analysis of the results in relevant literature and summarized the mechanisms, enhancement effects, and latest research progresses of nanofluids for enhancing heat transfer. Ebrahimnia-Bajestan et al. [18] measured the thermal conductivity at different temperatures and different particle volume fractions and established a calculation model. Results showed that nanofluids have higher convective heat transfer coefficients than the base fluid. Malvandi and Ganji [19] studied the fully developed flow of nanofluid laminar convection in vertical annular tubes, focusing on the migration of nanoparticles and its effect on heat transfer and pressure drop. It is found that the asymmetric thermal boundary conditions can change the direction of particle migration. At low heat flux, nanoparticles tend to aggregate toward the wall. When the heat flux ratio of the inner and outer walls is less than 1, there is an optimal inner-to-outer diameter ratio (0.6) to obtain the optimal heat transfer performance. Sheremet et al. [20] found that the average Nusselt number near the hot wall increases with an increase in Rayleigh number and Brownian diffusion coefficient and decreases with a decrease in Lewis number, buoyancy coefficient, and thermophoretic diffusion coefficient. The average Sherwood number and the Nusselt number have exactly the opposite relationship in Brownian diffusion and thermophoretic diffusion. It was also found that the low Rayleigh number, the low Lewis number, and the high thermophoretic diffusion coefficient cause the nanoparticles to be distributed uniform, so using a non-uniform model is more appropriate.

Many scholars have begun to pay attention to the application of metal foam to enhance the heat transfer. Through experimental research, the accurate physical properties and empirical correlations of the metal foam can be obtained. When numerical and theoretical studies based on volume averaging method are conducted on transport in porous metals, the energy equations can be classified as local thermal equilibrium (LTE) model and local thermal non-equilibrium (LTNE) model. Lu et al. [21] and Zhao et al. [22] found that increasing the pore density and porosity will increase the Nusselt number. They also found that the use of metal foam can obviously enhance the effect of heat transfer, but also bring a lot of pressure loss. This can be overcome by partially filling the heat transfer channel with the metal foam, which is proven to be with a much smaller pressure loss [23–26]. Umavathi et al. [27] use the conventional perturbation method to analyze the Poiseuille-Couette flow and heat in a channel filled with two porous media sintered on the upper and lower sides of the channel centerline. Hong and Herling [28] conducted an experimental study on the effect of the aluminum foam specific surface area on the thermal conductivity of the PCM composite. The results show that the pore diameter has a significant effect on the time of heating and cooling for PCM composite. Zhao et al. [29] and Tian and Zhao [30] conducted an experimental/numerical study of the heat transfer of paraffin phase materials in copper foams with pore sizes ranging from 0.8 mm to 2.5 mm. This study considered the effects of non-equilibrium heat transfer in the phase transition and liquid phase regions. The coupling of heat transfer and natural convection was considered, and the experimental results were compared with the numerical results. Results show that using metal foams or increasing thermal conductivity can enhance the heat transfer, but the large flow resistance of metal foam significantly constrains natural convection. The overall heat conduction promotion exceeds the weakening of natural convection [30]. Metal foams can be well used for promoting the thermal performance of compact heat exchangers [31,32]. Xu [33] performed a thermal analysis for the micro heat exchangers filled with micro metal foams with the slip boundary condition and the LTNE effect and found that the presence of metal foam is favorable for enhancing heat transfer. The asymmetrical thermal boundary conditions have obvious influence on the distributions of velocity and temperature of the forced convection in metal foams [34–36]. Dukhan et al. [37] studied the effect of convective heat transfer during the filling of open-cell aluminum foam in circular tubes and measured the average temperatures of the wall surface and the inlet and outlet. In the fully developed area, the Nusselt number and the Reynolds number strongly correlate with the power law. Siavashi et al. [38] adopted a two-phase model and used a new parameter, PN (performance number), to represent the ratio of heat transfer enhancement to pressure drop. At low permeability, the PN will become smaller and the best thickness will correspond to the PN at medium-sized penetration.

Recently, the use of nanofluids and porous media has received widespread attention, and scholars have begun extensive research in this area. The porous media increases the contact area between the liquid and the solid, while on the other hand, the dispersed nanoparticles increase the effective thermal conductivity of the nanofluid. So, the use of porous media and nanofluids can greatly increase the heat transfer rate of typical thermal systems. Bhadauria et al. [39] studied the steady and unsteady heat transfer of nanofluids in metal foams using the Brinkman model and the LTNE effect that nanofluids are only affected by Brownian and thermophoretic motions. A three-temperature model of particles, liquids, and solids was established. Sun et al. [40] numerically studied the natural convection of nanofluids in an enclosure filled with a porous medium. Results show that the increase of the volume fraction of nanoparticles at low Rayleigh number can increase the average Nusselt number, but the average Nusselt number decreases with an increase in the volume fraction of nanoparticles at high Rayleigh number. Hajipour and Molaei Dehkordi [41] found that the heat transfer from nanofluid to the cold wall increases with an increase in viscous dissipation and buoyancy, but the heat transfer from the hot wall to the fluid decreases with an increase in the viscous force and buoyancy. There is a large difference between different viscous dissipative models. Subsequently, they conducted experimental and numerical studies [42] and measured the outlet temperature and pressure drop at different flow rates. Xu and Xing [43] presented a mathematical model for the nanofluid natural convective heat transfer in a metal-foam enclosure with the lattice Boltzmann method by considering the LTE assumption. Sakai et al. [44] used the volume average method to establish macroscopic equations describing heat transfer of nanofluids in metal foams. The analytical results show that the combination of metal foam and nanofluid can give very good heat transfer performance, but the model they established is based on the LTE model, ignoring the temperature difference between the solid and the fluid. Zhang et al. [45] established a system of equations for the convection heat transfer of nanofluids in metal foams based on a LTNE model and analyzed the numerical treatments in modeling Brownian diffusion, thermophoretic diffusion, and the heat transfer between the metal skeleton and the fluid. Results show that the Nusselt number between the pores has a great influence on the temperature difference between the fluid and the solid. Xu et al. [46] obtained the distribution of velocity field and temperature field of convective heat transfer in nanofluids in metal foams by numerical simulation using LTNE model and analyzed the influence of various important parameters on flow and heat transfer performance. Kasaeian et al. [1] analyzed and summarized the latest research progress of convective heat transfer in nanofluids in metal foams. Xu et al. [47] summarized the correlations for thermophysical properties, mathematical models for nanofluid, and the recent progress for nanofluid flow and heat transfer in metal foams and suggested some aspects for future research, including turbulent flow of nanofluids flow in metal foams, new mathematical models for heat transfer, the micro effect of nanofluid, the nanofluid non-Newtonian effect, and the slip effect. Ting et al. [48] considered the effects of fluid viscous dissipation, internal heat sources in the solid phase, asymmetric heating, and nanoparticle size on nanofluid heat transfer characteristics in porous media. It is found that the average heat transfer coefficient with symmetrical heat sources increases by a maximum of 47% compared with asymmetrical heat sources. The heat source from the solid phase makes the heat transfer effect better, and when this heat source is large, the results obtained with a simple LTE model are erroneous. Thus, it is necessary to use a LTNE model.

Overall, there are many research results that both metal foams and nanofluids have good heat transfer enhancement effects, but the specific heat transfer enhancement mechanisms are not yet in depth. At present, there is still relatively little related research that combines the two. The effects of some key factors such as nanoparticle size and shape, Brownian diffusion, and thermophoretic diffusion are not well defined. Interactions among nanoparticles, base fluid, and foam metals need to be thoroughly investigated. The complex interplay between phases in nanofluids and metal foams is an important factor limiting further research. The lattice Boltzmann (LB) method, developed very rapidly in recent years, has a good computational efficiency, easy programming procedure, and easy processing of boundary conditions. The purpose of this chapter is to introduce the procedure of LB method for predicting the natural and forced convection of nanofluids in porous metals.

9.2 LB Modeling on the Nanofluid Natural Convection

9.2.1 Introduction to the Application of LB Method

The earliest LB model was proposed by McNamara and Zanetti [49] in 1988. But it cannot solve the problem of exponent complexity of the collision operator, and it is very complicated for too many particle types. Zou and He [50] derived the boundary conditions of D2Q9 and D3Q15 for the incompressible Lattice Bhatnagar-Gross-Krook (LBGK) model and verified it with the Poiseuille flow at the velocity and pressure boundary. By comparing the results with the analytical solution, the 3D model has a second-order accuracy. It is also found that using the half-step boundary to simulate the flow in the pipeline has a second-order accuracy, which can be considered in the LBGK model. With the development of LB method, many scholars began to try to apply it to the simulation of nanofluids and porous media.

Guo et al. [51] used LB method to simulate the convection heat transfer in porous media and introduced a new distribution function to simulate the temperature field. Nemati et al. [52] simulated the top cap driven flow of nanofluids and found that LB method is suitable for the simulation of nanofluids. The results show that the type of nanofluids has a significant effect on the flow and heat transfer. Mehriziet al. [53] used the LB method to study the heat dissipation of a cooling plate device filled with porous media based on the LTE model and found that porosity has a significant effect on the heat transfer (especially at high Reynolds number and Prandtl number). The Nusselt number at the inlet is lower than that at the outlet for a low porosity, and the porous medium at the outlet does not have a significant enhancing effect on the heat transfer. Sheikholeslami et al. [54] used LB method to simulate the natural convection of nanofluids in an elliptical cavity and reasonably processed the distribution function at the boundary. The influences of various factors on the Nusselt number were analyzed. Sheikholeslami et al. [55] studied the natural convection of circular inner boundary in a square cavity and analyzed the heat transfer effects of various nanofluids and multiple flow states. Sidik et al. [56] studied the nanofluid forced convection using LB method. Results show that increasing the Reynolds number and increasing the volume fraction of nanoparticles can lead to a better heat transfer. The same conclusions were drawn by using the traditional method, but the simulation process was greatly simplified with the LB method. Ata et al. [57] used the LB method to study the effect of a uniform vertical magnetic field on the nanofluid forced convection in a channel filled with a porous medium. It has been found that increasing the volume fraction of nanoparticles increases the outlet temperature and velocity, and the Nusselt number significantly increases. The effect of the magnetic field on the Nusselt number is gradually weakened. When the Hagen number increases, the outlet temperature and velocity slightly decrease, but the Nusselt number still increases slowly.

The LB method is a new research and development tool in the field of computational fluid dynamics and plays an important role in various industrial and academic applications. As a mesoscopic simulation tool, the LB method has good numerical stability and versatility and plays a crucial role in understanding mesoscopic flow and solving incompressible and compressible flow and heat transfer problems. The LB method is very suitable for dealing with conjugating transport problem of nanofluids within the metal foam [58]. Thus, we use LB method to numerically simulate the convection heat transfer of the nanofluid in highly conductive metal foams.

9.2.2 Nanofluid Thermophysical Properties

The thermophysical properties of nanofluid are related to the properties of the base fluid, nanoparticles, and the volume fraction of the nanoparticles. Among them, the kinematic viscosity, density, specific heat capacity, and thermal conductivity need to be determined for the nanofluid convection heat transfer. Density and specific heat capacity can be directly calculated based on nanoparticle volume fraction [16]:

$$\rho_{nf} = \varphi\rho_{np} + (1-\varphi)\rho_{bf} \tag{9.1}$$

$$c_{nf} = [\varphi\rho_{np}c_{np} + (1-\varphi)\rho_{bf}c_{bf}] / \rho_{nf}. \tag{9.2}$$

The thermal conductivity and kinematic viscosity need to be determined experimentally or calculated by empirical correlations. Tables 9.1 and 9.2, respectively, give the correlations for calculating the thermal conductivity and kinematic viscosity. Correlations in Lee et al. [59] and Einstein [60] are widely used for the higher accuracy, which is used to calculate the properties of nanofluids.

TABLE 9.1

The Semi-empirical Correlations of the Thermal Conductivity of Nanofluid

Year	Calculation Correlation	References
1873	$k_{nf} = \dfrac{k_{np} + 2k_{bf} + 2\left(k_{np} - k_{bf}\right)\varphi}{k_{np} + 2k_{bf} - \left(k_{np} - k_{bf}\right)\varphi} k_{bf}$	[61]
1999	$k_{nf} = \dfrac{k_{np} + (n-1)k_{bf} - (n-1)\left(k_{bf} - k_{np}\right)\varphi}{k_{np} + (n-1)k_{bf} + \left(k_{bf} - k_{np}\right)\varphi} k_{bf}$	[59]
2011	$\dfrac{k_{nf}}{k_{bf}} = 1 + 4.4 \left(\dfrac{T}{T_{fr}}\right)^{10} \left(\dfrac{k_{np}}{k_{bf}}\right)^{0.03} Pr^{0.66} Re^{0.4}$	[10]
2002	$k_{nf} = (1-\varphi)k_{bf} + \varphi k_{bf} + 3C \dfrac{d_{bf}}{d_{np}} k_{bf} Re_{d_{np}}^2 Pr \varphi$	[62]
2005	$\dfrac{k_{nf}}{k_{bf}} = 1 + 64.7\varphi^{0.746} \left(\dfrac{d_{bf}}{d_{np}}\right)^{0.369} \left(\dfrac{k_{np}}{k_{bf}}\right)^{0.7476} Pr^{0.9955} Re^{1.2321}$	[63]
2011	$\dfrac{k_{nf}}{k_{bf}} = 1 + 1.011\varphi + 2.438\varphi \left(\dfrac{47}{d_{np}}\right) - 0.025\varphi \left(\dfrac{k_{np}}{0.613}\right)$	[13]
2013	$k_{nf} = \begin{cases} k_{bf}(1 + 7.47\varphi), \text{ alumina} \\ k_{bf}(1 + 2.92\varphi - 11.99\varphi^2), \text{ titanian} \end{cases}$	[16]

TABLE 9.2

The Semi-empirical Correlations of the Kinematic Viscosity of Nanofluid

Year	Calculation Correlation	References
1906	$v_{nf} = \left(1 + 2.5\varphi\right)v_{bf}$	[60]
2005	$v_{nf} = \left(1 + 1.73\varphi + 123\varphi^2\right)v_{bf}$	[64]
2013	$v_{nf} = \begin{cases} v_{bf}(1 + 39.11\varphi + 533.9\varphi^2), \text{ alumina} \\ v_{bf}(1 + 5.45\varphi + 108.2\varphi^2), \text{ titanian} \end{cases}$	[16]

9.2.3 LB Modeling the Nanofluid Transport

Figure 9.1 shows the natural convection heat transfer of nanofluids in a square cavity. The length and width of the cavity are both H. The upper and lower walls are insulated, and the left and right boundaries maintain a constant temperature of $T_{c,h}$ and $T_{c,l}$, respectively. It is assumed that there is no chemical reaction between the nanoparticles and the base fluid, and they have the same temperature and velocity. The nanofluids are stable and the nanoparticles do not deposit.

Based on the assumptions, the above problem can be modeled with the Coupled Lattice Bhatnagar-Gross-Krook (CLBGK) model proposed by Guo et al. [65] to simulate the nanofluid natural convection heat transfer. The model includes the velocity distribution function for the velocity field and the temperature distribution function for the temperature field, respectively, corresponding to evolution equations are Eqs. (9.3) and (9.4).

$$f_i\left(\mathbf{r} + \mathbf{e}_i \delta t, t + \delta t\right) - f_i\left(\mathbf{r}, t\right) = -\frac{1}{\tau}\left[f_i\left(\mathbf{r}, t\right) - f_i^{eq}\left(\mathbf{r}, t\right)\right] + \delta t F_i \tag{9.3}$$

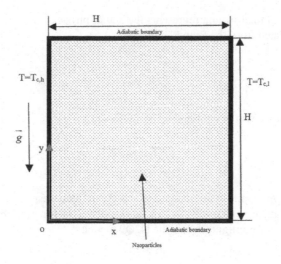

FIGURE 9.1   Schematic diagram of nanofluid natural convective heat transfer in a cavity.

$$g_i\left(\mathbf{r}+\mathbf{e}_i\delta t,t+\delta t\right)-g_i\left(\mathbf{r},t\right)=-\frac{1}{\tau_\mathrm{T}}\left[g_i\left(\mathbf{r},t\right)-g_i^{\mathrm{eq}}\left(\mathbf{r},t\right)\right]. \tag{9.4}$$

The above equations are based on the D2Q9 model, i.e., two-dimensional and nine-velocity directions. The subscript i (i = 0, 1,..., 8) in the evolution equation represents nine different directions. The parameters f_i and g_i are the velocity distribution function and the temperature distribution function in each direction. The corresponding equilibrium distribution functions are shown in Eqs.(9.5) and (9.6), respectively. *Fi* is the force distribution function (Eq. [9.7]).

$$f_i^{\mathrm{eq}} = \rho\omega_i\left[1+\frac{\mathbf{e}_i\cdot\mathbf{u}}{c_\mathrm{s}^2}+\frac{\left(\mathbf{e}_i\cdot\mathbf{u}\right)^2}{2c_\mathrm{s}^4}-\frac{\mathbf{u}^2}{2c_\mathrm{s}^2}\right] \tag{9.5}$$

$$g_i^{\mathrm{eq}} = T\omega_i\left[1+\frac{\mathbf{e}_i\cdot\mathbf{u}}{c_\mathrm{s}^2}+\frac{\left(\mathbf{e}_i\cdot\mathbf{u}\right)^2}{2c_\mathrm{s}^4}-\frac{\mathbf{u}^2}{2c_\mathrm{s}^2}\right] \tag{9.6}$$

$$F_i = \rho\omega_i\left(1-\frac{1}{2\tau}\right)\left[\frac{\mathbf{e}_i\cdot\mathbf{F}}{c_\mathrm{s}^2}+\frac{\mathbf{u}\mathbf{F}:(\mathbf{e}_i\mathbf{e}_i-c_\mathrm{s}^2\mathbf{I})}{c_\mathrm{s}^4}\right], \tag{9.7}$$

where ρ, **u**, and *T* are the density, velocity, and temperature in lattice unit, respectively. The LB method has its own dimensionless unit system. The dimensionless numbers in the lattice unit, such as Reynolds number and Rayleigh number, must be equal to those in the actual physical unit. When using LB method, the viscosity (υ) and the lattice number (*N*) in the lattice unit should be generally determined. The viscosity should be of a proper value. The velocity in the lattice unit can be calculated from the Reynolds number, shown as:

$$u = \mathrm{Re}\cdot\frac{v}{N} = \frac{u_\mathrm{ap}H_\mathrm{ap}}{v_\mathrm{ap}}\cdot\frac{v}{N}, \tag{9.8}$$

where the subscript "ap" in the formula represents the actual physical unit. The thermal diffusivity in lattice units can be calculated from the Prandtl number:

$$a = \frac{v}{Pr} = \frac{v}{v_{ap}/a_{ap}}. \tag{9.9}$$

The temperature in the lattice unit is obtained by normalization:

$$T = \frac{T_{ap} - T_{c,l}}{T_{c,h} - T_{c,l}}. \tag{9.10}$$

The lattice velocity $c = \delta_x/\delta_t$, δ_x and δ_t are the grid step and time step, respectively, and $c = 1$ in general. The parameter c_s is the lattice velocity, $c_s^2 = c^2/3$, ω_i is the weight function, $\omega_0 = 4/9$, $\omega_{1-4} = 1/9$, and $\omega_{5-8} = 1/36$. e_i are the velocity configuration of the D2Q9 model:

$$\mathbf{e}_i = \begin{cases} (0,0) & i = 0 \\ c\left(\cos\left[(i-1)\frac{\pi}{2}\right], \sin\left[(i-1)\frac{\pi}{2}\right]\right) & i = 1, 2, 3, 4 \\ \sqrt{2}c\left(\cos\left[(2i-1)\frac{\pi}{4}\right], \sin\left[(2i-1)\frac{\pi}{4}\right]\right) & i = 5, 6, 7, 8. \end{cases} \tag{9.11}$$

The parameters τ and τ_T are, respectively, the relaxation times for the velocity equation and the temperature equation, which can be calculated from the kinematic viscosity (v) and thermal diffusion coefficient (a) of the nanofluid, shown as:

$$\tau = \frac{v\delta t}{c_s^2} + \frac{1}{2} \tag{9.12}$$

$$\tau_T = \frac{a\delta t}{c_s^2} + \frac{1}{2}. \tag{9.13}$$

The macro physical quantities of nanofluids, such as density, velocity, and temperature, can be calculated from the distribution functions:

$$\rho = \sum_{i=0}^{8} f_i \tag{9.14}$$

$$\rho\mathbf{u} = \sum_{i=0}^{8} \mathbf{e}_i f_i + \frac{\delta t}{2}\rho\mathbf{F} \tag{9.15}$$

$$T = \sum_{i=0}^{8} g_i. \tag{9.16}$$

$\mathbf{F}$ is the force in Eq. (9.5), and it is the force caused by the density difference in natural convection ($\mathbf{F}_G$, Buoyancy force):

$$\mathbf{F}_G = \mathbf{g}\beta(T - T_m), \tag{9.17}$$

where:

  g is the gravitational acceleration

  β is the coefficient of thermal expansion

  Tm is the average temperature of the left and right walls.

In the physical model in Figure 2.1, the velocities on the four walls of the square cavity are all non-slip, and no heat is transferred on the upper and lower boundaries. The temperature on the left and right boundaries does not change, as shown in Eq. (9.16):

$$\begin{cases} x = 0: \mathbf{u} = 0, T = T_{c,h} \\ x = H: \mathbf{u} = 0, T = T_{c,l} \\ y = 0: \mathbf{u} = 0, \partial T/\partial y = 0 \\ y = H: \mathbf{u} = 0, \partial T/\partial y = 0. \end{cases} \tag{9.18}$$

The boundary condition in the LB model refers to the calculation method of the distribution function on the given boundary. The non-equilibrium extrapolation scheme [65] used in this chapter has a second-order accuracy and good numerical stability. The calculation formulas for the velocity and temperature distribution functions of the non-equilibrium extrapolation scheme are shown in Eqs. (9.17) and (9.18), respectively, as:

$$f_i(\mathbf{x}_b) - f_i^{eq}(\mathbf{x}_b) = f_i(\mathbf{x}_f) - f_i^{eq}(\mathbf{x}_f) \tag{9.19}$$

$$g_i(\mathbf{x}_b) - g_i^{eq}(\mathbf{x}_b) = g_i(\mathbf{x}_f) - g_i^{eq}(\mathbf{x}_f), \tag{9.20}$$

where $\mathbf{x}_b$ and $\mathbf{x}_f$ are, respectively, the lattice points on the wall and lattice points adjacent to the wall in the fluid area.

9.2.4 Analysis

The natural convective heat transfer of nanofluids in the cavity is simulated using the LB method. In order to verify the correctness of the program, the calculation results are compared with those of He et al. and Hortmann et al. [66–68]. The results of the study have been compared and verified. Figure 9.2 shows the comparison of velocity field and temperature field with Rayleigh number Ra = 10^5, and Figure 9.2a and b are the natural convection of the cavity in the cavity simulated by He et al. [67]. The streamlines and isotherms, Figure 9.2c, are the streamline and temperature fields for the calculations in this chapter's program. From Figure 9.2c it is shown that the velocity field and temperature field obtained by simulation in this chapter are the same as those in the literature, which can explain the correctness of the program. By defining the Nusselt number at the wall,

$$Nu_{ave} = -\frac{1}{\Delta T} \int_0^H \frac{\partial T}{\partial x}\bigg|_{x=0} dy. \tag{9.21}$$

Table 9.3 shows the maximum Nusselt number (Nu_{max}) at the left wall surface and its position (y_{Nu}/H) and the average Nusselt number at the left wall surface (Nu_{ave}, as defined by Eq. [21]). Comparing with the results in the literature, some of the results are almost identical to the baseline solution. The maximum error of some values is only 1.2%. This verifies the accuracy of the present results. It also shows that LB method is feasible to simulate the natural convection of nanofluids. Hot feasibility. Figure 9.3

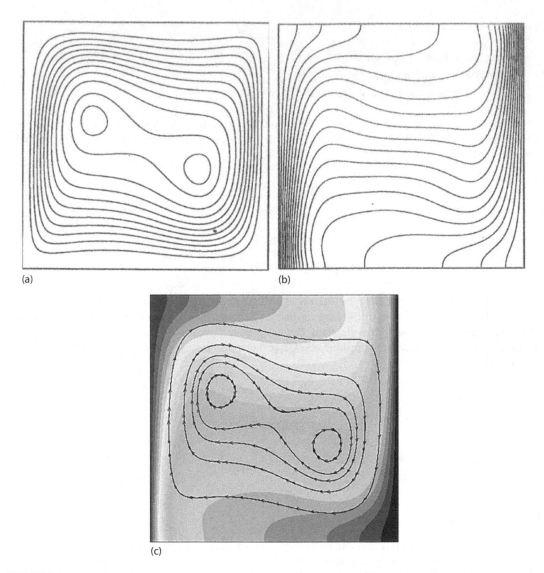

(a)   (b)

(c)

FIGURE 9.2 Validation of the present velocity and temperature results with results. (a) stream function, (b) isothermin, and (c) the present results. (From Hortmann, M. et al., *Int. J. Numer. Methods Fluids*, 11, 189–207, 1990.)

TABLE 9.3

The Comparison of the Result Between Present Study and Others' Work

Rayleigh Number	Comparison	Nu_{max}	y_{Nu}/H	Nu_{ave}
10^5	Present	7.6633	0.9218	4.4936
	He et al. [67]	7.7500	0.9219	4.5219
	Hortmann et al. [66]	7.7201	0.9180	4.5216
10^6	Present	17.4374	0.9648	8.7165
	He et al. [67]	17.6140	0.9648	8.7926
	Hortmann et al. [66]	17.5360	0.9608	8.8251

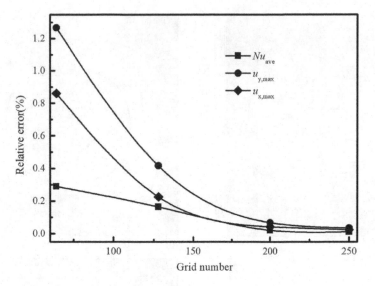

FIGURE 9.3   Relative error of the numerical solution.

shows the grid independence assessment. The results of Hortmann et al. [66] are used to give the relative sizes of the results obtained by applying this program under different grid numbers and in the literature. From Figure 9.3, it can be seen that as the number of grids increases, the program results are closer to the correct results, and when the number of grids is greater than 200, the calculation results almost do not change. Therefore, we use 250 × 250 as the grid density.

Figure 9.4 shows the relationship between the Rayleigh number and the mean velocity (u_{ave}, calculated by Eq. [22]), and the numerical results with Al_2O_3/H_2O nanoparticle volume fractions of 5% and 10% were compared with each other. With an increase in Ra, the average velocity rapidly increases, and the average velocity is very small when Ra is small. The average velocity slightly increases when the volume fraction of nanoparticles increases from 5% to 10%.

$$u_{ave} = \frac{\int\limits_0^H \int\limits_0^H \left(u_x^2 + u_y^2\right) dy dx}{\int\limits_0^H \int\limits_0^H \sqrt{u_x^2 + u_y^2} dy dx}. \qquad (9.22)$$

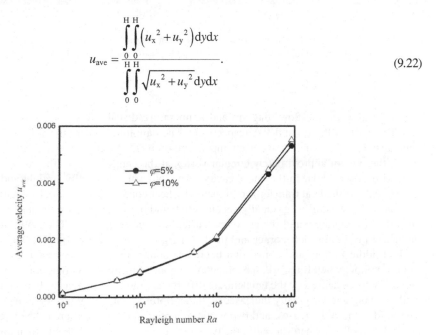

FIGURE 9.4   Effect of Ra on the average velocity.

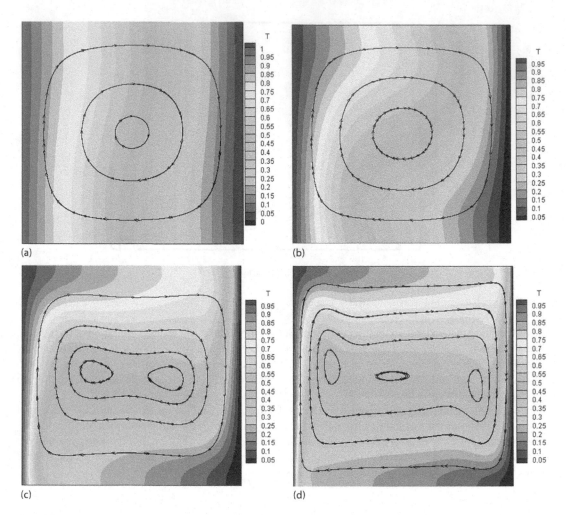

FIGURE 9.5 The streamline and temperature distribution diagram of natural convective heat transfer in nanofluid for different Rayleigh number (a) Ra = 10^3, (b) Ra = 10^4, (c) Ra = 10^5, and (d) Ra = 10^6.

Figure 9.5a–d are the flow diagram and temperature distribution of natural convection heat transfer with Ra = 10^3, 10^4, 10^5, and 10^6, respectively. The nanofluid is with the volume fraction of Al_2O_3/H_2O nanoparticles of 5% and an average temperature of 30°C. As can be seen from Figure 9.5a, when Ra is small, there is an approximately circular vortex at the central area of the cavity, and the streamline is approximately circular in the entire cavity. The isotherm is basically straight and evenly distributed in the cavity, and the heat transfer is dominated by heat conduction for small velocity. When Ra gradually increases, the streamline gradually becomes flat, and its shape is approximately an ellipse. The velocity of flow is accelerated, and the convection effect gradually appears. The isotherm begins to bend, and the heat in the upper left corner and the lower right corner of the square cavity is quickly transferred to the middle region, and convection begins to dominate the heat transfer, as shown in Figure 9.5b. In Figure 9.5c, when Ra = 10^5, the temperature non-uniformity becomes apparent, and accordingly, the buoyancy force caused by the temperature difference gradually increases. Especially in the area near the wall, the buoyancy is greatly increased, and therefore the rotational motion of nanofluid is very obvious. The fluid velocity increases, and two vortices appear in the central area. The streamline becomes flatter. In Figure 9.5d, when Ra = 10^6, the two vortices in the flow field begin to approach the left and right

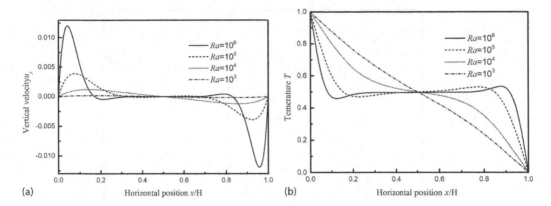

(a)

(b)

FIGURE 9.6   The distribution of velocity and temperature at $y = 0.5$ H. (a) velocity and (b) temperature.

walls, and a third vortex appears in the central area of the enclosure. The isotherm at the center becomes straight and flat, indicating that the convective heat transfer is more intense with an increase in Rayleigh number, and the natural convection in the central area is also gradually enhanced.

Figure 9.6a shows the comparison of the vertical velocity (u_y) distribution for different Rayleigh numbers with the nanoparticle volume fraction 5% at $y = 0.5$ H. From Figure 9.6a, compared with other situations, the vertical velocity for Ra = 10^3 is very small, which means that the flow is very weak. Compared with the isotherm in Figure 9.5a, the temperature difference in vertical direction is almost zero, so the buoyancy due to temperature difference is very small. Figure 9.6b shows the comparison of temperature distribution for different Rayleigh numbers at $y = 0.5$ H. The temperature at Ra = 10^3 is approximately linear, indicating that heat conduction dominates the heat transfer. A large Rayleigh number corresponds to the large coefficient of thermal expansion, large temperature difference, or the small viscosity. Therefore, an increase in Rayleigh number facilitates fluid flow and heat diffusion. From Figure 9.6, as Ra increases, the velocity gradually increases, and the temperature profile and the isotherm begin to bend. Due to the intensified flow, the temperature gradient near the wall is large, and heat transfer is enhanced. For Ra = 10^5, the vertical velocity is negative in the left half of x < 0.5, and the temperature is smaller than the average temperature, indicating that the vortex brought by vortex at large Rayleigh number corresponds to the intensified local flow. This is conducive to enhancing natural convective heat transfer.

Figure 9.7 shows the relationship between the Rayleigh number and the average Nusselt number on the left wall with the volume fraction of nanoparticles 5% and 10%, respectively. From Figure 9.7, with an increase in Ra, the variation trend of the average Nusselt number is similar to that of the average velocity, but the effect of nanoparticle volume fraction on the average Nusselt number is more obvious. This is because the addition of highly conductive nanoparticles increases the effective thermal conductivity and viscosity of nanofluids, leading to the increased Nusselt number. When Ra is small, the velocity and the average Nusselt number are very small. For example, for Ra = 10^3, $u_{ave} = 1.24 \times 10^{-4}$, $Nu_{ave} = 1.2308$, the fluid is almost stationary, and the heat transfer is also dominated by heat conduction. Figure 9.8 shows the relationship between the nanoparticle volume fraction and the average Nusselt number of Al_2O_3/H_2O nanofluids for Rayleigh numbers of 10^5 and 10^6, respectively. From Figure 9.8, as the nanoparticle volume fraction increases, Nu_{ave} gradually increases, and the increment is more obvious. It is also shown that Rayleigh number has a more significant effect on the natural convection heat transfer of nanofluids. For Ra = 10^5, as the nanoparticle volume fraction increases from 0% to 10%, Nu_{ave} increases by 70%. For Ra = 10^6, Nu_{ave} increases by 40%. When Ra is higher, the convection plays a more significant role, so increasing the nanoparticle volume fraction and the thermal conductivity for high Rayleigh number is not as obvious as the case for the low Rayleigh number.

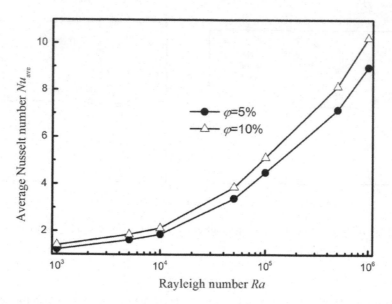

FIGURE 9.7   Effect of Ra on Nu.

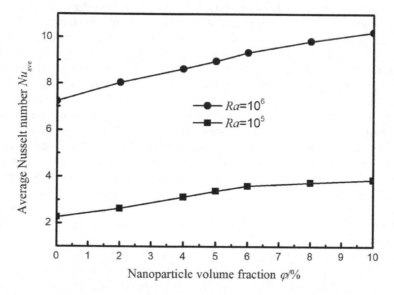

FIGURE 9.8   Effect of nanoparticle volume fraction on Nu.

9.3 Natural Convection of Nanofluids in Metal Foams

9.3.1 Physical Problem and Mathematical Model

Figure 9.9 shows the natural convection heat transfer model of the nanofluid in the cavity filled with metal foam. The upper and lower walls of the cavity are insulated, and the left and right walls are kept at constant temperatures $T_{c,h}$ and $T_{c,l}$, and the width of the cavity is H. Assuming that there is no chemical reaction between the nanoparticles and the base fluid, the nanofluid is stable, and the nanoparticles do not deposit on the surfaces of the wall and solid ligaments. The nanoparticle and the base fluid velocities are considered as the same, and the same temperature between the nanoparticle, the base fluid, and the metal foam, are assumed, i.e., the LTE assumption.

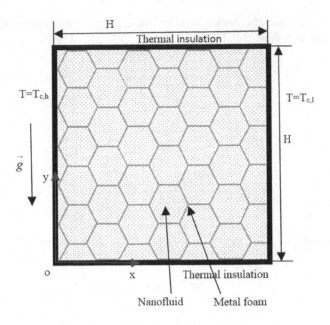

FIGURE 9.9   The physical model of natural convection of nanofluid in metal foams.

For the above problem, a mathematical model for the convective heat transfer of nanofluids in metal foams can be established [44], including the continuity equation (Eq. 9.23), the momentum equation (Eqs. 9.24 and 9.25), and energy equation (Eq. 9.26), shown as:

$$\frac{\partial(\rho_{nf}u_x)}{\partial x}+\frac{\partial(\rho_{nf}u_y)}{\partial y}=0 \tag{9.23}$$

$$\rho_{nf}\left[\frac{\partial(u_xu_x)}{\partial x}+\frac{\partial(u_yu_x)}{\partial y}\right]=-\frac{dp}{dx}+\frac{\mu_{nf}}{\varepsilon}\left(\frac{\partial^2u_x}{\partial x^2}+\frac{\partial^2u_x}{\partial y^2}\right)-\frac{\mu_{nf}}{K}u_x-\rho_{nf}\frac{F_\varepsilon}{\sqrt{K}}\sqrt{u_x^2+u_y^2}u_x \tag{9.24}$$

$$\rho_{nf}\left[\frac{\partial(u_xu_y)}{\partial x}+\frac{\partial(u_yu_y)}{\partial y}\right]=-\frac{dp}{dy}+\frac{\mu_{nf}}{\varepsilon}\left(\frac{\partial^2u_y}{\partial x^2}+\frac{\partial^2u_y}{\partial y^2}\right)-\frac{\mu_{nf}}{K}u_y-\rho_{nf}\frac{F_\varepsilon}{\sqrt{K}}\sqrt{u_x^2+u_y^2}u_y+G_y \tag{9.25}$$

$$\rho_{nf}c_{nf}\left[\frac{\partial(u_xT)}{\partial x}+\frac{\partial(u_yT)}{\partial y}\right]=\frac{\partial}{\partial x}\left[(k_{ne}+k_d)\frac{\partial T}{\partial x}\right]+\frac{\partial}{\partial y}\left[(k_{ne}+k_d)\frac{\partial T}{\partial y}\right]. \tag{9.26}$$

For the physics problem in this section, the LBGK model based on the D2Q9 model [46] can be used to simulate the flow and heat transfer process. The evolution equations of the velocity and temperature distribution functions are, respectively [43]:

$$f_i(\mathbf{r}+\mathbf{e}_i\delta t,t+\delta t)-f_i(\mathbf{r},t)=-\frac{1}{\tau_{nf}}\left[f_i(\mathbf{r},t)-f_i^{eq}(\mathbf{r},t)\right]+\delta tF_i \tag{9.27}$$

$$g_i(\mathbf{r}+\mathbf{e}_i\delta t,t+\delta t)-g_i(\mathbf{r},t)=-\frac{1}{\tau_T}\left[g_i(\mathbf{r},t)-f_i^{eq}(\mathbf{r},t)\right]. \tag{9.28}$$

For calculating the equilibrium distribution function, the effect of porosity should be considered compared to the LBGK model with no metal foam:

$$f_i^{eq} = \rho \omega_i \left(1 + \frac{e_i \cdot u}{c_s^2} + \frac{(e_i \cdot u)^2}{2\varepsilon c_s^4} - \frac{u^2}{2\varepsilon c_s^2} \right) \tag{9.29}$$

$$g_i^{eq} = T w_i \left(1 + \frac{e_i \cdot u}{c_s^2} + \frac{(e_i \cdot u)^2}{2 c_s^2} - \frac{u^2}{2 c_s} \right) \tag{9.30}$$

$$F_i = \rho \omega_i \left(1 - \frac{1}{2\tau_{nf}} \right) \left[\frac{e_i \cdot F}{c_s^2} + \frac{uF : (e_i e_i - c_s^2 I)}{\varepsilon c_s^4} \right]. \tag{9.31}$$

The force F in Eq. (9.31) includes the buoyancy F_G and the flow resistance due to metal foam F_R, expressed as:

$$F = F_G + F_R \tag{9.32}$$

$$F_R = -\frac{\varepsilon v}{K} u - \frac{\varepsilon F_\varepsilon}{\sqrt{K}} |u| u \tag{9.33}$$

$$F_G = g\beta(T - T_m) \tag{9.34}.$$

The density, velocity, and temperature of the nanofluid can be calculated based on velocity and temperature distribution functions, shown as:

$$\rho = \sum_{i=0}^{8} f_i \tag{9.35}$$

$$u = \frac{V}{d_0 + \sqrt{d_0^2 + d_1 |V|}} \tag{9.36}$$

$$\rho V = \sum_{i=0}^{8} e_i f_i, \quad d_0 = \frac{1}{2} \left(1 + \varepsilon \frac{\delta t}{2} \frac{v}{K} \right), \quad d_1 = \varepsilon \frac{\delta t}{2} \frac{F_\varepsilon}{\sqrt{K}} \tag{9.37}$$

$$T = \sum_{i=0}^{8} g_i. \tag{9.38}$$

9.3.2 Analysis

In order to verify the calculation procedure of the LB model, the results are compared with the results in the literature. Figure 9.10 compares the temperature distribution at the middle position ($y = 0.5$ H) of the square cavity in the Reference [51] for Da = 10^{-2}, Ra = 10^5, and Da = 10^{-4}, Ra = 10^7. Table 9.4 compares the average Nusselt number in this chapter with that of References [68,69] for Pr = 1. From Figure 9.10 and Table 9.4, it can be seen that the results in this chapter are basically the same as those in the literature. This shows that the LB model based on the LTE assumption is suitable for the natural convection of nanofluids in metal foams. The simulation also shows that the results of the present simulation are accurate.

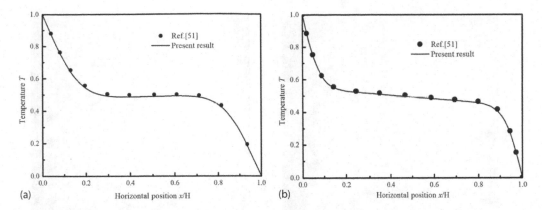

FIGURE 9.10 Contrast of temperature distribution of nanofluid natural convection in metal foam (a) Da = 10^{-2}, Ra = 10^5 and (b) Da = 10^{-4}, Ra = 10^7.

TABLE 9.4

Contrast of Average Nusselt Number of Nanofluid Natural Convection in Metal Foam

			Nu_{ave}		
Da	**Ra**	**References**	$\varepsilon = 0.4$	$\varepsilon = 0.6$	$\varepsilon = 0.9$
10^{-4}	10^5	[68]	1.067	1.071	1.072
		[69]	1.063	1.066	1.067
		Present	1.065	1.068	1.069
	10^6	[68]	2.550	2.725	2.740
		[69]	2.554	2.610	2.637
		Present	2.530	2.610	2.651
10^{-2}	10^3	[68]	1.010	1.015	1.023
		[69]	1.007	1.012	1.017
		Present	1.015	1.019	1.023
	10^4	[68]	1.408	1.530	1.640
		[69]	1.362	1.493	1.633
		Present	1.364	1.493	1.636
	10^5	[68]	2.983	3.555	3.910
		[69]	2.992	3.433	3.902
		Present	2.988	3.451	3.901
	5×10^5	[68]	4.990	5.740	6.700
		[69]	4.923	5.637	6.336
		Present	4.967	5.721	6.499

Figure 9.11 shows the flow and temperature distribution of the natural convection heat transfer in the cavity of nanofluids filled with metal foams with Rayleigh numbers Ra = 10^4, 10^5, 10^6, Darcy number 10^{-4}, the porosity 0.6, and the volume fraction of Al_2O_3 nanoparticles 5%. Figure 9.12 shows the temperature distribution at y = 0.5 H, y = 0.75 H, and y = H for Rayleigh numbers Ra = 10^4, Ra = 10^5, and Ra = 10^6.

From Figure 9.11a, it can be seen that when Ra = 10^4, the temperature distribution is very uniform and the heat transfer is dominated by heat conduction. From the streamline point of view, the vertical velocity is larger and the horizontal velocity is smaller. In Figure 9.12a, it can be clearly seen that the temperature is linearly distributed in the x direction, and the temperatures at the three positions y = 0.5 H,

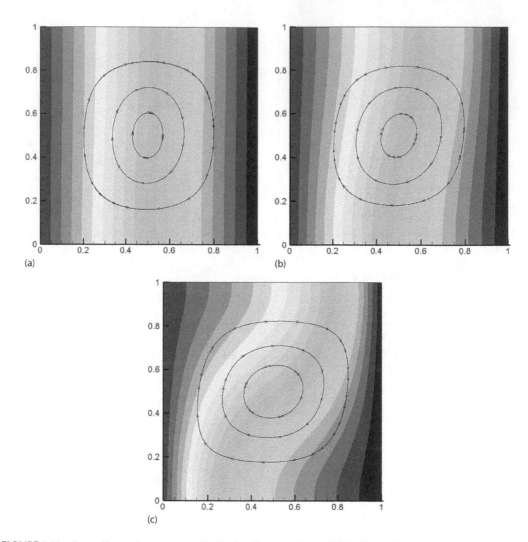

FIGURE 9.11   Streamline and temperature distribution diagram (a) Ra = 10^4, (b) Ra = 10^5, and (c) Ra = 10^6.

$y = 0.75$ H, and $y =$ H are substantially the same. It also shows that the temperature distribution between the left and right wall surfaces at this time approximately satisfies the law of Fourier heat conduction, and the convection heat transfer can be ignored. In Figure 9.11b and c, the fluid for the high temperature area in the upper left corner of the cavity and that for the low temperature area in the lower right corner are rapidly transferred to other areas and fluid flow becomes stronger as Ra increases. Temperature diffusion becomes faster and buoyancy gradually plays a more significant role in flow and heat transfer. The non-linearity of the temperature distribution in Figure 9.11c is more pronounced, which means that the convection heat transfer has occupied a dominant position. Accordingly, in Figure 9.12c, it can clearly be seen that there are different temperature distributions at different locations. The obvious difference is that the temperature gradients near the left and right walls also become larger. Figures 9.11 and 9.12 clearly show the heat transfer transition from the heat conduction to the convection in the natural convection with an increase in Ra. The non-linear effect for convection is enhanced with an increase in Rayleigh number.

Figure 9.13 shows the vertical velocity (u_y) distribution at Rayleigh number Ra = 10^4, Ra = 10^5, and Ra = 10^6 at $y = 0.5$ H, and $y = 0.75$ H, corresponding to Figure 9.11a–c, respectively. In Figure 9.13a, the maximum value of u_y is only 6×10^{-5}, and u_y is distributed uniformly in the x direction. In Figure 9.13b and c, u_y increases rapidly and begins to become uneven. The flow is stronger in the region near the left and right walls, and the flow is weaker in the middle area.

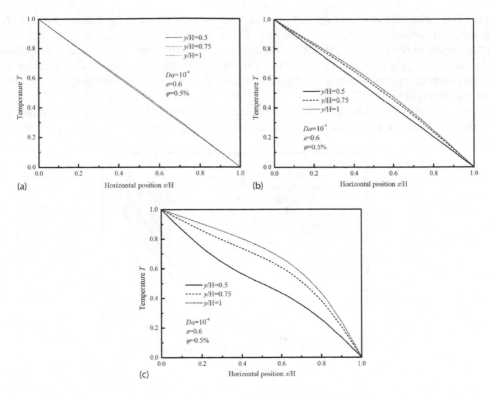

FIGURE 9.12 The temperature distribution diagram at $y = 0.5\,H$, $y = 0.75\,H$ and $y = H$ (a) Ra = 10^4, (b) Ra = 10^5, and (c) Ra = 10^6.

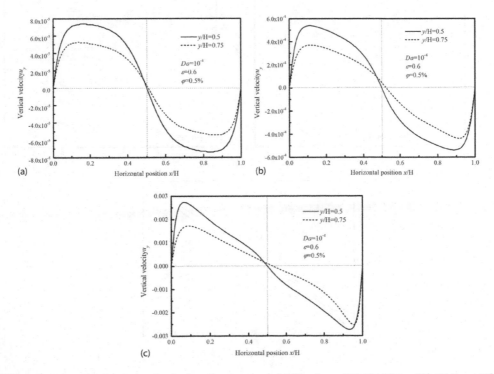

FIGURE 9.13 The vertical velocity distribution diagram at $y = 0.5\,H$ and $y = 0.75\,H$ (a) Ra = 10^4, (b) Ra = 10^5, and (c) Ra = 10^6.

Figure 9.14 shows the relationship between Rayleigh number and average Nusselt number (Nu_{ave}) under different porosities for Da = 10^{-2} and Pr = 1. This figure shows the comparison with the corresponding results in the literature [68,69]. As Ra increases, Nu_{ave} increases with a smaller amplitude for smaller Ra and sharply increases for larger Ra. This variation pattern is related to temperature difference, fluid volume expansion ability, and fluid fluidity, which corresponds to the changes in the flow and temperature fields in Figure 9.11.

Figure 9.15a shows the relationship between the Darcy number and the average Nusselt number. The Rayleigh numbers are Ra = 10^6 and Ra = 10^7, the porosity is 0.6, and the Al_2O_3 nanoparticle volume

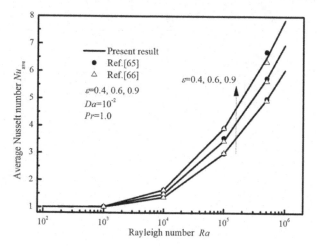

FIGURE 9.14 Effects of porosity and Rayleigh number on average Nusselt number.

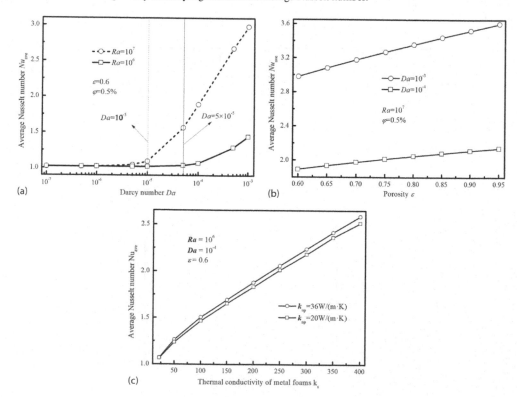

FIGURE 9.15 Effects of metal foam on average Nusselt number (a) Darcy number, (b) porosity, and (c) metal foam thermal conductivity.

fraction is 0.5%. When Da is large, the fluid has a strong penetration ability in the metal foam, so when Da increases, the flow strengthens and the average Nusselt number increases. The effect of Ra on Nu_{ave} is more obvious when Da is larger. When Ra increases from 10^6 to 10^7, the average Nusselt number corresponding to Da = 10^{-3} and Da = 10^{-4} increases to 2.06 and 1.77, respectively. Times when Da is small, the flow resistance due to the solid skeleton of the metal foam is too large and the fluid velocity is small, so the gain brought by the increase of Ra to the velocity is not significant. The average Nusselt number almost never changes at Da < 10^{-5}. When Da increases from 10^{-7} to 10^{-6}, the average Nusselt number only increases by 0.03% (Ra = 10^6) and 0.14% (Ra = 10^7), while the temperature distribution at this time is similar to Figures 9.11a and 9.12a, so Da = 10^{-5} can be considered as the starting point of natural convection, before that can be considered only heat conduction. Figure 9.15b shows the relationship between the porosity (ε) and the average Nusselt number. In this figure, the Darcy numbers are 10^{-3} and 10^{-4}, the Rayleigh number is 10^7, and the Al_2O_3 nanoparticle volume fraction is 0.5%. An increase in the porosity can lead to the decreased volume of the solid skeleton in the metal foam, the increased fluid space, and the decreased flow resistance, and thus fluid flow and macro convective heat transfer are enhanced. Figure 9.15c shows the relationship between the coefficient of thermal conductivity (k_s) of metal foam and Nu_{ave}, for two different nanoparticle thermal conductivities (36 W/[m·K] and 20 W/[m·K]). In this figure, the Darcy number is 10^{-4}, the Rayleigh number is 10^7, and the porosity is 0.6. From this figure, the average Nusselt number also increases significantly as the thermal conductivity of the metal foam increases, whereas the thermal conductivity of the nanoparticles (k_{np}) has a less significant effect on the average Nusselt number. When the pore size of nanoparticles is comparable to the pore size of metal foams, the fluid flow will be very small due to the too small pore size and permeability of metal foams.

Figure 9.16a shows the relationship between nanoparticle thermal conductivity (k_{np}) and nanoparticle volume fraction (φ) and the average Nusselt number. The volume fraction of nanoparticles has a great influence on the thermophysical properties of nanofluids. With the increase of φ, the thermal conductivity of nanofluids also increases gradually, which is beneficial to enhance heat transfer and therefore Nu_{ave} also increases. Figure 9.16b shows the relationship between the thermal conductivity of nanoparticles and Nu_{ave} for two thermal conductivities of metal foams (k_s = 22.2 and 100 W/[m·K]). The increase in the thermal conductivity of the nanoparticles increases the average Nusselt number, but the larger the k_{np}, the slower the increase of the Nu_{ave}. This is related to the relationship between the thermal conductivity of nanoparticles and nanofluids. The calculation formula given above shows that when k_{np} is small, there is a large change in k_{nf}.

Figure 9.17 shows the relationship between the Darcy number and the medium thermal conductivity (k_m, Eq. [39]). The thermal conductivity of the medium refers to the equivalent thermal conductivity of the entire cavity as a solid with only heat conduction. When Da or Ra is high, natural convection is greatly strengthened and heat transfer is accelerated, so the thermal conductivity of the medium is increased. This shows that the natural convection of the nanofluid in the square cavity of the metal foam

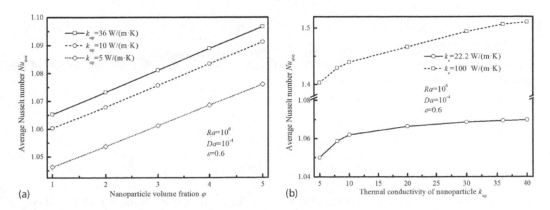

FIGURE 9.16 Effects of nanoparticle on average Nusselt number (a) Nanoparticle volume fraction and (b) Nanoparticle thermal conductivity.

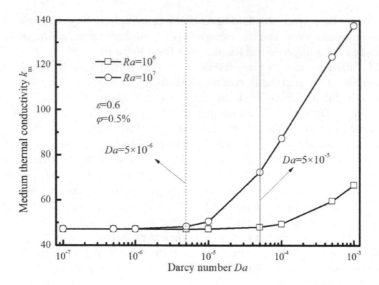

FIGURE 9.17   Effect of Da on medium thermal conductivity.

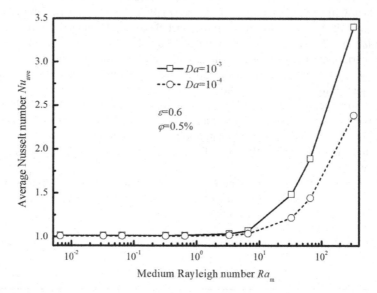

FIGURE 9.18   Effect of Da on medium Ra.

can be regarded as a high thermal conductivity solid medium. The medium thermal conductivity of the hypothetical solid can be evaluated with:

$$k_m = \frac{Q}{A\Delta T} = \frac{\int_0^H \left(-k_{ne} \frac{\partial T}{\partial x}\right)\Big|_{x=H} dy}{H(T_{c,h} - T_{c,l})}. \tag{9.39}$$

Figure 9.18 shows the relationship between the medium Rayleigh number (Ram, Eq. [40]) and Nu_{ave}. The medium Rayleigh number is a combination of metal foam and nanofluid. The characteristics of fluid flow and heat transfer, as well as the permeability of metal foam, thermal conductivity, and other

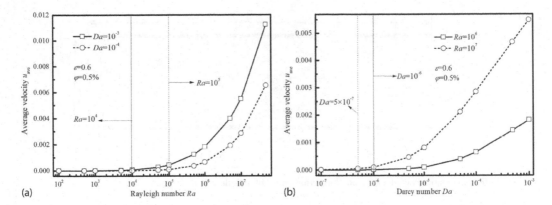

FIGURE 9.19   The effect of Rayleigh number and Darcy number on average velocity (a) Rayleigh number and (b) Darcy number.

natural convection heat transfer relationships are considered. It can be seen from Figure 9.18 that Nu_{ave} is increased and natural convection is enhanced with an increase in Ram defined as:

$$Ra_{m} = Ra \cdot Da \frac{k_{nf}}{k_{ne}}. \tag{9.40}$$

Figure 9.19a and b show the average velocity influenced by the Darcy and Rayleigh numbers, respectively. It can be seen from this figure that the average velocity is very low at low Darcy or low Rayleigh numbers, in which condition the fluid is very weak and only heat conduction is at work. With the increase of Da and Ra, u_{ave} gradually increases and convective heat transfer can no longer be ignored. When Da or Ra exceeds a certain value, the increment in the average velocity will be enhanced, and the convection heat transfer starts to dominate the natural convection.

In this section, the convective heat transfer of nanofluids in an enclosure filled with metal foams is numerically studied. The corresponding LB model is established based on LTE and LTNE models. The effects of porosity, nanoparticle concentration, thermal conductivity, Darcy number, etc., on natural convective heat transfer are analyzed. For the natural convection heat transfer of the nanofluid in the square cavity filled with metal foams, an increase in the Raleigh number or Darcy number will increase the Nusselt number. When Da < 10^{-5}, the natural convection can be neglected and only heat conduction should be considered. The Nusselt number also increases with an increase in the thermal conductivity of the nanoparticle or that of the metal foam. It should be noted that the transition boundary between conductive and convective regimes is hard to define, since the conduction in natural convective heat transfer of nanofluid in porous foams is always very strong.

9.4   Forced Convection of Nanofluids in Metal Foams

9.4.1   Physical Problem and Mathematical Model

Figure 9.20 is a metal-foam flat-plate channel filled with the length L and the height H. The upper and lower walls are with the condition of a constant temperature T_w. A nanofluid flows from the left side with the velocity u_{in} and temperature T_{in}, and convective heat transfer occurs inside the channel. In order to establish a suitable mathematical model, it is assumed that the nanofluid is incompressible with no chemical reactions considered, external forces ignored (such as gravity), and viscous dissipation and radiant heat transfer neglected. The flow and heat transfer at the outlet are treated as hydraulically and thermally fully developed, and the fluid flow is in laminar state. Thermal equilibrium state between the

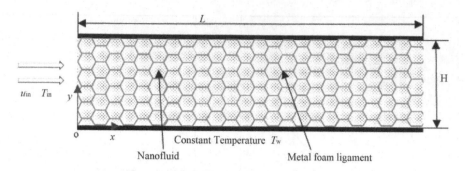

FIGURE 9.20 Schematic diagram of forced convection model of nanofluid in metal foams.

nanoparticles and the base fluid are assumed, and the temperature difference between the nanofluid and the solid framework of the metal foam is involved.

Based on the basic assumptions, a mathematical model for the forced convection heat transfer of nanofluids in metal foams is established using the volume average method, including mass conservation equations, momentum conservation equations, and energy conservation equations, shown as:

$$\frac{\partial(\rho_{nf} u_x)}{\partial x} + \frac{\partial(\rho_{nf} u_y)}{\partial y} = 0 \tag{9.41}$$

$$\rho_{nf}\left[u_x \frac{\partial u_x}{\partial x} + u_y \frac{\partial u_x}{\partial y}\right] = \frac{\mu_{nf}}{\varepsilon}\left(\frac{\partial^2 u_x}{\partial x^2} + \frac{\partial^2 u_x}{\partial y^2}\right) - \frac{dp}{dx} - \frac{\mu_{nf}}{K} u_x - \rho_{nf} \frac{F_\varepsilon}{\sqrt{K}}\sqrt{u_x^2 + u_y^2}\, u_x \tag{9.42}$$

$$\rho_{nf}\left[u_x \frac{\partial u_y}{\partial x} + u_y \frac{\partial u_y}{\partial y}\right] = \frac{\mu_{nf}}{\varepsilon}\left(\frac{\partial^2 u_y}{\partial x^2} + \frac{\partial^2 u_y}{\partial y^2}\right) - \frac{dp}{dy} - \frac{\mu_{nf}}{K} u_y - \rho_{nf} \frac{F_\varepsilon}{\sqrt{K}}\sqrt{u_x^2 + u_y^2}\, u_y \tag{9.43}$$

$$0 = k_{se}\left(\frac{\partial^2 T_s}{\partial x^2} + \frac{\partial^2 T_s}{\partial y^2}\right) - h_{s,nf} a_{s,nf}\left(T_s - T_{nf}\right) \tag{9.44}$$

$$\rho_{nf} c_{nf}\left[\frac{\partial(u_x T_{nf})}{\partial x} + \frac{\partial(u_y T_{nf})}{\partial y}\right] = k_{fe}\left(\frac{\partial^2 T_{nf}}{\partial x^2} + \frac{\partial^2 T_{nf}}{\partial y^2}\right) + h_{s,nf} a_{s,nf}\left(T_s - T_{nf}\right), \tag{9.45}$$

where the subscripts "nf" and "s" denote nanofluids and metal foams, respectively. The parameters $h_{s,nf}$, $a_{s,nf}$, and k_e are the local convection heat transfer coefficients between the nanofluid and the metal foam, the specific surface area of the metal foam, and the effective thermal conductivity, respectively. The calculation correlations for $h_{s,nf}$, $a_{s,nf}$, and k_e can be referred from Reference [70].

Compared with LTE-LB method, there is no force term in the evolution equation for the velocity distribution function in the LTNE-LB model, but there are two evolution equations for the nanofluid temperature distribution function and the metal foam temperature distribution function, shown as:

$$g_{i,nf}\left(\mathbf{r} + \mathbf{e}_i \delta t, t + \delta t\right) - g_{i,nf}\left(\mathbf{r}, t\right) = -\frac{1}{\tau_{T,nf}}\left[g_{i,nf}\left(\mathbf{r}, t\right) - g_{i,nf}^{eq}\left(\mathbf{r}, t\right)\right] + \delta t S_{i,nf} \tag{9.46}$$

$$g_{i,s}\left(\mathbf{r} + \mathbf{e}_i \delta t, t + \delta t\right) - g_{i,s}\left(\mathbf{r}, t\right) = -\frac{1}{\tau_{T,s}}\left[g_{i,s}\left(\mathbf{r}, t\right) - g_{i,s}^{eq}\left(\mathbf{r}, t\right)\right] + \delta t S_{i,s}. \tag{9.47}$$

The equilibrium state equations of the nanofluid temperature distribution function and the metal foam temperature distribution function are as follows:

$$g_{i,\text{nf}}^{\text{eq}} = T_{\text{nf}}\omega_i \left[1 + \frac{\mathbf{e}\cdot\mathbf{u}}{c_s^2} + \frac{(\mathbf{e}\cdot\mathbf{u})^2}{2c_s^4} - \frac{\mathbf{u}^2}{2c_s^2} \right]$$ (9.48)

$$g_{i,s}^{\text{eq}} = \omega_i T_s.$$ (9.49)

The source terms in the governing equations represent the heat exchange between the nanofluid and the solid in a LTNE model, shown as:

$$S_{i,\text{nf}} = \omega_i \left[1 + \left(1 - \frac{1}{2\tau_{\text{T,nf}}} \right) \frac{\mathbf{e}\cdot\mathbf{u}}{\varepsilon c_s^2} \right] \cdot \frac{h_{\text{sf}} a_{\text{sf}}}{\varepsilon \rho_{\text{nf}} c_{\text{nf}}} (T_s - T_{\text{nf}})$$ (9.50)

$$S_{i,s} = \omega_i \frac{h_{\text{sf}} a_{\text{sf}}}{(1-\varepsilon)\rho_s c_{ss}} (T_{\text{nf}} - T_s),$$ (9.51)

where c_{ss} represents the specific heat capacity of the metal foam.

The nanofluid flows in the inlet with u_{in} and T_{in}. The upper and lower wall temperatures are kept as T_w, and there is no slip at the wall surface. The flow and heat transfer are fully developed at the outlet. The boundary conditions can be expressed as:

$$x = 0: \mathbf{u} = (u_{\text{in}}, 0), T_s = T_{\text{nf}} = T_{\text{in}}$$ (9.52)

$$x = \text{L}: \frac{\partial u_x}{\partial x} = 0, u_y = 0, \frac{\partial T_s}{\partial x} = \frac{\partial T_{\text{nf}}}{\partial x} = 0$$ (9.53)

$$y = 0 \text{ or } y = H: \frac{\partial \mathbf{u}}{\partial y} = 0, T_s = T_f = T_w.$$ (9.54)

Guo and Zhao [51] presented the calculation formula of the LB method to deal with this kind of boundary conditions. For a boundary with a known velocity, the fluid density at the boundary is equal to the density of the fluid-side point that is adjacent to it, expressed as:

$$\rho(\mathbf{x_b}) = \rho(\mathbf{x_f})$$ (9.55)

$$f_i(\mathbf{x_b}) - f_i^{\text{eq}}(\mathbf{x_b}) = f_i(\mathbf{x_f}) - f_i^{\text{eq}}(\mathbf{x_f}).$$ (9.56)

When the temperature on the wall is known, the nanofluid temperature distribution function and the metal foam temperature distribution function are as follows:

$$g_{i,\text{nf}}(\mathbf{x_b}) - g_{i,\text{nf}}^{\text{eq}}(\mathbf{x_b}) = g_{i,\text{nf}}(\mathbf{x_f}) - g_{i,\text{nf}}^{\text{eq}}(\mathbf{x_f})$$ (9.57)

$$g_{i,s}(\mathbf{x_b}) - g_{i,s}^{\text{eq}}(\mathbf{x_b}) = g_{i,s}(\mathbf{x_f}) - g_{i,s}^{\text{eq}}(\mathbf{x_f}).$$ (9.58)

When the temperature gradient on the wall is already known, $\overline{g}_i^{eq}(\mathbf{x}_b)$ in the following equations can be used to replace the term $g_i^{eq}(\mathbf{x}_b)$ in Eqs. 59 and 60:

$$\overline{g}_{i,\mathrm{nf}}^{eq}(\mathbf{x}_b) = \omega_i \left[T_{\mathrm{nf}}(\mathbf{x}_f) - (\mathbf{x}_f - \mathbf{x}_b) \cdot \nabla T_{\mathrm{nf}}(\mathbf{x}_b) \right] \left[1 + \frac{\mathbf{e} \cdot \mathbf{u}}{c_s^2} + \frac{(\mathbf{e} \cdot \mathbf{u})^2}{2c_s^4} - \frac{\mathbf{u}^2}{2c_s^2} \right] \tag{9.59}$$

$$\overline{g}_{i,\mathrm{s}}^{eq}(\mathbf{x}_b) = \omega_i \left[T_{\mathrm{s}}(\mathbf{x}_f) - (\mathbf{x}_f - \mathbf{x}_b) \cdot \nabla T_{\mathrm{s}}(\mathbf{x}_b) \right] \left[1 + \frac{\mathbf{e} \cdot \mathbf{u}}{c_s^2} + \frac{(\mathbf{e} \cdot \mathbf{u})^2}{2c_s^4} - \frac{\mathbf{u}^2}{2c_s^2} \right]. \tag{9.60}$$

The boundary condition scheme in Eqs. (9.59) and (9.60) is called the non-equilibrium extrapolation. This scheme divides the distribution function at the boundary nodes into the equilibrium part and the non-equilibrium part, and there is also another simple boundary scheme, known as the equilibrium extrapolation which only considers the equilibrium part, shown as:

$$g_{i,\mathrm{nf}}(\mathbf{x}_b) = g_{i,\mathrm{nf}}(\mathbf{x}_f) \tag{9.61}$$

$$g_{i,\mathrm{s}}(\mathbf{x}_b) = g_{i,\mathrm{s}}(\mathbf{x}_f). \tag{9.62}$$

Compared with the extrapolated scheme, the non-equilibrium relationship between the boundary and the internal nodes was taken into account in the non-equilibrium extrapolation scheme, which has a second-order accuracy both in time and space. A non-equilibrium extrapolation scheme for processing boundary conditions is employed in this section.

9.4.2 Analysis

In this section, the above LB model is used to numerically study the forced convective heat transfer of nanofluids in a flat-plate channel filled with metal foam. The effects of Reynolds number, Darcy number, porosity, nanoparticle volume fraction, thermal conductivity of metal foam, and thermal conductivity of nanofluid on forced convective heat transfer are analyzed. The Nusselt number representing the thermal performance can be defined as:

$$\mathrm{Nu} = -\frac{1}{T_{\mathrm{in}} - T_{\mathrm{w}}} \int_0^L \left. \frac{\partial T_{\mathrm{nf}}}{\partial y} \right|_{y=0} dx. \tag{9.63}$$

To verify the numerical procedure, the present numerical result is compared with the numerical result in Reference [46] and the experimental result in Reference [71]. The comparison is shown in Figure 9.21. In Figure 9.21a, the velocity agrees well with the result obtained by the finite volume method (FVM) method. Figure 9.21b shows the comparison of the average Nusselt number with the present LB model with the experimental data. It can be seen from Figure 9.21b that the trend of Nu is basically the same as the experimental result. Since the present configuration (a flat-plate channel) is different from the circular tube in Nazari et al. [71], the Nu values are slightly different from each other, but the difference is acceptable, showing that the LB model established is feasible. From Figure 9.21, it can be seen that the velocity of nanofluids from the wall surface rapidly reaches its maximum from 0, and its velocity is very uniform throughout the channel. Because of the great velocity gradient near the wall, the heat transfer coefficient between the nanofluid and the metal foam is very high, and the heat transfer is very sufficient. The thermal conductivity of the metal foam is much larger than that of the nanofluid, and more heat is transferred in the solid ligament of the metal foam than in the fluid. The extending surface of the metal foam can efficiently enhance the local heat transfer between solid and fluid.

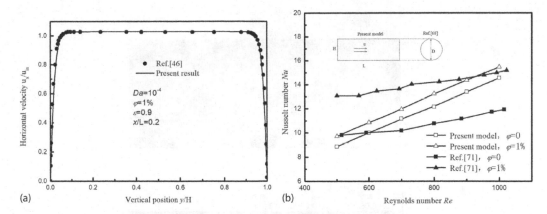

(a)

(b)

FIGURE 9.21 Validation of the velocity and Nusselt number: (a) horizontal velocity and (b) Nusselt number.

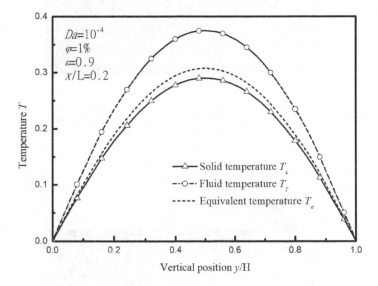

FIGURE 9.22 The temperature distribution diagram at $x = 0.2$ L.

Figure 9.22 shows the temperature profiles of nanofluids at a cross-section of the metal-foam channel with an equivalent temperature defined as $T_e = T_s + T_f \cdot k_{se} / k_{fe}$. The working fluid is Al_2O_3-H_2O nanofluid, and the metal foam is made of a copper alloy. It can be seen from Figure 9.22 that T_{nf} is higher than T_s and T_e is in between. Since the wall temperature is lower than the inlet temperature, the fluid is cooled in the channel. At this time, the temperature of the solid is lower than the temperature of the fluid, which indicates that the heat is transferred more quickly in solid. Equivalent temperature is closer to solid temperature, which also reflects the promotion of heat transfer by the metal foam.

Figure 9.23 shows the temperature distributions of the nanofluid and solid in the metal-foam channel. From the figure, the fluid is quickly cooled after entering the channel and is very close to the wall temperature when it passes through the channel at $x = 0.5$ L. Figure 9.24 shows the results obtained by using the equilibrium extrapolation scheme to process the boundary condition at the outlet. In this figure, the temperature at the outlet rises, but the convective heat transfer in the channel without metal foam does not occur when using the extrapolation scheme. This phenomenon indicates that the equilibrium extrapolation scheme is not suitable for the treatment of the metal foam temperature boundary. Therefore, the non-equilibrium extrapolation scheme should be used to ensure the accuracy and correctness of the calculation results.

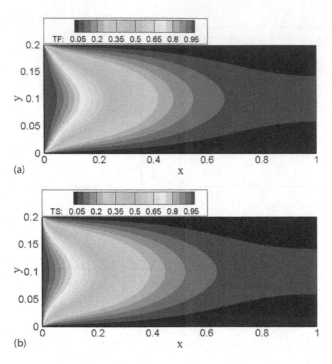

FIGURE 9.23 The temperature distribution using with non-equilibrium extrapolation scheme: (a) fluid temperature and (b) solid temperature.

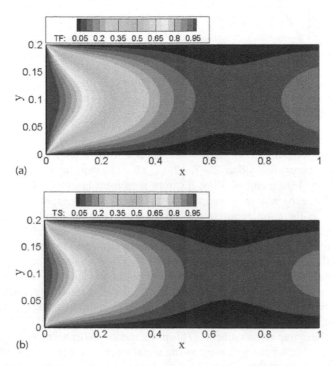

FIGURE 9.24 The temperature distribution using with equilibrium extrapolation scheme: (a) fluid temperature and (b) solid temperature.

The local surface heat transfer coefficient can be defined as:

$$h_x = -\left.\frac{\partial T_{nf}}{\partial y}\right|_{y=0} \frac{1}{\left(T_w - T_{nf,b}\right)_x}.$$

(9.64)

Figure 9.25a and b are, respectively, the local surface heat transfer coefficients h_x and the average nanofluid temperature along the flow direction. As the nanofluid flows forward, the temperature of the fluid gradually decreases and ultimately remains unchanged. The change of the local surface heat transfer coefficient shows that the convection heat transfer effect at the inlet is very strong, and the temperature rapidly decreases and approaches the wall temperature just after the fluid enters the metal foam, so the local surface heat transfer coefficient also rapidly decreases.

Figure 9.26 shows the effect of Reynolds number on Nusselt number for different nanoparticle volume fractions, and Figure 9.27 shows the effect of Darcy number on Nusselt number for different nanoparticle

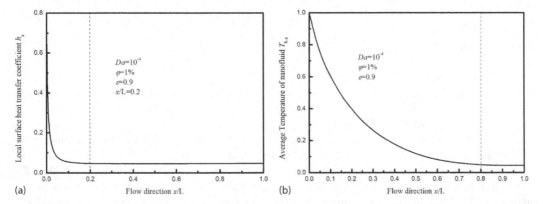

FIGURE 9.25   Local convective heat transfer coefficient and cross-sectional bulk temperature: (a) local heat transfer coefficient and (b) bulk fluid temperature.

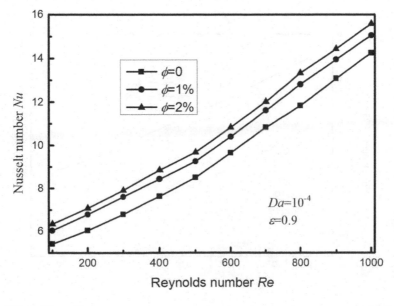

FIGURE 9.26   Effect of Re on Nu.

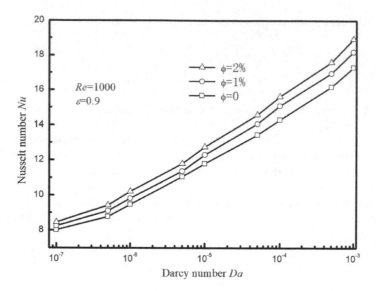

FIGURE 9.27   Effect of Da on Nu.

volume fractions. Figures 9.26 and 9.27 present the dependence of Nu with Re and Da when the volume fractions of nanoparticles are 0%, 1%, and 2%, respectively. It can be seen that when Re and Da increase, Nu increases. Since the Reynolds number reflects the flow intensity, the fluid velocity is fast or the viscosity of the fluid is low for a larger Reynolds number. The stronger flow capability of fluid makes the fluid easily carry heat in the heat convection process and thus increases the Nusselt number. The greater Darcy number leads to a larger permeability of metal foam, which is conducive to the fluid flow in the metal foam. This also has a strengthening effect on convection heat transfer. Incorporating solid nanoparticles of high thermal conductivity into base fluid significantly increases the thermal conductivity of the fluid and thus has a certain role in promoting heat transfer.

Figure 9.28 shows the relationship between the Nusselt number and the porosity. The volume fraction of solids in the metal foam is high for small porosity. This will lead to the heat transfer enhancement in

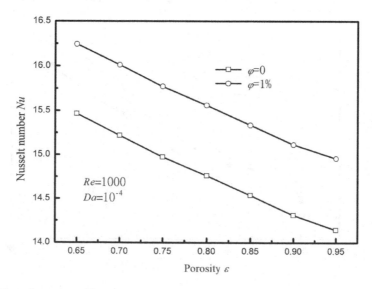

FIGURE 9.28   Effect of porosity on Nu.

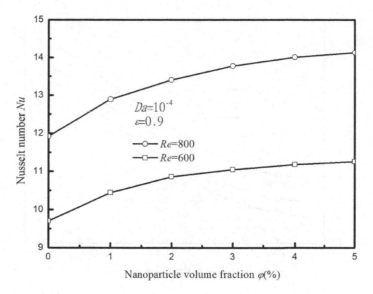

FIGURE 9.29   Effect of nanoparticle volume fraction on Nu.

the metal foam, but the flow resistance and pressure drop will also increase. When the porosity is high, the metal foam has fewer solid skeletons, and its heat transfer ability is not as strong as that at low porosity, but the flow resistance is weak and the pressure drop is much smaller. Figure 9.29 shows the effect of nanoparticle volume fraction on Nusselt number. From Figure 9.29, it can be seen that *Nu* increases as the nanoparticle volume fraction increases. When the nanoparticle volume fraction is larger, the effect of the nanoparticle volume fraction on Nu is weakened because the metal foam thermal conductivity increases with an increase in the nanoparticle volume fraction. But, its viscosity, density, and specific heat capacity also increase, which is not conducive to enhancing heat transfer. Therefore, the effect of nanoparticle volume fraction on Nu is not as significant as those of Re and Da.

Figure 9.30 shows the relationship between the Nusselt number and the thermal conductivities of the metal foam and the nanofluid. From Figure 9.30, it can be seen that the Nu increases when the thermal

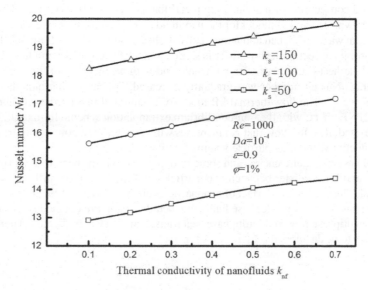

FIGURE 9.30   Effect of nanofluid thermal conductivity on Nu.

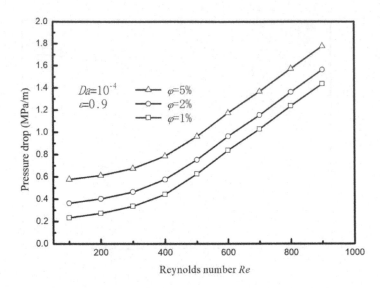

FIGURE 9.31   Effect of Re on the press drop.

conductivities of the metal foam and the nanofluid increase. The effect of foam thermal conductivity on Nusselt number is more significant than that of nanofluid. The thermal conductivity of metal foam is generally much larger than the nanofluid thermal conductivity, so it has a greater influence on the effective thermal conductivity and a more obvious effect on the enhancement of heat transfer. Figure 9.31 shows the relationship between Re and pressure drop for different nanoparticle volume fractions. When Re is larger, the flow rate is generally larger, which needs a larger pressure drop to maintain the fluid flow. The presence of metal foam will produce a large flow resistance. As the nanoparticle volume fraction increases, the nanofluid viscosity increases slightly, so its effect on pressure drop is relatively small.

9.5 Summary

Natural and forced convections of nanofluids in metal foams were studied with the LB method. Some conclusions are drawn. (1) The LB model for the nanofluid natural convection phenomenon in the square cavity of metal foam was established and numerically studied based on the LTE model. The results show that nanofluid natural convection heat transfer is significantly affected by Ra and Da. With the addition of nanoparticles, the heat transfer was strengthened, and with an increase in the nanoparticle volume fraction, the average Nusselt number was gradually increased, indicating that nanofluid has good heat transfer performance. (2) The flow/thermal LB model of a nanofluid in a porous medium is established by considering LTNE effect, with the non-equilibrium extrapolation scheme for nanofluid outlet boundary conditions revised. Results show that Nu increases with a decrease in porosity and an increase in Re, Da, thermal conductivity, and nanoparticle volume fraction.

This work still has many deficiencies, and some in-depth aspects still need to be further studied. The interaction force and heat transfer between nanoparticles and the base fluid still need a more accurate calculation model. The microscopic effect of nanoparticles on the local heat transfer requires an in-depth study. The three phases, nanoparticle, base fluid, and metallic foam, are coupled with each other in heat transfer, and the multiphase flow and multiphase heat transfer are very complicated. Therefore, the three-phase model needs to be developed.

ACKNOWLEDGMENT

This work is financially supported by the National Natural Science Foundation of China (51876118, 51406238) and the Initiative Program of New Youth Teachers in SJTU (WF220428001).

NOMENCLATURE

$\tilde{a}$ specific surface area, m^{-1}

c specific heat, $J \cdot kg^{-1} \cdot K^{-1}$

Da Darcy number

$\vec{e}_i$ the particle streaming velocity

f friction factor

f_i the particle distribution function

F_i the distribution function of resisting force comes from porous media

F_I Inertial coefficient

FFI the entropy generation rate due to flow, $W \cdot m^{-3} \cdot K^{-1}$

g_i the particle distribution function of temperature

h_v volumetric heat transfer coefficient, $W \cdot m^{-3} \cdot K^{-1}$

H length of the channel

k thermal conductivity, $W \cdot m^{-1} \cdot K^{-1}$

L width of the channel

Nu   Nusselt number

p pressure, $N \cdot m^{-2}$

Po Poiseuille number

Pr Prandtl number

q heat flux, $W \cdot m^{-2}$

Re Reynolds number

S the source term

S_i the distribution of source term

T temperature, K

$\vec{u}$ velocity, $m \cdot s^{-1}$

V a local averaging volume, m^3

$\vec{v}$ a temporal velocity, $m \cdot s^{-1}$

w_i the weight function

GREEK SYMBOLS

ϕ volume fraction of nanoparticles in nanofluid

ε porosity

ρ density, $kg \cdot m^3$

μ dynamic viscosity, $kg \cdot m^{-1} \cdot s^{-1}$

δt lattice time step

τ_f the single relaxation time of fluid

τ_t the relaxation time of the temperature distribution

SUBSCRIPTS AND SUPERSCRIPTS

ave	average
bf	base fluid
e	effective
eq	equilibrium state
f	flow
nf	nanofluid
np	nanoparticle
s	solid
w	wall

REFERENCES

1. Kasaeian, A., R. Daneshazarian, O. Mahian, et al. (2017). Nanofluid flow and heat transfer in porous media: A review of the latest developments. *International Journal of Heat and Mass Transfer*. 107: 778–791.
2. Buongiorno, J., D.C. Venerus, N. Prabhat, et al. (2009). A benchmark study on the thermal conductivity of nanofluids. *Journal of Applied Physics*. 106(9): 094312.
3. Das, S.K., S.U.S. Choi, W. Yu, et al. *Nanofluids: Science and Technology*. Hoboken, NJ: Wiley-Interscience; 2007.
4. Buongiorno, J. (2005). Convective transport in nanofluids. *Journal of Heat Transfer*. 128(3): 240–250.
5. Kim, S.J., T. McKrell, J. Buongiorno, et al. (2010). Subcooled flow boiling heat transfer of dilute alumina, zinc oxide, and diamond nanofluids at atmospheric pressure. *Nuclear Engineering and Design*. 240(5): 1186–1194.
6. Shafahi, M., V. Bianco, K. Vafai, et al. (2010). Thermal performance of flat-shaped heat pipes using nanofluids. *International Journal of Heat and Mass Transfer*. 53(7): 1438–1445.
7. Wu, S., D. Zhu, X. Zhang, et al. (2010). Preparation and melting/freezing characteristics of Cu/paraffin nanofluid as phase-change material (PCM). *Energy & Fuels*. 24(3): 1894–1898.
8. Zeng, J.L., Z. Cao, D.W. Yang, et al. (2010). Thermal conductivity enhancement of Ag nanowires on an organic phase change material. *Journal of Thermal Analysis and Calorimetry*. 101(1): 385–389.
9. Corcione, M. (2011). Rayleigh-Bénard convection heat transfer in nanoparticle suspensions. *International Journal of Heat and Fluid Flow*. 32(1): 65–77.
10. Corcione, M. (2011). Empirical correlating equations for predicting the effective thermal conductivity and dynamic viscosity of nanofluids. *Energy Conversion and Management*. 52(1): 789–793.
11. Barber, J., D. Brutin, and L. Tadrist. (2011). A review on boiling heat transfer enhancement with nanofluids. *Nanoscale Research Letters*. 6(1): 280.
12. Guo, C.X. (2011). Application study of nanoparticle-enhanced phase change material in ceiling board. *Advanced Materials Research*. 150–151: 723–726.
13. Khanafer, K. and K. Vafai. (2011). A critical synthesis of thermophysical characteristics of nanofluids. *International Journal of Heat and Mass Transfer*. 54(19): 4410–4428.
14. Lee, J.-H., S.-H. Lee, C. Choi, et al. (2010). A review of thermal conductivity data, mechanisms and models for nanofluids. *International Journal of Micro-Nano Scale Transport*. 1(4): 269–322.
15. Wu, S.Y., H. Wang, S. Xiao, et al. (2011). An investigation of melting/freezing characteristics of nanoparticle-enhanced phase change materials. *Journal of Thermal Analysis and Calorimetry*. 110(3): 1127–1131.
16. Yang, C., W. Li, and A. Nakayama. (2013). Convective heat transfer of nanofluids in a concentric annulus. *International Journal of Thermal Sciences*. 71: 249–257.
17. Hadadian, M., S. Samiee, H. Ahmadzadeh, et al. (2013). Nanofluids for heat transfer enhancement—A review. *Physical Chemistry Research*. 1(1): 1–33.
18. Ebrahimnia-Bajestan, E., M. Charjouei Moghadam, H. Niazmand, et al. (2016). Experimental and numerical investigation of nanofluids heat transfer characteristics for application in solar heat exchangers. *International Journal of Heat and Mass Transfer*. 92: 1041–1052.

19. Malvandi, A. and D.D. Ganji. (2016). Mixed convection of alumina–water nanofluid inside a concentric annulus considering nanoparticle migration. *Particuology*. 24: 113–122.
20. Sheremet, M.A., I. Pop, and M.M. Rahman. (2015). Three-dimensional natural convection in a porous enclosure filled with a nanofluid using Buongiorno's mathematical model. *International Journal of Heat and Mass Transfer*. 82: 396–405.
21. Lu, W., C.Y. Zhao, and S.A. Tassou. (2006). Thermal analysis on metal-foam filled heat exchangers. Part I: Metal-foam filled pipes. *International Journal of Heat and Mass Transfer*. 49(15): 2751–2761.
22. Zhao, C.Y., W. Lu, and S.A. Tassou. (2006). Thermal analysis on metal-foam filled heat exchangers. Part II: Tube heat exchangers. *International Journal of Heat and Mass Transfer*. 49(15): 2762–2770.
23. Xu, H.J., Z.G. Qu, and W.Q. Tao. (2011). Analytical solution of forced convective heat transfer in tubes partially filled with metallic foam using the two-equation model. *International Journal of Heat and Mass Transfer*. 54(17): 3846–3855.
24. Xu, H.J., Z.G. Qu, and W.Q. Tao. (2011). Thermal transport analysis in parallel-plate channel filled with open-celled metallic foams. *International Communications in Heat and Mass Transfer*. 38(7): 868–873.
25. Xu, H.J., Z.G. Qu, T.J. Lu, et al. (2011). Thermal modeling of forced convection in a parallel-plate channel partially filled with metallic foams. *Journal of Heat Transfer*. 133(9): 092603–092603-9.
26. Xu, H., C. Zhao, and K. Vafai. (2018). Analysis of double slip model for a partially filled porous microchannel—An exact solution. *European Journal of Mechanics—B/Fluids*. 68: 1–9.
27. Umavathi, J.C., J.P. Kumar, and K.S.R. Sridhar. (2010). Flow and heat transfer of Poiseuille-Couette flow in an inclined channel for composite porous medium. *International Journal of Applied Mechanics and Engineering*. 15(1): 249–266.
28. Hong, S.T. and D.R. Herling. (2007). Effects of surface area density of aluminum foams on thermal conductivity of aluminum foam-phase change material composites. *Advanced Engineering Materials*. 9(7): 554–557.
29. Zhao, C.Y., W. Lu, and Y. Tian. (2010). Heat transfer enhancement for thermal energy storage using metal foams embedded within phase change materials (PCMs). *Solar Energy*. 84(8): 1402–1412.
30. Tian, Y. and C.Y. Zhao. (2011). A numerical investigation of heat transfer in phase change materials (PCMs) embedded in porous metals. *Energy*. 36(9): 5539–5546.
31. Xu, H.J., Z.G. Qu, and W.Q. Tao. (2014). Numerical investigation on self-coupling heat transfer in a counter-flow double-pipe heat exchanger filled with metallic foams. *Applied Thermal Engineering*. 66(1): 43–54.
32. Xu, H., L. Gong, S. Huang, et al. (2014). Non-equilibrium heat transfer in metal-foam solar collector with no-slip boundary condition. *International Journal of Heat and Mass Transfer*. 76: 357–365.
33. Xu, H. (2017). Performance evaluation of multi-layered porous-medium micro heat exchangers with effects of slip condition and thermal non-equilibrium. *Applied Thermal Engineering*. 116: 516–527.
34. Xu, H., C. Zhao, and K. Vafai. (2017). Analytical study of flow and heat transfer in an annular porous medium subject to asymmetrical heat fluxes. *Heat and Mass Transfer*. 53(8): 2663–2676.
35. Xu, H. (2017). Convective heat transfer in a porous-medium micro-annulus with effects of the boundary slip and the heat-flux asymmetry: An exact solution. *International Journal of Thermal Sciences*. 120: 337–353.
36. Xu, H.J., C.Y. Zhao, and Z.G. Xu. (2016). Analytical considerations of slip flow and heat transfer through microfoams in mini/microchannels with asymmetric wall heat fluxes. *Applied Thermal Engineering*. 93: 15–26.
37. Dukhan, N., Ö. Bağcı, and M. Özdemir. (2015). Thermal development in open-cell metal foam: An experiment with constant wall heat flux. *International Journal of Heat and Mass Transfer*. 85: 852–859.
38. Siavashi, M., H.R. Talesh Bahrami, and H. Saffari. (2015). Numerical investigation of flow characteristics, heat transfer and entropy generation of nanofluid flow inside an annular pipe partially or completely filled with porous media using two-phase mixture model. *Energy*. 93: 2451–2466.
39. Bhadauria, B.S. and S. Agarwal. (2011). Convective transport in a nanofluid saturated porous layer with thermal non equilibrium model. *Transport in Porous Media*. 88(1): 107–131.
40. Sun, Q. and I. Pop. (2011). Free convection in a triangle cavity filled with a porous medium saturated with nanofluids with flush mounted heater on the wall. *International Journal of Thermal Sciences*. 50(11): 2141–2153.

41. Hajipour, M. and A. Molaei Dehkordi. (2012). Analysis of nanofluid heat transfer in parallel-plate vertical channels partially filled with porous medium. *International Journal of Thermal Sciences*. 55: 103–113.

42. Hajipour, M. and A. Molaei Dehkordi. (2014). Mixed-convection flow of Al_2O_3–H_2O nanofluid in a channel partially filled with porous metal foam: Experimental and numerical study. *Experimental Thermal and Fluid Science*. 53: 49–56.

43. Xu, H. and Z. Xing. (2017). The lattice Boltzmann modeling on the nanofluid natural convective transport in a cavity filled with a porous foam. *International Communications in Heat and Mass Transfer*. 89: 73–82.

44. Sakai, F., W. Li, and A. Nakayama (2014). A rigorous derivation and its applications of volume averaged transport equations for heat transfer in nanofluid saturated metal foams. *The 15th International Heat Transfer Conference*: 8575.

45. Zhang, W., W. Li, and A. Nakayama. (2015). An analytical consideration of steady-state forced convection within a nanofluid-saturated metal foam. *Journal of Fluid Mechanics*. 769: 590–620.

46. Xu, H., L. Gong, S. Huang, et al. (2015). Flow and heat transfer characteristics of nanofluid flowing through metal foams. *International Journal of Heat and Mass Transfer*. 83: 399–407.

47. Xu, H.J., Z.B. Xing, F.Q. Wang, et al. (2018). Convective transport characteristics of nanofluids in lightweight metal foams with high porosity. In *Nanomaterials*. Kyzas G. and Mitropoulos A. C. (Eds.), Intech Publisher; 2018.

48. Ting, T.W., Y.M. Hung, and N. Guo. (2015). Viscous dissipative nanofluid convection in asymmetrically heated porous microchannels with solid-phase heat generation. *International Communications in Heat and Mass Transfer*. 68: 236–247.

49. McNamara, G.R. and G. Zanetti. (1988). Use of the Boltzmann equation to simulate Lattice-Gas automata. *Physical Review Letters*. 61(20): 2332–2335.

50. Zou, Q. and X. He. (1997). On pressure and velocity boundary conditions for the lattice Boltzmann BGK model. *Physics of Fluids*. 9(6): 1591–1598.

51. Guo, Z. and T.S. Zhao. (2005). A lattice Boltzmann model for convection heat transfer in porous media. *Numerical Heat Transfer, Part B*. 47(2): 157–177.

52. Nemati, H., M. Farhadi, K. Sedighi, et al. (2010). Lattice Boltzmann simulation of nanofluid in lid-driven cavity. *International Communications in Heat and Mass Transfer*. 37(10): 1528–1534.

53. Abouei Mehrizi, A., M. Farhadi, K. Sedighi, et al. (2013). Lattice Boltzmann simulation of heat transfer enhancement in a cold plate using porous medium. *Journal of Heat Transfer*. 135(11): 111006–111006-9.

54. Sheikholeslami, M., M. Gorji-Bandpy, S.M. Seyyedi, et al. (2013). Application of LBM in simulation of natural convection in a nanofluid filled square cavity with curve boundaries. *Powder Technology*. 247: 87–94.

55. Sheikholeslami, M., M. Gorji-Bandpy, and G. Domairry. (2013). Free convection of nanofluid filled enclosure using lattice Boltzmann method (LBM). *Applied Mathematics and Mechanics*. 34(7): 833–846.

56. Sidik, N.A.C., M. Khakbaz, L. Jahanshaloo, et al. (2013). Simulation of forced convection in a channel with nanofluid by the lattice Boltzmann method. *Nanoscale Research Letters*. 8(1): 178.

57. Ata, S.V.A., K. Javaherdeh, and H.R. Ashorynejad. (2014). Magnetic field effects on force convection flow of a nanofluid in a channel partially filled with porous media using lattice Boltzmann method. *Advanced Powder Technology*. 25(2): 666-675.

58. Arumuga Perumal, D. and A.K. Dass. (2015). A Review on the development of lattice Boltzmann computation of macro fluid flows and heat transfer. *Alexandria Engineering Journal*. 54(4): 955–971.

59. Lee, S., S.U.S. Choi, S. Li, et al. (1999). Measuring thermal conductivity of fluids containing oxide nanoparticles. *Journal of Heat Transfer*. 121(2): 280–289.

60. Einstein, A. (1906). Eine neue Bestimmung der Moleküldimensionen. *Annalen der Physik*. 324(2): 289–306.

61. Maxwell, J.C. (1873). *Treatise on electricity and magnetism*. Oxford, UK: Clarendon Press.

62. Keblinski, P., S.R. Phillpot, S.U.S. Choi, et al. (2002). Mechanisms of heat flow in suspensions of nano-sized particles (nanofluids). *International Journal of Heat and Mass Transfer*. 45(4): 855–863.

63. Chon, C.H., K.D. Kihm, S.P. Lee, et al. (2005). Empirical correlation finding the role of temperature and particle size for nanofluid (Al_2O_3) thermal conductivity enhancement. *Applied Physics Letters*. 87(15): 153107.

64. Maïga, S.E.B., S.J. Palm, C.T. Nguyen, et al. (2005). Heat transfer enhancement by using nanofluids in forced convection flows. *International Journal of Heat and Fluid Flow*. 26(4): 530–546.

65. Guo, Z., B. Shi, and C. Zheng. (2002). A coupled lattice BGK model for the Boussinesq equations. *International Journal for Numerical Methods in Fluids*. 39(4): 325–342.

66. Hortmann, M., M. Perić, and G. Scheuerer. (1990). Finite volume multigrid prediction of laminar natural convection: Bench-mark solutions. *International Journal for Numerical Methods in Fluids*. 11(2): 189–207.

67. He, Y.L., Y. Wang, and Q. Li. (2009). *Lattice Boltzmann Method: Theory and Applications*. Beijing, China: Science Press.

68. Nithiarasu, P., K.N. Seetharamu, and T. Sundararajan. (1997). Natural convective heat transfer in a fluid saturated variable porosity medium. *International Journal of Heat and Mass Transfer*. 40(16): 3955–3967.

69. Seta, T., E. Takegoshi, and K. Okui. (2006). Lattice Boltzmann simulation of natural convection in porous media. *Mathematics and Computers in Simulation*. 72(2): 195–200.

70. Xu, H.J., L. Gong, C.Y. Zhao, et al. (2015). Analytical considerations of local thermal non-equilibrium conditions for thermal transport in metal foams. *International Journal of Thermal Sciences*. 95: 73–87.

71. Nazari, M., M. Ashouri, M.H. Kayhani, et al. (2015). Experimental study of convective heat transfer of a nanofluid through a pipe filled with metal foam. *International Journal of Thermal Sciences*. 88: 33–39.

10

CNT-Water Nanojet Impingement Cooling of a Sinusoidally Moving Isothermal Hot Wall

Fatih Selimefendigil and Hakan F. Oztop

CONTENTS

10.1 Introduction

Nanofluid technology was successfully used in a variety of thermal engineering applications such as in heat exchangers, refrigeration, renewable energy systems, thermal storage, and many others [1,2,4,6,8,12,16,17,19,21,22,24,26,31,33,39,44–49,51,53,54,56–60]. Jet impingement cooling is another field where nanofluid technology was successfully implemented. The impinging jet cooling plays an important role in obtaining higher heat and mass transfer coefficients in various applications such as in glass annealing, drying, thermal treatment of surfaces, and many others. Combining nanofluid technology in jet impingement cooling provides a novel solution for achieving enhanced heat and mass transfer characteristics. An experimental study with nanofluid impingement was performed for the heat transfer characteristics onto a flat circular disc in [18], and it was observed that nano solid particle concentration beyond a certain value was not effective for heat transfer augmentation. Convective heat transfer characteristics of water, nanofluid, and nano-encapsulated phase-change material jet impingement onto a heated copper plate were examined by Rehman et al. [37]. Nanofluid was found to enhance the cooling performance when it is used as a coolant. Nakharintr et al. [30] performed an experimental study for nanojet impingement problem with TiO_2 nanoparticles and effects of jet-plate spacing, jet diameter ratios, and nanofluid particle concentration on the heat transfer and pressure drop were analyzed. It was found that the Nusselt number enhances for higher nanofluids concentration and for lower jet-plate spacing. Swirling impinging jet heat transfer characteristics with TiO_2 nanoparticles were experimentally analyzed with thermochromic liquid crystal technique in reference [61]. It was observed that for similar operational conditions, swirling impinging jets give higher heat transfer performance as compared to conventional impinging jets and inclusion of nanoparticles further enhances the heat transfer coefficient. In a recent study, Selimefendigil and Oztop [42] performed numerical and optimization study of jet impingement cooling of a curved isothermal surface with CuO nanoparticles by using finite element method. They observed 20% of average heat transfer enhancement at the highest volume fraction of nanoparticle and similar trends for the average heat transfer versus solid particle volume fraction (a linear trend) both for flat and curved surface. In another study, Selimefendigil and Oztop [43] numerically examined the effects of using a rotating obstacle and nanofluid for the jet impingement cooling problem. When SiO_2-water nanofluid was used at the highest particle volume fraction, only 8% in the average heat transfer coefficient is achieved and rotating cylinder was found to be more effective in changing the heat transfer coefficient as compared to inclusion of the nanoparticle. In a recent study, Selimefendigil and Oztop [50] analyzed the jet impingement cooling of an elastic surface with nanofluid considering various shapes of the particles (spherical, cylindrical, blade). It was observed that heat transfer enhances with higher values of elastic modulus and higher nanoparticle solid volume fraction. Among various shapes of the particles, cylindrical ones performed the best in terms of heat transfer enhancement. Numerical analysis of cooling of a corrugated hot surface with jet impingement was performed in reference [49]. Effects of various nanoparticle solid volume fraction (between 0 and 0.04) and various particle shapes (spherical, blade, brick, and cylindrical) on the convective heat transfer performances were examined. It was noted that cylindrical particles perform the best heat transfer characteristics in terms of stagnation and average Nusselt number and corrugated surface results in higher heat transfer rates as compared to a flat surface.

As compared to some other conventional particles carbon nanotubes (CNTs) have significantly higher thermal conductivity. Studies with CNT nanofluid revealed significant enhancements in the heat transfer coefficients [7,35,36]. Due to their higher aspect ratio and complex morphology, heat transfer mechanism using CNT-nanofluid becomes very complex. A review for the heat and mass transfer characteristics of CNT can be found in reference [63]. In the study by Murshed and de Castro [29], recent applications and advancements for the carbon-nanotube nanofluids were discussed. In this review study, some features such as thermal conductivity enhancement, convective heat transfer coefficient, and boiling critical heat flux and how they were affected by the carbon nanotube concentration and temperature were discussed in detail. Thermal conductivity enhancement of CNT is even more at the elevated temperatures, and it gives a good opportunity to use them in high temperature applications [29]. Lotfi et al. [27] performed experimental study for a horizontal shell and tube heat exchanger with multi-walled carbon nanotube/water nanofluid. To increase the stability of the nanofluid, COOH functional groups were used, and they observed significant heat transfer enhancement with nanofluids. In the experimental study by Ettefaghi et al. [13], multi-walled carbon nanotubes were used with engine oil to enhance their thermal conductivity and flash point of nano-lubricants. For the stabilization of the nanotube in the engine oil, carbon nanotubes functionalization method and planetary ball mill were used. They improved the thermal conductivity and flash point of nano-lubricant by 13.2% and 6.7% with only 0.1%wt. In a recent review study, CNT nanofluids applications in solar collectors and factors affecting thermal conductivity and optical properties of nanofluid are discussed in detail.

In many applications, the cooling surface with jet impingement is not stationary and surface movement brings additional complexity in the fluid flow and convective heat transfer characteristics. Aldabbagh and Mohamad [3] performed 3D numerical study of fluid flow and heat transfer for a jet array impingement of a moving heated surface. The numerical simulations were performed for various jet-to-jet spacing, moving plate to the jet velocity ratios. It was observed that the streamwise Nusselt shows periodic oscillations, and they were not affected with the velocity of the moving plate. In the study by Solliec [55], fluid flow characteristics for a slot jet impinging on a moving surface were examined for air with particle image velocimetry. Various nozzle-to-plate spacing, Reynolds numbers and surface-to-jet velocity ratios were considered. It was observed that fluid flow characteristics are greatly affected for higher surface-to-jet velocity ratios. Chattopadhyay and Saha [10] performed numerical study of fluid flow characteristics in turbulent regime for an impinging jet emerging from a rectangular slot onto a moving surface. Large eddy simulation technique was used, and it was observed that the surface velocity has significant effects on turbulence production rate and heat transfer. In the study by Kadiyal and Chattopadhyay [20], a numerical analysis of heat transfer due to impinging slot jets onto a moving surface in laminar flow was conducted. An optimization was also included by using the micro-genetic algorithm. In this study, surface velocity was found to significantly affect the fluid flow and heat transfer characteristics. Confined slot jet impingement heat transfer for a moving isothermal surface was numerically examined in Sharif and Banerjee [52] in a 2D turbulent flow configuration. Average heat transfer coefficient was found to enhance with plate velocity and jet exit velocity, and the skin friction was found to significantly change with hot surface velocity.

Introducing periodic oscillation of the moving lid was found to affect the flow mixing in heat transfer characteristic in cavities. Anderson et al. [5] performed numerical study of fluid flow mixing in a lid driven cavity which has a pulsating lid velocity by using spectral element method. They observed chaotic mixing when a pulsating component is added to the steady lid velocity. In the numerical study of Nishimura and Kunitsugu [32], finite element method was used to examine the heat and mass transfer in lid driven cavity with a time-periodic lid velocity. They obtained a certain frequency for the best mixing and noted that fluid mixing depends on aspect ratio and oscillation amplitude. Fluid flow and heat transfer characteristics within a triangular cavity due to an oscillating lid and natural convection were examined in reference [11]. Interaction of the natural frequency of periodic flow and oscillation lid frequency was investigated, and they were shown to appear as the major frequencies in the spectrum. In the study of Selimefendigil and Oztop [48], mixed convection in a cavity due to an oscillation lid was examined under the effects of magnetic field and nanoparticle inclusion by using finite element method. It was observed that higher oscillation frequency of the lid results in heat transfer deterioration, and average heat transfer enhancement of 28.96% was obtained with nanoparticle inclusion at the highest particle volume fraction when compared to base fluid.

Based on the above literature survey and to the best of authors' knowledge, jet impingement cooling of a moving isothermal surface that has an oscillating velocity component with nanoparticle inclusion has never been studied despite its importance in various thermal engineering applications. In practice, cooling of such a surface that has a sinusoidally varying component may be needed with jet impingement or oscillating velocity component can be superimposed on the steady velocity component to affect the fluid flow and thermal characteristics. Inclusion of nanotechnology to the jet impingement cooling problem has been outlined above in the light of recent literature survey. In the present study, nanofluid with single-walled carbon nanotube which has a higher thermal conductivity as compared to other convectional nanoparticles was considered. The results of this present investigation could be utilized for design and optimization of such systems and to obtain best performance in terms of fluid flow and heat transfer characteristics.

10.2 Numerical Modeling

Schematic view of the considered problem is shown in Figure 10.1a with boundary conditions. Two cold jets with temperature of T_c and uniform velocity $v = v_0$ are emerged from rectangular slot of width w. The twin jets are separated by a distance $d = 5$ w while the separating distance between the plates is $H = 10$ w. Bottom plate is isothermal at constant hot temperature of T_h. The bottom plate is also moving sinusoidally with velocity of $u = U (1 + A \sin(2\pi \text{ft}))$, and its length is $L = 100$ w. Nanofluid with water as base fluid and single-walled CNTs as nanoparticles were used for various solid particle volume fractions. Table 10.1 shows the thermophysical properties of base fluid and nanoparticle. The fluid was assumed to be Newtonian and incompressible. The flow is 2D, laminar, and unsteady. Effects of viscous dissipation, radiation, and natural convection were neglected.

Conservation equations of mass, momentum, and energy are written as:

$$\frac{\partial u}{\partial x} + \frac{\partial u}{\partial y} = 0 \tag{10.1}$$

$$u\frac{\partial u}{\partial x} + v\frac{\partial u}{\partial y} = -\frac{1}{\rho_{nf}}\frac{\partial p}{\partial x} + \upsilon_{nf}\left(\frac{\partial^2 u}{\partial x^2} + \frac{\partial^2 u}{\partial y^2}\right) \tag{10.2}$$

$$u\frac{\partial v}{\partial x} + v\frac{\partial v}{\partial y} = -\frac{1}{\rho_{nf}}\frac{\partial p}{\partial y} + \upsilon_{nf}\left(\frac{\partial^2 v}{\partial x^2} + \frac{\partial^2 v}{\partial y^2}\right) \tag{10.3}$$

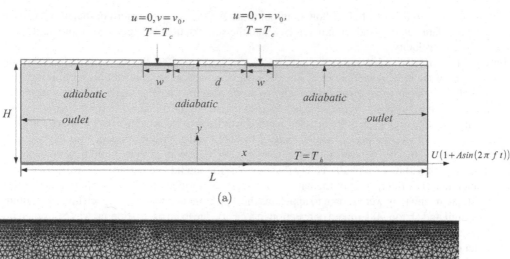

(a)

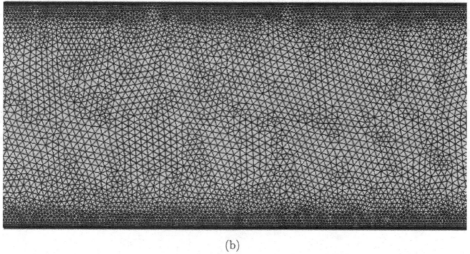

(b)

FIGURE 10.1 Schematic description of the physical model (a) and mesh distribution with triangular elements (b).

TABLE 10.1

Thermophysical Properties of Base Fluid and Nanoparticles

Property	Water	Single-walled carbon nanotube	Multi-walled carbon nanotube
ρ (kg/m³)	997.1	2600	1600
Cp (J/kg K)	4179	425	796
k (W/mK)	0.61	6600	3000

Source: Khan, W.A. et al., *Appl. Nanosci.*, 4, 633–641, 2014.

$$u\frac{\partial T}{\partial x}+v\frac{\partial T}{\partial y}=\alpha_{nf}\left(\frac{\partial^2 T}{\partial x^2}+\frac{\partial^2 T}{\partial y^2}\right). \qquad (10.4)$$

Dimensional form of the boundary conditions can be stated as:

- At the inlet where the jets are emerged, isothermal cold temperature and uniform velocity are used:

$$u=0, \quad v=v_0, \quad T=T_c.$$

- Gradients in the x-direction are set to 0 at the side surfaces:

$$\frac{\partial u}{\partial x} = \frac{\partial v}{\partial x} = \frac{\partial T}{\partial x} = 0.$$

- Bottom plate is at isothermal temperature and moving with a sinusoidally varying component:

$$u = U(1 + A\sin(2\pi ft)), \quad v = 0, \quad T = T_h.$$

- Top plate is adiabatic and stationary:

$$u = v = 0, \quad \frac{\partial T}{\partial y} = 0.$$

Single-walled carbon nanotube was used as an additive to water. The effective density and specific heat of the nanofluid are given in terms of base fluid, carbon nanotube properties, and solid volume fraction as:

$$\rho_{nf} = (1-\phi)\rho_f + \phi\rho_{CNT} \tag{10.5}$$

$$(\rho C_p)_{nf} = (1-\phi)(\rho C_p)_f + \phi(\rho C_p)_{CNT}. \tag{10.6}$$

Dynamic viscosity of the nanofluid was defined according to the Brinkman model [9]:

$$\mu_{nf} = \frac{\mu_f}{(1-\phi)^{2.5}}. \tag{10.7}$$

Even though this model is widely used in the computational nanofluid studies, it does not take into account temperature and particle size effects.

For accurate prediction of the thermal conductivity of the CNT-nanofluid, many studies were performed. Patel et al. [34] used an existing theoretical model for predicting the thermal conductivity of CNT-nanofluid which matches with the experimental data. In the study by Sastry et al. [40], a theoretical model for the thermal conductivity of the CNT-nanofluid using 3D CNT chain formation in the liquid which considers the CNT orientation and CNT-CNT interaction was used. The predicted model was found to be in good agreement with the experimental one. Gao et al. [15] developed a theoretical model for the thermal conductivity of carbon nanotubes-nanofluids by generalizing the Bruggeman effective medium theory. The developed model predicts the nonlinear behavior of thermal conductivity versus volume fraction and gives good agreement with experimental data.

The model of thermal conductivity of water-CNT nanofluid developed by Xue [62] takes into account the axial ratio and the space distribution of the carbon nanotubes, and it is defined as:

$$\frac{k_{nf}}{k_f} = \frac{(1-\phi) + 2\phi \dfrac{k_{CNT}}{k_{CNT} - k_f} \ln \dfrac{k_{CNT} + k_f}{2k_f}}{(1-\phi) + 2\phi \dfrac{k_f}{k_{CNT} - k_f} \ln \dfrac{k_{CNT} + k_f}{2k_f}}. \tag{10.8}$$

10.3 Solution Methodology

Finite volume method was used to solve the governing equations along with the boundary conditions. Unsteady convection-diffusion equation for a scalar transport variable Ψ can be written as:

$$\frac{\partial(\rho\psi)}{\partial t} + \nabla \cdot (\rho\mathbf{u}\psi) = \nabla \cdot (\Gamma\nabla\psi) + b, \tag{10.9}$$

with velocity u, source term b, and diffusion coefficient Γ. Integration of this partial differential equation over a control volume and after using suitable discretization schemes for convective and diffusion terms, algebraic equation at the node point p surrounded by neighboring relevant nodes (subscript n) can be written as:

$$a_p \varphi_p = \sum a_n \varphi_n + s. \tag{10.10}$$

For the discretization of the convective terms in the momentum and energy equations, Quadratic Upstream Interpolation for Convective Kinematics (QUICK) scheme was utilized and Semi-Implicit Method for Pressure-Linked Equations (SIMPLE) algorithm was used for velocity-pressure coupling. Second order implicit time stepping with a time step size of $\Delta t = \tau/50$ was used. Gauss-Seidel point-by-point iterative method and algebraic multi-grid method were used for the solution of the resulting system of algebraic equations. The normalized residual is calculated as:

$$R^\phi = \frac{\sum_{all\,cells} \left| a_p \phi_p - a_n \phi_n - s \right|}{\sum_{all\,cells} \left| a_p \phi_p \right|}. \tag{10.11}$$

Iterative solution is stopped when the residuals for all dependent variables become less than 10^{-6}. Local and average Nusselt number along the isothermal hot wall is calculated as:

$$\mathrm{Nu}_x = \frac{h_x D_h}{k_{nf}} = -\frac{D_h}{T_h - T_c} \left. \frac{\partial T}{\partial y} \right|_{y=0} \, , \quad \mathrm{Nu}_m = \frac{1}{L} \int_0^L \mathrm{Nu}_x dx, \tag{10.12}$$

where h_x is the local heat transfer coefficient, k_{nf} represents the thermal conductivity of nanofluid and $D_h = 2\,w$ is the characteristic length based on the slot width.

Grid and time step independence tests were performed. Table 10.2 shows grid independence test results for average Nusselt number along the hot surface considering various number of triangular elements (Re = 400, $U = 2v_0$). In the present study, G4 with 16568 triangular elements was used in the subsequent computations. Time step independence study was performed for two different solid particle volume fractions and for various time step sizes (Re = 400, St = 2). Average Nusselt number results for various time step sizes are shown in Table 10.3, and time step size of $\Delta t = \tau/50$ was chosen. The present code was validated by using the numerical results of Sahoo and Sharif [38]. Figure 10.2 shows the comparison results of average Nusselt number along the hot surface for various aspect ratios for the problem of jet impingement cooling of a heated surface (Re = 500 and Ri = 1).

TABLE 10.2

Grid Independence Study (Re = 400, $U = 2v_0$)

Grid Name	Number of Elements	Number ($\varphi = 0$)	Number ($\varphi = 0.05$)
G1	3132	5.071	11.604
G2	5744	5.327	12.105
G3	9059	54.609	12.272
G4	16568	5.483	12.276
G5	38724	5.490	12.279

TABLE 10.3

Time Step Size Independence Study (Re = 400, St = 2)

Time Step Size (Δt)	$\varphi = 0$	$\varphi = 0.05$
$\tau/25$	6.124	14.343
$\tau/50$	6.318	14.890
$\tau/100$	6.324	14.896
$\tau/200$	6.330	14.905

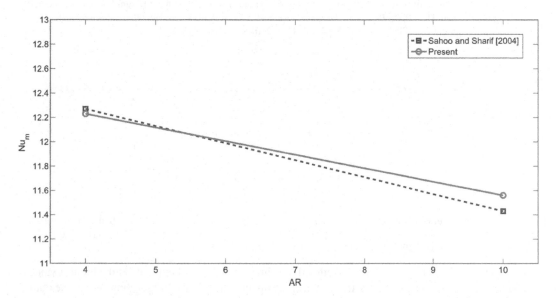

FIGURE 10.2 Comparison of the average Nusselt number variation at Reynolds number of 500 and Richardson number of 1. (Present study and results of Sahoo, D. and Sharif, M.A.R., *Int. J. Therm. Sci.* 43, 877–887, 2004.)

10.4 Results and Discussions

In this study, nanojet impingement cooling of a hot moving surface was numerically analyzed. Numerical simulations were performed for various values of Reynolds number (between 100 and 400), solid particle volume fraction (between 0 and 0.05), velocity of the moving hot wall (between 0 and $2v_0$), and Strouhal number (between 0.1 and 1). Reynolds number is defined based on the hydraulic diameter of the slot jet and jet exit velocity (Re = $v_0 D_h/\upsilon$). Simulations are first performed for steady configurations where various moving wall velocities are considered, then unsteady simulations were performed where moving wall velocity was kept fixed at v_0 and different sinusoidally varying components were added to the moving wall considering different frequencies. The value of amplitude for the sinusoidally varying velocity component was kept fixed at $A = 0.5$. In the steady configuration, U_w denotes the ratio of moving plate velocity to the jet exit velocity as $U_w = U/U_0$ and for the oscillating velocity configuration this value was kept fixed at $U_w = 1$.

10.4.1 Steady Simulation Results

Figures 10.3 and 10.4 demonstrate the effects of Reynolds number on the distribution of streamlines and isotherms for various moving wall velocity ($\varphi = 0.05$). When the hot wall is stationary, recirculating regions are established due to the entrainment and confining effects. As the value of Reynolds

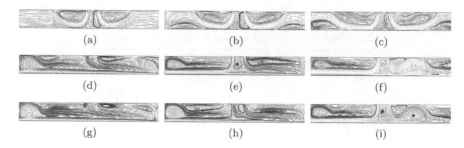

FIGURE 10.3 Streamline distribution for various values of Reynolds numbers and for different bottom wall velocities ($\varphi = 0.05$). (a) Re = 100, $U_w = 0$, (b) Re = 200, $U_w = 0$, (c) Re = 400, $U_w = 0$, (d) Re = 100, $U_w = 1$, (e) Re = 200, $U_w = 1$, (f) Re = 400, $U_w = 1$, (g) Re = 100, $U_w = 2$, (h) Re = 200, $U_w = 2$, (i) Re = 400, $U_w = 2$.

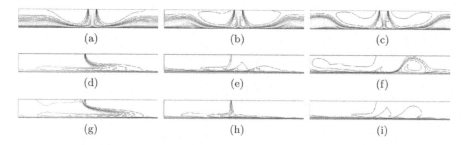

FIGURE 10.4 Isotherm distribution for various values of Reynolds numbers and for different bottom wall velocities ($\varphi = 0.05$). (a) Re = 100, $U_w = 0$, (b) Re = 200, $U_w = 0$, (c) Re = 400, $U_w = 0$, (d) Re = 100, $U_w = 1$, (e) Re = 200, $U_w = 1$, (f) Re = 400, $U_w = 1$, (g) Re = 100, $U_w = 2$, (h) Re = 200, $U_w = 2$, (i) Re = 400, $U_w = 2$.

number is enhanced, vortices are established in the vicinity of the lower hot plate. At the highest value of Reynolds number, two vortex centers are seen near the upper plate of the configuration. When the bottom wall moves in the positive x direction, the streamlines are skewed toward the moving direction. At low value of Reynolds number, enhancing the moving plate velocity inserts higher drag on the fluid and more skewness of the streamlines are observed. When the value of Reynolds number increases, at the highest value of moving wall speed, flow field becomes more complex and multi-cellular flow structure is established near the jet inlet region which can be contributed to the higher values of jet momentum for higher values of Reynolds number. Isotherms are steeper near the stagnation point and become less clustered toward the outlet when the bottom plate is stationary. Due to the presence of the secondary vortices near the bottom wall, the highest value of Reynolds number are here, additional location except from the stagnation point for the steep temperature gradients are observed on the bottom wall. When the hot plate is started to move, isotherms show very similar characteristics to that of streamlines, and steep temperature gradient locations move toward the left opening and become more closely dense for higher wall velocity.

Influence of hot plate velocity on the distribution of average Nusselt number along the hot wall for various Reynolds number and for two solid particle volume fractions is shown in Figure 10.5. When the plate is stationary ($U_w = 0$), average Nusselt number rises with higher Reynolds number. As the value of U_w increases, average heat transfer enhances and discrepancy between the average Nusselt number for different Reynolds numbers diminishes which is due to the dominance of shear driven flow both for base fluid and for nanofluid with the highest solid particle volume fraction. When the hot wall is stationary, average Nusselt number enhancements are 167.92% and 160.94% when Reynolds number is increased from Re = 100 to Re = 400 for base fluid and for nanofluid with $\varphi = 5\%$. On the other hand, when velocity of the hot plate is increased from $U_w = 0$ to $U_w = 2$, 164.66% and 144.38% of average Nusselt number increment are obtained for base fluid and nanofluid at the highest solid particle volume fraction. Use of nanoparticles is not advantageous for the average heat transfer enhancement as compared to pure fluid for the shear dominated flow configuration due to the increasing effect of drag exerted on the nanofluid. Average Nusselt number is a linear increasing function of solid particle volume fraction for all

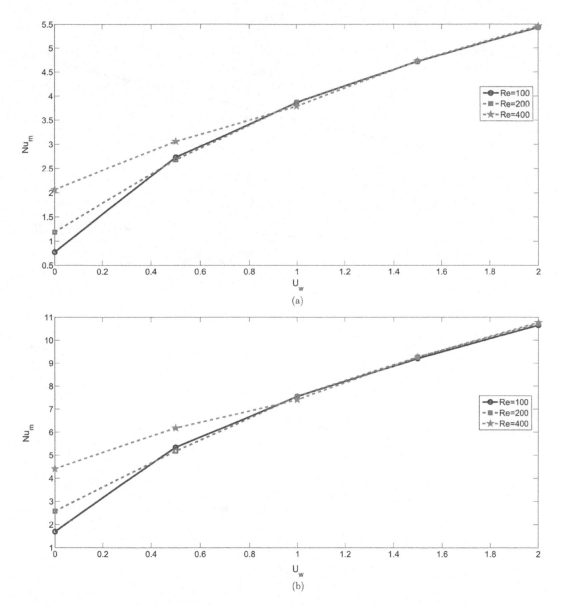

FIGURE 10.5 Influence of bottom wall velocity on the variation of average Nusselt number for different Reynolds numbers (a) base fluid and (b) nanofluid with $\varphi = 0.05$.

considered velocity of the moving wall. The slope of the linear curve is higher for the highest velocity of the moving wall when the shear driven flow becomes dominant. The trend of the curves is similar for Re = 100 and for Re = 400, but the slope becomes highest for stationary wall at Re = 100 as compared to case at Re = 400. The average Nusselt number increases 122.85% and 124.90% at Reynolds number of 100 and 400 when nanoparticles with the highest volume fraction are included in the base fluid. A significant amount of average heat transfer enhancement is obtained which is due to the higher thermal conductivity of the carbon nanotubes as compared to other nanoparticles (Figure 10.6).

10.4.2 Simulation Results for Oscillating Velocity of Moving Plate

Particular attention was given to the configuration when an oscillating component was added to the moving wall velocity of the hot plate. Sinusoidally varying component of the hot wall velocity was considered for

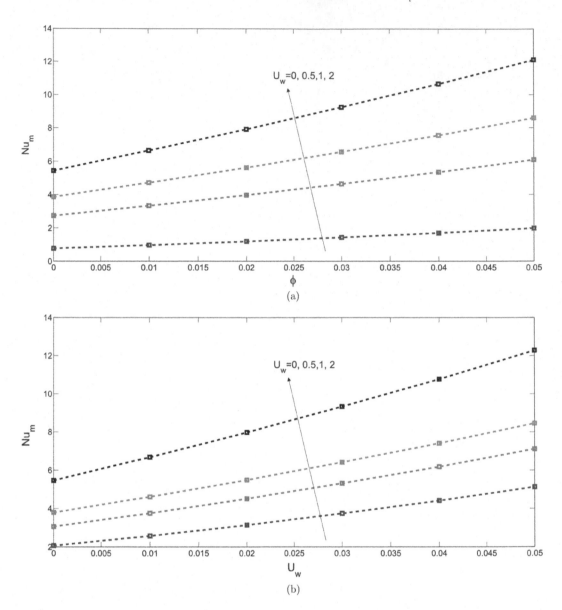

FIGURE 10.6   Average Nusselt number along the hot wall versus nanoparticle volume fraction for various wall velocities (a) Re = 100, (b) Re = 400.

different frequencies. The amplitude of the oscillating component was kept at $A = 0.5$ and moving wall velocity of $U = U_w (1 + A \sin(2\pi ft)$ with oscillating frequency f were considered. Figure 10.7 depicts the variation of spatial average Nusselt number along the hot wall for different Reynolds number for base fluid and for nanofluid with $\varphi = 5\%$. Strohual number was kept fixed at St = 0.1. A higher value of Reynolds number represents a higher velocity of jet velocity and thus corresponds to a higher frequency of the oscillation. A number of cycles is needed to reach the steady periodic oscillations for all Reynolds numbers. Nusselt number is higher for higher values of Reynolds number, and phase shift between time dependent Nusselt number signal is seen for different Reynolds numbers. One dominant oscillation frequency is noticeable from the signal, and the trends in the signals are similar when base fluid and nanofluid with $\varphi = 5\%$ are compared.

Wavelet analysis method can be used to extract transient features of time series data. As compared to Fast Fourier Transform, wavelet transformation of a time series signal reveals frequency and time portrait which is called a spectrogram and important transient features can be obtained [25,28].

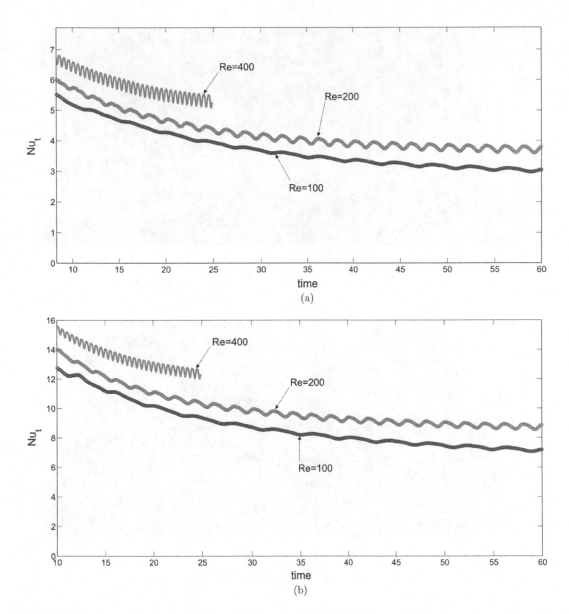

FIGURE 10.7   Time evolution of spatial averaged Nusselt number along the hot wall for various values of Reynolds number, (a) $\varphi = 0$, (b) $\varphi = 0.05$.

The wavelet transformation is obtained by using dilations and translations of a mother function $\psi(t)$ having finite energy and zero average. The wavelet transform of a function $f(t)$ can be represented as [41]:

$$\tilde{F}(a,b) = \int_t f(t)\Psi_{a,b}(t)dt \quad a,b \in R, \tag{10.13}$$

where Ψa, $b(t)$, a, and b represent the wavelet basis, scale, and position parameters. These parameters are related to the frequency and time of the function $f(t)$. Even though a lot of mother wavelet form is available, in the application of the fluid mechanics, Mexican hat and complex Morlet wavelets are utilized [14,41]. Figure 10.8 shows the spectrogram of the time dependent spatial averaged Nusselt number for two values of Reynolds numbers. One dominant frequency is seen in the wavelet map which has

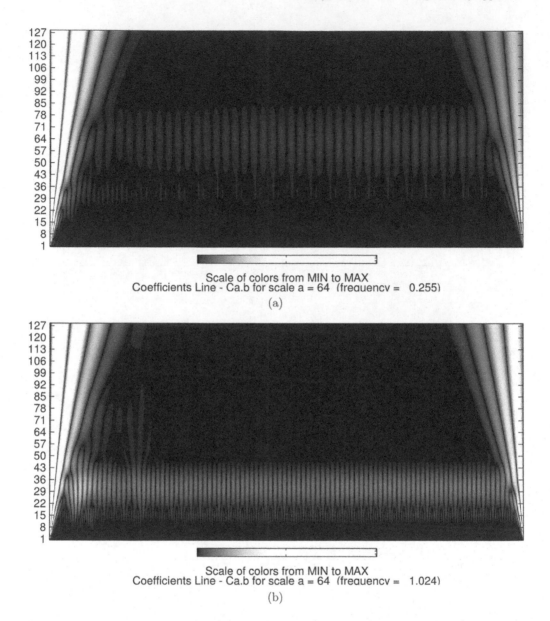

FIGURE 10.8   Spectrogram of wavelet analysis for two different Reynolds numbers, (a) Re = 100, (b) Re = 400.

the high energy density, and it can be identified from the given scale. Figure 10.9 presents the wavelet transformation of Nusselt number time series data for various Strouhal numbers. In these maps, one can also see periodicity of the signals and dominance of one frequency component which is proportional to the Strouhal number. In the low frequency oscillation of the moving hot wall, initial transients are seen to be small as compared to higher frequency oscillation velocity of the hot wall.

Time-spatial average Nusselt number enhances almost linearly with solid particle volume fraction for all Strouhal numbers and Reynolds numbers in the case of sinusoidal varying hot wall which is shown in Figure 10.10. Highest values of the average Nusselt number is achieved for the highest value of Strouhal number. For base fluid when Strouhal number is increased from St = 0.1 to St = 2, 23.68% of the average Nusselt number enhancement is achieved, whereas this value is 28.50% when nanofluid with the highest solid particle volume fraction is considered.

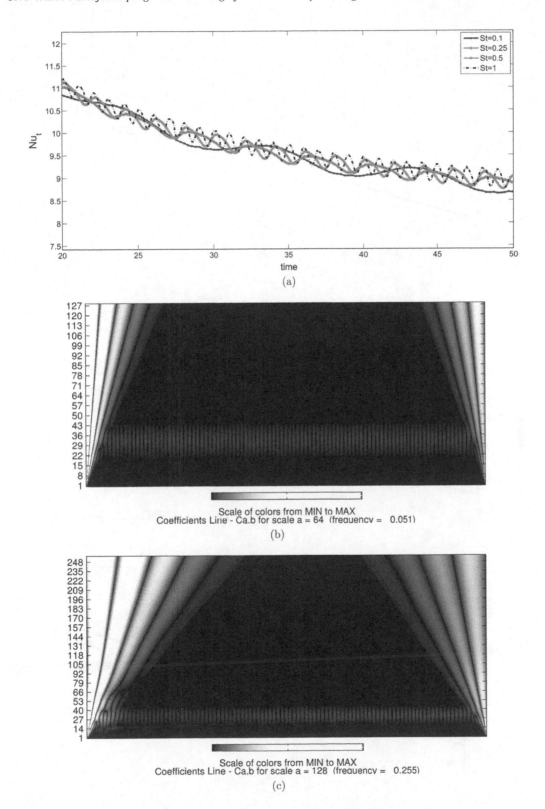

FIGURE 10.9 (a) Time evolution of spatial averaged Nusselt number along the hot wall for different Strouhal numbers and (b, c) wavelet analysis spectrogram for various Strouhal numbers (Re = 200, $\varphi = 0.02$).

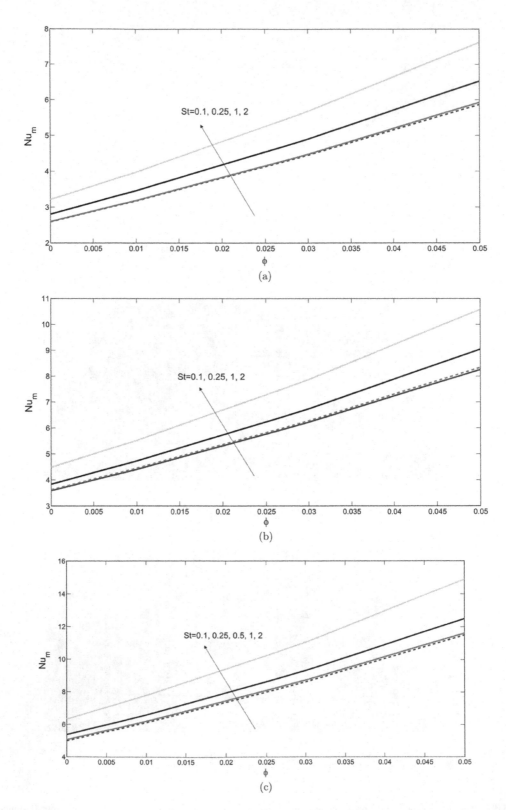

FIGURE 10.10 Spatial-temporal averaged Nusselt number along the hot wall versus solid particle volume fraction for different values of Strouhal number and Reynolds number, (a) Re = 100, (b) Re = 200, (c) Re = 400.

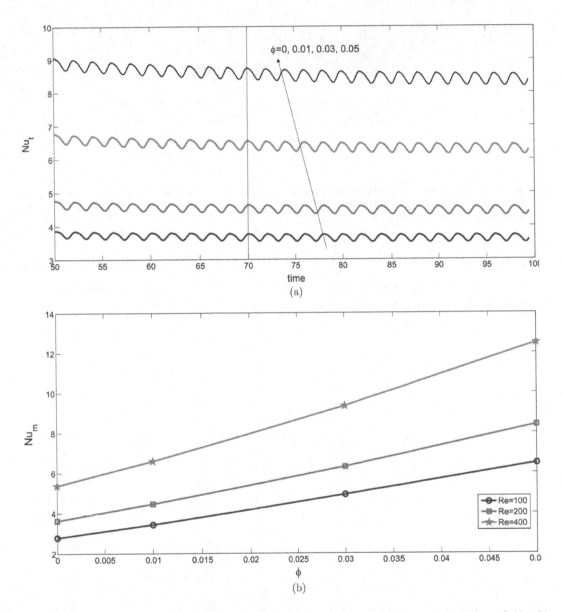

FIGURE 10.11   (a) Time evolution of spatial averaged Nusselt number along the bottom wall for different fluids with various nanoparticle volume fractions and (b) spatial-temporal average Nusselt number along the hot bottom versus solid particle volume fraction for different Reynolds number, (a) Re = 200, St = 0.5 (b) St = 0.5.

When nanoparticles are added to the base fluid, due to the thermal conductivity enhancement, significant increase in the time dependent and average Nusselt number is obtained which is indicated in Figure 10.11. There is no shift in the location where maximum value of the Nusselt number is obtained in the time domain signal when nanofluid with different nanoparticles are considered.

In the configuration with sinusoidally moving hot wall, average heat transfer enhancements are 132.68% and 132.86% when nanoparticles with the highest volume fraction are added to the base fluid at Reynolds numbers of 100 and 400 (St = 0.5).

Figure 10.12 demonstrates the variation of flow patterns for steady and oscillating velocity configurations (Re = 400, U_w = 1, and St = 2). When the hot wall moves with a sinusoidally varying component, the appearance of flow structure is different than that of a steady configuration.

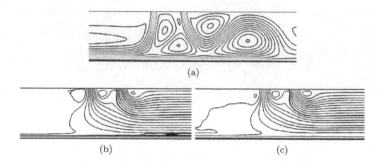

(a)

(b)   (c)

FIGURE 10.12 Streamline distribution for steady configuration with constant moving wall velocity and unsteady configuration with sinusoidally varying wall velocity (Re = 400, $Uw = 1$, $\varphi = 0.05$), t^*: total time of simulation, Δt: time step size, (a) Steady, (b) St =2, $t^*-20\Delta t$, (c) St = 2, t^*.

TABLE 10.4

Average Nusselt Number Enhancements with Oscillating Lid as Compared to Steady Moving Lid for Various Reynolds Number, Solid Particle Volume Fraction, and Strouhal Numbers

Re	φ	St	Enchancement (%)	Re	φ	St	Enchancement (%)
100	0	0.1	−32.99	200	0.03	0.1	−5.13
100	0	1	−27.51	200	0.03	1	2.73
100	0	2	−16.94	200	0.03	2	19.95
100	0.01	0.1	−32.44	200	0.05	0.1	−4.10
100	0.01	1	−26.65	200	0.05	1	5.14
100	0.01	2	−15.74	200	0.05	2	10.58
100	0.03	0.1	−32.10	400	0	0.1	31.47
100	0.03	1	−25.25	400	0	1	41.63
100	0.03	2	−13.33	400	0	2	66.70
100	0.05	0.1	−31.70	400	0.01	0.1	32.71
100	0.05	1	−24.10	400	0.01	1	42.25
100	0.05	2	−11.28	400	0.01	2	69.02
200	0	0.1	−7.33	400	0.03	0.1	34.76
200	0	1	−0.95	400	0.03	1	45.55
200	0	2	15.68	400	0.03	2	72.40
200	0.01	0.1	−6.61	400	0.05	0.1	36.11
200	0.01	1	0.23	400	0.05	1	47.90
200	0.01	2	17.1	400	0.05	2	76.19

The vortices established in between the inlet jets diminish in size, and their number is reduced. Thus, shear driven flow due to the moving hot wall becomes effective as compared to jet flow for an oscillating velocity case.

Table 10.4 presents the summary of the results with comparison of the constant moving hot wall velocity. Enhancement values in percentage that are indicated in the last column of Table 10.4 denote the increment for the same flow condition when oscillating part of the moving velocity is neglected. Negative values represent deterioration of the average heat transfer which are obtained for lower values of Reynolds number, solid particle volume fraction, and Strouhal number. At Reynolds number of 100, even at the highest value of φ and St number, 11.28% deterioration of the average Nusselt number is seen. At Reynolds number of 400, irrespective of the nanoparticle volume fraction and Strouhal number values, positive values are seen and at the highest value of φ and St combination, 76.19% in the average heat transfer is obtained.

10.5 Conclusions

This study examines fluid flow and heat transfer characteristics for carbon nanotube-water nanofluid double jet impingement cooling of a hot surface which has a sinusoidally varying velocity component. Following, important conclusions can be drawn from the numerical simulation results as:

- Flow and thermal fields are highly affected by the variation of Reynolds number based on the jet velocity and moving wall velocity. In the steady configuration, higher values of hot wall velocity result in higher drag exerted on the fluid and more skewness and distortion of the streamlines in the moving wall direction. For higher values of Reynolds number, higher values of jet momentum and shear driven flow interaction result in more complex multi-cellular flow structures.

- Average Nusselt number increases as the moving wall velocity enhances. Influence of Reynolds number based on the jet velocity becomes negligible when the hot wall velocity increases.

- The average Nusselt number is linearly varying with the solid particle volume fraction nanofluid for all moving wall velocities. Significant enhancements in the average heat transfer (up to 124.90%) are obtained with carbon-nanotube nanofluid at the highest particle volume fraction which is due to the higher thermal conductivity of this nanoparticle as compared to other commonly used nanoparticles such as alumina, copper, copper-oxide, etc.

- When moving wall with sinusoidally varying velocity component was considered, wavelet spectrogram of the time dependent spatial averaged Nusselt number indicates one dominant frequency in the wavelet map which has the high energy density for each of the frequency of oscillation and for different Reynolds number.

- As compared to steady flow configuration, the average heat transfer deteriorates for lower values of Reynolds number, solid particle volume fraction, and Strouhal number. For Reynolds number of 400, average heat transfer enhances in oscillation velocity configuration as compared to steady moving velocity irrespective of nanoparticle volume fraction and Strouhal number. At the highest value of Reynolds number, solid particle volume fraction, and Strouhal number, 76.19% of average heat transfer enhancement is achieved with oscillating velocity case.

NOMENCLATURE

A amplitude of oscillation

b source term

d separating distance between twin jets

f frequency of oscillation

h local heat transfer coefficient

k thermal conductivity

H plate distance

L plate length

n unit normal vector

Nux   local Nusselt number

Num   average Nusselt number

p pressure

Pr   Prandtl number

Re   Reynolds number based on jet exit velocity

St   Strouhal number

U	moving wall velocity
U_w	nondimensional moving wall velocity
T	temperature
t	time
u, v	x-y velocity components
w	slot width
x, y	Cartesian coordinates

GREEK CHARACTERS

α	thermal diffusivity
φ	solid volume fraction
ν	kinematic viscosity
θ	non-dimensional temperature
ρ	density of the fluid
Ψ	scalar transport variable

SUBSCRIPTS

c	cold
CNT	carbon nanotube
h	hot

REFERENCES

1. Abu-Nada, E. (2008). Application of nanofluids for heat transfer enhancement of separated flows encountered in a backward facing step. *International Journal of Heat and Fluid Flow* 29, 242–249.
2. Abu-Nada, E. and A. J. Chamkha (2010). Mixed convection flow in a lid-driven inclined square enclosure filled with a nanofluid. *European Journal of Mechanics B/Fluids* 29, 472–482.
3. Aldabbagh, L. B. Y. and A. A. Mohamad (2009). A three-dimensional numerical simulation of impinging jet arrays on a moving plate. *International Journal of Heat and Mass Transfer* 52, 4894–4900.
4. Ali, H. M. and W. Arshad (2015). Thermal performance investigation of staggered and inline pin fin heat sinks using water based rutile and anatase TiO_2 nanofluids. *Energy Conversion and Management* 106, 793–803.
5. Anderson, P. D., O. S. Galaktionov, G. W. Peters, F. N. van de Vosse and H. E. Meijer (2000). Chaotic fluid mixing in non-quasi-static time-periodic cavity flows. *International Journal of Heat and Fluid Flow* 21, 176–185.
6. Arshad, W. and H. M. Ali (2017a). Experimental investigation of heat transfer and pressure drop in a straight minichannel heat sink using TiO_2 nanofluid. *International Journal of Heat and Mass Transfer* 110, 248–256.
7. Arshad, W. and H. M. Ali (2017b). Graphene nanoplatelets nanofluids thermal and hydrodynamic performance on integral fin heat sink. *International Journal of Heat and Mass Transfer* 107, 995–1001.
8. Aswadi, A., H. Mohammed, N. Shuaib and A. Campo (2010). Laminar forced convection flow over a backward facing step using nanofluids. *International Communications in Heat and Mass Transfer* 37, 950–957.
9. Brinkman, H. (1952). The viscosity of concentrated suspensions and solutions. *Journal of Chemical Physics* 20, 571–581.
10. Chattopadhyay, H. and S. K. Saha (2003). Turbulent flow and heat transfer from a slot jet impinging on a moving plate. *International Journal of Heat and Fluid Flow* 24, 685–697.

11. Chen, C.-L. and C.-H. Cheng (2009). Numerical study of the effects of lid oscillation on the periodic flow pattern and convection heat transfer in a triangular cavity. *International Communications in Heat and Mass Transfer* 36, 590–596.

12. Cho, C. C. (2014). Heat transfer and entropy generation of natural convection in nanofluid-filled square cavity with partially-heated wavy surface. *International Journal of Heat and Mass Transfer* 77, 818–827.

13. Ettefaghi, E., H. Ahmadi, A. Rashidi, A. Nouralishhi and S. S. Mohtasebi (2013). Preparation and thermal properties of oil-based nanofluid from multi-walled carbon nanotubes and engine oil as nano-lubricant. *International Communications in Heat and Mass Transfer* 46, 142–147.

14. Feng, L. H. and L. K. Ran (2018). Wavelet multi-scale analysis of the circular cylinder wake under synthetic jets control. *International Journal of Heat and Fluid Flow* 69, 73–82.

15. Gao, L., X. Zhou and Y. Ding (2007). Effective thermal and electrical conductivity of carbon nanotube composites. *Chemical Physics Letters* 434, 297–300.

16. Gherasim, I., G. Roy, C. T. Nguyen and D. Vo-Ngoc (2009). Experimental investigation of nanofluids in confined laminar radial flows. *International Journal of Thermal Sciences* 48, 1486–1493.

17. Hajialigol, N., A. Fattahi, M. H. Ahmadi, M. E. Qomi and E. Kakoli (2015). MHD mixed convection and entropy generation in a 3-D microchannel using Al_2O_3-water nanofluid. *Journal of the Taiwan Institute of Chemical Engineers* 46, 30–42.

18. Jaberi, B., T. Yousefi, B. Farahbakhsh and M. Saghir (2013). Experimental investigation on heat transfer enhancement due to Al_2O_3-water nanofluid using impingement of round jet on circular disk. *International Journal of Thermal Sciences* 74, 199–207.

19. Jajja, S. A., W. Ali, H. M. Ali and A. M. Ali (2014). Water cooled minichannel heat sinks for microprocessor cooling: Effect of fin spacing. *Applied Thermal Engineering* 64, 76–82.

20. Kadiyal, P. K. and H. Chattopadhyay (2014). Neuro-genetic optimization of laminar slot jets impinging on a moving surface. *International Communications in Heat and Mass Transfer* 59, 143–147.

21. Kamyar, A., R. Saidur and M. Hasanuzzaman (2012). Application of computational fluid dynamics (CFD) for nanofluids. *International Journal of Heat and Mass Transfer* 55, 4104–4115.

22. Kasaeian, A., S. Daviran, R. D. Azarian and A. Rashidi (2015). Performance evaluation and nanofluid using capability study of a solar parabolic trough collector. *Energy Conversion and Management* 89, 368–375.

23. Khan, W. A., Z. H. Khan and M. Rahi (2014). Fluid flow and heat transfer of carbon nanotubes along a flat plate with Navier slip boundary. *Applied Nanoscience* 4, 633–641.

24. Kherbeet, A., H. Mohammed, K. Munisamy and B. Salman (2013). Combined convection nanofluid flow and heat transfer over microscale forward-facing step. *International Journal of Nanoparticle* 6, 350–365.

25. Latemeh, H. D., V. D. Narasimhamurthy, H. I. Andersson and B. Pettersen (2017). Turbulent wake behind a t-shaped plate: Comparison with a cross-shaped plate. *International Journal of Heat and Fluid Flow* 65, 127–140.

26. Lin, Y., Y. Jia, G. Alva and G. Fang (2018). Review on thermal conductivity enhancement, thermal properties and applications of phase change materials in thermal energy storage. *Renewable and Sustainable Energy Reviews* 82, 2730–2742.

27. Lotfi, R., A. M. Rashidi and A. Amrollahi (2012). Experimental study on the heat transfer enhancement of MWNT-water nanofluid in a shell and tube heat exchanger. *International Communication in Heat and Mass Transfer* 39, 108–111.

28. Mittal, R. C. and S. Pandit (2017). Numerical simulation of unsteady squeezing nanofluid and heat flow between two parallel plates using wavelets. *International Journal of Thermal Sciences* 118, 410–422.

29. Murshed, S. S. and C. A. N. de Castro (2014). Superior thermal features of carbon nanotubes-based nanofluids—A review. *Renewable and Sustainable Energy Reviews* 37, 155–167.

30. Nakharintr, L., P. Naphon and S. Wiriyasart (2018). Effect of jet-plate spacing to jet diameter ratios on nanofluids heat transfer in a mini-channel heat sink. *International Journal of Heat and Mass Transfer* 116, 352–361.

31. Nielda, D. and A. Kuznetsov (2014). Forced convection in a parallel-plate channel occupied by a nanofluid or a porous medium saturated by a nanofluid. *International Journal of Heat and Mass Transfer* 70, 430–433.

32. Nishimura, T. and Koji Kunitsugu (1997). Fluid mixing and mass transfer in two-dimensional cavities with time-periodic lid velocity. *International Journal of Heat and Fluid Flow* 18, 497–506.

33. Oztop, H. F. and E. Abu-Nada (2008). Numerical study of natural convection in partially heated rectangular enclosures filled with nanofluids. *International Journal of Heat and Fluid Flow* 29, 1326–1336.

34. Patel, H. E., K. B. Anoop, T. Sundararajan and S. K. Das (2008). Model for thermal conductivity of CNT-nanofluids. *Bulletin of Materials Science* 31, 387–390.

35. Piratheepan, M. and T. Anderson (2014). An experimental investigation of turbulent forced convection heat transfer by a multi-walled carbon-nanotube nanofluid. *International Communications in Heat and Mass Transfer* 57, 286–290.

36. Rahman, M., S. Mojumder, S. Saha, S. Mekhilef and R. Saidur (2014). Effect of solid volume fraction and tilt angle in a quarter circular solar thermal collectors filled with CNT-water nanofluid. *International Communications in Heat and Mass Transfer* 57, 79–90.

37. Rehman, M. M. U., Z. Qu, R. Fu and H. Xu (2017). Numerical study on free-surface jet impingement cooling with nanoencapsulated phase-change material slurry and nanofluid. *International Journal of Heat and Mass Transfer* 109, 312–325.

38. Sahoo, D. and M. A. R. Sharif (2004). Numerical modeling of slot-jet impingement cooling of a constant heat flux surface confined by a parallel wall. *International Journal of Thermal Sciences* 43, 877–887.

39. Sarkar, S., S. Ganguly and G. Biswas (2014). Buoyancy driven convection of nanofluids in an infinitely long channel under the effect of a magnetic field. *International Journal of Heat and Mass Transfer* 71, 328–340.

40. Sastry, N., A. Bhunia, T. Sundararajan and S. Das (2008). Predicting the effective thermal conductivity of carbon nanotube based nanofluids. *Nanotechnology* 19, 055704.

41. Seena, A. and H. J. Sung (2011). Wavelet spatial scaling for reducing dynamic structures in turbulent open cavity flows. *Journal of Fluids and Structures* 27, 962–975.

42. Selimefendigil, F. and H. F. Oztop (2017a). Jet impingement cooling and optimization study for a partly curved isothermal surface with CuO-water nanofluid. *International Communications in Heat and Mass Transfer* 89, 211–218.

43. Selimefendigil, F. and H. Oztop (2018). Analysis and predictive modeling of nanofluid-jet impingement cooling of an isothermal surface under the influence of a rotating cylinder. *International Journal of Heat and Mass Transfer* 121, 233–245.

44. Selimefendigil, F. and H. F. Oztop (2013). Identification of forced convection in pulsating flow at a backward facing step with a stationary cylinder subjected to nanofluid. *International Communications in Heat and Mass Transfer* 45, 111–121.

45. Selimefendigil, F. and H. F. Oztop (2014a). Numerical study of MHD mixed convection in a nanofluid filled lid driven square enclosure with a rotating cylinder. *International Journal of Heat and Mass Transfer* 78, 741–754.

46. Selimefendigil, F. and H. F. Oztop (2014b). Pulsating nanofluids jet impingement cooling of a heated horizontal surface. *International Journal of Heat and Mass Transfer* 69, 54–65.

47. Selimefendigil, F. and H. F. Oztop (2015b). Mixed convection in a two-sided elastic walled and SiO_2 nanofluid filled cavity with internal heat generation: Effects of inner rotating cylinder and nanoparticle's shape. *Journal of Molecular Liquids* 212, 509–516.

48. Selimefendigil, F. and H. F. Oztop (2016). Mixed convection of nanofluid filled cavity with oscillating lid under the influence of an inclined magnetic field. *Journal of the Taiwan Institute of Chemical Engineers* 63, 202–215.

49. Selimefendigil, F. and H. F. Oztop (2017b). Effects of nanoparticle shape on slot-jet impingement cooling of a corrugated surface with nanofluids. *Journal of Thermal Science and Engineering Application* 9, 021016.

50. Selimefendigil, F. and H. F. Oztop (2018). Cooling of a partially elastic isothermal surface by nanofluids jet impingement. Journal of Heat Transfer 140, 042205.

51. Selimefendigil, F., H. F. Oztop and N. Abu-Hamdeh (2016). Mixed convection due to rotating cylinder in an internally heated and flexible walled cavity filled with SiO_2water nanofluids: Effect of nanoparticle shape. *International Communications in Heat and Mass Transfer* 71, 9–19.

52. Sharif, M. A. R. and A. Banerjee (2009). Numerical analysis of heat transfer due to confined slot-jet impingement on a moving plate. *Applied Thermal Engineering* 29, 532–540.

53. Sharma, K., L. S. Sundar and P. Sarma (2009). Estimation of heat transfer coefficient and friction factor in the transition flow with low volume concentration of Al_2O_3 nanofluid flowing in a circular tube and with twisted tape insert. *International Communications in Heat and Mass Transfer* 36, 503–507.

54. Sheikholeslami, M., M. Gorji-Bandpy, D. Ganji, S. Soleimani and S. Seyyedi (2012). Natural convection of nanofluids in an enclosure between a circular and a sinusoidal cylinder in the presence of magnetic field. *International Communications in Heat and Mass Transfer* 39, 1435–1443.

55. Solliec, J. S. C. (2007). Flow field analysis of a turbulent slot air jet impinging on a moving flat surface. *International Journal of Heat and Fluid Flow* 28, 708–719.

56. Suresh, S., K. Venkitaraj, P. Selvakumar and M. Chandrasekar (2012). Effect of Al_2O_3-Cu/water hybrid nanofluid in heat transfer. *Experimental Thermal and Fluid Science* 38, 54–60.

57. Teamah, M. A. and W. M. El-Maghlany (2012). Augmentation of natural convective heat transfer in square cavity by utilizing nanofluids in the presence of magnetic field and uniform heat generation/absorption. *International Journal of Thermal Sciences* 58, 130–142.

58. Teng, T. P., Y. H. Hung, T. C. Teng, H.-E. Mo and H.-G. Hsu (2010). The effect of alumina/water nanofluid particle size on thermal conductivity. *Applied Thermal Engineering* 30, 2213–2218.

59. Tiwari, R. and M. K. Das (2007). Heat transfer augmentation in a two-sided lid-driven differentially heated square cavity utilizing nanofluids. *International Journal of Heat and Mass Transfer* 50, 2002–2018.

60. Wen, D. and Y. Ding (2004). Experimental investigation into convective heat transfer of nanofluids at the entrance region under laminar flow conditions. *International Journal of Heat and Mass Transfer* 47, 5181–5188.

61. Wongcharee, K., V. Chuwattanakul and S. Eiamsaard (2017). Heat transfer of swirling impinging jets with TiO_2-water nanofluids. *Chemical Engineering and Processing: Process Intensification* 114, 16–23.

62. Xue, Q. Z. (2005). Model for thermal conductivity of carbon nanotube—Based composites. *Physica B* 368, 302–307.

63. Yazid, M. N. A. W. M., N. A. C. Sidika and W. J. Yahya (2017). Heat and mass transfer characteristics of carbon nanotube nanofluids: A review. *Renewable and Sustainable Energy Reviews* 80, 914–941.

Section IV

Nanofluids in Solar Applications

11

An Insight of Ionanofluids Flow and Heat Transfer Behavior for Solar Energy Applications

Alina Adriana Minea and S.M. Sohel Murshed

CONTENTS

11.1 Introduction

Ionanofluids (INFs) are suspensions of nanoparticles in ionic liquids (ILs), and they are a new class of heat transfer fluids. They are also a new type of nanofluids. The interesting and potential applications of these new fluids can be in the area of both nanofluids and ILs. Ionic liquids are entirely made of ions and have the melting point lower than 100°C [1]. Recently, the development of room temperature ionic liquids have attracted tremendous interest from researchers and industrial people due to their low melting temperatures (<30°C) which allow these host fluids for INFs to be used in wide range of applications [2]. Ionic liquids exhibit several unique features that allow to develop and synthesize by tailoring of the cation-anion structure for desired physiochemical properties and thus for the targeted applications. As these liquids are not combustible or volatile at ambient conditions and also are recyclable, they are considered as environmental-friendly fluids [3,4]. Ionic liquids also have extremely low vapor pressure, high thermal stability, as well as high heat capacity, and the combination of these features makes ILs and thus ionanofluids better heat transfer fluids.

Some studies [e.g., 5] also suggested that, due to their very low vapor pressure preventing them to be cooled by evaporations, ILs can be used for thermal energy storage in an open system. Thus, ionic liquids and their suspensions of nanoparticles (INFs) can be good media for thermal storage systems as well as heat transfer fluids in solar power generation applications.

França [6] demonstrated that the combination of nanomaterials with ILs show great potential as heat transfer fluids through the enhancement of the thermal properties and heat transfer.

Since the first work on carbon nanotubes (CNT) mixed in ionic liquids (i.e., INFs) reported by Fukushima and Aida [7] in 2007, numbers of research works [8–28] have been performed in this new area of ionanofluids. However, the research on various properties and heat transfer features of these new heat transfer fluids are still at very early stage and some findings are carefully reviewed in later sections.

In terms of real-life applications, few research groups (e.g., [14–16,24]) demonstrated that ionanofluids could be very well suited for solar energy applications, especially for the efficient absorption of the solar radiation and for its transmission to heating/cooling systems. This chapter aims to provide an insight of these new fluids thermal properties and convective heat transfer behavior particularly focusing on their solar energy applications.

11.2 Preparation and Stability of Ionanofluids

Preparation of ionanofluids and ensuring their stability are the first step for their properties characterization and applications. Although conventional nanofluids are prepared in two methods (i.e., one-step and two-steps) [29], ionanofluids are mostly prepared by two-step methods as demonstrated in Figure 11.1.

It is of great importance to make sure that the added nanoparticles are properly (if possible homogenously) dispersed in base ILs and ionanofluids have long stability. However, it is very challenging to achieve long-term stability of INFs as many factors such as nanoparticles types, size shapes, purity, and degree of agglomerations, as well as properties of host fluids (ILs) are involved in this process. As for nanofluids, various techniques including sonication, surfactant addition, agitation, and surface treatment of nanoparticles are commonly employed for the dispersion and stability of ionanofluids. However, in most cases, researchers used sonication and addition of surfactants for better stability. Both of these means need to be carefully studied and understood before applying them to the preparation of nanofluids for their better stability and properties without changing the chemistry of nanofluids and the original structures of nanoparticles [30]. Although ultrasonication is most widely used in nanofluids as well as ionanofluids studies, there is a lack of adequate knowledge on the effects of its various parameters and duration of use. Also, no standard dispersion procedure or protocol for ultrasonication is available either for nanofluids or for ionanofluids, and therefore proper dispersion of nanoparticles and long-term stability of these new fluids remain very challenging.

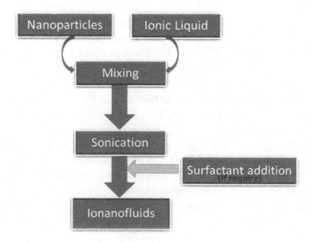

FIGURE 11.1   Flow chart of preparation of ionanofluids by two-step method.

11.3 Ionanofluids Thermal Conductivity

In the literature, thermal conductivity of INFs is the most studied property under different temperature and concentrations of nanoparticles [20–28,31]. In this section, influences of both of these parameters on this key property of ionanofluids are discussed besides summarizing some representative results.

11.3.1 Effect of Temperature on Thermal Conductivity

In general, temperature has mixed effect on thermal conductivity of these new fluids. For instance, Franca et al. [20] found that the thermal conductivities of their $[C_4mim][(CF_3SO_2)_2N]$ and $[C_2mim]$ $[EtSO_4]$-based multi-walled carbon nanotubes (MWCNTs) ionanofluids remained almost constant with increasing temperature. Ferreira et al. [23] also reported that temperature has almost no influence on the thermal conductivity of several INFs used in their study. Whereas several other studies in the literature [21,25,28,32] reported even a decrease of thermal conductivity of INFs with temperature increase. Some studies also found slight increase in the enhanced thermal conductivity of INFs with increasing temperature [e.g., 25]. Ribeiro et al. [27] measured thermal conductivity of several INFs at a fixed 1 wt.% concentration of MWCNT at various temperatures between 20°C and 80°C and at pressure of 0.1 MPa. Except $[C_4mim][PF_6]$/MWCNT ionanofluids, all others exhibited moderate increase in the thermal conductivity with temperature. Very recently, França et al. [31] reported a comprehensive study on thermal conductivity and heat transfer feasibility assessment for various ionanofluids. Thermal conductivity of their ionanofluids was found to be higher than those of the base ILs, and the enhancement of this property was found further increased with increasing temperature. Temperature dependence of thermal conductivity of various ionanofluids from literature are shown in Figure 11.2a–c which confirmed the mixed influence of temperature on the enhanced thermal conductivity of ionanofluids. It can also be noticed that some ionanofluids showed significant increase in thermal conductivity and no noticeable effect of temperature on this property. For example, Ribeiro et al. [21] found mostly moderate increase in thermal conductivity of $[C_nmim][(CF_3SO_2)_2N]$-based MWCNT (1 wt.%) ionanofluids (Figure 11.2b). However, highest thermal conductivity enhancement was for $[C_4mim][(CF_3SO_2)_2N]$-based INF and effect of temperature was found

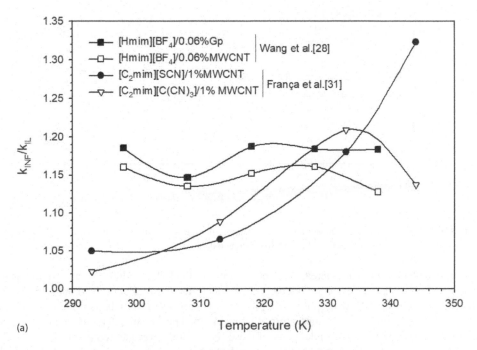

FIGURE 11.2 Effect of temperature on the enhanced thermal conductivity of various ionanofluids. a) Research from Wang et al. [28] and Franca et al. [31]. *(Continued)*

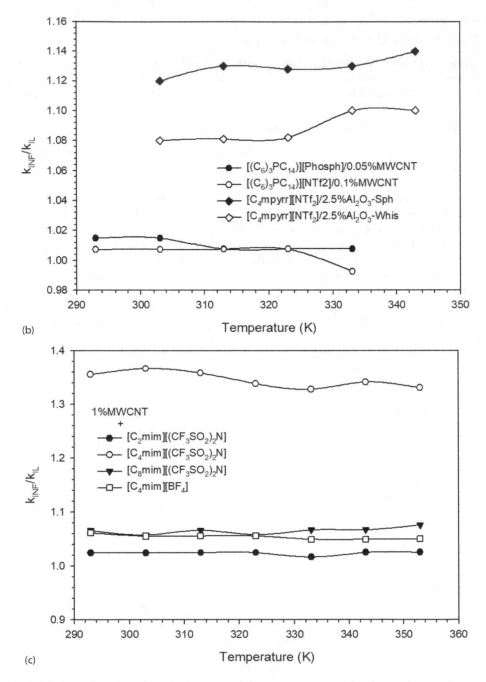

FIGURE 11.2 (Continued) Effect of temperature on the enhanced thermal conductivity of various ionanofluids. b) research on MWCNT and alumina INF and c) adding MWCNT to different ionic liquids.

slightly negative (decreasing with increasing temperature) for this INF. In addition, shape of the nanoparticles also reported to have an effect on the thermal conductivity enhancement. Paul et al. [25] reported thermal conductivities of INFs with spherical (Sph)- and whiskers (Whis)-shaped Al_2O_3 nanoparticles in [C4mpyrr][NTf2] as a function of temperature (Figure 11.2c). Their results demonstrated mostly slight increase in the enhanced thermal conductivity of these INFs with increasing temperature and INF with spherical-shaped nanoparticle showed higher thermal conductivity compared to that of the whiskers-shaped Al_2O_3 nanoparticles (Figure 11.2c).

Similarly, Wang et al. [28] measured [Hmim][BF$_4$]/Graphene (Gp) as well as MWCNT ionanofluids, and their results (in Figure 11.2a) showed no obvious effect of temperature. However, ionanofluids with GP was found to have higher thermal conductivity compared to INFs with MWCNT.

11.3.2 Effect of Concentration of Dispersed Nanoparticles on Thermal Conductivity

Like nanofluids, effect of concentration of nanoparticles on the thermal conductivity of INFs is the most obvious factor, and it was reported to increase with increasing the loading of nanoparticles. Ribeiro et al. [21] observed an increase in thermal conductivities of three different ionic liquids when adding MWCNT in to them. The highest enhancement of thermal conductivity (about 35%) was obtained for [C$_4$mim][(CF$_3$SO$_2$)$_2$N]-based INF at 1 wt.% loading of MWCNT at room temperature.

França et al. [20] measured thermal conductivity of [C$_4$mim][(CF$_3$SO$_2$)$_2$N] and [C$_2$mim][EtSO$_4$] ionic liquids at different concentrations of MWCNT at a fixed temperature of 20°C. In all considered cases, the thermal conductivity of their INFs increased with adding higher concentrations of nanoparticles (Figure 11.3).

Effect of nanoparticles concentration and shape of Al$_2$O$_3$ (<50 nm) nanoparticles on the thermal conductivity of different types of ILs including [C$_4$mpyrr][NTf$_2$]-based INFs were reported by Paul and co-workers [24,25]. Their results showed (in Figure 11.4) that thermal conductivity of these INFs increased considerably with increasing the concentration of both nanoparticles, and spherical-shaped Al$_2$O$_3$ nanoparticles were found to have higher enhancement of thermal conductivity compared to their whiskers counterparts. While authors delivered some explanations supporting their results, such a phenomenon can occur mostly because of the whiskers orientation, and we assume that was influenced also by the dispersion uniformity.

Very recently, França et al. [31] reported thermal conductivity enhancement of different INFs at two different concentrations of MWCNT. They found an increase in thermal conductivity of [C$_4$mim][SCN] IL about 7.13% due to adding 1 w/w% of MWCNT. Other INFs in their study [31] also showed moderate increase in thermal conductivity. Interestingly, except some speculations and presumptions, no studies were able to elucidate the underlying mechanisms of the observed enhancement of thermal conductivity. Very little research has also been made on the theoretical determination or model development of the thermal conductivity of these new nanofluids. Some efforts were, however, made to assess the applicability of classical and recent models for nanofluids in order to predict the thermal conductivity of INFs.

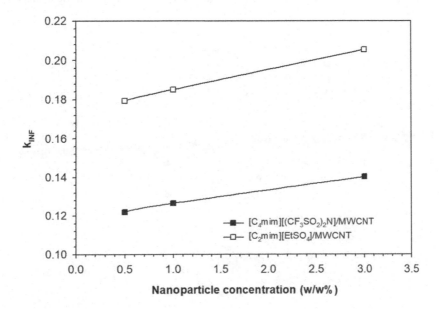

FIGURE 11.3   Thermal conductivity variation of two INFs as a function of loading of MWCNT.

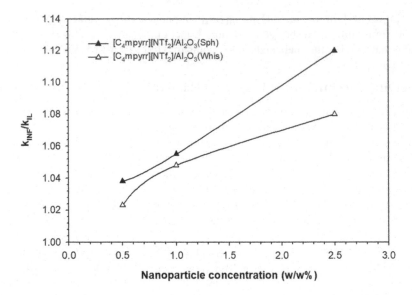

FIGURE 11.4 Thermal conductivity enhancement of $[C_4mpyrr][NTf_2]$-based ionanofluids as a function of concentration of spherical (Sph)- and whiskers (Whis)-shaped Al_2O_3 nanoparticles at 40°C.

11.4 Ionanofluids Heat Capacity

As another important property, specific heat of INFs was also investigated by a handful of researchers. For example, Wang et al. [28] measured the specific heats of Gp and MWCNT dispersed INFs, and a slight increase of specific heat with temperature and a minor decrease while adding the nanoparticles to the ionic liquid was reported as shown in Figure 11.5. Overall, the variations of specific heat remain between 0% and 3% decrease while increasing temperature and concentration as well.

Effects of temperature and concentration of nanoparticles on specific heat of INFs with Al_2O_3 of whiskers (2 nm–6 nm × 200 nm–400 nm and aspect ratio > 100) as well as spherical shapes dispersed

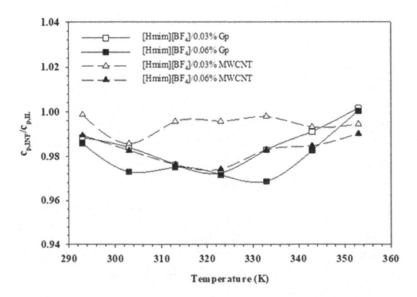

FIGURE 11.5 Specific heat ratio of $[Hmim][BF_4]$-based graphene (G_p) and MWCNT ionanofluids.

in [C$_4$mpyrr][NTf$_2$] were investigated by Paul et al. [25]. Interestingly, their results demonstrated a small increase (fluctuating as well) in specific heat with increasing temperature for INF with whiskers-shaped Al$_2$O$_3$, while temperature has apparently no effect on this property for spherical-shaped Al$_2$O$_3$ as shown in Figure 11.6 [25]. No physics-based explanation behind such results was provided. However, such phenomenon can appear mostly because of the whiskers orientation and probably the dispersion uniformity as mentioned before. On the other hand, the effect of concentration for both shapes of this nanoparticle was obvious as substantial increase in specific heats of INFs with increasing loading of both nanoparticles was reported as can also be seen from Figure 11.7 [25]. Nevertheless, any increase in specific heat of these new fluids is desirable for their applications particularly in energy storage and harvesting.

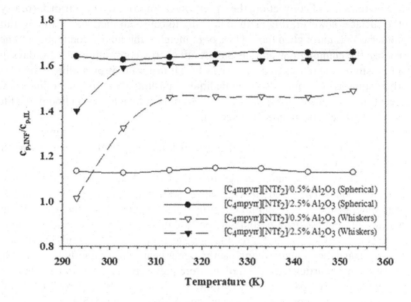

FIGURE 11.6 Enhancement of specific heat of INFs with whiskers- and spherical-shaped Al$_2$O$_3$ nanoparticles as a function of temperature.

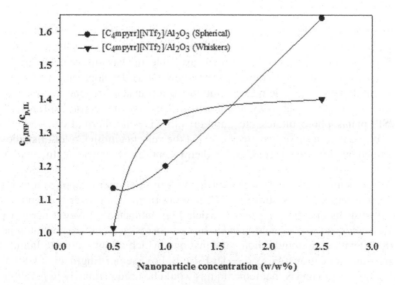

FIGURE 11.7 Enhancement of specific heat of [C$_4$mpyrr][NTf$_2$]-based INFs containing spherical- and whiskers-shaped Al$_2$O$_3$ nanoparticles as a function of their concentration.

11.5 Ionanofluids Convective Heat Transfer

There are numbers of studies in the archived literature on the probable applications of ionanofluids. Most of these are concentrated on their application as thermal fluids due to the improved thermophysical properties of ionanofluids, in comparison with the corresponding base fluids. Possible applications are related to: industrial cooling, refrigeration, energy extraction, solar energy, etc. Nevertheless, the most important issue, besides the determination of thermophysical properties remains their comportment in convective heat transfer, and this section aims to make an outline of published research on this specific topic.

According to Newton's law of convection, the easier way to improve the heat transferred by convection is to increase the surface, however, this method is going to cause an increase of the overall size of the equipment. Another way is to roughen the surface or to increase the fluid velocity, but all these methods are going to increase the pumping power, thus the overall energy consumption. Nowadays, heat transfer practice built a new strategy and tries to develop new heat transfer fluids by adding nanoparticles to base fluids. One of these approaches refers also to ionic liquids, pushing them to ionanofluids development.

Convective heat transfer studies are very limited at this moment, and one can find in the literature only few numerical studies while experimental is ongoing.

Few recent studies [33–36] are discussed further on with reference to the numerical approach and results are summarized in Table 11.1 which also provides the proposed correlations for heat transfer coefficients from those numerical studies.

In regard to experimental studies, Paul et al. [15] investigated the heat transfer performance of [C4mim] [NTf2]+1%Al_2O_3 ionanofluid. They studied heat transfer performances under forced convection in a circular tube with an inner diameter of 3.86 mm in laminar flow and found an increase of about 20% of the convective heat transfer by adding nanoparticles to the base fluid. Another experimental study performed by the same group (Paul et al. [37]) considered the same test set-up and ionanofluid type as was used in their earlier study [15], and they found a significantly higher heat transfer coefficient than base ionic liquid, depending also on nanoparticles concentration. More precisely, the heat transfer enhancement goes from 15% to 31% while two nanoparticles concentrations were used: 0.18%vol. and 0.36%vol. alumina in base fluid. However, no specific correlation was reported in these studies [15,37].

11.6 Solar Energy Applications

Most of the currently used sources of energy are damaging the environment, and, thus, much research is directed nowadays toward some better alternatives like green and renewables energies. One of the most studied alternative is solar energy that is freely available and has almost no environmental impact. Research in solar energy is strongly connected with new fluids development, as a new way to minimize the equipment dimensions while maximizing the heat transfer efficiency [33]. Actually, there are two important ways to harvest solar energy: one is directed to solar-electric conversion (concentrating solar power (CSP) plants, photovoltaics, etc.), and the other is solar-thermal conversion by using solar collectors. Even if research in these two areas is not at the very beginning, a lot of studies are ongoing, as were also outlined by Muhammad et al. [38] in their recent review article dealing with nanofluids use in solar collectors.

On the other hand, ionic liquids have major advantages in solar energy applications mainly because they do not freeze at temperatures above 100°C, making them stable over the temperature range at which many CSP systems operate. In a recent article [39] published on New Energy Update website quoted as: "Ionic liquids can easily be designed to be non-corrosive and there are potentially up to 10 to the power of 18 different cation and anion combinations, which allow the ionic liquids properties to be tailored to specific applications. Fox's (i.e., Dr Elise B. Fox, Senior Engineer at US-based Savannah River National Laboratory) view is that ionic liquids would be used primarily in parabolic troughs and linear Fresnel systems, although they may also have the potential to be 'usefully integrated' in tower-based systems."

TABLE 11.1

Numerical Research Results on Convection Heat Transfer of Ionanofluids

References	Ionanofluids	Flow Conditions/CFD Code	Findings and/Correlations
El-Maghlany and Minea [33]	$[C_4mim][NTf_2]/Al_2O_3$	• Fully developed under steady state conditions laminar flow a tube, with longitudinal and radial flow (no swirl flow). • Computational fluid dynamics (CFD) code: in-house developed	Developed correlations for Nu number: • $Nu = 0.558 (Re\ Pr\ D/L)^{0.376}$ for $\phi = 0.0$ • $Nu = 0.6 (Re\ Pr\ D/L)^{0.372}$ for $\phi = 0.5$ • $Nu = 0.63 (Re\ Pr\ D/L)^{0.369}$ for $\phi = 1.0$ • $Nu = 0.696 (Re\ Pr\ D/L)^{0.361}$ for $\phi = 2.5$ • $Nu^{2.702} = 0.226 Re\ Pr^{1/L}(\alpha_f/\alpha_{ionano})$ valid in laminar flow for all studied fluids. The performance evaluation criteria is 1–1.5 that sustain the advantages of using ionanofluids in solar applications.
Minea and Murshed [34]	$[C_4mim][NTf_2]/Al_2O_3$; $[C_2mim][EtSO_4/MWCNT$; $[Hmim][BF_4]/MWCNT$; $[Hmim][BF_4]/Graphene$	• Steady laminar forced flow combined with heat transfer inside a tube (0.12 m diameter) with two zones: isothermal zone of 5.76 m at the entrance followed by a uniformly heated area of 2.88 m • CFD Code: Ansys Fluent	With Re number increase, the heat transfer coefficient increases significantly. The maximum escalation in heat transfer coefficient is reached by $[C_4mim]$ $[NTf_2]$ + alumina. Also, adding MWCNT or graphene to base ionic liquids, the convective heat transfer coefficient rises 1.7%–12.1%.
Chereches et al. [35]	$[C_4mim][NTf_2]/Al_2O_3$; $[C_4mpyrr][NTf_2]/Al_2O_3$	• Laminar flow in a uniformly heated tube of 1.75 m length and 0.014 m inner diameter • CFD Code: Ansys Fluent	Authors proposed a correlation valid under the laminar flow for $500 < Re < 2000$ and total weight concentration ranging from 0% to 2.5% $Nu = 4.15 Re^{0.09} Pr^{0.195} (1 - \phi - 200\ \phi^2)$ Plus, heat transfer enhancement by adding nanoparticles to the base ionic liquid was noticed and is depending on Re and nanoparticles concentration.
Minea and El-Maghlany [36]	$[C_4mim][NTf_2]/Al_2O_3$	• Natural convection in a square enclosure. • CFD code: in-house developed	Two correlations were proposed which are valid for $10^4 \leq Ra \leq 10^6$ and $0\% \leq \phi \leq 2.5\%$ • For the hot element at the bottom wall: $Nu = 81.663\ \phi + 0.555\ (Ra-4614.793)^{0.226} - 3710.366\ \phi^2$ • For the hot element at the left wall: $Nu = 116.173\ \phi + 0.484\ Ra^{0.245} - 5001.894\ \phi^2$ For both situations, a clear heat transfer enhancement was noticed.

With ILs as based fluids, ionanofluids were identified as a new possible heat transfer fluid with high performances in solar energy applications, as was also described in some numerical and experimental studies outlined in the previous section (see [15,33–37]). Nevertheless, the major issue remains the ionanofluids´ costs and their stability in long-run service [38].

In this idea, the ionic liquids enhancement by adding nanoparticles is a subject worth to be further investigated extensively. Nevertheless, experimental as well as numerical studies on ionanofluids effective behavior in CSP plants or solar collectors are very limited in the literature.

Some research [15,33–35,37] demonstrated the benefits of using ionanofluids in thermal-solar applications, and all results indicated a beneficial use of ionanofluids in comparison with their base ionic liquids. In some cases (see for example [35]), the heat transfer coefficient is increased as high as 6.2 times, at high Reynolds number and higher nanoparticle loading (i.e., 2.5% alumina).

Zhang et al. [16] were perhaps the pioneers in measuring the radiative properties of ionanofluids and their base ionic liquids. They demonstrated that the ionanofluid with carbon coated Ni nanoparticles displayed the lowermost transmittance and uppermost extinction coefficient if compared with ionanofluids with Ni and Cu nanoparticles at volume fraction of 10 ppm.

Nevertheless, the beneficial effects of adding solid nanoparticles to a basic fluid was demonstrated for nanofluids [40] and can be extended also to ionanofluids. The idea of adding nanoparticles to fluids goes undoubtedly to reducing the overall size of the solar collector by 25.5%–21.5% and increasing the collector efficiency by at least 40% [40].

Another review published by Murshed and Nieto de Castro [41] outlines the importance of using CNT-based ionanofluids in solar energy due to their superior features in regard to thermal conductivity.

On the other hand, Chen et al. [42] affirm that ionic liquids application as heat transfer fluids is limited due to low absorption of sunlight and adding nanoparticles will increase their potential. Hence, their studies are directed to evaluation of [Hmim]BF$_4$/SiC ionanofluids' thermophysical and optical properties, while their conclusion was that adding 0.06% wt.% of SiC nanoparticles in this IL the optical properties were enhanced and the thermal conductivity was increased by 10.2%. These improvements in properties made the [Hmim]BF$_4$/SiCionanofluids better candidates for direct absorption solar collectors.

Wang et al. [28] studied graphene-based ionanofluids and reported their results on thermophysical properties enhancement, affirming that these new fluids have unique properties for solar energy collection at high temperatures due to liquid temperature, low vapor pressure, and high thermal stability.

As a conclusion, even if ionic liquids applications in solar area are intensively acknowledged, intensively since many years [43–45] ionanofluids studies are at their very early stage and oriented only to thermophysical properties. Thus, more application-oriented research is needed in this area to implement these new fluids in CSP or solar collectors.

11.7 Conclusion

Ionanofluids are complex dispersions of nanomaterials in ionic liquids, and their development is in the early stages, as was outlined in this chapter. The nature of their structure and thermophysical properties have encouraged their possible usage in many applications, while some of them are already being used in the chemical industry while some attempts exist also for the solar energy applications. A brief framework on the nature, preparation, properties, and applications as heat transfer fluids especially is solar energy area is given in this chapter.

As a final conclusion, it can be surmised that ionanofluids are suitable for solar application due to their unique and enhanced properties. This chapter sets a frame and outlines the studied ionic liquids with several nanoparticles enhanced their thermophysical properties and convective heat transfer behavior targeting toward their applications in solar energy areas.

Ionanofluids research has many challenges to proceed in systematic and efficient ways, however, it can be a future support for environmentally safe and biodegradable fluids, contributing to a healthier and ecological world. Nevertheless, more intensive studies are highly needed and to be performed, especially in terms of stability, thermophysical properties estimation, and real-life applications of these new class of fluids.

NOMENCLATURE

cP   specific heat
D hydraulic diameter
k thermal conductivity
L channel length
Pr   Prandtl number
Nu   Nusselt number
Re   Reynolds number
Ra   Rayleigh number

GREEK SYMBOLS

α Thermal diffusivity
ϕ volume fraction of particles

ABBREVIATIONS

CFD   computational fluid dynamics
IL   ionic liquid
INF   ionanofluid

REFERENCES

1. Wasserscheid, P. and T. Welton, *Ionic Liquids in Synthesis.* VCH-Wiley: Weinheim, Germany; 2008.
2. Marsh, K. N., J. A. Boxall and R. Lichtenthaler (2004) Room temperature ionic liquids and their mixtures—A review. *Fluid Phase Equilibria* 219, 93–98.
3. Earle, M. J. and K. R. Seddon (2000) Ionic liquids. *Green Solvents for the Future Pure and Applied Chemistry* 72, 1391–1398.
4. Ghandi, K. (2014) A review of ionic liquids, their limits and applications. *Green and Sustainable Chemistry* 4, 44–53.
5. Zhang, M. and R. Reddy (2006) Application of [C$_4$min][Tf$_2$N] ionic liquid as thermal storage and heat transfer fluids. *ECS Transactions* 1(4).
6. França J. M. P. (2010) *Thermal properties of ionanofluids*, MSc Thesis, Faculdade de Ciências da Universidade de Lisboa, Portugal.
7. Fukushima, T. and T. Aida (2007) Ionic liquids for soft functional materials with carbon nanotubes. *Chemistry-A European Journal* 13, 5048–5058.
8. Murshed, S. M. S., C. A. Nieto de Castro, M. J. V. Lourenço, M. L. M. Lopes and F. J. V. Santos (2012) Current research and future applications of nano- and ionano-fluids. *Journal of Physics Conference Series* 395, 012117.
9. Nieto de Castro C. A., A. P. C. Ribeiro, S. I. C. Vieira, J. P. M. França, M. J. V. Lourenço, F. V. Santos, S. M. S. Murshed, P. Goodrich and C. Hardacre (2013) Synthesis, properties and physical applications of ionanofluids, In *Ionic Liquids—New Aspects for the Future.* Kadokawa J. I. (Ed.), Rijeka, Croatia: InTech.
10. França, J. M. P., F. Reis, S. I. C. Vieira, M. J. V. Lourenço, F. J. V. Santos, C. A. Nieto de Castro and A. H. Pádua (2014) Thermophysical properties of ionic liquid dicyanamide (DCA) nanosystems. *The Journal of Chemical Thermodynamics* 79, 248–257.
11. Shevelyova, M. P., Y. U. Paulechka, G. J. Kabo, A. V. Blokhin, A. G. Kabo and T. M. Gubarevich (2013) Physicochemical properties of imidazolium-based ionic nanofluids: Density, heat capacity, and enthalpy of formation. *The Journal of Physical Chemistry C* 117, 4782–4790.
12. Wang, B., X. Wang, W. Lou and J. Hao (2011) Gold-ionic liquid nanofluids with preferably tribological properties and thermal conductivity. *Nanoscale Research Letters* 6, 259.

13. Visser, A. E., N. J. Bridges, B. L. Garcia-Diaz, J. R. Gray and E. B. Fox (2012) Al_2O_3-based nanoparticle-enhanced ionic liquids (NEILs) for advanced heat transfer fluids. *Ionic Liquids: Science and Applications, ACS Symposium Series* 1117, 259–270.

14. Bridges, N. J., A. E. Visser and E. B. Fox (2011) Potential of nanoparticle-enhanced ionic liquids (NEILs) as advanced heat-transfer fluids. *Energy Fuels* 22, 4862–4864.

15. Paul, T. C., A. M. Morshed and J. A. Khan (2013) Nanoparticle enhanced Ionic liquids (NEILs) as working fluid for the next generation solar collector. *Procedia Engineering* 56, 631–636.

16. Zhang, L., J. Liu, G. He, Z. Ye, X. Fang and Z. Zhang (2014) Radiative properties of ionic liquid-based nanofluids for medium-to-high-temperature direct absorption solar collectors. *Solar Energy Materials & Solar Cells* 130, 521–528.

17. Nieto de Castro, C. A., X. Paredes, S. I. C. Vieira, S. M. S. Murshed, M. J. V. Lourenço and F. V. Santos (2017) Ionanofluids: Innovative agents for sustainable development, In *Nanotechnology for Energy Sustainability*, Raj, B., M. Van de Voorde, Y. Mahajan (Eds.), Chapter 37, Weinheim, Germany: Wiley-VCH.

18. Atashrouz, S., M. Mozaffarian and G. Pazuki (2015) Modeling the thermal conductivity of ionic liquids and ionanofluids based on a group method of data handling and modified Maxwell model. *Industrial & Engineering Chemistry Research* 54, 8600–8610.

19. He, Z. and P. Alexandridis (2015) Nanoparticles in ionic liquids: Interactions and organization, *Physical Chemistry Chemical Physics* 17, 18238–18261.

20. França, J. M. P., S. I. C. Vieira, M. J. V. Lourenço, S. M. S. Murshed and C. A. Nieto de Castro (2013) Thermal conductivity of ionanofluids of $[C_4mim][NTf_2]$ and $[C_2mim][EtSO_4]$ with carbon nanotubes Experiment and theory. *Journal of Chemical and Engineering Data* 58, 467–476.

21. Ribeiro, A. P. C., S. I. C. Vieira, P. Goodrich, C. Hardacre, M. J. V. Lourenço and C. A. Nieto de Castro (2013) Thermal conductivity of $[C_nmim][(CF_3SO_2)_2N]$ and $[C_4mim][BF_4]$ ionanofluids with carbon nanotubes—Measurement, theory and structural characterization. *Journal Nanofluids* 2, 55–62.

22. Patil, V., A. Cera-Manjarre, D. Salavera, C. Rode, K. Patil, C. A. Nieto de Castro and A. Coronas (2016) Ru-Imidazoliumhalide ionanofluids: Synthesis, structural, morphological and thermophysical properties. *Journal of Nanofluids* 5, 191–208.

23. Ferreira, A. G. M., P. N. Simoes, A. F. Ferreira, M. A. Fonseca, M. S. A. Oliveira and A. S. M. Trino (2013) Transport and thermal properties of quaternary phosphonium ionic liquids and IoNanofluids. *The Journal of Chemical Thermodynamics* 64, 80–92.

24. Paul, T. C. Investigation of thermal performance of nanoparticle enhanced ionic liquids (NEILs) for solar collector applications. PhD Thesis, University of South Carolina; 2014.

25. Paul, T. C., A. K. M. M. Morshed, E. B. Fox and J. A. Khan (2015) Experimental investigation of natural convection heat transfer of Al_2O_3 nanoparticle enhanced ionic liquids (NEILs). *International Journal of Heat and Mass Transfer* 83, 753–761.

26. Nieto de Castro, C. A., S. M. S. Murshed, M. J. V. Lourenço, F. J. V. Santos, M. L. M. Lopes, and J. M. P. França (2012) Enhanced thermal conductivity and heat capacity of carbon nanotubes-ionanofluids. *International Journal of Thermal Sciences* 62, 34–39.

27. Ribeiro A. P. C., S. I. C. Vieira, J. M. P. França, C. S. Queirós, E. Langa, M. J. V. Lourenço, S. M. S. Murshed and C. A. Nieto de Castro (2011) Thermal properties of ionic liquids and ionanofluids. In *Ionic Liquids: Theory, Properties, New Approaches.* Kokorin, A. (Ed.), Rijeka, Croatia: InTech.

28. Wang, F., L. Han, Z. Zhang, X. Fang, J. Shi and W. Ma (2012) Surfactant-free ionic liquid-based nanofluids with remarkable thermal conductivity enhancement at very low loading of graphene. *Nanoscale Research Letters* 7, 314–319.

29. Murshed, S. M. S. and C. A. Nieto de Castro (2014) *Nanofluids: Synthesis, Properties and Applications*, New York: Nova Science Publishers.

30. Nieto de Castro, C. A., S. I. Vieira, M. J. V. Lourenço and S. M. S. Murshed (2017) Understanding stability, measurements, and mechanisms of thermal conductivity of nanofluids. *Journal of Nanofluids* 6, 804–811.

31. França, J. M. P., M. J. V. Lourenço, S. M. S. Murshed, A. A. H. Pádua and C. A. Nieto de Castro (2018) Thermal conductivity of ionic liquids and ionanofluids and their feasibility as heat transfer fluids. *Industrial & Engineering Chemistry Research* 57, 6516–6529.

32. Nieto de Castro, C. A., M. J. V. Lourenço, A. P. C. Ribeiro, E. Langa, S. I. C. Vieira, P. Goodrich and C. Hardacre (2000) Thermal properties of ionic liquids and ionanofluids of imidazolium and pyrrolidinium liquids. *Journal of Chemical and Engineering Data* 55, 65–69.

33. El-Maghlany, W. M. and A. A. Minea (2018) Novel empirical correlation for ionanofluid PEC inside tube subjected to heat flux with application to solar energy. *Journal of Thermal Analysis and Calorimetry*, 10, doi:10.1007/s10973-018-7461-y.

34. Minea, A. A. and S. M. S. Murshed (2018) A review on development of ionic liquid based nanofluids and their heat transfer behavior. *Renewable and Sustainable Energy Reviews* 91, 584–599.

35. Chereches, E. I., K. V. Sharma and A. A. Minea (2018) A numerical approach in describing ionanofluids behavior in laminar and turbulent flow. *Continuum Mechanics and Thermodynamics* 30, 657–666.

36. Minea, A. A. and W. M. El-Maghlany (2017) Natural convection heat transfer utilizing ionic nanofluids with temperature-dependent thermophysical properties. *Chemical Engineering Science* 174, 13–24.

37. Paul, T. C., A. K. M. Morshed, E. B. Fox and J. A. Khan (2015) Thermal performance of Al$_2$O$_3$ nanoparticle enhanced ionic liquids (NEILs) for concentrated solar power (CSP) applications. *International Journal of Heat and Mass Transfer* 85, 585–594.

38. Muhammad, M. J., I. A. Muhammad, N. A. CheSidik, M. N. Yazid, R. Mamat and G. Najafi (2016) The use of nanofluids for enhancing the thermal performance of stationary solar collectors: A review. *Renewable and Sustainable Energy Reviews* 63, 226–236.

39. Williams, A. (2011) Ionic liquids: The future heat transfer fluid of choice? New Energy Update: CSP, (Jan 28, 2011) available: http://www.newenergyupdate.com/csp-today/technology/ionic-liquids-future-heat-transfer-fluid-choice.

40. Bozorgan, N. and M. Shafahi (2015) Performance evaluation of nanofluids in solar energy: A review of the recent literature. *Micro and Nano Systems Letters* 3, 5–20.

41. Murshed, S. M. S. and C. A. Nieto de Castro (2014) Superior thermal features of carbon nanotubes-based nanofluids—A review. *Renewable and Sustainable Energy Reviews* 37, 155–167.

42. Chen, W., C. Zou and X. Li (2017) An investigation into the thermophysical and optical properties of SiC/ionic liquid nanofluid for direct absorption solar collector. *Solar Energy Materials and Solar Cells* 163, 157–163.

43. Plechkova, N. V. and K. R. Seddon (2008) Applications of ionic liquids in the chemical industry. *Chemical Society Reviews* 37, 123–150.

44. Siddharth, P. (2006) Analytical applications of room-temperature ionic liquids: A review of recent efforts. *Analytica Chimica Acta* 556, 38–45.

45. Wang, P., B. Wenger, R. Humphry-Baker, J. F. Moser, J. Teuscher, W. Kantlehner, J. Mezger, E. V. Stoyanov, S. M. Zakeeruddin and M. Grätzel (2005) Charge separation and efficient light energy conversion in sensitized mesoscopic solar cells based on binary ionic liquids. *Journal of the American Chemical Society* 127, 6850–6856.

12

Nanofluid-Based Direct Absorption Solar Collectors

Gary O'Keeffe, Sarah Mitchell, Tim Myers, and Vincent Cregan

CONTENTS

12.1 Introduction

Energy consumption is a significant driver of economic growth. Technological improvements, political pressure, and global gross domestic product (GDP) growth are all affecting the global energy consumption (Omri 2013). Global solar energy production is predicted to rise at a rate of 8.9% annually from 2012 to 2040, making it the fastest growing form of energy generation in the coming decades (U.S. Energy Information Administration 2017). The solar energy industry has experienced phenomenal growth in recent years due to both technological improvements resulting in cost reductions and government policies supportive of renewable energy development and utilisation (Timilsina et al. 2012). Currently, solar-thermal technologies produce more energy than solar-voltaic technologies (Li et al. 2016). Solar-thermal technologies provide hot water, to heat and cool space, and to generate high-temperature heat for industrial processes (REN21 2016). Whilst the global capacity of solar-thermal technology continues to rise, recently the rate of capacity increase has seen a decline, falling 14% in 2015 (REN21 2016). More innovation is needed if solar-thermal technologies are to see this recent trend reversed.

Several types of solar-thermal collectors have been used to harness solar energy in both residential, electricity generation, and industrial settings. Taylor et al. (2011) outline the three ideal characteristics of a solar-thermal collector. (1) An ideal collector should efficiently absorb solar radiation. (2) An ideal collector must minimise heat loss due to convection and radiation. (3) An ideal collector should have minimal pumping and maintenance costs. In the following paragraphs, we discuss the most common types of solar collectors.

We classify solar collectors based on whether or not the incoming radiation is concentrated. In the non-concentrating type of solar collector, the aperture area is the same as the area absorbing the radiation—this type of solar collector is typically used in a residential setting to heat water or buildings. The most common residential solar collectors are flat-plate black-surface absorbers and evacuated tubes. Flat-plate collectors consist of a dark flat-plate absorber and a working fluid which circulates through the system, extracting heat from the absorber plates. Flat-plate collectors can be cheap relative to other collectors due to the simplicity of their design—they can also be integrated into roofs or façades, however, they do not perform as well in wet/cold environments (Lamnatou et al. 2016). Average annual flat-plate collector efficiency in Dublin, Ireland, was found to be 46.1% (Ayompe et al. 2011).

Evacuated tube collectors come in various design layouts, one of the most common forms is composed of multiple evacuated glass tubes, each containing an absorber plate fused to a heat pipe (Mahjouri 2004). The working fluid extracts heat from the top of these heat pipes as it flows through a manifold. Insulation covers this manifold, and a vacuum surrounds the glass tubes, reducing the convective and conductive heat losses, and this feature makes evacuated tube collectors ideal for solar-thermal conversion in cold, windy, and humid conditions. Average annual vacuum tube collector efficiency in Dublin, Ireland, was found to be 60.7% (Ayompe et al. 2011), which is significantly higher than the flat-plate collector efficiency. In fact, vacuum tube collectors can perform up to 30% better than flat-plate collectors, however, they are more expensive to manufacture (Comerford 2011).

Concentrating collectors have a larger aperture area than absorber area—they can be used in many industrial processes including: sterilising, pasteurising, drying, hydrolysing, distillation and evaporation, washing and cleaning, and polymerisation (Kalogirou 2003). In an industrial heat process, the operating temperature needs to be consistent and predictable, models for accurately predicting the temperature at the outlet of a collector are of vital importance to the viability of this solar-thermal application. Concentrating solar collectors are also used to power turbines and generate electricity. A parabolic trough solar collector is one example of a solar collecting system that can be used to generate electricity or heat for an industrial process. In this design, a parabolic mirror reflects incoming sunlight onto a receiver located on its focal line. A working fluid flows through the receiver, and as the receiver heats up, so does the working fluid. The parabolic mirror tracks the sun, similarly to the heliostats in a solar-power tower.

All of the examples of concentrating and non-concentrating solar collectors mentioned above are 'surface-based' absorbers—that is, in each of these designs, the receiver's surface absorbs the incoming sunlight. For example, parabolic trough collectors incorporate an opaque metal surface coated with a selective thin film to efficiently convert solar radiation into thermal energy (Li et al. 2016). Then, a working fluid flows through the receiver and absorbs thermal energy from this heated surface. Heat loss occurs via convection and radiation at the surface of the absorber. Unsurprisingly, the surface is the hottest part of a 'surface-based' absorber. However, this is not the only way to capture and convert electromagnetic radiation into usable thermal energy—direct absorption solar collectors (DASCs) offer a different solution (Taylor et al. 2011). DASCs were first proposed in the mid-1970s as an alternative to surface absorbers. A DASC does not contain an absorbing surface, instead, the working fluid absorbs incoming solar energy directly. This leads to lower temperatures at the surface of the receiver and potentially less energy lost to the surrounding environment. Standard fluids are inefficient at absorbing sunlight due to their low absorptive properties, for example, Otanicar et al. (2009) show that water only absorbs 13% of the available solar energy in a DASC of depth 1 cm. Therefore, DASCs initially failed to take off as they were not a commercially viable solar collector design (Parvin et al. 2014). In the past, people have added particles/dyes to liquids in an attempt to enhance the thermophysical and/or optical properties of the working fluid in a DASC. Large particles can settle and clog systems, and this escalates the cost of maintenance, however, nanoparticle-laden fluids can serve as the absorbing medium in DASCs to overcome the poor absorption properties of conventional fluids (Taylor et al. 2011).

Unlike traditional DASCs, nanofluid-based direct absorption solar collectors (NDASCs) use nanofluids to absorb incident sunlight. A nanofluid is a colloidal suspension of nanoparticles in a liquid medium. Nanoparticles do not settle as quickly as larger particles, however, incorrect nanofluid preparation can lead to nanoparticle agglomeration, and pockets of slow-moving or stationary flow in the system can lead to nanoparticle sedimentation (Colangelo et al. 2013). High particle volume fractions in a nanofluid can significantly change the viscosity and overall fluid dynamics in the system, potentially increasing the pump's workload, and so usually nanofluids appear in NDASCs at very low particle volume fractions (<0.01%). Nanofluids with low nanoparticle volume fractions have similar thermophysical properties to their base fluid, however, they exhibit enhanced optical properties. These optical enhancements show a potential for volumetric absorption to be a more efficient solar collector design than surface-based absorption (Lenert and Wang 2012, Otanicar et al. 2010).

Numerous NDASC designs have been proposed and studied in previous literature. There are three related (but different) approaches to modelling and predicting the performance of an NDASC: statistical, numerical, and analytical. Statistical approaches to NDASC modelling rely on fitting parameters to experimental data. This approach is limited as it does not allow a modeller to make accurate performance predictions outside the parameter space that has been tested experimentally. Numerical approaches to NDASC modelling generally simulate the system via a computational fluid dynamics (CFD) method. This approach is generally more useful than the statistical approach, however, it can be computationally expensive (and prohibitive) to conduct an exhaustive parameter space exploration on a CFD-simulated model. Analytical approaches to NDASC modelling generally employ applied mathematical techniques to simplify the system's governing physical equations such that they can be solved analytically. Analytic expressions for the performance of NDASCs are desirable as they allow a system to be simulated and optimised with relative computational ease. Thus, we focus the discussion in this chapter towards analytic modelling approaches.

Tyagi et al. (2009) present the first theoretical model of an NDASC. Their model consists of a system of two differential equations: a radiative transport equation (RTE), describing the propagation of solar radiation through the nanofluid, and a conservation of energy equation. The model neglects thermal re-emission effects by assuming low operating temperatures. They simulate the model numerically and demonstrate up to a 10% efficiency enhancement over conventional flat-plate collectors by using water-based Al nanofluids. The collector efficiency increased rapidly with volume fraction and reached a maximum value of ~80% at a volume fraction of ~2%. Otanicar et al. (2010) use a reflective surface to coat the bottom of an NDASC and numerically and experimentally test the collector performance. A two-dimensional representation of the Otanicar et al. (2010) experimental set-up is detailed in Figure 12.1. NDASC efficiency is very much dependent on receiver height, Otanicar et al. report relatively high collector efficiencies with a small collector height ($H = 0.00015$ m). No existing analytic model has been proposed for the NDASC detailed in Otanicar's study. In this chapter, we discuss the general process of modelling an NDASC's performance—we achieve this whilst describing a novel analytic modelling approach for predicting the performance of the NDASC explored in Otanicar et al. (2010).

We discuss the literature surrounding analytic modelling approaches in Section 12.1.1. In Section 12.2, we present a two-dimensional model for the conservation of energy in a parallel-plate NDASC under laminar flow. In this system, the base panel of the collector is coated with aluminium foil which reflects

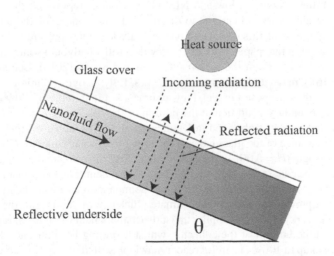

FIGURE 12.1 Nanofluid flow through a channel with a glass cover and a reflective underside.

incoming radiation back out of the receiver, therefore, in Section 12.2.4, we introduce a term describing the source of heat which accounts for reflected radiation. Whilst NDASCs are designed to absorb sunlight, sometimes they are tested using heat lamps to generate incoming radiation. Heat lamps can produce a constant and predictable source of radiation making them ideal for testing solar collectors in controlled environments. Therefore, it is important to better understand how the optical properties of nanofluids change with different sources of radiation. In Section 12.2.6, we propose a case study to consider two alternative sources of black-body radiation, the first is a heat lamp at 3158 K, and the second is the sun at 5800 K. In Section 12.2.6, we apply the realistic parameter values to the model and discuss the collector performance further in Section 12.3.1.

12.1.1 Analytic Modelling Approaches

Analytical models for the performance of concentrating NDASCs are rare in the literature. Table 5 from Gorji and Ranjbar (2017) provides a chronological summary of the literature on concentrating NDASCs. Of the seven studies mentioned, only one proposes a fully analytic solution for the temperature of the nanofluid as it flows through a receiver—Veeraragavan et al. (2012). Veeraragavan et al. assume laminar and plug flow through a two-dimensional channel of height 1 cm and apply a zero-flux boundary condition at the bottom of the receiver. This model excludes the effect of scattering due to the nanoparticles and also assumes that the base fluid is a non-absorbing medium, moreover, this model does not describe an NDAPSC, different types of concentrating systems must be modelled differently. The other six studies discussed in Table 5 from Gorji and Ranjbar (2017) are numerical and/or experimental. Further to Table 5 from Gorji and Ranjbar (2017), O'Keeffe et al. (2018b) propose an analytic expression for the efficiency of a large parabolic trough NDASC assuming turbulent flow through a three-dimensional pipe. In that study, the authors explore the relationship between heat-mirrors and collector performance. A heat-mirror is a selectively transmissive/reflective material that is highly transparent at short wavelengths, but highly reflective at long wavelengths, and was introduced for use in solar-thermal energy conversion applications in the 1970s by Fan and Bachner (1976). Previous research had suggested that heat-mirror coatings could improve the efficiency of an NDAPSC (Li et al. 2016, Taylor et al. 2011), however, O'Keeffe et al. (2018b) show that this is not always the case: for lower temperatures, an uncoated system may be more efficient. Also, as the solar concentration ratio increases, an uncoated parabolic trough NDASC becomes more efficient than a parabolic trough NDASC coated with a heat-mirror, at higher inlet temperatures, the concentration ratio required for an uncoated parabolic trough NDASC to be more efficient than a coated parabolic trough NDASC increases.

Gorji and Ranjbar (2017) also offer an extensive review of the literature surrounding the use of nanofluids in non-concentrating DASCs. Table 4 from Gorji and Ranjbar (2017) presents a chronological summary of the literature around on low-flux NDASCs, this table reports on the findings of 18 previous studies. Since the aim of this chapter is to discuss analytic models for the temperature (and efficiency) of an NDASC, we focus this discussion on the analytic studies presented by Gorji and Ranjbar. Of the articles mentioned in that paper, only three provide a fully analytical solution for the temperature of the nanofluid as it flows through the receiver: Cregan and Myers (2015), Lee and Jang (2015), and Turkyilmazoglu (2016). Cregan and Myers present the first fully analytic solution for the temperature of a low-flux NDASC, in their proposed model, they assume a laminar and plug flow through the receiver and apply a zero-flux boundary condition at the bottom of the receiver. Cregan and Myers (2015) also analytically reproduce the results obtained in Otanicar et al. (2010), however, they achieve this by setting an unrealistic reflectance of the glass panel at the top of their solar collector ($\Upsilon_T = 0.65$) and using a much larger receiver height ($H = 0.0012$ m). Lee and Jang (2015) investigate the efficiency of an NDASC, treating flow in the receiver as steady, laminar, and depth-dependent. They obtain an analytic solution for the temperature in the system after applying a zero-flux boundary condition at the bottom of the receiver. This study shows that previous methods overestimate efficiency by modelling the collector subject to plug-flow. Meanwhile, Turkyilmazoglu applies an isothermal boundary condition rather than a zero-flux boundary condition at the bottom of the collector, whilst assuming laminar and plug flow and reports collector efficiencies of up to 100% (a careful reader may look sceptically on such high efficiency values, we discuss this further in Section 12.3).

12.2 Modelling

An NDASC must be modelled in such a way that is appropriate to its: system geometry, source of incoming radiation, velocity field, operating conditions, etc. In this chapter, we model a low-profile NDASC consisting of nanofluid flowing between parallel plates on an inclined plane under a laminar flow regime. Li et al. (2016) define a low-profile to be a collector <15 cm in height.) In this design, incoming solar radiation is absorbed as it propagates in through the system and also back out of the system after it is reflected by the base panel. We model the system as a two-dimensional channel on an inclined plane at an angle of θ to the horizontal, see Figure 12.1 for an illustrated schematic of the system. The system geometry is defined such that x^* is the coordinate in the downstream direction and z^* is the perpendicular distance downwards from the upper surface (* denotes a dimensional variable). The underside of the collector ($z^* = H$) is completely insulated and so heat is lost due to Newton cooling at the upper panel ($z^* = 0$). The temperature of the surrounding air, T_A^*, remains constant, also, the nanofluid enters the collector with initial temperature T_I^* and heats up as it absorbs solar radiation volumetrically.

12.2.1 Conservation of Energy

Generally when modelling NDASCs, conservation of heat energy in the system is governed by the continuity equation:

$$\rho_{nf} C_{p,nf} T_{t^*}^* + \nabla \cdot \mathbf{j} = q, \tag{12.1}$$

where $c_{p, nf}$ is the nanofluid heat capacity, ρ_{nf} is the nanofluid density, T^* is the nanofluid temperature, j is the total flux, and q is the volumetric source. In this chapter, we only consider the steady-state case where the fluid is incompressible, and $\mathbf{j}$ is the sum of the advective and diffusive fluxes. Furthermore, we define our system in a two-dimensional Cartesian co-ordinate system and therefore the conservation of energy equation in this system is:

$$\rho_{nf} C_{p,nf} \mathbf{u} \cdot \nabla T^* = k_{nf} \nabla^2 T^* + q, \tag{12.2}$$

where $T^*(x^*, z^*)$ is the nanofluid temperature, and k_{nf} is the nanofluid thermal conductivity. We note that these NDASCs have also been studied in alternative co-ordinate systems, see for example O'Keeffe et al. (2018a, 2018b) for an expression of (2) in cylindrical co-ordinate systems. Also, in this collector, there is a zero-flux boundary condition at $z^* = H$ due to the underside of the collector being insulated, and a Newton cooling (or heating) boundary condition at $z^* = 0$, the top of the collector. More explicitly these boundary conditions are:

$$\left. \frac{\partial T^*}{\partial z^*} \right|_{z^* = H} = 0, \quad \left. k_{nf} \frac{\partial T^*}{\partial z^*} \right|_{z^* = 0} = h_s \left(T^* \big|_{z^* = 0} - T_{Amb}^* \right), \tag{12.3}$$

and the inlet and outlet conditions are:

$$T^* \big|_{x^* = 0} = T_I^* \quad \left. \frac{\partial T^*}{\partial x^*} \right|_{x^* = L} = 0. \tag{12.4}$$

We note that these boundary conditions can change depending on collector design. For example, Turkyilmazoglu (2016) models an NDASC with an isothermal boundary condition at $z^* = H$.

12.2.2 Nanofluid Properties

The nanofluid properties, such as density, heat capacity, and thermal conductivity, depend on the associated base fluid and nanoparticles properties, as well as the particle volume fraction, f_v. We note that, in reality, these properties are also temperature dependent, however, they do not vary much across small temperature fluctuations. Hence, for simplicity, we calculate them using the inlet temperature and assume they remain constant as the nanofluid flows through the collector. The nanofluid density and specific heat capacity are calculated using classical mixing theory:

$$\rho_{nf} = f_v \rho_{np} + (1 - f_v)\rho_{bf}, \quad C_{p,nf} = \frac{f_v \rho_{np} C_{p,np} + (1 - f_v)\rho_{bf} C_{p,bf}}{\rho_{nf}}, \quad (12.5)$$

where the subscripts *bf* and *np* denote base fluid and nanoparticle, respectively. It is well-known that the thermal conductivity of a nanofluid may be significantly higher than that predicted by the classical Maxwell theory. This has led to many models and mechanisms to explain the 'enhanced thermal conductivity'. Yu and Choi (2003) add a nanolayer around the particle, with unknown thickness and conductivity. Other models are either multi-component or include clustering. In each case, the correct choice of the introduced unknowns permits better agreement with experiment. Myers et al. (2013), point out that the Maxwell model is derived from the assumptions of steady-state heat flow on an infinite domain, so it is only valid for highly disperse nanofluids (with effectively an infinite distance between particles). Hence, it should not be expected to hold for typical nanoparticle volume fractions. Myers et al. (2013) perform the analysis for unsteady heat flow over a finite fluid volume and obtain

$$k_{nf} = \frac{k_{bf}}{\left(1 - f_v^{1/3}\right)^2}\left[(1 - f_v) + f_v \frac{\rho_{np} C_{p,np}}{\rho_{bf} C_{p,bf}}\right]\frac{n-1}{2(n+1)}\left[\frac{1 + f_v^{1/3}}{2} - \frac{1}{n+1}\right]^{-1}, \quad (12.6)$$

where $n = 2.233$ results from the solution technique. Eq. (12.6) is accurate for higher volume fractions than Maxwell's model and has no fitting parameters. It also shows excellent agreement with experimental results, as f_v increases, Maxwell's model under-predicts thermal conductivity, see for example Figure 12.4 in Myers et al. (2013) which shows that at $f_v = 0.01$, Eq. (12.6) outperforms Maxwell's model.

We note that typical NDASC particle volume fractions are in the range of $0 \leq f_v \leq 0.01$, a nanofluid's thermophysical properties approach those of its base fluid at such low particle concentrations. For example, at 50°C the thermophysical properties of pure water are $\rho = 0.988 \times 10^3$ kg m^{-3}, $c_p = 4.181 \times 10^3$ J kg^{-1} K^{-1}, and $k = 0.644$ W m^{-1} K^{-1}, meanwhile a water/Aluminium nanofluid at $f_v = 0.01$ has $\rho = 0.998 \times 10^3$ kg m^{-3}, $c_p = 4.128 \times 10^3$ J kg^{-1} K^{-1}, and $k = 0.65$ W m^{-1} K^{-1}, the density, heat capacity, and thermal conductivity change by only 1%, −1.3%, and 0.9%, respectively, as f_v increases from 0 to 0.01.

Nanofluids also have different optical properties compared to their base fluids. In particular, adding nanoparticles to a base fluid dramatically increases the amount of light that is absorbed and scattered, which leads to an increase in the efficiency of a DASC. Attenuation of the solar radiation takes place through scattering and absorption in the nanofluid and is accounted for by the extinction coefficient:

$$K_e = K_a^{bf} + K_s^{bf} + K_a^{np} + K_s^{np}, \quad (12.7)$$

where K_a and K_s are the absorption and scattering coefficients, respectively. We assume independent scattering in the nanofluid and thus the intensities may be added (Tien 1988). For pure fluids, light scattering is negligible and only the attenuation due to absorption need be considered (Ladjevardi et al. 2013). Hence, the spectral absorption and scattering coefficients for the base fluid are:

$$K_a^{bf} = (1 - f_v)\frac{4\pi k_a^{bf}}{\lambda}, \quad K_s^{bf} \approx 0, \quad (12.8)$$

where κ_a^{bf} is the fluid index of absorption, and λ is the wavelength (O'Keeffe et al. 2018a). The dominant wavelengths of incoming solar radiance are greater than 250 nm and are thus at least ten times larger than the mean diameter (*D*) of typical NDASC nanoparticles (<25 nm). Hence, the higher order spectral components associated with Mie scattering theory may be neglected, and the Rayleigh scattering approximation is applicable (Taylor et al. 2011). The extinction coefficient for a spherical nanoparticle (Bohren and Huffman 2008) is:

$$K_a^{bf} = (1 - f_v)\frac{4\pi k_a^{bf}}{\lambda}, \quad K_s^{bf} \approx 0, \tag{12.9}$$

From Tyagi et al. (2009), Taylor et al. (2011), Cregan and Myers (2015), Saidur et al. (2012), the nanoparticle absorption and scattering efficiencies are:

$$Q_{np}^a = 4\alpha \operatorname{Im}\left\{\frac{m^2 - 1}{m^2 + 2}\left[1 + \frac{\alpha 2}{15}\left(\frac{m^2 - 1}{m^2 + 2}\right)\frac{m^4 + 27m^2 + 38}{2m^2 + 3}\right]\right\}, \tag{12.10}$$

$$Q_{np}^s = \frac{8}{3}\alpha^4\left|\frac{m^2 - 1}{m^2 + 2}\right|^2, \tag{12.11}$$

where $\alpha = (\pi D)/\lambda$ is the size parameter, n_{np} and n_{bf} are the refractive indices, κ_{np} is the nanoparticle absorption index, and $m = (n_{np} + i\kappa_{np})/n_{bf}$ is the relative complex refractive index of particles to the fluid. The values for n_{np}, n_{bf} and κ_{np} depend on the nanoparticle and base fluid. We note that, although most NDASC studies model light scattering via the Rayleigh approximation, several studies include larger nanoparticles and thus require the Mie scattering approximation (Lee et al. 2012, Otanicar et al. 2013, Menbari and Alemrajabi 2016).

12.2.3 Nanofluid Flow

Nanofluid flow can vary from collector to collector. In this chapter, we express the conservation of momentum in the system via the steady-state Navier-Stokes equations for laminar, incompressible fluid flow down an inclined plane under the action of gravity:

$$(\mathbf{u} \cdot \nabla)\mathbf{u} + \frac{1}{\rho_{nf}}\nabla p^* - \frac{\mu_{nf}}{\rho_{nf}}\nabla^2\mathbf{u} + \mathbf{g}, \tag{12.12}$$

where μ_{nf} is the viscosity of the nanofluid, $\mathbf{u} = (u^*, v^*)$ is the fluid velocity, p^* is the pressure, and $\mathbf{g}$ is the forces exerted on the system due to gravity. Naturally, this expression for the conservation of momentum would not be accurate in different system geometry or if the flow was turbulent. See for example O'Keeffe et al. (2018a), which details an expression of the conservation of momentum for flow in a pipe, or see O'Keeffe et al. (2018b) which details an analytic approach to NDASC modelling when flow is turbulent. Acheson (1990) uses the following argument to express Eq. (12.12) in a simpler form: since $\mathbf{u} = 0$ on $z^* = 0, H$ (due to the no-slip conditions at these boundaries), $\mathbf{u}$ needs to depend on z^*. Furthermore, since we have no reason to believe that $\mathbf{u}$ also depends on anything else, let's explore the case where $\mathbf{u} = (u^*(z^*), v^*(z^*))$. Since the liquid in this system is incompressible, mass continuity implies:

$$\frac{\partial v^*}{\partial z^*} = 0, \tag{12.13}$$

for example, v^* is constant. Moreover, for v^* to be constant and also satisfy the no-slip condition, then $v^* = 0$. Therefore, substituting $\mathbf{u} = (u^*(z^*), 0)$ into Eq. (12.12) yields:

$$-\frac{\partial p^*}{\partial x^*} + \mu_{nf}\frac{\partial^2 u^*}{\partial z^{*2}} + \rho_{nf}\,g\sin\theta = 0, \quad -\frac{\partial p^*}{\partial z^*} + \rho_{nf}\,g\cos\theta = 0. \tag{12.14}$$

We integrate the latter of these equations and apply the boundary condition $p^*(0) = p_a{}^*$ (where p_a is the ambient pressure) to obtain $p^* = p_a{}^* + \rho_{nf}gz^*\cos\theta$, and this result implies that $\partial p^*/\partial x^* = 0$. Therefore, a solution for u^* is found by integrating the first partial differential equation in Eq. (12.14) twice and applying the no-slip boundary conditions at the upper and lower surfaces of the collector yielding:

$$u^* = \frac{\rho_{nf}gH^2\sin\theta}{2\mu_{nf}}\left(\frac{z^*}{H} - \frac{z^{*2}}{H^2}\right). \tag{12.15}$$

12.2.4 Heat Source

Generally when one models a NDASC, the heat source, q, from Eq. (12.2) is obtained via an energy balance, whereby the change in the normally incident solar spectral flux due to attenuation of the nanofluid is dissipated as heat is released. This heat source term can vary from collector to collector and depends on several factors: whether or not incoming radiation is concentrated, whether or not the incoming radiation is reflected off the base panel, the spectral intensity of the incoming radiation, the temperature of the nanofluid, the optical properties of the nanofluid, etc. In this chapter, we model a collector wherein the incoming radiation is not concentrated, and the radiation is reflected off the collector's base panel. The spectral intensity at the top of the receiver, J_0, is approximated via Planck's black-body relation and can be expressed as:

$$J_0 = \frac{2hc^2\Omega_S S_{Att}\Upsilon_T}{\lambda^5\left(\exp\left(\dfrac{hc}{\lambda k_B T_{Sun}^*}\right) - 1\right)}, \tag{12.16}$$

where h is Planck's constant, c the speed of light, Ω_S the solid angle of the sun, S_{Att} the attenuation of sunlight through the Earth's atmosphere, k_B the Boltzmann constant, T_{Sun}^* the temperature of the sun, and Υ_T the transmittance of the glass cover.

We note that in this chapter, the system operates at temperatures below 750 K, and thus we follow the approach of Veeraragavan et al. (2012) where thermal re-emission in the fluid is presumed negligible since it constitutes less than 5% of the total radiative heat loss. Recall, Otanicar et al. (2010) model a flat-plate NDASC with a highly reflective bottom surface. In their system, incoming radiation is reflected back at the bottom of the collector, and so they also split up the solar intensity into its incoming and outgoing directional components. We follow the approach taken by Otanicar et al. (2010) and assume that the incoming radiation is reflected back off the aluminium tape in the direction that is normal to the receiver. The change in normally incident solar spectral flux due to attenuation of the nanofluid is dissipated as heat release. Mathematically, it is given by:

$$\frac{J_{A\lambda}^*}{z^*} = -K_e J_{A\lambda}^*, \quad \frac{J_{B\lambda}^*}{z^*} = -K_e J_{B\lambda}^* \tag{12.17}$$

where the subscripts A and B denote the incoming and outgoing directional components, respectively. This system has solutions:

$$J_{A\lambda}^*(z^*) = J_0 e^{-K_e z^*}, \quad J_{B\lambda}^*(z^*) = J_0 e^{-K_e H} e^{K_e z^*}. \tag{12.18}$$

which are obtained after applying the boundary conditions:

$$J_{A\lambda}^*(z^* = 0) = J_0, \quad J_{A\lambda}^*(z^* = H) = J_{B\lambda}^*(z^* = H). \tag{12.19}$$

Next, we integrate Eq. (12.18) over the entire wavelength spectrum to determine the corresponding radiative fluxes:

$$P_A^*(z^*) = \int_0^\infty J_0 e^{-K_e z^*} \lambda, \quad P_B^*(z^*) = \int_0^\infty J_0 e^{-K_e H} e^{K_e z^*} \lambda. \tag{12.20}$$

The equations in Eq. (12.20) are highly non-linear with respect to wavelength due to the wavelength-dependent J_0 and K_e terms; so, to make analytic progress, we follow the approach of Cregan and Myers (2015) and use the method of least squares to approximate this radiative flux integral. They show that the radial flux is well approximated for a nanofluid by the power law:

$$\int_0^\infty J_0 e^{-K_e z^*} \lambda \approx \frac{G_s^*}{\left(1 + \dfrac{\beta_0}{H} z^*\right)^{\beta_1}}, \tag{12.21}$$

where β_0 and β_1 are dimensionless fitting parameters, and G_s^* is the solar intensity at $z^* = 0$. We use Eq. (12.21) to approximate Eq. (12.20) and obtain:

$$P_A^*(z^*) = \frac{G_s^*}{\left(1 + \dfrac{\beta_0}{H} z^*\right)^{\beta_1}}, \quad P_B^*(z^*) = \frac{G_s^*}{\left(1 + \dfrac{\beta_0}{H}(2H - z^*)\right)^{\beta_1}} \tag{12.22}$$

The qualitative behaviours of P_A^* and P_B^* are understood with the aid of Figure 12.2. We observe that P_A^* decreases as z^* increases, this is because incoming radiation enters the receiver at $z^* = 0$ and is attenuated due to the scattering and absorption as it passes through the receiver. Meanwhile, P_B^* decreases as z^* decreases because at $z^* = H$ the solar radiation is reflected and continues to get attenuated until it exits the receiver at $z^* = 0$. As f_v increases, the incoming and outgoing radiative fluxes are attenuated more quickly. Only 15% of the incoming radiation is absorbed by the pure base fluid ($f_v = 0$), whilst the nanofluid with $f_v = 0.006$ absorbs 96% of the incoming radiation. We note that 100% of the incoming solar radiation at $z^* = H$ is reflected back out of the collector in order to generate Figure 12.2, however, most

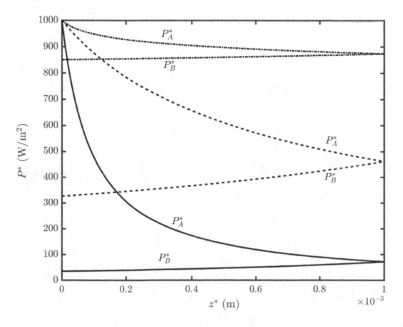

FIGURE 12.2 The solar intensities, $P_A^*(z^*)$, and $P_B^*(z^*)$ in the collector for $f_v = 0.006$ (solid lines), $f_v = 0.0005$ (dashed lines), and $f_v = 0$ (dot-dashed lines). The nanofluid is Water/Aluminium, $H = 0.001$ m, and G_s^* 1000 Wm^{-2}.

surfaces are not perfect reflectors so we introduce R_A, the reflectivity of the base panel. For example, an aluminium base panel has $R_A = 0.73$, and so it only reflects 73% of the incoming radiation (Janecek 2012).

The solar intensities given by Eq. (12.22) are differentiated to obtain the heat source term:

$$q^*(z^*) = \left(-\frac{P_A^*}{z^*} + \frac{P_B^*}{z^*} \right) \tag{12.23}$$

$$= \left(\frac{G_s^* \beta_0 \beta_1}{H\left(1 + \frac{\beta_0}{H} z^*\right)^{\beta_1+1}} + R_A \frac{G_s^* \beta_0 \beta_1}{H\left(1 + \frac{\beta_0}{H}\left(2H - z^*\right)\right)^{\beta_1+1}} \right). \tag{12.24}$$

12.2.5 Model Reduction

We rewrite the conservation of energy in the system using the velocity profile, Eq. (12.15), and the heat source term, Eq. (12.24), to obtain:

$$\frac{C_{p,nf}\rho_{nf}^2 g \sin\theta}{2\mu_{nf}}\left(Hz^* - z^{*2}\right)\frac{\partial T^*}{\partial x^*} = k_{nf}\left(\frac{\partial^2 T^*}{\partial z^{*2}} + \frac{\partial^2 T^*}{\partial x^{*2}}\right) +$$

$$\Upsilon_T \left(\frac{G_s^* \beta_0 \beta_1}{H\left(1 + \frac{\beta_0}{H} z^*\right)^{\beta_1+1}} + R_A \frac{G_s^* \beta_0 \beta_1}{H\left(1 + \frac{\beta_0}{H}\left(2H - z^*\right)\right)^{\beta_1+1}} \right), \tag{12.25}$$

with the boundary conditions:

$$\left.\frac{\partial T^*}{\partial z^*}\right|_{z^*=H} = 0, \text{ and } k_{nf} \left.\frac{\partial T^*}{\partial z^*}\right|_{z^*=0} = h_s\left(T^*\big|_{z^*=0} - T_{Amb}^*\right). \tag{12.26}$$

For consistency, we rescale and non-dimensionalise Eq. (12.25) choosing the same dimensionless parameters that were used in Cregan and Myers (2015). Therefore, Eq. (12.25) (in dimensionless form) is:

$$(z - z^2)\frac{\partial T}{\partial x} = \gamma\frac{\partial^2 T}{\partial z^2} + \frac{1}{Pe}\frac{\partial^2 T}{\partial x^2} + \frac{1}{(1+\beta_0 z)^{\beta_1+1}} + \frac{R_A}{(1+\beta_0(2-z))^{\beta_1+1}} \tag{12.27}$$

a second order inhomogeneous partial differential equation with an initial condition $T(x = 0) = 0$, outlet condition $T_x(x = L) = 0$, and the boundary conditions:

$$\left.\frac{\partial T}{\partial z}\right|_{z=1} = 0, \text{ and } \left.\frac{\partial T}{\partial z}\right|_{z=0} = Nu\left(T\big|_{z=0} - \hat{T}\right). \tag{12.28}$$

Since $1/Pe \ll 1$, one can ignore the boundary layer at $x^* = L$ (and drop the boundary condition at the outlet), thus approximating Eq. (12.28) with:

$$(z - z^2)\frac{\partial T}{\partial x} = \gamma\frac{\partial^2 T}{\partial z^2} + \frac{1}{(1+\beta_0 z)^{\beta_1+1}} + \frac{R_A}{(1+\beta_0(2-z))^{\beta_1+1}} \tag{12.29}$$

12.2.6 Case Study

As a case study for exploring this model further, we apply the parameter values used by Otanicar et al. (2010), these parameter values are detailed in Table 12.1.

We compare this case study to the case study explored in O'Keeffe et al. (2016) where a similar collector is modelled, but without a reflective base panel. For convenience, the collector studied in O'Keeffe et al. will be referred to as Collector 1, and the collector studied in Otanicar ct al. will be referred to as Collector 2. Some of the key differences between parameter values from the case study in O'Keeffe et al. (2016) and the parameter values from the case study in Otanicar et al. (2010) are detailed in Table 12.2. We discuss these parameter values further in Section 12.3.1, highlighting the differences between the two case studies.

12.2.7 Solution Method

The system described by Eq. (12.29) is a second-order, linear partial differential equation. Therefore, we obtain a solution for T via the method of separation of variables. The presence of non-homogeneous terms in the system suggests a solution of the form $T(x, z) = v(x, z) + w(z)$, and we choose $w(z)$ such that:

$$\gamma w_{zz} + \frac{1}{(1+\beta_0 z)^{\beta_1+1}} + \frac{R_A}{(1+\beta_0(2-z))^{\beta_1+1}} = 0, \tag{12.30}$$

TABLE 12.1

Solar Collector and Nanofluid Parameters and Physical Constants

Quantity	Symbol	Value	Units
Collector length/width/height	$L/W/H$	1,1,0.00015	m
Average velocity downstream	u	0.0259 m s^{-1}	—
Viscosity (H$_2$O)	μ_{bf}	10^{-3}	kg m^{-1} s^{-1}
Surface heat transfer coefficient	h_s	23	W m^{-2} K^{-1}
Air/inlet nanofluid temperature	T_{Amb}^{*}/T_I^{*}	298.15/308.15	K
Incident solar radiation	G_S	1000	W m^{-2}
Gravity	g	9.8	m s^{-2}
Density water/aluminium	ρ_{bf}/ρ_{np}	1000/2700	kg m^{-3}
Conductivity water/aluminium	k_{bf}/k_{np}	0.609/247	W m^{-1} K^{-1}
Specific heat capacity water/aluminium	$c_{p,bf}/c_{p,np}$	4187/900	J kg^{-1} K^{-1}
Fitting parameters (for $\phi = 0.006$)	β_0/β_1	3.4135/0.4014	—
Dimensionless parameters	$\gamma/\mathrm{Nu}/$	12.62/0.0056/2.63	—

Source: Otanicar, T.P. et al., *J. Renew. Sustain. Ener.*, 2, 033102, 2010.

TABLE 12.2

Some Key Differences Between the Case Study

	O'Keeffe et al. (Collector 1)	Otanicar et al. (Collector 2)
Temperature of source	5800 K	3158 K
Receiver depth	0.0012 m	0.00015 m
Material at $z^* = 0$	Non-reflective insulator	Aluminium tape ($R_A = 0.73$ from Janecek [2012])
Nanoparticles	Aluminium	Carbon nanotubes, graphite, and silver

Source: Otanicar, T.P. et al., *J. Renew. Sustain. Ener.*, 2, 033102, 2010; O'Keeffe, G.J. et al., *Solar Energy*, 159, 44–54, 2018b.

which has the solution:

$$w(z) = \frac{1}{\gamma\beta_1(\beta_1-1)\beta_0^2}\left(\frac{1}{(1+\beta_0 z)^{\beta_1-1}} + \frac{R_A}{(1+\beta_0(2-z))^{\beta_1-1}}\right) + K_1 z + K_2. \tag{12.31}$$

We apply the boundary conditions to Eq. (12.31) and solve for the integration constants:

$$K_1 = \frac{R_A-1}{\gamma\beta_1\beta_0(1+\beta_0)^{\beta_1}},$$

$$K_2 = \frac{1}{Nu}\left(\frac{1}{\gamma\beta_0\beta_1} - \frac{R_A}{\gamma\beta_0\beta_1(1+2\beta_0)^{\beta_1}} + K_1\right)$$

$$+ \frac{1}{\gamma\beta_0^2\beta_1(\beta_1-1)}\left(1 + \frac{R_A}{(1+2\beta_0)^{\beta_1-1}}\right) + \check{T}.$$

Returning our attention to $v(x, z)$, the homogeneous part of $T(x, z)$, we assume $v(x, z) = f(x)g(z)$, therefore, Eq. (12.29) is rearranged to obtain:

$$\frac{f_x}{\gamma f} = \frac{g_{zz}}{(z-z^2)g} = -p_n^2 \tag{12.32}$$

for some constant p_n. We find expressions for $f(x)$ and $g(z)$ by solving the ordinary differential equations:

$$\frac{f_x}{\gamma f} = -p_n^2 \tag{12.33}$$

and

$$\frac{g_{zz}}{(z-z^2)g} = -p_n^2. \tag{12.34}$$

The solution to Eq. (12.33) is $f(x) = K_1 e^{-\gamma p_n^2}$ for some constant K_1. Meanwhile, we obtain $g(z)$ via a change of variables: first we let:

$$z = \frac{\bar{z}}{\sqrt{2p_n}} + \frac{1}{2}, \tag{12.35}$$

which enables us to rewrite Eq. (12.34) as:

$$\bar{g}_{\bar{z}\bar{z}} + \left(v + \frac{1}{2} - \frac{1}{4}\bar{z}^2\right)\bar{g} = 0, \tag{12.36}$$

where $v = (p_n-4)/8$. Eq. (12.36) has solutions of the form:

$$\bar{g}(\bar{z}) = K_2 D_v(\bar{z}) + K_3 D_v(-\bar{z}) \tag{12.37}$$

for some constants K_2 and K_3, where $D_a(b)$ is the parabolic cylinder function (we are using the notion described in Whittaker (1902) for this expression). However, Eq. (12.37) is still defined in terms of z, and so we rewrite it in terms of its original variable z:

$$g(z) = K_2 D_v\left(\sqrt{2p_n}\left[z - \frac{1}{2}\right]\right) + K_3 D_v\left(-\sqrt{2p_n}\left[z - \frac{1}{2}\right]\right) \tag{12.38}$$

which suggests a solution to our model of the form:

$$T(x,z) = w(z) + \sum_{n=1}^{\infty}\left[C_n \exp\left(-\lambda\, p_n^2 x\right)\left[D_v\left(\sqrt{2p_n}\left[z - \frac{1}{2}\right]\right) + K_3 D_v\left(-\sqrt{2p_n}\left[z - \frac{1}{2}\right]\right)\right]\right]$$

The weight function in the Sturm-Liouville problem, defined by Eq. (12.34), is $(z - z^2)$ and so orthogonality requires:

$$\int_0^1 (z - z^2) g_n g_m z = 0 \tag{12.39}$$

for $n \neq m$. The series coefficient C_n is determined after imposing the initial condition, and the constant K_3 is obtained by applying the boundary conditions:

$$K3 = \frac{v D_{v-1}\left(\sqrt{\frac{p_n}{2}}\right) - D_{v+1}\left(\sqrt{\frac{p_n}{2}}\right)}{v D_{v-1}\left(-\sqrt{\frac{p_n}{2}}\right) - D_{v+1}\left(-\sqrt{\frac{p_n}{2}}\right)}, \tag{12.40}$$

$$C_n = -\frac{\int_0^1 w(z)\left(z - z^2\right)\left[D_v\left(\sqrt{2p_n}\left[z - \frac{1}{2}\right]\right) + K_3 D_v\left(-\sqrt{2p_n}\left[z - \frac{1}{2}\right]\right)\right] dz}{\int_0^1 \left(z - z^2\right)\left[D_v\left(\sqrt{2p_n}\left[z - \frac{1}{2}\right]\right) + K_3 D_v\left(-\sqrt{2p_n}\left[z - \frac{1}{2}\right]\right)\right]^2 dz}, \tag{12.41}$$

where the eigenvalues, p_n. satisfy:

$$\left[v D_{v-1}\left(-\sqrt{\frac{p_n}{2}}\right) - D_{v+1}\left(-\sqrt{\frac{p_n}{2}}\right) - K_3\left[v D_{v-1}\left(\sqrt{\frac{p_n}{2}}\right) - D_{v+1}\left(\sqrt{\frac{p_n}{2}}\right)\right]\right]$$

$$= \sqrt{\frac{2Nu^2}{p_n}}\left[D_v\left(-\sqrt{\frac{p_n}{2}}\right) + K_3 D_v\left(\sqrt{\frac{p_n}{2}}\right)\right]. \tag{12.42}$$

We acknowledge that some software programs do not have inbuilt parabolic cylinder functions, and so this hypergeometric function-based solution is not universally practical. Should this be an issue, we direct the reader's attention to Lee and Jang (2015) where a similar system is expressed via an infinite series.

12.3 Performance Evaluation

Collector efficiency is defined as the ratio of usable thermal energy to incident solar energy. Otanicar et al. (2010) and Tyagi et al. (2009) calculate the efficiency, η, of an NDASC with:

$$\eta = \frac{\dot{m}\rho_{nf} C_{p,nf}\left(\hat{T}_O^* - T_I^*\right)}{G_S A}, \tag{12.43}$$

where $\dot{m}$ is the mass flow rate of the fluid through the collector, $\hat{T}_O^*$ is the mean fluid outlet temperature, and A is the top surface area of the collector—this is a three-dimensional calculation of efficiency. Cregan and Myers (2015) calculate the solar collector efficiency in two-dimensional space with:

$$\eta = \frac{\dot{m}\rho_{nf}C_{p,nf}\left(\hat{T}_{O}^{*}-T_{I}^{*}\right)}{G_{S}L},$$ (12.44)

However, these analytic expressions are limited since they do not consider depth-dependence in the nanofluid velocity profile. Lee and Jang (2015) alter this efficiency calculation to be a true measure of the ratio of usable thermal energy to incident solar energy, i.e.:

$$\eta = \frac{\rho_{nf}C_{p,nf}}{G_{S}L}\int_{0}^{H}\left(T_{O}^{*}(z^{*})-T_{I}^{*}\right)u(z^{*})dz^{*},$$ (12.45)

or in its non-dimensional form:

$$\eta = \beta_{0}\beta_{1}\int_{0}^{1}T_{O}(z)\,(z-z^{2})dz.$$ (12.46)

We note that sometimes the ratio of usable thermal energy to incident solar energy is a bad measure of collector efficiency. For example, Turkyilmazoglu (2016) uses this definition of efficiency whilst modelling an NDASC with an isothermal (fixed temperature) base panel (rather than the insulated base panel which is used in this chapter's model). Turkyilmazoglu reports 100% efficiency, however, this result only occurs when the base panel is sufficiently hotter than the liquid flowing through the receiver. In fact, Turkyilmazoglu might have even reported $\eta > 100\%$ had the base panel been heated further. Of course, efficiencies exceeding 100% are unphysical. Since the base of Turkyilmazoglu's collector is at a constant (in this case elevated) temperature, the temperature gradient of the nanofluid at $z^{*} = H$ is non-zero, and the isothermal base panel is a source of heat. In Figure 12.7c (from Turkyilmazoglu [2016]), for example, the nanofluid's initial temperature rise is predominantly due to the fixed temperature boundary condition at the base panel. This additional heating phenomenon is not included in an efficiency expression defined by Eq. (12.45). We propose, that in the case of a non-zero-flux boundary condition, this additional source/sink of heat should be accounted for, i.e.:

$$\eta = \frac{\rho_{nf}C_{p,nf}\int_{0}^{H}\left(T_{O}^{*}(z^{*})-T_{I}^{*}\right)u(z^{*})dz^{*}}{Gs^{*}L+k_{nf}\int_{0}^{L}\frac{T^{*}}{z^{*}}\Big|_{z^{*}=H}x^{*}},$$ (12.47)

the second term in the denominator represents the energy entering the system across the boundary at $z^{*} = H$ along the length of the collector.

12.3.1 Results

Although the case study in this chapter closely resembles the case study that is explored in Otanicar et al. (2010), the nanoparticles used in this chapter differ to the nanoparticles used in that study. Therefore, both nanofluids have different extinction coefficients, and we cannot do a like-for-like comparison between these two studies. Nonetheless, in this section, we do make several indirect comparisons between the research in Otanicar et al. (2010) and the work in this chapter.

Numerous studies have explored the relationship between NDASC efficiency and nanofluid particle volume fraction across several collector designs and found that generally, as particle volume fraction increases, NDASC efficiency also increases (Cregan and Myers 2015, Lee and Jang 2015, O'Keeffe et al. 2016, 2018a, 2018b, Otanicar et al. 2010, Tyagi et al. 2009). Figure 12.3 shows collector efficiency versus particle volume fraction across two different inlet temperatures for an NDASC with a reflective base panel (dashed lines) and an NDASC without a reflective base panel (dot-dashed lines). The nanofluid heats up as it absorbs the incoming radiation ($P_A^*(z^*)$), this incoming radiation is reflected back out at

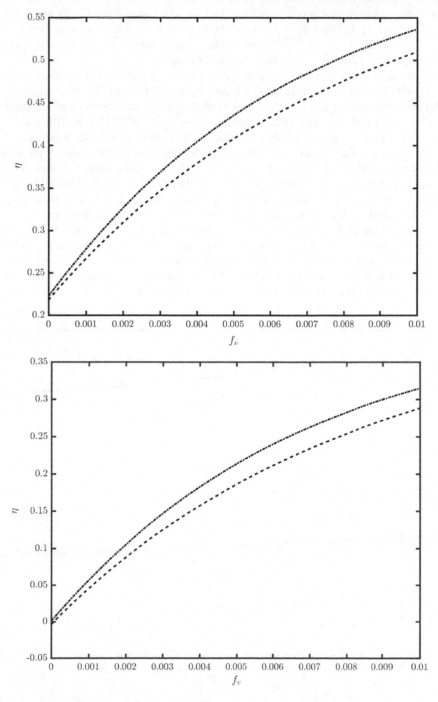

FIGURE 12.3 Efficiency versus particle volume fraction for Collector 1 (dot-dashed line), and Collector 2 (dashed line), (a) $T_I = T_{Amb}$, and (b) $T_I = T_{Amb} + 10°C$.

$z^* = H$, and so the nanofluid also absorbs outgoing radiation, $P_B^*(z^*)$, thus, heating up further. Therefore, an NDASC coated with a reflective base panel absorbs more radiation than an uncoated NDASC. In Figure 12.3, the coated NDASC performs better than the uncoated NDASC under all circumstances due to this increased amount of absorbed radiation. Both collectors perform similarly when $f_v = 0$ and little incoming radiation is absorbed, however, the performance differences between the two collectors become more pronounced as the nanofluid's particle volume fractions are increased—the efficiency of a collector with a reflective base panel rises faster than the efficiency of a collector without a reflective base panel. In Figure 12.3a, when $T_A^* = T_I^*$, both NDASCs are more efficient than in Figure 12.3b when $T_A^* = T_I^* - 10°C$, as the temperature of the nanofluid in the system increases, more energy is lost via Newton cooling at $z^* = 0$, this decreases collector performance.

As one might expect, as nanoparticle volume fraction increases (and the collector becomes more efficient), an NDASC's outlet temperature also increases. Figure 12.4 shows the outlet temperature versus receiver depth for Collector 1, an NDASC without a reflective base panel (dot-dashed line), and Collector 2, an NDASC with a reflective base panel (dashed line), when $T_I^* = 25°C$, $f_v = 0.01$, and $\Upsilon_T = 1$. The nanofluid is hotter at the outlet of Collector 2, than at the outlet of Collector 1 because Collector 2 absorbs more radiation. The temperature profile of the nanofluid as it flows through the receiver does not vary much with receiver depth because heat is diffused in the z^*-dimension at a much faster rate than it is absorbed volumetrically. Since the depth of the receiver is so small ($H = 0.00015$ m), $\gamma = 12.96$, and the diffusion term dominates the model in Eq. (12.29). If the receiver depth was larger, γ would be smaller, and we would observe larger gradients in the temperature profile at the outlet.

Recall, we use Planck's black-body relation, Eq. (12.16), to calculate the spectral intensity of the incoming electromagnetic radiation. The spectral intensity is highly dependent on the temperature of the source. Sunlight was the source of incoming radiation in the case study used to explore Collector 1, however, in this case study, we follow Otanicar et al.'s experimental trials and use a heat lamp at 3158 K as the source of incoming radiation (Otanicar et al. 2010). We note that the surface of the sun, at 5800 K,

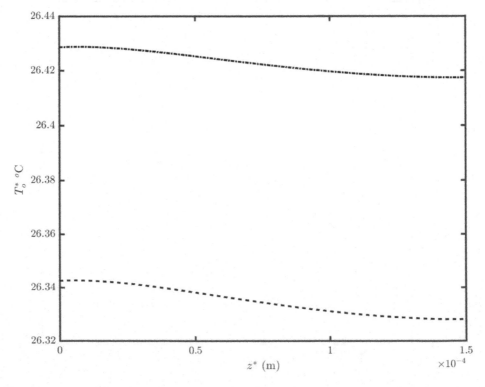

FIGURE 12.4 Temperature at the outlet versus receiver depth for Collector 1 (dot-dashed line), and Collector 2 (dashed line).

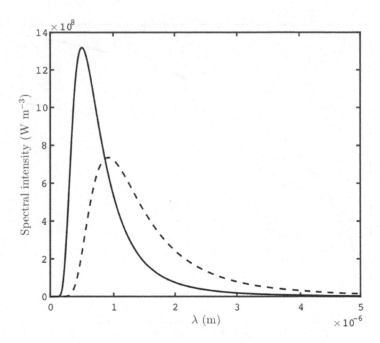

FIGURE 12.5 Spectral dependence of black-body radiation for two bodies at 5800 K and 3158 K.

is much hotter than the heat lamp used in Otanicar et al.'s experimental trials, Figure 12.5 illustrates the spectral distribution of Planck's black-body relation for two objects, one at the temperature of the sun, and the other at the temperature of the heat lamp. Whilst the spectral intensities have been scaled such that the area under both curves is 1000 w/m², the radiative intensities are distributed differently across the electromagnetic spectrum. The radiation emitted from the heat lamp is noticeably more intense at longer wavelengths and less intense at shorter wavelengths.

The qualitative performance of an NDASC depends on the spectral power distribution of the incoming radiation, and so it is important to consider this phenomenon when designing and testing NDASCs. Water is better at absorbing radiation at longer wavelengths, therefore, one might expect the water-based NDASC to always perform better when incoming radiation comes from the heat lamp rather than from the sun. However, we observe in Figure 12.6 that this is not always the case. When $f_v = 0$, Collector 2 is more efficient when the incoming radiation is emitted by the heat lamp, however, when $f_v = 0.01$, Collector 2 is more efficient when the incoming radiation is emitted by the sun. This is because the water is better at absorbing the heat lamp generated radiation, but nanoparticles are generally better at absorbing solar radiation (Taylor et al. 2011). As the nanoparticle volume fraction in the nanofluid increases, the efficiency of the sunlight-absorbing NDASC increases faster than the efficiency of the NDASC which is absorbing radiation from the heat lamp.

Figure 12.7 shows the efficiency of the solar collector versus particle volume fraction for $H = 0.0006$ (solid line), $H = 0.0012$ (dashed line), and $H = 0.0018$ (dot-dashed line) using parameter values from the case study detailed in O'Keeffe et al. (2016). This plot demonstrates Collector 2's performance when its parameter values are equivalent to the ones that produced the optimal efficiency of Collector 1 (81.2%). In Figure 12.7, we observe a sharp increase in collector efficiency across each of the three height configurations when nanoparticles are first added to the base fluid and particle volume fractions are quite low. The presence of nanoparticles increases the rate at which incoming radiation is absorbed into the working fluid. However, as the nanoparticle volume fraction continues to rise, we observe different qualitative behaviours in all three of the collector configurations. This initial increase is sharpest at larger collector depths and is progressively more gradual as the collector depth decreases. The efficiencies of all three collector configurations plateaus with further increase in f_v, after the system has absorbed all of the available incoming radiation any additional increase in particle volume fraction does not lead to any further efficiency enhancement. In Figure 12.7, the maximum efficiency value is (81.3%), and so Collector 2

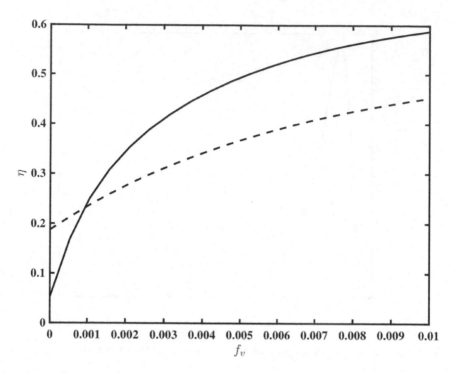

FIGURE 12.6   Efficiency of the solar collector versus nanofluid particle fraction. The source of radiation is a black-body at 5800 K (solid line) 3158 K (dashed line).

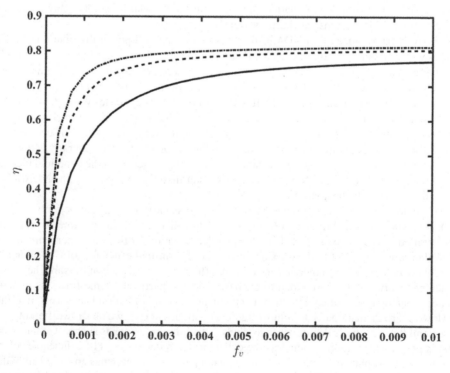

FIGURE 12.7   Efficiency of the solar collector versus particle volume fraction for $H = 0.0006$ (solid line), $H = 0.0012$ (dashed line), and $H = 0.0018$ (dot-dashed line).

performs marginally better than Collector 1. Furthermore, the efficiency of Collector 2 plateaus at lower particle volume fractions across all three heights than the efficiency of Collector 1—the reflective base panel always increases efficiency across all of the operating conditions.

12.3.2 General Findings

In this chapter, we modelled the efficiency of a parallel plate NDASC on an inclined plane with a reflective base panel. This type of collector has been studied experimentally and modelled numerically, however, this particular system configuration has not been studied analytically. The modelling approach taken in this chapter closely followed the modelling methodology outlined in previous studies (Cregan and Myers 2015, Lee and Jang 2015, O'Keeffe et al. 2016). These models have many similarities: the system geometries are defined using the same coordinate system, conservation of momentum takes the same form, the thermophysical properties of the nanofluid are calculated using the same method, and there are various other system parallels. However, there are also some fundamental differences. One major difference is the heat source term, in Section 12.2.4, we showed how the reflective base panel alters the form of the radiative flux integral and thus the partial differential equation (PDE) describing the conservation of energy. Other significant differences between the two collectors are highlighted in Table 12.2 from Section 12.2.6, where we detailed the parameter values used in this chapter's case study. Following dimensionless analysis, we observed that $\gamma \sim O(10)$, and so diffusion has much more relative importance in this system than it would in a larger NDASC design. The dimensionless conservation of energy PDE describing this collector was solved via a separation of variables, the homogeneous component of this solution was equivalent to the homogeneous solution from Lee and Jang (2015). Meanwhile, the inhomogeneous part of the solution was new, and we outlined its solution method in Section 12.2.7.

We plotted and discussed various parameter configurations and how they affect collector efficiency in Section 12.3.1. Figure 12.3 shows that the reflective base panel does indeed improve collector efficiency, as does nanoparticle volume fraction. The efficiency in this figure reaches 54% and does not plateau like the efficiency expression in other studies (Cregan and Myers 2015, Lee and Jang 2015, O'Keeffe et al. 2016) because the channel was much more shallow. We compared the efficiencies of a collector with and without a reflective underside in Figure 12.7, showing that collector with a reflective underside was more efficient and can reach efficiency values of up to 81.3%. A collector with a reflective underside was universally more efficient, however, at lower channel heights the reflective underside was a more significant determinant of collector efficiency. In Figure 12.4, we highlighted the difference that a reflective base panel makes to the temperature at the outlet, we also noted that even though this collector has lower efficiencies than the collector studied in (Cregan and Myers 2015, Lee and Jang 2015, O'Keeffe et al. 2016), it has a higher temperature at the outlet due to its decreased cross-sectional area—although this system does not absorb as much of the incoming radiation, there is a lower volume of nanofluid to heat, and so the overall temperature rise across the system is greater. Figure 12.6 shows how the temperature of the heat source significantly affects collector efficiency, this is important to note as it suggests that NDASCs will perform differently when they are in some testing environments than they will in exposure to real sunlight.

The reflective base panel certainly improves collector performance over the non-reflective base panel, and the temperature at the outlet improves with decreasing channel height, however, at higher operating temperatures, this solar collector becomes less efficient, and the temperature rise across the system is still too low for most real-world applications. In the following section, we discuss alternative avenues of research with an aim to increase the viability of NDASCs in real-world settings.

12.4 Future Work

Nanofluid-based direct absorption solar collectors are a relatively new technology, and the scientific literature surrounding them is evolving at a rapid pace. Although this chapter discusses many contributions to the literature, future complementary work could be extremely valuable to the field of NDASC design. We note that even though one-dimensional parameter studies are generally conducted on NDASC models, the results of these studies are somewhat limited since they only focus on optimising receiver efficiency and outlet temperature. We anticipate that mathematical models for NDASCs could be used within formal multi-objective optimisation processes which seek to identify optimal collector designs. (Multi-objective design optimisation has previously been implemented in the design of solar-powered irrigation systems, see for example Ganesan et al. [2013]). Before looking for optimal collector designs, one would need to identify the characteristics which contribute the most to the overall value of the design. For example, a good design may involve multiple criteria/objectives such as manufacturing cost, operating cost, expected life-cycle, receiver efficiency, outlet temperature, etc. Also, we note that some of these objectives are conflicting: As an NDASC's operating temperature increases, its receiver efficiency typically decreases. Furthermore, if an NDASC was to be used in conjunction with a Carnot heat engine to generate electricity, its Carnot efficiency would increase and its receiver efficiency would decrease, as its operating temperatures increase (see Figure 9 from Veeraragavan et al. [2012] for an illustration of the relationship between receiver efficiency and Carnot efficiency). Therefore, optimisation search algorithms such as conjugate gradients or combinations of design of experiments and meta-modelling approaches may be needed to find optimal NDASC designs. This chapter solely focuses on calculating receiver efficiency and outlet temperature, therefore, in order to find optimal collector designs, further work would be needed to define and calculate: manufacturing cost, operating cost, expected life-cycle, etc.

Looking forwards, we identify the following as opportunities for further research in the field of NDASC modelling:

- Improved solar collector designs,
- Experimental validation/invalidation of existing models,
- Time-dependent modelling of NDASCs under laminar flow,
- Further improvements to numerical solution methods, and
- Better heat-mirrors.

ACKNOWLEDGEMENTS

G. J. O'Keeffe acknowledges the support of the Irish Research Council (GOIPG/2014/887). S. L. Mitchell and V. Cregan acknowledge the support of the Mathematics Applications Consortium for Science and Industry funded by the Science Foundation Ireland (12/IA/1683). T. G. Myers acknowledges the support of a Ministerio de Ciencia e Innovaciòn (MTM2014-56218).

REFERENCES

Acheson, D. J. (1990), *Elementary Fluid Dynamics*. Oxford, UK: Clarendon Press.

Ayompe, L., Duffy, A., Mc Keever, M., Conlon, M. and McCormack, S. (2011), Comparative field performance study of flat plate and heat pipe evacuated tube collectors for domestic water heating systems in a temperate climate. *Energy* 36, 3370–3378.

Bohren, C. F. and Huffman, D. R., *Absorption and Scattering of Light by Small Particles*. New York: John Wiley & Sons; 2008.

Colangelo, G., Favale, E., de Risi, A. and Laforgia, D. (2013), A new solution for reduced sedimentation flat panel solar thermal collector using nanofluids. *Applied Energy* 111, 80–93.

Comerford, W. (2011), *Dispelling the solar myth—Evacuated tube versus flat plate panels*. Lecture presented at European Solar Days in EPA Headquarters, Johnstown Castle, Wexford.

Cregan, V. and Myers, T. G. (2015), Modelling the efficiency of a nanofluid direct absorption solar collector. *International Journal of Heat and Mass Transfer* 90, 505–514.

Fan, J. C. C. and Bachner, F. J. (1976), Transparent heat mirrors for solar-energy applications. *Applied Optics* 15, 1012–1017.

Ganesan, T., Elamvazuthi, I., Shaari, K. Z. K. and Vasant, P. (2013), Hypervolume-driven analytical programming for solar-powered irrigation system optimization, In *Nostradamus 2013: Prediction, Modeling and Analysis of Complex Systems*. I. Zelinka, G. Chen, O. E. Rössler, V. Snasel, A. Abraham (Eds.), Heidelberg, Germany: Springer, pp. 147–154.

Gorji, T. B. and Ranjbar, A. (2017), A review on optical properties and application of nanofluids in direct absorption solar collectors. *Renewable and Sustainable Energy Reviews* 72, 10–32.

Janecek, M. (2012), Reflectivity spectra for commonly used reflectors. *IEEE Transactions on Nuclear Science* 59, 490–497.

Kalogirou, S. (2003), The potential of solar industrial process heat applications. *Applied Energy* 76, 337–361.

Ladjevardi, S., Asnaghi, A., Izadkhast, P. and Kashani, A. (2013), Applicability of graphite nanofluids in direct solar energy absorption. *Solar Energy* 94, 327–334.

Lamnatou, C., Cristofari, C., Chemisana, D. and Canaletti, J. (2016), Building-integrated solar thermal systems based on vacuum-tube technology: Critical factors focusing on life-cycle environmental profile. *Renewable and Sustainable Energy Reviews* 65, 1199–1215.

Lee, B. J., Park, K., Walsh, T. and Xu, L. (2012), Radioactive heat transfer analysis in plasmonic nanofluids for direct solar thermal absorption. *Journal of Solar Energy Engineering* 134, 021009.

Lee, S. H. and Jang, S. P. (2015), Efficiency of a volumetric receiver using aqueous suspensions of multi-walled carbon nanotubes for absorbing solar thermal energy. *International Journal of Heat and Mass Transfer* 80, 58–71.

Lenert, A. and Wang, E. N. (2012), Optimization of nanofluid volumetric receivers for solar thermal energy conversion. *Solar Energy* 86, 253–265.

Li, Q., Zheng, C., Mesgari, S., Hewkuruppu, Y. L., Hjerrild, N., Crisostomo, F., Rosengarten, G., Scott, J. A. and Taylor, R. A. (2016), Experimental and numerical investigation of volumetric versus surface solar absorbers for a concentrated solar thermal collector. *Solar Energy* 136, 349–364.

Mahjouri, F. (2004), Vacuum tube liquid-vapor (heat-pipe) collectors, in *Proceedings of the Solar Conference. American Solar Energy Society; American Institute of Architects*, 341–346.

Menbari, A. and Alemrajabi, A. A. (2016), Analytical modeling and experimental investigation on optical properties of new class of nanofluids (Al2O3–CuO binary nanofluids) for direct absorption solar thermal energy. *Optical Materials* 52, 116–125.

Myers, T. G., MacDevette, M. M. and Ribera, H. (2013), A time-dependent model to determine the thermal conductivity of a nanofluid. *Journal of Nanoparticle Research* 15, 1–11.

Omri, A. (2013), CO2 emissions, energy consumption and economic growth nexus in MENA countries: Evidence from simultaneous equations models. *Energy Economics* 40, 657–664.

Otanicar, T. P., Hoyt, J., Fahar, M., Jiang, X. and Taylor, R. A. (2013), Experimental and numerical study on the optical properties and agglomeration of nanoparticle suspensions. *Journal of Nanoparticle Research* 15, 2039.

Otanicar, T. P., Phelan, P. E. and Golden, J. S. (2009), Optical properties of liquids for direct bsorption solar thermal energy systems. *Solar Energy* 83, 969–977.

Otanicar, T. P., Phelan, P. E., Prasher, R. S., Rosengarten, G. and Taylor, R. A. (2010), Nanofluid-based direct absorption solar collector. *Journal of Renewable and Sustainable Energy* 2, 033102.

O'Keeffe, G. J., Mitchell, S. L., Myers, T. G. and Cregan, V. (2016), The effect of depth-dependent velocity on the performance of a nanofluid-based direct absorption solar collector, In *European Consortium for Mathematics in Industry*. Quintela et al. (Eds.), European Consortium for Mathematics in Industry. Springer, pp. 327–334.

O'Keeffe, G. J., Mitchell, S. L., Myers, T. G. and Cregan, V. (2018a), Modelling the efficiency of a low-profile nanofluid-based direct absorption parabolic trough solar collector. *International Journal of Heat and Mass Transfer* 126, 613–624.

O'Keeffe, G. J., Mitchell, S. L., Myers, T. G. and Cregan, V. (2018b), Modelling the efficiency of a nanofluid-based direct absorption parabolic trough solar collector. *Solar Energy* 159, 44–54.

Parvin, S., Nasrin, R. and Alim, M. A. (2014), Forced convection through nanofluid-based direct absorption solar collector. *Proceedings of the 15th Annual Paper Meet* 7, 08, The Institution of Engineers, Bangladesh.

REN21 (2016), Renewables 2016 global status report, Technical report, Paris: REN21 Secretariat.

Saidur, R., Meng, T., Said, Z., Hasanuzzaman, M. and Kamyar, A. (2012), Evaluation of the effect of nanofluid-based absorbers on direct solar collector. *International Journal of Heat and Mass Transfer* 55, 5899–5907.

Taylor, R. A., Phelan, P. E., Otanicar, T. P., Adrian, R. and Prasher, R. (2011), Nanofluid optical property characterization: Towards efficient direct absorption solar collectors. *Nanoscale Research Letters* 6, 1–11.

Tien, C. L. (1988), Thermal radiation in packed and fluidized beds. *Journal of Heat Transfer* 110, 1230–1242.

Timilsina, G. R., Kurdgelashvili, L. and Narbel, P. A. (2012), Solar energy: Markets, economics and policies. *Renewable and Sustainable Energy Reviews* 16, 449–465.

Turkyilmazoglu, M. (2016), Performance of direct absorption solar collector with nanofluid mixture. *Energy Conversion and Management* 114, 1–10.

Tyagi, H., Phelan, P. and Prasher, R. (2009), Predicted efficiency of a low-temperature nanofluid-based direct absorption solar collector. *Journal of Solar Energy Engineering* 131, 041004.

U.S. Energy Information Administration (2017), International Energy Outlook 2017.

Veeraragavan, A., Lenert, A., Yilbas, B., Al-Dini, S. and Wang, E. N. (2012), Analytical model for the design of volumetric solar flow receivers. *International Journal of Heat and Mass Transfer* 55, 556–564.

Whittaker, E. (1902), On the functions associated with the parabolic cylinder in harmonic analysis. *Proceedings of the London Mathematical Society* 1, 417–427.

Yu, W. and Choi, S. (2003), The role of interfacial layers in the enhanced thermal conductivity of nanofluids: A renovated Maxwell model. *Journal of Nanoparticle Research* 5, 167–171.

13

Applications of Nanofluids in Solar Thermal Systems

Kalyani K. Chichghare, Divya P. Barai, and Bharat A. Bhanvase

CONTENTS

13.1 Introduction

The increased demand of energy due to the ever-growing population and rapid expansion has always given the push to the minds of researchers to try and find newer technologies that meet the requirements of mankind. As we are surrounded by plenty of resources and sources of energy, there has always been utilization of these resources for different purposes along with timely exploitation of the sources.

One should be aware that the consumption of the non-renewable conventional fossil fuels is increasing, while they are running short expeditiously. The time has come that these sources can no longer fulfill the prodigious demand of the world. Thus, the attention has now diverted toward exploiting the non-conventional energy sources such as solar, wind, or geothermal energy [1].

Inexhaustible solar energy is considered as one of the most convenient renewable energy sources [2]. Due to the sun being naturally and abundantly available source of heat, harnessing energy from it is clean and non-polluting [3]. Figure 13.1 summarizes various ways to utilize the solar energy. Radiations and heat received from the sun that can be used up are sufficiently available at most places on earth. Mankind has already found many techniques to harness the energy that the sun provides. Harnessing the heat energy from the sun involves use of a fluid that absorbs the heat and then can be used directly or indirectly to transfer the heat to another place or fluid. Water is the most common fluid that is used in solar thermal techniques. The only factor that shall increase the efficiency of the liquid to absorb the heat is the amount of heat present in the atmosphere which depends on the climatic conditions. This being uncontrollable by humans, one can think of increasing the heat absorbance of the fluid itself. A little enhancement in the absorbance of heat by the fluid may cause a huge positive impact on the process of harnessing the sun's energy. Since the discovery of Maxwell [4], that the solids possess higher thermal conductivity than the liquids, there have been tremendous studies to enhance the thermal properties of the base liquids by use of different solids particles. But these types of fluids containing micro- or macro-sized particles also possess many disadvantages like agglomeration, clogging, erosion, etc. [5]. Nanofluids, a new type of fluid developed in 1995 by Choi and Eastman [6] are made by dispersing nano-sized particles in base liquids. These fluids are found to have superior properties than the conventional fluids like water, ethylene glycol, etc. Along with possession of high thermal conductivity, nanofluids have outstanding absorption of light and excellent thermal transport properties as well which makes them the superb working fluids for solar thermal systems [7]. Utilization of nanofluids in the solar technologies has attracted much attention recently [3].

Figure 13.2 depicts graphical illustration of the nanofluids applications in various solar thermal systems.

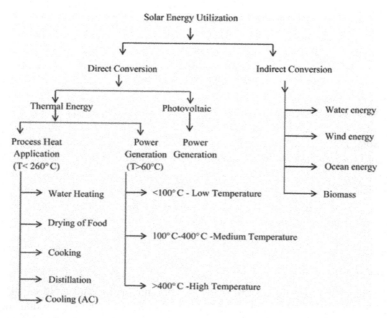

FIGURE 13.1 Categorization of solar energy utilization for different purposes. (From Reddy, K.S. et al., *Int. J. Low Carbon Tech.*, 12, 1–23, 2017.)

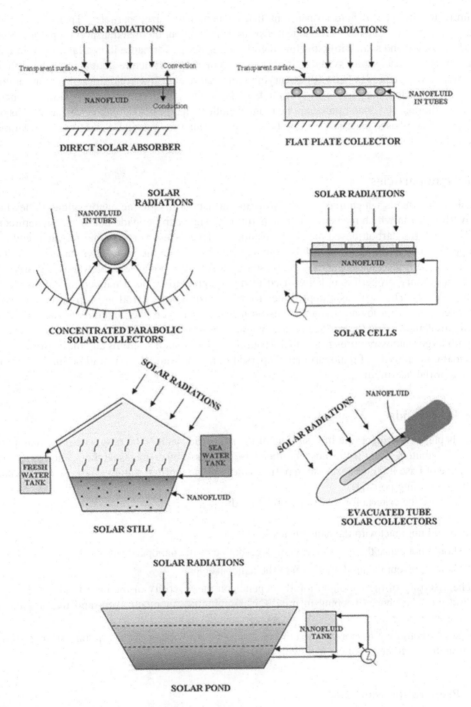

FIGURE 13.2 Solar applications of nanofluids.

13.2 Nanofluids

Nanofluids are the type of fluid suspensions that contains nano-sized particles. Their thermophysical properties have been studied to a great extent. Due to their excellent thermal transport properties, researchers were motivated and started investigating their heat transfer performance in natural convection, forced convection, and condensation system. Nanofluids have a simple structure containing a base fluid and nanoparticles, yet a complex behavior at varying temperature and flow conditions. There are numerous factors that may affect the performance of the nanofluids which include their composition, preparation, physical conditions, and conditions at application. Nanofluids have outstanding thermal as well as optical properties. Carbon-based materials are also being used nowadays as nanofluid additives for solar thermal collectors [7].

13.2.1 Nanoparticles

The basic components which make up the structure of nanofluids are the nanoparticles. They have at least one dimension which measures less than 100 nm. They can be made out of metal, semiconductor, polymeric, and carbon-based materials. Nanoparticles possess very distinct physical and chemical properties as compared to their bulk counterparts. They become more advantageous as one can benefit from the individual properties possessed by a single nanoparticle. Nowadays, graphene due to its extraordinary properties is being used in many applications as a nanocomposite with different materials [9]. The two basic approaches to synthesize nanomaterial are top-down and bottom-up approach [10]. Top-down approach deals with reducing larger size materials to nanometer-sized materials, thus increasing the surface area to volume ratio. While the bottom-up approach deals with making up larger nanostructures from smaller building blocks such as atoms and molecules. The more the thermal conductivity of nanoparticles dispersed in a nanofluid, more would be the heat transport properties of the nanofluid.

13.2.2 Base Fluids

Base fluids play a dominant role in designing of the nanofluids as it is the base where the nanoparticles would be distributed. Water is the most common base fluid used by many researchers. Factors considered in the choice of base fluid for best system performance are the thermal stability, viscosity, heat capacity, freezing, and boiling point [11].

Following are the criteria for selection of base fluid:

- It should not react with the nanoparticles.
- It should not cause the deterioration or degradation of the nanoparticles.
- It should prevent the agglomeration of the nanoparticles.
- The viscosity of the nanofluid totally depends on the viscosity of the base fluid. So the base fluid must be chosen keeping in mind the pumping power requirement and that amount of pumping cost is low.
- Also, it should not be corrosive to the material of construction of the equipment or pipelines through which the nanofluid is flowing.

13.2.3 Preparation Methods

The preparation of nanofluids is the most important step which may affect its thermal properties [12]. Two types of methods that have been investigated in producing nanofluids are the two-step method and the one-step method. Additives like surfactants or other dispersing agents that are added during the preparation of the nanofluids play a crucial role in disturbing the thermal transport properties of nanofluids [13].

13.2.3.1 Two-Step Method

The two-step method is an extensively used method for the preparation of nanofluids [14]. Nanoparticles, nanofibers, nanotubes, or any other nanomaterials employed in this method are first processed as dry powders by physical or chemical methods. In the second processing step, the nano-sized powder is mixed into a fluid with the cooperation of magnetic force agitation, ultrasonic agitation, high shear mixing, homogenizing, and ball milling. As a result, agglomeration of nanoparticles may take place in any step and this may lower down the thermal conductivity. Ultrasonic agitation technique or adding surfactants to the fluids are mainly used to lower the particle aggregation and thus help enhance the dispersion behavior [12]. Due to the complications in preparing stable nanofluids by two-step method, several different techniques have been implemented to produce nanofluids which includes single-step method [14].

13.2.3.2 One-Step Method

The one-step method was introduced to reduce the agglomeration of nanoparticles [14]. In this method, nanoparticles are directly formulated by physical vapor deposition technique or liquid chemical method [12]. In single-step process, preparing and simultaneously dispersing the particles in the fluid takes place rapidly. The process of drying, storage, transportation, and dispersion of nanoparticles are neglected, so that the agglomeration of nanoparticles is minimized [14]. The only disadvantage of this method is that low vapor pressure fluids are well-suited with the process [12]. Table 13.1 summarizes some nanofluids systems that have been studied previously as collected by Li et al. [12].

TABLE 13.1

Some Nanofluids and Their Properties Reported in Literature

System	Synthesis Process	Particle Loading (vol.%)	Particle Size (nm)	Enhancement in Thermal Conductivity (%)	References
Cu/EG	Single-step	0.3	10	40	[15]
Cu/H_2O	Single-step	0.1	75–100	23.8	[16]
Cu/H_2O	Two-step	7.5	100	78	[17]
Fe/EG	Single-step	0.55	10	18	[18]
Ag90/toluene	Two-step	0.001	60–80	16.5	[19]
Au/toluene	Two-step	0.0026	10–20	21	[19]
Au/ethanol	Two-step	0.6	4	1.3 ± 0.8	[20]
Fe_3O_4/H_2O	Single-step	4	10	38	[21]
TiO_2/H_2O	Two-step	5	15	30–33	[22]
Al_2O_3/EG	Two-step	0.05	60	29	[23]
CuO/H_2O	Two-step	5	33	11.5	[24]
Sic/H_2O	Two-step	4.2	25	15.9	[25]
NCTs/engine oil	Two-step	2	20–50	30	[26]
NCTs/poly oil	Two-step	1	25 nm × 50 μm	160	[27]
NCTs/EG	Two-step	1	15 × 30 μm	19.6	[28]
NCTs/H_2O	Two-step	1	15 × 30 μm	7	[28]
NCTs/decene	Two-step	1	15 × 30 μm	12.7	[28]
H_2O/FC-72	Two-step	12	9.8 nm	52	[29]
Al_2O_3/H_2O	Two-step	5	20	20	[23]

Source: Li, Y. et al., *Powder Technol.*, 196, 89–101, 2009.

13.3 Properties of Nanofluids

Nanofluid being a combination of a base fluid and a nano-sized solid material, it is bound to have a complex participation of different properties that will affect its overall performance. Properties of both the individual components play role in determining the properties of the nanofluid. The nanoparticles in the fluid are in a continuous Brownian motion which is one of the characteristics of a nanofluid. There are many factors that affect this motion of the nanoparticles within the fluid which include the type of nanoparticles, their size and shape, type of fluid, its viscosity, density, etc. All these factors also affect the marked properties of the nanofluid some of which are discussed below.

13.3.1 Thermal Conductivity

Thermal conductivity of a nanofluid is the most important property which certifies its heat transfer performance. It is the degree of heat conduction that a nanofluid provides. To measure the thermal conductivity of nanofluids, various methods like transient hot wire (THW) method, temperature oscillation, and steady-state parallel plate method have been employed [17,18,22,30–43]. Solids have much higher thermal conductivity than the base liquids and thus enhance the heat transfer performance of the fluids. This happens in a nanofluid due to the suspended solids/particles [12]. Many researchers use the well-known KD2 Pro thermal property analyser to determine the thermal conductivity of the nanofluid which can be set up as given in Figure 13.3 [44].

> *Influence of nanoparticles*: Thermal conductivity of nanofluids is being influenced by the nanoparticles, i.e., the additives of nanofluids. Most results from the investigations reveal that the thermal conductivity of the nanofluids increases drastically with an increase in the volume fraction of nanoparticles [12]. Fe/ethylene glycol nanofluids thermal conductivity enhances non-linearly with that of the volume fraction of Fe nanoparticles [18].
>
> In the nano-suspensions, with the increase in volume fraction of Al_2O_3 nanoparticles, the thermal conductivity ratios also increase [23]. Liu et al. [16,26] and Xie et al. [28] carried out the investigations for the enhancement of thermal conductivity, and then they compared the effects of carbon nanotubes (CNT) and CuO, CNT and Al_2O_3 nanoparticles. It was found that the CNT suspensions show higher thermal conductivity than that shown by oxide nanoparticle

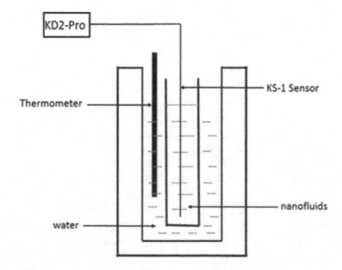

FIGURE 13.3 The schematic diagram of experimental set up for thermal conductivity measurement using KD2 Pro thermal property analyser (Decagon Devices, United States). (From Chen, W. et al., *Int. J. Heat Mass Transf.*, 107, 264–270, 2017.)

with same volume fraction. This may be due to difference in the morphology between oxide nanoparticles and in their thermal conductivity. Hong et al. [18] observed that Cu nanoparticles show higher thermal conductivity than that compared with the Fe nanoparticles.

Influence of base fluids: Base fluids being a major component of the nanofluid, their properties affect the thermal conductivity of nanofluids which are their thermal conductivity, temperature, and viscosity [12]. Xie et al. [23] carried out experiments and concluded that for suspensions with same nanoparticles, with the increase of the thermal conductivity of nanofluids of the base fluid, the thermal conductivity ratio is reduced.

Influence of temperature: Temperature of a nanofluid has the most significant impact on its thermal conductivity. An increase in temperature is found to increase the thermal conductivity of the nanofluid. The reasons behind can be the increase in Brownian motion, decrease in base fluid viscosity and thus a decrease in agglomeration of the nanoparticles [33]. The nanoparticles are found to carry the heat along with them through the nanofluid and dissipate it and act as stirrers to induce convection within the fluid [34]. High temperature decreases the viscosity of the base fluid thus increasing the random motion of the particles within it which is called the Brownian motion. Also, a decrease in the surface energy of the particles at higher temperatures causes a decrease in the agglomeration of the particles leading to more enhanced Brownian motion of the particles and thus more enhanced nanofluid thermal conductivity.

Influence of the liquid-solid interface: Liquid-solid interface also affects the thermal conductivity. As the crystalline solids have higher thermal transport properties than liquids, so this type of liquid layering shows higher thermal conductivity. Figure 13.4 shows a spherical particle along with the interfacial layer present in a fluid medium.

Many investigations showed that nano-layer behaves as a thermal bridge between a solid nanoparticle and a bulk fluid which plays crucial role in increasing the thermal conductivity of nanofluids. Yu and Choi [35] stated that the thermal conductivity is only affected when the particle diameter of the nano-layer is less than 10 nm. Xie et al. [36] summarized that with increasing thickness of the nano-layer, the thermal conductivity increases when the particle size is small, manipulating nano-layer structure would be more pronounced to enhance the thermal conductivity. Leong et al. [37] stated that interfacial layer plays a significant role in enhancing the thermal conductivity. Later, Xue et al. [38] summarized that layering of the liquid atoms at the liquid-solid interface did not have pronounced effect on the thermal conductivity. A nice compilation of the data available in literature has been done by Li et al. [12] for comparing the thermal conductivity possessed by different nanofluids as given in Figure 13.5.

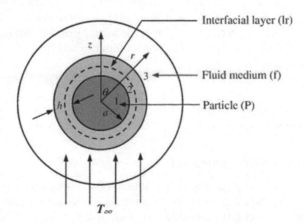

FIGURE 13.4 Single spherical particle with interfacial layer in a fluid medium. (From Li, Y. et al., *Powder Technol.*, 196, 89–101, 2009.)

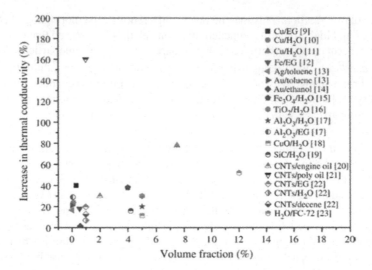

FIGURE 13.5 Comparison of experimental data on thermal conductivity of nanofluids. (From Li, Y. et al., *Powder Technol.*, 196, 89–101, 2009.)

13.3.2 Optical Properties

As far as the application of nanofluids in solar thermal systems is concerned, the optical properties become the most relevant. A nanofluid along with having extraordinary thermal properties also possesses good absorption properties as compared to the base fluids [39].

Kameya et al. [40] have demonstrated that the suspension of Ni nanoparticles in a base fluid increases the absorption coefficient over the visible wavelength and near-infrared wavelength of the solar radiation. A detailed review on optical properties possessed by the nanofluids in solar collector applications has been done by Ahmed et al. [41]. Lee et al. [42] investigated a Monte Carlo algorithm with Mie scattering theory to examine the optical property of nanofluids. The sharp absorption peak can be elaborated by using plasmonic nanoparticles and is examined by surface plasmon. Hence, the efficiency (absorption) of direct absorption solar collectors (DASCs) is enhanced. Taylor et al. [43] developed nanofluid-based optical fibers for photovoltaic thermal (PV/T) system. The nanofluid-based filters show higher solar-weight efficiency as compared with the optical fibers and pure fluids. Saidur et al. [39] studied the optical properties of nanofluids by using Rayleigh scattering theory. Extinction coefficient is affected by the particle size. As the particle volume fraction increases, the extinction coefficient also increases. Hong et al. [31] summarized on the optical properties of single particle and excluded the multiple scattering effects between particles. The results revealed that many of the nanofluids are agglomerated. It was also noted that particle aggregates also affected the thermal conductivity of nanofluids. Optical properties of different concentrations of CNT nanofluids can be seen in Figure 13.6 [2].

13.3.3 Specific Heat

Specific heat of a material is capacity to hold and carry the heat along with it. Materials of high specific heat can take up larger amounts of heat with lower changes in its temperature.

13.3.3.1 Effect of Volume Concentration

Many researchers have investigated about the effect of volume concentration on specific heat of nanofluids [45], and some proved that with the enhancement of volume fraction, the specific heat of nanofluid decreases [46–48]. Zhou and Ni [49] summarized that specific heat of Al_2O_3 water-based nanofluid decreased till 21% for enhancement of nanoparticle vol.% from 0% to 25% based on mixing theory and 45% based on thermal equilibrium. Namburu et al. [50] noted that specific heat falls to 12% for SiO_2 nanoparticle concentration

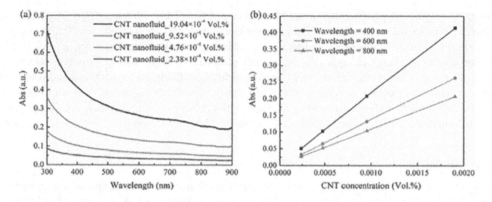

FIGURE 13.6 Optical properties of CNT nanofluids: (a) UV-Vis spectra of CNT nanofluids at different concentrations. (b) absorbance of CNT nanofluids versus CNT concentration at three different wavelengths (400, 600, and 800). (From Wang, X. et al., *Solar Energy Mater. Sol. Cells*, 130, 176–183, 2006.)

of 10 vol.% in (ethylene glycol [EG]/W):(60/40) mixture. Murshed [51] carried out experimentations with various volume fractions (0–0.05) and shape of nanoparticles of various nanofluids to study the specific heat of TiO_2-EG, Al_2O_3-EG, Al-EG, and Al-EO by using double hot-wire technique. The results revealed that with the enhancement of volume fraction of nanoparticles, specific heat of nanofluids decreases.

13.3.3.2 Effect of Temperature

Most of the investigations reveal how the temperature affects the specific heat of nanofluids. De Robertis et al. [52] investigated experimentally the specific heat of Cu-EG nanofluid and noted a decrease in the specific heat due to the nanoparticles and also that it enhances with the increase in temperature. Later, Saeedinia et al. [53] investigated by suspending CuO in pure engine oil (PEO) and it was found that with the enhancement of volume concentration of CuO nanoparticles, the specific heat of nanoparticles decreases, and with the increase of temperature, the specific heat decreases. Kumaresan and Velraj [54] studied the specific heat of multi-walled carbon nanotube (MWCNT) nanofluid, and it was reported that specific heat enhanced with enhancement of temperature till 30°C, but it decreased with enhancement of volume (%) of MWCNT as it is higher than that of base fluid. It was also noted that for a nanofluid having 0.3 vol.% of CNT, the increase of temperature above 30°C, specific heat showed a decreasing trend.

Zhou and Ni [49] investigated the specific heat of Al_2O_3 water-based nanofluid with differential scanning calorimeter. It was found that when the nanoparticle volume concentration increased, the specific heat of nanofluid decreased. Mansour et al. [55] located functions as ratio to the specific heat of base fluid with the nanoparticle concentration in the nanofluid. It was not confirmed which relation should be used, so they have considered both relations to be correct, but Zhou and Ni studied both relations carefully and found that it was valid. In their investigation. specific heat is calculated as below:

$$C_{pnf} = \frac{\phi\rho(C_p)_p + (1+\phi)_p(C_p)_f}{P_{nf}}$$ (13.1)

where:

$(\rho c_p)_p$ is the density and specific heat of particle
$(\rho c_p)_f$ is the density and specific heat of fluid
$(\rho c_p)_{nf}$ is the density and specific heat of nano fluid

Sekhar and Sharma [56] carried out the study over a wide range of nanoparticle sizes with combination of concentration and studied the temperature-dependence of specific heat capacity. Teng and Hung [57] studied the difference between experimental values and the calculated values of specific heat and density of Al_2O_3-water nanofluids having weight percentage of nanoparticles as 0.5, 1.0, and 15 with added 0.2 wt.% of chitosan in all mixtures. For 0.5 wt.% of Al_2O_3 nanoparticles in water, they found good results with experimental and calculated value. The variation of specific heat was over the range of -0.07% to 5.88% and -0.35% to 4.95%. As the concentration of nanofluid increases, the variation of calculated results of specific heat and density also increases. It was also found that various sizes of nanoparticles, base fluids, and temperature and volume fractions have significant effect on the specific heat of nanofluids. With the enhancement of volume fraction, the specific heat of nanoparticle decreases, and it increases with the increase of temperature. To know the nanofluid with desired specific heat capacity, knowledge of specific heat of base fluid and nanoparticles is needed.

13.3.4 Viscosity

Viscosity is one of the most crucial properties of nanofluids. It is observed that viscosity is as important as thermal conductivity in most of the engineering systems as the pumping costs are related to it. And it is strived to always have a nanofluid with low viscosity as the pumping requirements should not nullify the profits made by the high thermal properties of the nanofluid. Various investigations have been done to study the factors affecting the viscosity of nanofluids. Some factors include concentration, size, temperature, and shear rate of the nanoparticles [12]. Also, an increase in viscosity of the nanofluid will increase the pumping power requirements. Mostly, there are reports where they have found a negligible increase in the viscosity, thus certifying the nanofluid as an economical heat transfer fluid that does not require greater pumping costs than the respective base fluid [58–60].

Prasher et al. [61] studied the viscosity of alumina-based nanofluids and observed that viscosity of these nanofluids is even though not strong function of nanoparticle diameter and temperature, but is definitely a strong function of concentration of the nanoparticles. Li et al. [62] calculated the viscosity by using capillary viscometer of water with CuO nanoparticle suspensions. The results revealed that with the enhancement of temperature, the viscosity decreases, and with increasing concentration, the effect is not so much pronounced as that of temperature. It was observed that at lower temperatures, capillary diameter plays a significant role in affecting the viscosity. Einstein [63] obtained the formula for a linearly viscous fluid containing dilute, suspended, and spherical particles.

$$\mu_{nf} = \mu_{bf}\left(1 + 2.5\varphi\right) \tag{13.2}$$

μ_{nf} is the viscosity of suspension
μ_{bf} is the viscosity of base fluid
φ is the volume fraction of particles in base fluid.

The above formula is only valid for low particle volume fractions $\varphi \leq 0.02$.

Brinkman [64] has further developed Einstein's formula for the use with moderate particle concentrations as follows:

$$\mu_{nf} = \mu_{bf}\frac{1}{\left(1-\varnothing\right)^{2.5}} \tag{13.3}$$

Batchelor [65] again modified the Einstein equation as given in Eq. 13.4. It takes into account the Brownian motion of the particles.

$$\mu_{nf} = \mu_{bf}\left(1 + 2.5\varnothing + 6.5\varnothing^2\right) \tag{13.4}$$

Lundgren [66] in 1972 developed an equation under the form of a Taylor series, and that if the terms of second or higher order are neglected, then it reduces to that of Einstein:

$$\mu_{nf} = \mu_{bf}\left[1 + 2.5\varnothing + \frac{25}{4}\varnothing^2 + f\left(\varnothing^3\right)\right] \tag{13.5}$$

13.3.5 Other Properties

As we know that nanofluids are not only simple mixtures of liquid and solid particles, but also have a complex tendency to form aggregates as the time passes due to its high surface forces and surface activity. These agglomerations lead to settling and clogging of micro-channels which ultimately leads to decrease of thermal conductivity of nanofluids. So stability also affects the properties of nanofluids. Many methods have been investigated to study the stability of nanofluids. Sedimentation is the simplest method for the evaluation of stability [12].

Peng et al. [67] summarized the factors which affect the stability of nanofluids. The results showed that nanoparticles concentration, dispersant, viscosity of base liquid, and pH value are the factors which influence the stability of suspensions. Hwang et al. [68] investigated the stability of nanofluids by using UV-V is spectrophotometer. They observed that characteristics of suspended particle and base fluids affect the stability of nanofluids and that addition of a surfactant to the nanofluid may help improving its stability. Wang [69,70] summarized that stability of suspension is also affected by equivalent diameter of nanoparticle and dynamic viscosity of nanofluids. Many investigators [71,72] also observed that pH of nanofluids and mass fraction of dispersant affect the stability of the nanofluids.

13.4 Applications of Nanofluids in Solar Thermal Systems

13.4.1 Direct Solar Absorber Systems

To enhance the efficiency of the conventional indirect absorption-based solar collectors, the direct absorption solar collectors were developed [73–77]. A typical direct solar absorber system schematic is presented in Figure 13.7 [78].

In the newer studies, nanometer scale particles have been developed by the methods of nanotechnology. Nanoparticles are being used to enhance the thermal conductivity of the base fluid. Many investigations found enhanced thermal properties of a base fluid by addition of metal or metal oxide nanoparticles specifically for the direct absorption solar collector. The optical properties are changed from transparent to absorbing over the range of solar spectrum by the addition of nanoparticles [79–85].

Non-concentrating solar collectors are found to have lower conversion of the thermal energy and are not used in high temperature applications. Many investigations have reported that they achieved lower degree of capturing solar energy, lower heat transfer, and increased heat losses in the collector due to the working fluid absorbing heat very well [86]. To improve the absorbance, the absorbers are coated with chromium black, but they may pose environmental hazard as most of the coatings are toxic.

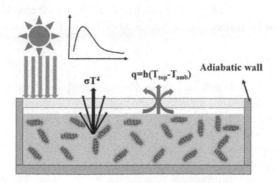

FIGURE 13.7 Schematic of nanofluid based direct absorption solar collector. (From Zhuocheng, J.L. et al., *Solar Energy Mater. Sol. Cells*, 136, 177–186, 2015.)

The high thermal energy transfer to generate high temperatures and pressures required for electricity generation are provided by concentrating solar thermal collectors. These types of collectors are expensive and are economic for a large scale. However, to overcome these gaps, an alternative to non-concentrating solar thermal collector was developed, i.e., direct solar absorption collector. The feature of this collector was using the conventional solar collector configurations in which large energy can be transferred [87]. The efficiency of the traditional solar collector is affected by various designs. Various factors that must be considered while designing a solar collector are size of the collector, material used in the construction of collector, and properties of the collector. The convective heat flow rate (Q) can be evaluated:

$$Q = hA\Delta T, \tag{13.6}$$

where:
   h is the heat transfer coefficient
   A is the surface area of heat transfer region
   ΔT is the temperature difference due to heat flow [88].

Area (A) cannot be enlarged in case of solar collectors as the size of the collector becomes large and bulky. If ΔT is increased, there is a need to improve the solar concentration ratio which is not effective due to collector design. To overcome these shortcomings, researchers developed the direct solar absorption collectors so as to make use of the working fluid as both a solar radiation absorber as well as a heat transfer medium.

Taylor et al. [89] investigated that nanofluids have outstanding potential for power tower and solar thermal power plants. By using the nanofluid in the receiver side, efficiency can be improved on the order of 5%–10%. More additional improvements can be made by changing the optical and thermal properties of the nanofluid more closely with the specific system conditions.

Li et al. [90] introduced studies analogous to Taylor et al. [89] by using three distinct nanofluids (Al_2O_3/water, ZnO/water, & MgO/water) on the tubular solar collectors. Experimental results show that 95% of the sunlight can be absorbed effectively by nanofluid containing nanoparticles less than 10 ppm. Yousefi et al. [91] experimentally investigated the efficiency of the flat-plate solar collector using particle size of 15 nm and Al_2O_3/water nanofluid with weight concentrations of 0.2% and 0.4%. The experimentation was carried out with Triton X-100 as surfactant as well as without it. The results obtained show that the efficiency increases by 28.3% by using nanofluid having concentration of 0.2 wt.%. Increasing the mass flow rate with the use of a surfactant increases the efficiency more by 15.63%. Khullar et al. [92] summarized their experimental investigation on aluminium-based nanofluid on concentrating parabolic solar collector (CPSC). The results showed a 5%–10% of increase in thermal efficiency as compared to conventional concentric parabolic solar collector with a nanofluid of concentration of 0.05 vol.%.

Graphite nanofluids and silver-based nanofluids performance has been studied experimentally and numerically by Octanicar et al. [93] on DASC. Using the nanofluids as an absorption medium, the efficiency improvement up to 5% was observed. The results showed that by using nanoparticles of size 30 nm, 3% increase in efficiency can be noted, using silver nanoparticles of size 20 nm–40 nm, 5% increase in efficiency was observed. When the particle size was reduced to its half, 6% efficiency enhancement is noted.

13.4.1.1 Flat-Plate Collectors

There are various applications of nanofluids among which applications in harnessing renewable energies using nanofluids have been paid more attention. Compared to other common fluids, nanofluids possess lower heat capacity, so they have gained popularity. There are various studies on solar collectors in which nanofluid is being used as a working fluid [94].

The widespread use of solar energy for heating and cooling buildings is affected by the cost of the flat-plate collector [74]. Flat-plate solar collectors are classified as a special class of heat exchangers. These devices are used for the conversion of heat energy by absorbing the solar radiation [95]. They usually have a fixed position, so they need to be arranged properly. A flat-plate solar collector basically consists of glazing covers and insulation layers to decrease losses of heat, absorber plates to absorb the heat energy, recuperating tubes to contain the liquid which carries the heat, and other accompanying parts. The glazing covers are made with materials having high transmissivity of short-wave radiation and low transmissivity of long-wave radiation, like glass, these are single or multiple sheets. It reduces irradiation losses from the collector due to the greenhouse effect and also decreases the convection losses from the absorber plate [96]. A typical schematic along with its various components has been shown in Figure 13.8. Their applications have been found in solar water-heating systems in homes and in solar space heating.

The pictorial view of the flat-plate collector is shown below Figure 13.9. As observed from Figure 13.9a, a flat-plate collector consists of a glass- or plastic-coated insulated metal box and also a dark colored absorber plate. All the incoming solar radiations falling on the absorber plate are absorbed by it and are circulated through the collector in tubes by a medium. The isometric view of a flat-plate solar collector

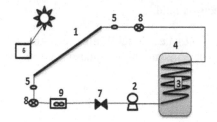

FIGURE 13.8 General experimental set up of solar collector. 1. collector 2. pump 3. heat exchanger 4. tank 5. thermometer 6. solar meter 7. control valve 8. automate valve 9. rotameter. (From Moghadam A.J. et al., *Exp. Therm. Fluid Sci.*, 58, 9–14, 2014.)

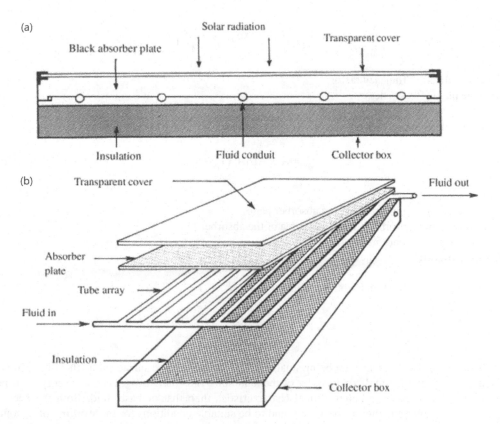

FIGURE 13.9 (a) Cross-sectional view and (b) isometric view of a flat-plate solar collector. (From Colangelo, G. et al., *Appl. Energy*, 111, 80–93, 2013.)

is depicted in Figure 13.9b [98]. Air is the circulating fluid in the air-based collector, while water is used in the liquid-based collector. These are mostly used where the temperature requirement is around 30°C–70°C [95].

Energy analysis of flat-plate collector:

The working fluid's heat gain (Q_u) can be given as:

$$Q_u = \dot{m}C_p\left(T_{out} - T_{in}\right),$$ (13.7)

where:

T_{in} is the fluid inlet temperature
T_{out} is the fluid outlet temperature
C_p is the heat capacity
$\dot{m}$ is the mass flow rate of the agent fluid.

The useful equation for heat gain (Q_u), i.e., Hottel-Whillier equation for a flat-plate collector system, in which the heat loss to from the solar collector to the atmosphere is given as [99],

$$Q_u = A_p F_R\left[S - U_1\left(T_{in} - T_a\right)\right].$$ (13.8)

T_a is the ambient temperature
F_R is the heat removal factor.

The heat removal factor is given as:

$$F_R = \frac{\dot{m}\cdot C_p}{U_1 A_p}\left[1 - \exp\left\{\frac{-F'U_1 A_p}{\dot{m}\cdot C_p}\right\}\right].$$ (13.9)

F is the collector efficiency factor
Φ is the plate effectiveness.

An energy balance on the absorber-plate can be given as [99]:

$$Q_u = A_p \rho - U_1 A_p\left(T_p - T_a\right).$$ (13.10)

From equations:

T_p is the average temperature of the absorber plate
S is the radiation absorbed flux by unit area of the absorber plate
A_p is the absorber plate area
U_1 is the overall loss coefficient.

Thermal efficiency of the solar collector can be given as:

$$\eta_{cn} = \frac{Q_u}{I_T}.$$ (13.11)

Due to the size of the nanoparticles being small and possession of large surface area, there is an increase in heat capacity and thermal conductivity of the nanofluid, thus increasing its solar energy absorption properties. Nanofluids have better optical characteristics than that of base fluid. Both the spectrum ranges, i.e., the solar and the infrared are found to be suitable conditions for the working of nanofluids as they show high absorption with low emittance of the radiation. They also are an excellent absorbing medium due to their good stability at a wide range of temperatures with large absorption coefficient.

The problems like sedimentation of solids, clogging due to choking and deposition of the solids, and fouling of pipes and pumps can be avoided by the use of nanofluids which is due to the smaller size of the solid particles compared with micrometer- or millimeter-sized particles. If the nanofluids are not very stable, they shall lose their thermal transport properties. This may be overcome by introducing dispersing agents or surfactants, but this addition of surfactants again reduces the ability of the nanofluids to conduct heat. As the heat transfer area of the solar thermal devices can be reduced by the nanofluids, it results in reduction of the total cost of the solar energy systems. Nanofluids have high convective heat transfer coefficient (HTC) and high density, and the specific heat of nanoparticles results in enhancement of the efficiency of the thermal devices [1].

Francia [100] investigated that insertion of honeycomb generated from a transparent material placed in the gap which is the airspace between the glazing layer and the absorber is advantageous for heat loss reduction. To prevent the system from overheating, the heat absorbed needs to be circulated to working fluids rapidly [101]. Kumar and Reddy used the porous insertions to increase the heat transfer of solar collectors, and they noted that enhancement in heat transfer up to 64.3% was obtained [102]. Sopian et al. [103] reported outstanding heat transfer rate with the use of a double-pass structure for solar receiver. The typical schematic of double-pass solar collector is given in Figure 13.10 [103].

Yousefi et al. [91] investigated performance of Al_2O_3 nanofluid in a flat-plate collector and found that mixing a surfactant will give maximum efficiency of 15.63%. The results reveal that a 0.2 wt.% Al_2O_3 nanofluid increases the efficiency of solar collector by 28.3%. Later, Yousefi et al. [104] used MWCNT/water nanofluids and found that the thermal efficiency of a flat-plate solar collector can be increased by increasing the weight fraction of the MWCNTs in the nanofluid. Yousefi et al. [105] found the optimum pH of MWCNT/water nanofluids that give maximum efficiency of a flat-plate collector.

Jamal-Abad et al. [106] experimentally showed that addition of Cu nanoparticles in water can improve the efficiency by 24%. Faizal et al. [107] incorporated four distinct water-based nanofluids using CuO, SiO_2, TiO_2, and Al_2O_3 nanoparticles in a flat-plate collector. They observed that the CuO/water nanofluid is the best choice as far as the economic and environmental considerations are concerned. Colangelo et al. [108] studied that by decreasing the sedimentation of Al_2O_3 nanoparticles distributed in water, the heat transfer coefficient in the flat-plate collector increases up to 25%. Parvin et al. [109] studied natural convection in a direct absorption solar collector using Cu/water and Ag/water nanofluids. They also developed correlations to calculate the Nusselt number and the collector efficiency. Chaji et al. [110] studied the performance of TiO_2/water nanofluids in a flat-plate collector and found an increase in the efficiency of the collector at nanoparticle concentration of 0.3 wt.%.

Zamzamian et al. [111] investigated that use of Cu-synthesized/EG nanofluid with volume fraction of 0.2 wt.% and 0.3 wt.% in a flat-plate collector enhances the efficiency. In another experimental work,

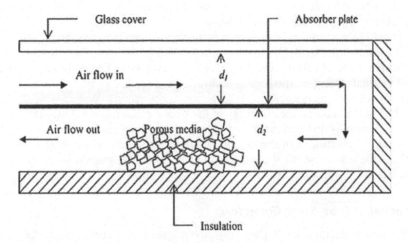

FIGURE 13.10 Schematic of double pass solar collector with porous media in second channel. (From Sopian, K. et al., *Renew. Energy*, 34, 640–645, 2009.)

Mohadam et al. [112] investigated performance of CuO/water nanofluid flow in a flat-plate collector and found an optimal value of mass flow rate to attain the optimal thermal efficiency. Hendrie et al. [113] recently reported that liquid collector performance could be enhanced by giving more attention toward design aspects like PV cell packing factor, thermal conductivities, and optical transmissivities. Natarajan et al. [114] summarized the increase in thermal conductivity of the base fluid by using carbon nanotube and recommended that these fluids increase the efficiency of the conventional solar water heater.

Tyagi [115] has theoretically investigated the potentiality of using a non-concentrating direct absorption solar collector (DAC) and compared with the conventional flat-plate collector. In his research, nanofluid which is a mixture of water and aluminium nanoparticles, which was used as an absorbing medium. The results from this study reveal that using nanofluid as the working fluid shows 10% increase in efficiency of a DAC than that of the flat-plate collector is obtained. Octanicar and Golden [116] investigated the environmental and economic impact by using nanofluids to increase solar collector efficiency compared with the conventional solar collectors. Octanicar et al. [117] summarized an efficiency enhancement up to 5% in solar thermal collectors by using the nanofluids as the absorption medium.

13.4.1.2 Concentrating Collectors

DASCs were developed so as to enhance the efficiency of the conventional indirect absorption solar collectors [73–77]. Octanicar et al. [117] investigated that using water in place of nanofluids can enhance the efficiency of solar collector by 5%. Sani et al. [118] and Mercaralli et al. [119] studied the absorption properties of carbon nanohorn-based nanofluids with base fluids as water and ethylene glycol. With the base fluids and the carbon black nanofluids, the absorption coefficient showed outstanding value over a wide range. Wang et al. [120] investigated that the maximum bulk temperature can be enhanced by 25% under natural sunlight by using 0.1% (in weight) of copper nanofluid.

Liu et al. [121] showed that by using CuO nanofluid, the solar collector has a better performance than the deionized water. Javadi et al. [122] studied that the photo thermal conversion efficiency of any nanofluid can get affected by the nano particle size. Lee et al. [42] summarized that using particle concentration of 0.05% by volume of the gold nanoparticle can increase the efficiency by 70% of a solar collector. Zhang et al. [123] found that no carbon-based nanoparticles can help achieve greater specific absorption rates than the gold nanoparticles.

Tyagi et al. [124] have studied the feasibility of using a non-concentrating DAC and compared with a conventional flat-plate collector. As the absorbing medium Al-H_2O nanofluid was used. They observed that the efficiency of a DAC using nanofluid as the working fluid showed 10% increase in efficiency than that of a flat-plate collector was noted. Octanicar et al. [117] studied the photo thermal properties for nanofluids made from different nanoparticles. By using the nanofluids as the absorption medium, 5% increase in efficiency in solar thermal collectors is observed.

Sani et al. [118,125] have studied the thermal and optical properties of a single wall carbon nanohorn (SWCNHs) using direct sunlight absorbers in solar thermal devices. The results reveals that nano SWCNHs have higher sunlight absorption. Han et al. [126] studied the carbon black aqueous nanofluids and their photo thermal properties, rheological behavior, and optical properties. The results revealed that nanofluids have better solar energy absorption properties as compared to water.

Yousefi et al. [127,128] studied the efficiency of flat-plate collector using Al_2O_3-H_2O and MWCNT-H_2O nanofluids. The results revealed that the efficiency of solar collector can be increased by 28.3% by using 0.2 wt.% Al_2O_3 nanofluid than that of pure water. The addition of surfactants in MWCNT-H_2O nanofluids effects the efficiency of solar collector. Saidur et al. [39] studied that optical properties of nanofluids are affected by the size of particle.

13.4.1.3 Evacuated Tube Solar Collectors

Evacuated tube solar collectors (ETSC) are very advantageous over conventional flat-plate solar collectors (FPSC). Many investigations have been done to study the thermal performance of both cases [129–134]. The evacuated tube solar collectors are lower in cost, make less heat losses and provide higher

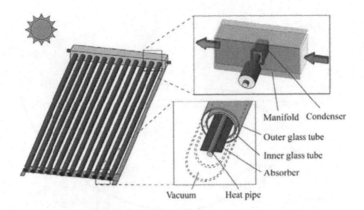

Manifold Condenser

Outer glass tube

Inner glass tube

Absorber

Vacuum   Heat pipe

FIGURE 13.11 Evacuated tube solar collector. (From Muhammad, M.J. et al., *Renew. Sust. Energy Rev.*, 63, 226–236, 2016.)

efficiencies as compared to the convectional flat plate solar collectors [129–132,135]. At higher tempera-ture conditions, ETSC proves to be a better option than the FPSC [132]. They have good applications at higher temperature like solar pre-heaters and solar tanks [1]. Figure 13.11 shows an evacuated tube solar collector [136].

The most crucial part in the evacuated tube solar collector is the parallel evacuated glass tube. It has an inner and outer tube. The inner tube is coated with a special absorptive coating so as to maximize the process of absorption of solar radiations, and the outer tube is kept transparent. Between the inner and the outer tube there is vacuum, which helps the solar radiations to be transferred, preventing the escape of heat. A heat pipe made out of copper which has a high thermal conductivity is welded along to the absorber plate and is placed inside the inner tube [1].

Numerous types of heat pipes are used with evacuated tubular solar collectors to enhance its solar collecting performance so as to gain better fluid temperature [121]. Chun et al. [137] experimented on various systems so as to select the most appropriate configuration of the system. They studied the performance of solar water heaters used for domestic purpose which includes heat pipes. Riffat et al. [138] theoretically studied a thin membrane heat pipe solar collector keeping in mind the high effi-ciency and lower capital cost. Choi and Eastman [6] have investigated the nanofluid which enhances the heat transfer. Khandekar et al. [139] summarized the thermal resistance of a closed two-phase thermosyphone by using different water-based nanofluids as working fluid with pure water.

Liu et al. [140] studied the heat transfer performance of CNT/water nanofluid in a miniature thermo-syphon and found that an optimum CNT mass concentration could achieve outstanding heat transfer increment. Later, Liu and co-workers wrote a review paper introducing applications of nanofluids in heat pipe [141,142]. Compound parabolic concentrator (CPC) has the ability to gain the incoming radiation coming from wide range of angles. The various internal reflectors can help any of the entering radia-tion to occupy its place at the bottom of the collector. As the CPC solar collectors are highly efficient as moderate temperature solar collector, they are booming in the design and analysis area of the solar collectors [143–146].

13.4.1.4 Parabolic Trough Collectors

Parabolic trough collectors are solar collectors that have U-shaped troughs to accumulate solar radiations in a small area called heat collector element and consists of water or oil as working fluid situated inside a pipe along the line of focus of the parabolic trough [147–153]. These types of collectors focus direct solar radiations on a focal line parallel to the collector axis. As shown in the Figure 13.12, the focal line has a receiver pipe in which the heat transfer fluid is moving which enhances the enthalpy by absorbing the solar energy concentrated on the pipe walls. The solar radiations must fall parallel to the axis of the collector which is provided with one-axis solar tracking system [154]. Figure 13.12 shows a schematic of parabolic-trough collector [154].

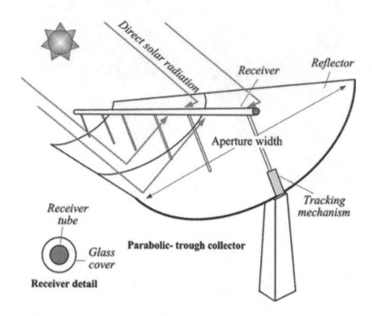

FIGURE 13.12   Schematic of a parabolic—trough collector. (From Cabrera, F.J. et al., *Renew. Sust. Energy Rev.*, 20, 103–118, 2013.)

The tracking system can be of one-axis or two-axis. Due to the lower angle of incidence on the two-axis moving system, the solar energy is collected over here [155,156]. Due to low cost and easy installation, one-axis solar tracking parabolic trough collectors (PTCs) are more preferred. The energy from the incident radiations received on the surface of the collector is higher, which ultimately results in higher efficiency of the collector and small auxiliary energy requirement when the PTC axis is North-South (N-S) (i.e., East-West [E-W] tracking system) [157]. PTC solar technique is lowest cost and best proved for a large scale solar power technology, due to which nine large scale power plants, solar electric generating systems (SEGS), with large scale (LS)-1, LS-2, and LS-3 PTC, were expanded by Luz International in California Mojave Desert (United States) [158].

The reliability of PTC was confirmed by other developed models. The feasibility of direct steam generation (DSG) has been studied in order to fill many gaps in PTC [159–161]. PTCs that are used in the solar thermal power plants at temperature around 400°C are very authentic. Other applications include industrial process heat and heat driven solar cooling, they work around 180°C and have space constraints due to the expensive land, so rooftop is used for installation [154]. To overcome the shortcomings, some companies have developed small, lightweight and low cost parabolic trough collectors (PTCs). These companies include: Industrial Solar Technology (IST) [162], Sopogy [163,164], CHAPS [165] in Australia, NEP Solar [166], and many others.

13.4.1.5 Concentrated Parabolic Solar Collectors

Harvesting of solar energy can be enhanced by use of nanofluid in CPSC [167]. Figure 13.13 shows a concentrating parabolic solar collector. A mathematical model of the collector was developed, after which the theoretical results were compared with the experimental results, and efficiency of CPSC using nanofluids was found to be enhancing by about 5%–10% as compared to the conventional. In CPSC, the nanofluid moves in a glass tube, making a direct contact of the working fluid with the solar irradiance. There is no direct contact between the solar irradiance and the working fluid in the conventional absorber tube as the solar radiations are initially absorbed by the absorber tube and then transmitted by the process of conduction and convection to the working fluid. While using nanofluids, a direct contact is occurred between the solar irradiance and the working fluid [1].

Monte Carlo ray-trace (MCRT) is a numerical method which is versatile, rigorous, and efficient method to simulate the collecting and concentrating characteristics of the concentrating solar collectors (CSCs).

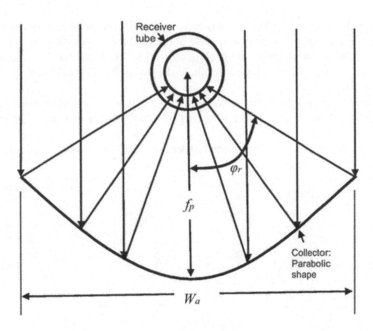

FIGURE 13.13 Parabolic trough solar collector with some of the incident and reflected rays. (From Mwesigye, A. et al., *Energy Convers. Manage.*, 120, 449–465, 2016.)

Daly studied flux distributions developed by parabolic and circular cylinder solar cylinders by using a backward ray tracing method [168]. Riveros and Oliva [169] found the optimum sizes and positions of parabolic collecting devices by using a computer-aided graphical method with modified ray tracing procedure. Shuai et al. [170] by using the MCRT method anticipated the performance of dish solar concentrator/cavity receiver systems.

He et al. [171], by using the MCRT method, replicated the solar energy conversion process in a parabolic trough solar collector system. They conjugated MCRT method with the finite volume method (FVM) to solve the complex heat transfer problems of heat conduction, radiation, and convection in the parabolic trough receiver subsystem [171].

There are various studies for heat transfer performance of absorber tube wall and working fluid. A descriptive thermal model was developed to calculate the heat loss of trough collector considering constant solar radiations which can be used to analyze the trough's thermal efficiency [172,173]. Martinez et al. [174] studied the finite difference method (FDW) to solve energy equations which uses sectional heat fluxes as boundary conditions. Eck et al. [175] determined the temperature profile for LS-3 type collector using finite element method (FEM) in ANSYS with a flux distribution boundary, then they have selected a Gaussian distribution assumption for finite element method (FEM) as the flux distribution boundary, and for a simplified analytical solution, a rectangular distribution assumption as the flux distribution boundary [176].

Hang et al. [177] have invented the possibility of concentrating solar power (CSP) systems in China. As per the study, this technology can be well used in northern and western parts of China, as large wasteland is available for these systems and there is better direct normal irradiation (DNI). Li [178] investigated a probability of development and distribution of CSP technologies in China, keeping in view both technical and socioeconomic perspectives. Fluri [179] investigated the possibility of CSP in South Africa by using geographical information (GIS) approach. Then it was revealed that South Africa has the ability for the installation of parabolic trough power plants with capacity of 547.6 GW.

13.4.1.6 Photovoltaic-Thermal Collectors

Solar thermal collection defines the performance of photo-thermal conversion. As the weather cannot be controlled artificially, experiments of photo-thermal conversion are carried out in solar stimulator rather in the natural sunlight. The collecting tube is prepared from organic glass. Four full spectrum solar stimulators were arranged properly at the end of the tube and for monitoring the temperature two thermocouples were also placed. Figures 13.14 [180] and 13.15 [181] depict photovoltaic/thermal systems.

Figure 13.16 depicts the concentrating photovoltaic/thermal (CPV/T) efficiencies for water and nanofluid cooling in a silicon solar cell.

Figure 13.17 shows a typical solar thermal collection experiment set up. In the collector, Chinese ink, Cu, CuO, and carbon black nanofluids were added with 0.05 wt.%, in which Cu and CuO nanofluids were dispersed ultrasonically for about 30 minutes.

Then for the next 2 hours, the tube was placed under the solar stimulator as shown in Figure 13.18.

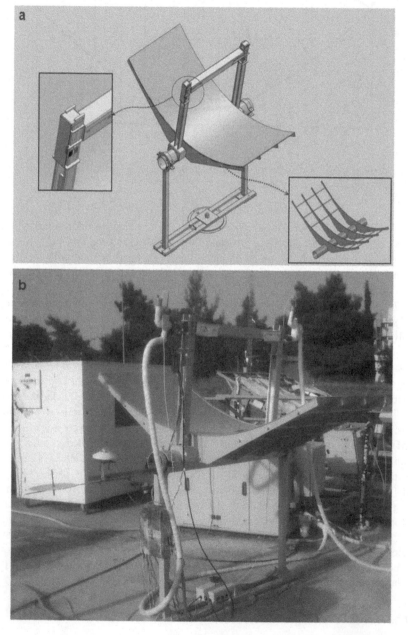

FIGURE 13.14 Metallic structure of the concentrating photovoltaic/thermal system: (a) CAD drawing and (b) constructed frame. (From Karathanassis, I.K. et al., *Renew. Energy*, 101, 467–483, 2017.)

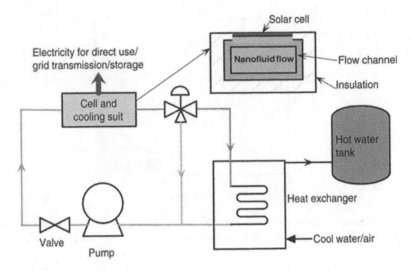

FIGURE 13.15 Schematic of photovoltaic/thermal system. (From Xu, Z. and Weinstreuer, C., *Energy Convers. Manage.*, 87, 504–512, 2014.)

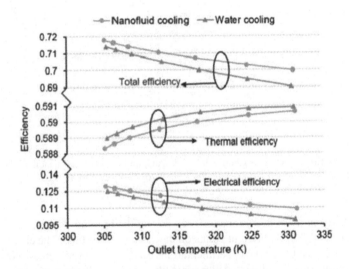

FIGURE 13.16 CPV/T efficiencies using water and nanofluid cooling for silicon solar cell. (From Xu, Z. and Weinstreuer, C., *Energy Convers. Manage.*, 87, 504–512, 2014.)

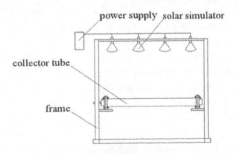

FIGURE 13.17 Solar thermal collection experiment platform. (From Wang, H. et al., *Solar Energy Materials and Solar Cells*, 176, 374–380, 2018.)

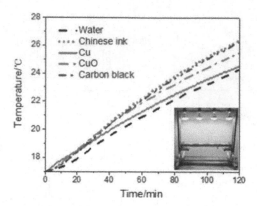

FIGURE 13.18 Temperature evolution of Chinese ink, Cu, CuO, and carbon black nanofluids with time, under solar simulators. (From Wang, H. et al., *Solar Energy Materials and Solar Cells*, 176, 374–380, 2018.)

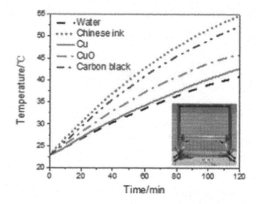

FIGURE 13.19 Temperature evolution of Chinese ink, Cu, CuO, and carbon black nanofluids with time, by sunlight. (From Wang, H. et al., *Solar Energy Materials and Solar Cells*, 176, 374–380, 2018.)

The temperature was noted at every minute. Then during the irradiation process, intensity of the light was measured by choosing seven different points on the surface of the tube. The effectiveness of the solar stimulators was also studied by similar experimentations carried out in sunlight as shown in Figure 13.19. Cu and CuO nanofluids were not dispersed ultrasonically. Chinese ink, Cu, CuO, and carbon black nanofluids were added with 0.05 wt.%. The experiment was carried out at 25°C, on a sunny day with average light intensity of 80,000 lx. Then the solar stimulators were shut down, and the system was placed in sunlight for the next 2 hours. The temperature was noted at both the ends of tubes with different working fluids. Later, to understand the properties of bone glue in Chinese ink, similar study was carried on bone glue solution, carbon black, Cu, and CuO nanofluids with or without addition of bone glue by using solar stimulator. Later, after 2 hours, temperature is evaluated [7].

Nilsson et al. [182] investigated the thermal and electrical properties of an asymmetric CPC which is made up from mono crystalline silicon cells built with high lattices. Assoa et al. [183] studied polycrystalline cells where air and water are used to cool a CPC by Conventry [184].

13.4.2 Solar Cells

The efficiency of solar cells can be improved by using cooling solar cells. The solution to this problem is by using nanofluids [185]. Elmir et al. [186] stimulated the cooling solar cell by using forced convection using nanofluid. They have solved the equations in Cartesian system by considering the solar panel as an inclined cavity. For the analysis, they have used Al_2O_3 and water nanofluid. Brinkman and Wasp models are used for calculating nanofluids thermal conductivity and viscosity. Rate of cooling can be improved by using nanofluids which ultimately increase the average Nusselt number. Many organic synthetic dyes have been studied to replace the high cost ruthenium compounds.

13.4.3 Solar Thermoelectric Devices

Figure 13.20 shows a solar thermoelectric device system. These devices are used for the conversion of heat energy into electricity and vice versa [188]. Thermoelectric devices have two semiconductors, *p*- and *n*-type, which are connected electrically and thermally in parallel. One end is kept at lower temperature while the other end is heated. This difference in temperature makes the electricity to flow in the circuit. The use of thermoelectric (TE) devices have been investigated by many researchers like electronic cooling and vehicle air conditioning [189–196].

Ahammed et al. [197] studied the feasibility by using the nanofluids in a thermoelectric cooling device (Figure 13.21). Volume concentration between 0.1% and 0.2% of Al_2O_3/water is used to study the feasibility as a working fluid in the heat exchanger to remove heat from hot side of the thermoelectric cooler. An enhancement of 9.15% in the difference in temperature between the hot and cold ends of the thermoelectric device was noted, using 0.2 vol.%.

Chang et al. [187] studied that the thermoelectric devices are used for the conversion of solar energy directly to electricity. CuO thin films were developed by the method of electrophoresis deposition and CuO nanofluid is prepared on a Cu plate which is later placed between dye-sensitized solar cells and TE devices. When solar radiation intensity of 100 mw/cm^2 was supplied, it was observed that 4.95 mw/cm^2 is generated by solar thermal electric devices. The TE efficiency increased by 10% by using CuO thin film coating on the solar thermal electric device (STED) surface with enhanced voltage of 14.8% and temperature of 2°C.

Mohammadian and Zhang [198] studied the parameters affecting Al_2O_3/Water nanofluid in a micro-pin-fin heat exchanger on TE module. Augmentation of the coefficient of performance at low Reynolds number was observed for low volume fraction of nanoparticles in the heat exchanger (hot surface). Using the nanoparticles in the heat exchanger (cold surface), it was observed that enhancement in coefficient of performance occurs at high Reynolds number. Ahammed et al. [199] found that graphene/water nanofluid increased the coefficient of performance of TE module by 72%.

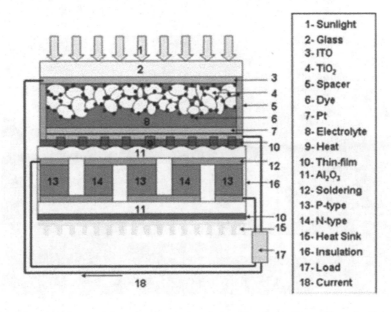

1- Sunlight
2- Glass
3- ITO
4- TiO$_2$
5- Spacer
6- Dye
7- Pt
8- Electrolyte
9- Heat
10- Thin-film
11- Al$_2$O$_3$
12- Soldering
13- P-type
14- N-type
15- Heat Sink
16- Insulation
17- Load
18- Current

FIGURE 13.20 Schematic diagram of a solar thermoelectric system. (From Chang, H. et al., *Curr. Appl. Phys.*, 11, S19–S22, 2011.)

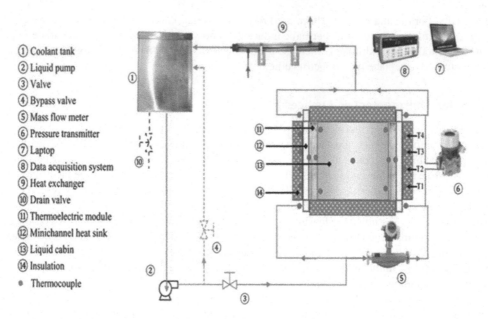

① Coolant tank
② Liquid pump
③ Valve
④ Bypass valve
⑤ Mass flow meter
⑥ Pressure transmitter
⑦ Laptop
⑧ Data acquisition system
⑨ Heat exchanger
⑩ Drain valve
⑪ Thermoelectric module
⑫ Minichannel heat sink
⑬ Liquid cabin
⑭ Insulation
• Thermocouple

FIGURE 13.21 Schematic of thermo-electric cooler set up. (From Ahammed, N. et al., *Exp. Therm. Fluid Sci.*, 74, 81–90, 2016.)

13.4.4 Solar Desalination (Solar Still)

A typical experimental set up of three single basin solar still is depicted in Figure 13.22 [200]. Further, a solar energy-based water desalination unit utilizing a solar still is as shown in Figure 13.22. It is cheap and is easy to operate. [200–203]. The only main disadvantage of this solar still is less productivity. One of the best methods studied in various investigations to overcome this problem is addition of nanoparticles to the brackish/salt water, which helps in improving the thermophysical properties of water which leads to increases in the heat transfer process inside the solar still basin. Nanofluids have good optical absorption

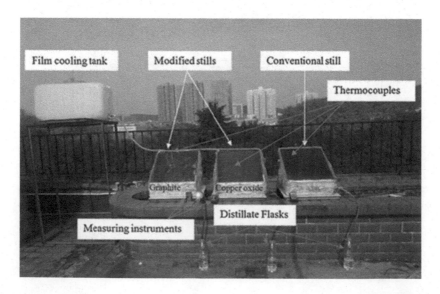

FIGURE 13.22 Experimental set up of three single basin solar still/desalination plant. (From Sharshir, S.W. et al., *Appl. Therm. Eng.*, 113, 684–693, 2017.)

capacity due to the plasmon resonance absorption capacity band arising in the infrared and visible spectrum present in nanoparticles. Nanoparticles have commendable ability for absorbing the solar radiation in the water as the spectrum of solar radiation matches the optical absorption of nanoparticles. With increase in temperature, the thermal conductivity increases, leading to better thermal energy transfer inside the basin.

More temperature difference is observed between the glass cover and basin fluid by using the nanofluids which ultimately increases the still productivity [1]. Sharshir et al. [200,204] studied a single basin solar still productivity at different conditions and also the effects of using nanofluids. Different concentrations of copper oxide and graphite were used for the study. It was observed that copper oxide and graphite micro-flakes enhance the solar still productivity by 44.9% and 53.95%, respectively, as compared to the conventional solar still without nanofluids.

Kabeel et al. [205] studied the performance of a single basin solar still by addition of different nano materials like cuprous and aluminium oxide with or without supplying vacuum inside the still. The results showed that higher productivity is obtained by using cuprous oxide/Water nanofluid than that of aluminium oxide/Water nanofluid as shown in Figure 13.23. The same authors continued the investigations by addition of aluminium-oxide and studied the effects on the single basin solar still with an external condenser. The productivity was increased by 116% when external condenser was attached with the still.

Sahota et al. [206] studied various designs of double slope solar still (DSSS) using different nanofluids and developed expression of the characteristic equation. These different designs include passive DSSS and an active DSSS having photovoltaic thermal flat-plate collectors which executed with and without helical heat exchanger as shown in Figures 13.24 and 13.25. Three different nanofluids were evaluated, i.e., TiO_2, Al_2O_3, and CuO in a passive DSSS. They have observed that thermal efficiency enhancement of TiO_2, Al_2O_3, and CuO were 46.10%, 50.34%, and 43.81%, respectively, compared to the base fluid which showed enhancement of 37.78%. The thermal exergy enhancement of TiO_2, Al_2O_3, and CuO is 12.38%, 14.10%, and 9.75% compared to base fluid (4.92%). Later, it was observed that maximum productivity of DSSS is observed by using Al_2O_3 nanoparticles which is then followed by TiO_2 and CuO. For the passive DSSS, the best performance was given by CuO nanoparticles followed by TiO_2 and Al_2O_3.

Later, to increase the performance of solar still using nanofluids, Mahian et al. [208] used nanofluids in the two-plate flat collector system which is connected in series, and then sent the nanofluid into the heat exchanger which is connected to the solar still as shown in Figure 13.26.

A comparative study between the yield of solar stills with and without heat exchanger was done and results obtained are as shown in Figure 13.27. They have also studied the effect of SiO_2/water and Cu/water nanofluids of concentrations between 0% and 2% and nanoparticle size of 7 nm and 40 nm on the

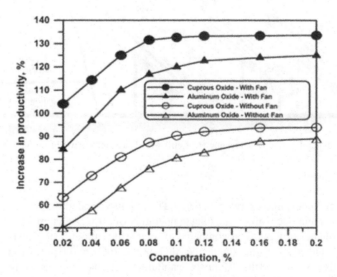

FIGURE 13.23 Productivity of solar still versus nanoparticles concentration using different nanofluids with and without fan. (From Kabeel, A.E. et al., *Energy Convers. Manag.*, 86, 268–274, 2014.)

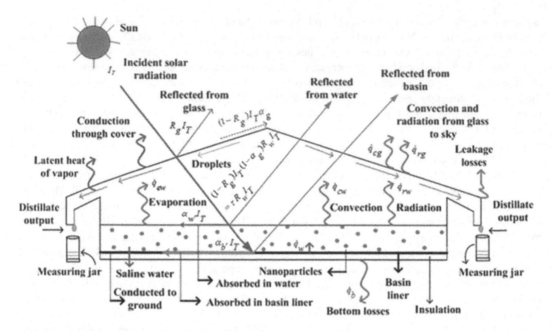

FIGURE 13.24 Schematic of passive double slope solar still (DSSS) with suspension of nanoparticles. (From Sahota, L. and Tiwari, G.N., *Desalination*, 388, 9–21, 2016.)

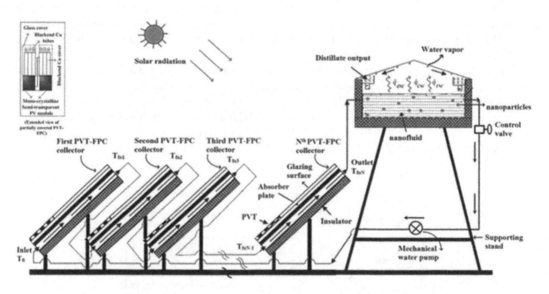

FIGURE 13.25 Schematic of active double slope solar still with heat exchanger. (From Sahota, L. and Tiwari, G.N., *Energy Convers. Manag.*, 135, 308–326, 2017.)

final yield, energy efficiency, and exergy efficiency. The results revealed that effect of heat exchanger is negligible on solar still final yield above the inlet temperature of 50°C. At a temperature of 70°C, the yield of the solar still is enhanced by 200% as compared to solar still without heat exchanger.

Elango et al. [209] studied solar stills employing Al_2O_3, ZnO, SnO_2, and Fe_2O_3/water nanofluids, and with the photographic method they found the stability of nanofluids to be 2 days, 3 days, 2 days, and 8 hours, respectively. Due to the bad stability of Fe_2O_3, it was eliminated from the study. The production of the still obtained by using Al_2O_3, ZnO, and SnO_2 nanofluids as compared to ordinary water was 29.59%, 12.67%, and 18.63%,

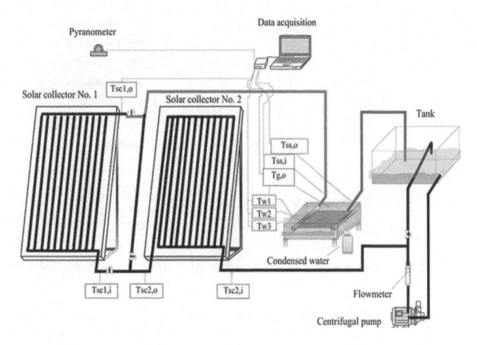

FIGURE 13.26 Schematic of experimental set up of solar still. (From Mahian, O. et al., *Nano Energy*, 36, 134–155, 2017.)

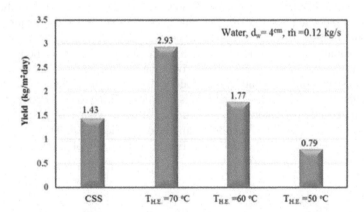

FIGURE 13.27 Comparison between the yield of solar stills with and without heat exchanger. (From Nijmeh, S. et al., *Int. Commun. Heat Mass Trans.*, 32, 565–572, 2005.)

respectively. To enhance the efficiency of solar stills, different methods have been studied [185]. Elango et al. [209] observed that solar stills' efficiency can be enhanced by using nanofluids. Further, they also investigated the effects of carbon nanotubes/water nanofluids in a single basin solar still. Figure 13.28 depicts a schematic of a vacuum solar still. Their results revealed that 50% increase in efficiency is obtained by addition of nanofluids. Solar stills efficiency can also be evaluated by addition of dyes in fluid. Nijmeh et al. [210] investigated that with the addition of violet dye to water, the efficiency of solar stills enhances by 29%. It involves lot of expenses using nanofluids in solar stills as they are costlier than dyes.

13.4.5 Solar Ponds

Salinity gradient solar ponds collect solar radiations and store them in the form of heat which are 2 m–5 m deep. For the industrial process heating, space heating, and power generation, heat is extracted from the solar

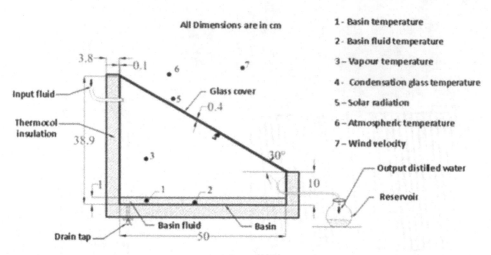

FIGURE 13.28 Schematic diagram of single basin single slope solar still. (From Elango, T. et al., *Desalination*, 360, 45–51, 2015.)

ponds at the temperature of about 50°C–60°C more than the upper surface of solar pond. It is expected that nanofluids can increase the rate of heat removed from the bottom of solar pond as shown in Figure 13.29 [211].

The low salinity brine is present in the upper convective zone (UCZ). Flushing of the fresh water is required continuously, or low salinity water is required to overcome evaporation and move the salt to non-convective zone (NCZ) by salt diffusion. Below the NCZ, the water close to the surface is less salty as compared to the water below it. As the natural convective currents are suppressed, the NCZ consists of a thermally insulating layer. The lower convective zone is also referred as storage zone, where heat is stored and extracted solid salt or high concentration brine must be regularly fed to the layer to lessen the losses of salt from lower convective zone (LCZ) and NCZ. Lower zone absorbs the solar energy, and it escapes only by conduction. Moderately low thermal conductivity is observed, and if the salinity gradient has less thickness, heat moves from lower zone to upper zone. This makes solar ponds to act as long term thermal storage systems and as efficient solar collectors. [212]. Solar ponds provide heat around temperature 50°C and 90°C and also have low cost. They can also be included into salinity mitigation schemes [213]. Table 13.2 presents a summary of some studies done on applications of nanofluids in solar collectors.

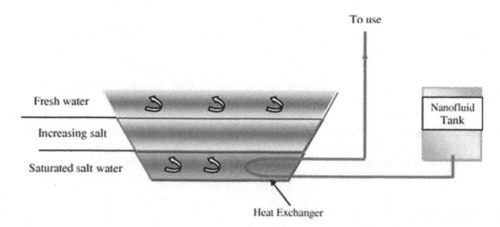

FIGURE 13.29 Experimental set up proposed for using nanofluids in solar ponds. (From Mahian, O. et al., *Int. J. Heat Mass Trans.*, 57, 582–594, 2013.)

TABLE 13.2

Summary of Some Studies Done on Applications of Nanofluids in Solar Collectors

Type of Solar Collector	Type of Nanofluid	Results	References
Flat-plate	Al_2O_3/water	The efficiency is enhanced by 28.3%. Using surfactant, it increased efficiency by 15.63%.	[214]
Flat-plate	Al_2O_3/water	25% increase in convective heat transfer coefficient.	[108]
Flat-plate	Al_2O_3/water	When the concentration was increased and nanoparticles size was decreased, the thermal conductivity was enhanced. 24.1% was the optimal efficiency.	[215]
Flat-plate	TiO_2/water	Increase in efficiency by 2.6%–7%.	[216]
Flat-plate	Cu/water	23.83% increase in efficiency.	[217]
Flat-plate	Cu/water	24% increase in efficiency.	[218]
Flat-plate	CuO/water	21.8% increase in efficiency.	[219]
Flat-plate	Cu synthesized/ ethylene glycol	Decrease in the volume flow rate decreases the efficiency. 37.5% enhancement of efficiency using nanofluids.	[111]
Flat-plate	SiO_2/ethylene glycol-water	8% efficiency enhancement by using nanofluids.	[94]
Flat-plate	Cu/water, SiO_2/ water, TiO_2/ water, and Al_2O_3/water	For obtaining maximum efficiency CuO/Water must be chosen.	[107]
Flat-plate	Cu/water, SiO_2/ water, TiO_2/ water, and Al_2O_3/water	SiO_2/water has lowest heat transfer coefficient, while Cu, TiO_2, and Al_2O_3 water-based have higher heat transfer coefficient than water.	[220]
Evacuated tube	Multi-walled carbon nanotube/ water	4% increase in the efficiency by using nanofluids.	[221]
Evacuated tube	Single-walled carbon nanotubes/ water	At 0.5 vol.% of nanofluid, the efficiency was 48.57%, and at 0.2 vol.% of nanofluid efficiency was 93.43%.	[222]
Parabolic	Carbon nanotube/oil	When 0.2% and 0.3% carbon nanotube/oil nanofluid is used, the efficiency is increased by 4%–5% and 5%–7%, respectively.	[223]
Parabolic	Cu/H_2O	At a mass concentration of 0.02%, highest efficiency of 93.4 was achieved.	[224]
Parabolic (direct absorption)	CuO/water	Enhancement of thermal efficiency from 18% to 52% can be obtained by using volume fraction from 0.02% to 0.008%.	[225]
Parabolic	Al_2O_3/synthetic oil	Using nanofluid with 0.05% volume fraction instead of oil, with the absorber plate bending decreased from 2.11 mm. to 0.54 mm.	[226]
Parabolic trough water heating system	Al_2O_3/water	Using a nanofluid-based parabolic trough for water heating application reduction of 2200 kg of CO_2/household/year is possible.	[227]
Parabolic trough solar collector	CNT/oil	Performance can be increased using 0.2% and 0.3% CNT/Oil nanofluids.	[223]
Parabolic trough solar collector	Al_2O_3/synthetic oil	At $\varphi = 5\%$ and $T = 300$ K, 14% average heat transfer coefficient enhances.	[228]

13.5 Challenges in Application of Nanofluids in Solar Thermal Systems

Many nanofluids have been developed, but the nanofluids with properties that can fulfill the practical engineering requirements have not been developed yet. For the scientific and practical applications, stability of nanofluids is very important. To date, the nanofluids which have been studied, their stability is not confirmed yet and more theoretical and experimental work is required to improve the stability of nanofluids. The factors affecting the thermal conductivity needs to be investigated properly. The mechanism which is affecting the thermal conductivity needs to be understood well. Thermal conductivity enhancement is abnormal or need not be understood properly [12]. Investigations on utilizing hybrid nanofluids in applications of solar energy are still underway. Phase change materials with different nanofluids must be used to increase storage [1].

13.6 Summary

Due to the increasing demand of energy among the human race, there is always a tremendous rush toward developing newer technologies to harness the energy from non-conventional energy sources. Nanofluids can prove as a better choice for absorbing heat from the sun and carry it through various solar thermal systems due to their superior properties. There have been already many research on the performance of nanofluids in the field of solar thermal energy absorbance. Specifically, the direct solar absorber systems have dominated the applications of nanofluids among other solar application.

ACKNOWLEDGMENT

This work was supported by the Science & Engineering Research Board (SERB), Department of Science and Technology (Government of India) [Start Up Research Grant (Young Scientists), Sanction order no. YSS/2014/000889, 2015].

REFERENCES

1. Elsheikh, A.H., Sharshir, S.W., Mostafa, M.E., Essa, F.A. and Ali, M.K.A. (2018). Application of nanofluids in solar energy. *Renewable and Sustainable Energy Reviews* 82, 3483–3502.
2. Wang, X., He, Y., Cheng, G., Shi, L., Lui X. and Zhu. J. (2006). Direct vapour generation through localized solar heating via carbon nanotube nanofluid. *Solar Energy Materials and Solar Cells* 130, 176–183.
3. Nagarajan, P.K., Subramani, J., Suyambazhahan. S. and Sathyamurthy. R. (2014). Nanofluids for solar collector applications: A review. *The 6th International Conference on Applied Energy* 61, 2416–2434.
4. Maxwell, J.C. (1881). *A Treatise On Electricity and Magnetism*, 2nd ed., Vol. 1, Clarendon Press, Oxford, UK, pp. 1–538.
5. Das, SK, Choi, S.U.S. and Patel, H.E. (2006). Heat transfer in nanofluids: A review. *Heat Transfer Engineering* 27, 3–19.
6. Choi, S.U.S. and Eastman, J.A. (1995). Enhancing thermal conductivity of fluids with nanoparticles. *ASME International Mechanical Engineering Congress Exposition* 66, 99–105.
7. Wang, H., Yang, W., Cheng, L., Guan, C. and Yan, H. (2018). Chinese ink: High performance nanofluids for solar energy. *Solar Energy Materials and Solar Cells* 176, 374–380.
8. Reddy, K.S., Kamnapure, N.R. and Srivastava, S. (2017). Nanofluid and nanocomposite applications in solar energy conversion systems for performance enhancement: A review. *International Journal of Low-Carbon Technologies* 12, 1–23.
9. Chang, H. and Wu, H. (2013). Graphene-based nanocomposite: Preparation, functionalization, and energy and environmental applications. *Energy & Environmental Science* 6, 3483–3507.
10. Wang, Y. and Xia, Y. (2004). Bottom-up top-down approaches to the synthesis of monodispersed spherical colloids of low melting-point metals. *Nano Letters* 4, 2047–2050.

11. Akilu, S., Baheta, A.T., Said, M.A.M., Minea, A.A. and Sharma, K.V. (2018). Properties of glycerol and ethylene glycol based SiO$_2$-CuO/C hybrid nanofluid for enhanced solar energy transport. *Solar Energy Materials and Solar Cells* 179, 118–128.

12. Li, Y., Zhou, J., Tung, S., Schneider, E. and Xi, S. (2009). A review on development of nanofluid preparation and characterization. *Powder Technology* 196, 89–101.

13. Sidik, N.A.C., Mohammed, H.A., Alawi, O.A. and Samion, S. (2014). A review on preparation methods and challenges of nanofluids. *International Communications in Heat and Mass Transfer* 54, 115–125.

14. Yu, W. and Xie, H. (2001). A review on nanofluids: Preparation, stability, mechanisms, and applications. *Journal of Nanomaterial's* 2012, 1–17.

15. Eastman, J.A. and Choi, S.U.S. (2001). Anomalously increased effective thermal conductivities of ethylene glycol-based nanofluids containing copper nanoparticles. *Applied Physics Letters* 78, 718–720.

16. Liu, M.S., Lin, M.C.C., Tsai, C.Y. and Wang, C.C. (2006). Enhancement of thermal conductivity with Cu for nanofluids using chemical reduction method. *International Journal of Heat Mass Transfer* 49, 3028–3033.

17. Xuan, Y. and Li, Q. (2000). Heat transfer enhancement of nanofluids. *International Journal of Heat Mass Transfer* 21, 58–64.

18. Hong, T.K., Yang, H.S. and Choi, C.I. (2005). Study of the enhanced thermal conductivity of Fe nanofluid. *Journal of Applied Physics* 97(064311), 1–4.

19. Patel, H.E., Das, S.K., Sundararagan, T., Nair, A.S., Geoge, B. and Pradeep, T. (2003). Thermal conductivities of naked and chemical effects. *Applied Physics Letters* 83, 2931–2933.

20. Putnam, S.A., Cahill, D.G. and Braun, P.V. (2006). Thermal conductivity of nanoparticle suspensions. *Applied Physics* 99(084308), 1–6.

21. Zhu, H., Zhang, C. and Liu, S. (2006). Effects of nanoparticle clustering and alignment on thermal conductivities of Fe$_3$O$_4$ aqueous nanofluids. *Applied Physics Letters* 89 (23123), 1–3.

22. Murshed, S.M.S., Leong, K.C. and Yang, C. (2005). Enhanced thermal conductivity of TiO$_2$-water based nanofluid. *International Journal of Thermal Sciences* 44, 367–373.

23. Xie, H., Wang, J., Xi, T., Liu, Y. and Ai, F. (2002). Thermal conductivity enhancement of suspensions containing nanosized alumina particle. *Journal of Applied Physics* 91, 4568–4572.

24. Zhang, X., Gu, H. and Fujii, M. (2006). Experimental study on the effective thermal conductivity and thermal diffusivity of nanofluid. *International Journal of Thermophysics* 27, 569–580.

25. Xie, H.Q. (2001). Study on the thermal conductivity of SiC nanofluids. *Journal of Chinese Ceramic Society* 29, 361–364 (in Chinese).

26. Liu, M.S., Lin, M.C., Huang, I.T. and Wang, C.C. (2005). Enhancement of thermal conductivity with carbon nanotube for nanofluids. *International Communications in Heat and Mass Transfer* 32, 1202–1210.

27. Choi, S.U.S., Zhang, Z.G., Yu, W., Lockwood, F.E. and Grulke, E.A. (2001). Anomalous thermal conductivity enhancement in nanotube suspensions. *Applied Physics Letters* 79, 2252–2254.

28. Xie, H., Lee, H., Youn, W. and Choi, M. (2003). Nanofluids containing multiwalled carbon nanotubes and their enhanced thermal conductivities. *Journal of Applied Physics* 954, 4967–4971.

29. Yang, B. and Han, Z.H. (2006). Thermal conductivity enhancement in water-in-FC72 nano emulsion fluids. *Applied Physics Letters* 88(261914), 1–3.

30. Lo, C.H., Tsung, T.T., Chen, L.C., Su, C.H. and Lin, H.M. (2005). Fabrication of copper oxide nanofluid using submerged arc nanoparticle synthesis system (SANSS). *Journal of Nanoparticle Research* 7, 313–320.

31. Hong, K.S., Hong, T.K. and Yang, H.S. (2006). Thermal conductivity of Fe nanofluids depending on the cluster size of nanoparticles. *Applied Physics Letters* 88(3), 031901.

32. Zhang, X., Gu, H. and Fujii, M. (2006) Effective thermal conductivity and thermal diffusivity of nanofluids containing spherical and cylindrical nanoparticles. *Journal of Applied Physics* 100(044325), 1–5.

33. Yu-Hua, L., Wei, Q. and Jian-Chao, F. (2008). Temperature dependence of thermal conductivity of nanofluids. *Chinese Physics Letters* 25, 3319–3322.

34. Yang, B., and Han, Z.H. (2006). Temperature-dependent thermal conductivity of nanorod-based nanofluids. *Applied Physics Letters* 89, 083111.

35. Yu, W. and Choi, S.U.S. (2003). The role of interfacial layers in the enhanced thermal of nanofluids: A renovated Maxwell model. *Nanoparticle Research* 5, 167–171.

36. Xie, H., Fujii, M. and Zhang, X. (2005). Effect of interfacial nanolayer on the effective thermal conductivity of nanoparticle-fluid mixture. *International Journal of Heat and Mass Transfer* 48, 2926–2932.

37. Leong, K.C., Yang, C. and Murshed, S.M.S. (2006). A model for the thermal conductivity of nanofluids—The effects of interfacial layer. *Journal of Heat Transfer* 125, 567–574.
38. Xue, L., Keblinski, P., Phillpot, S.R., Choi, S.U.S. and Eastman, J.A. (2004). Effect of liquid layering at the liquid–solid interface on thermal transport. *International Journal of Heat and Mass Transfer* 47, 4277–4284.
39. Saidur, R., Meng, T.C., Said, Z., Hasanuzzaman, M. and Kamyar, A. (2012). Evaluation of the effect of nanofluid-based absorbers on direct solar collector. *International Journal of Heat and Mass Transfer* 55, 5899–5907.
40. Kameya, Y. and Hanamura, K. (2011). Enhancement of solar radiation absorption using nanoparticle suspension. *Solar Energy* 85, 299–307.
41. Ahmad, S.H.A., Saidur, R., Mahbubul, I.M. and Al-Sulaiman, F.A. (2017) Optical properties of various nanofluids used in solar collector: A review. *Renewable Sustainable Energy Reviews* 73, 1014–1030.
42. Lee, B.J., Park, K., Walsh, T. and Xu, L. (2012). Radiative heat transfer analysis in plasmonic nanofluids for direct solar thermal absorption. *ASME Journal of Solar Energy Engineering* 134(2), 021009.
43. Taylor, R.A., Octanicar, T. and Rosengarten, G. (2012). Nanofluid-based optical filter optimization for PV/T systems. *Light: Science and Applications* 1(10), e34.
44. Chen, W., Zou, C., Li, X. and Li, L. (2017). Experimental investigation of Sic nanofluids for solar distillation system: Stability, optical properties and thermal conductivity with saline water-based fluid. *International Journal of Heat and Mass Transfer* 107, 264–270.
45. Kulkarni, D.P., Vajjha, R.S., Das, D.K. and Oliva, D. (2008). Application of aluminium oxide nanofluids in diesel electric generator as jacket water coolant. *Applied Thermal Engineering* 28, 1774–1781.
46. Pak, B.C. and Cho, Y.I. (1998). Hydrodynamic and heat transfer study of dispersed fluids with submicron metallic oxide particles. *Experimental Heat Transfer* 11, 151–170.
47. Fakoor Pakdaman, M., Akhavan-Behabadi, M.A. and Razi, P. (2010). An experimental investigation on thermo-physical properties and overall performance of MWCNT/heat transfer oil nanofluid flow inside vertical helically coiled tubes. *Experimental Thermal and Fluid Science* 40, 103–111.
48. Strandberg, R. and Das, D.K. (2010). Influence of temperature and properties variation on nanofluids in building heating. *Energy Conversation and Management* 51, 1381–1390.
49. Zhou, S.-Q. and Ni, R. (2008). Measurement of the specific heat capacity of water-based Al_2O_3 nanofluid. *Applied Physics Letters* 92, 093123.
50. Namburu, P.K., Kulkarni, D.P., Damdekar, A. and Das, D.K. (2007). Experimental investigation of viscosity and specific heat of silicon dioxide nanofluids. *Micro and Nano Letters IET* 2, 67–71.
51. Murshed, S.M.S. (2011). Determination of effective specific heat of nanofluids. *Journal of Experimental Nanoscience* 6(5), 539–546.
52. De Robertis, E., Cosme, E.H.H., Neves, R.S., Kuznetsov, A.Y., Campos, A.P.C., Landi, S.M. and Achete, C.A. (2012). Application of the modulated temperature differential scanning calorimetry technique for the determination of the specific heat of copper nanofluids. *Applied Thermal Engineering* 41, 10–17.
53. Saeedinia, M., Akhavan-Behabadi, M.A. and Nasr, M. (2012). Experimental study on heat transfer and pressure drop of nanofluid flow in a horizontal coiled wire inserted tube under constant heat flux. *Experimental Thermal Fluid Science* 36, 158–168.
54. Kumaresan, V. and Velraj, R. (2012). Experimental investigation of the thermo-physical properties of water-ethylene glycol mixture based CNT nanofluids. *Thermochimica Acta* 545, 180–186.
55. Mansour, R.B., Galanis, N. and Nguyen, C.T. (2007). Effect of uncertainties in physical properties on forced convection heat transfer with nanofluids. *Applied Thermal Engineering* 27, 240–249.
56. Sekhar, Y.R. and Sharma, K.V. (2013). Study of viscosity and specific heat capacity characteristics of water-based Al_2O_3 nanofluids at low particle concentrations. *Journal of Experimental Nanoscience* 1–17.
57. Teng, T.P. and Hung, Y.H. (2012). Estimation and experimental study of the density and specific heat for alumina nanofluid. *Journal of Experimental Nanoscience* 9, 1–12.
58. Sadri, R., Hosseini, M., Kazi, S.N., Bagheri, S., Ahmed, S.M., Ahmadi, G., Zubir, N., Sayuti, M. and Dahari, M. (2017). Study of environmentally friendly and facile functionalization of graphene nanoplatelet and its application in convective heat transfer. *Energy Conversation Management* 150, 26–36.
59. Yarmand, H., Zulkifli, N.W.B.M., Gharehkhani, S., Shirazi, S.F.S., Alrashed, A.A.A.A., Ali, M.A.B., Dahari, M. and Kazi, S.N. (2017). Convective heat transfer enhancement with graphene nanoplatelet/platinum hybrid nanofluid. *International Communications in Heat and Mass Transfer* 88, 120–125.

60. Sadri, R., Hosseini, M., Kazi, S.N., Bagheri, S., Abdelrazek, A.H., Ahmadi, G., Zubir, N., Ahmad, R. and Abidin, N.I.Z. (2018). A facile, bio-based, novel approach for synthesis of covalently functionalized graphene nanoplatelet nano-coolants toward improved thermo-physical and heat transfer properties. *Journal of Colloid and Interface Science* 509, 140–152.

61. Prasher, R. and Wang, D.D. (2006). Measurements of nanofluid viscosity and its implications for thermal applications. *Applied Physics Letters* 89(133108), 1–3.

62. Li, J.M., Li, Z.L. and Wang, B.X. (2002). Experimental viscosity measurements for copper oxide nanoparticle suspensions. *Tsinghua Science and Technology* 7, 198–201.

63. Einstein, A. (1906). Eine new Bestinmmung der Molekul- dimension (a new determination of the molecular dimensions). *Annalen der Physik* 19(2), 289–306.

64. Brinkman, H.C. (1952). The viscosity of concentrated suspensions and solution. *Journal of Chemical Physics* 20, 571–581.

65. Batchelor, G.K. (1977). The effect of Brownian motion on the bulk stress in a suspension of spherical particles. *Journal of Fluid Mechanics* 83, 97–117.

66. Lundgren, T.S. (1972). Slow flow through stationary random beds and suspensions of spheres. *Journal of Fluid Mechanics* 51, 273–299.

67. Peng, X. and Yu, X. (2007). Influence factors on suspension stability of nanofluid. *Journal of Zhejiang University: Engineering Science* 41, 577–580.

68. Hwang, Y., Lee, J.K., Lee, C.H., Jung, Y.M., Cheong, S.I., Lee, C.G., Ku, B.C. and Jang, S.P. (2007). Stability and thermal conductivity characteristics of nanofluids. *Thermochimica Acta* 455, 70–74.

69. Wang, B., Li, C. and Peng, X. (2003). Stability of nano-particle suspensions. *Journal of Basic Science and Engineering* 11, 169–173 (in Chinese).

70. Wang, B., Li, C. and Peng, X. (2003). Research on stability of nano-particle suspension. *Journal of University Shanghai Science Technology* 25, 209–212.

71. Liu, D.M. (2000). Influence of dispersant on powders dispersion and properties of Zirconia green compacts [J]. *Ceramics International* 26(3), 279–287.

72. Zhu, D., Li, X., Wang, N., Gao, J. and Li, H. (2009). Dispersion behaviour and thermal conductivity characteristics of Al_2O_3-H_2O nanofluids. *Current Applied Physics* 9, 131–139.

73. Gorji, T.B. and Ranjbar, A.A. (2017). A review on optical properties and application of nanofluids in direct absorption solar collectors (DASC). *Renewable Sustainable Energy Reviews* 72, 10–32.

74. Minardi, J.E. and Chuang, H.N. (1975). Performance of a "black" liquid flat-plate solar collector. *Solar Energy* 17, 179–183.

75. Bertocchi, R., Karni, J. and Kribus, A. (2004). Experimental evaluation of a non-isothermal high temperature solar particle receiver. *Energy* 29, 687–700.

76. Viskanta, R. (1987). Direct absorption solar radiation collection systems. In: Yuncu, H., Paykoc, E., Yener, Y., editors. *Solar Energy Utilization*, Dordrecht, the Netherlands: Springer, pp. 334–360.

77. Chen, M., He, Y., Zhu, J. and Wen, D. (2016). Investigating the collector efficiency of silver nanofluids based direct absorption solar collectors. *Applied Energy* 181, 65–74.

78. Zhuocheng, J.L., Zhang, Y.L., Fang, X. and Zhang, Z. (2015). A combined numerical and experimental study on graphene/ionic liquid nanofluid based direct absorption solar collector. *Solar Energy Materials and Solar Cells* 136, 177–186.

79. Singh, A. (2008). Thermal conductivity of nanofluids. *Defence Science Journal* 58, 600–607.

80. Lee, S., Choi, S.U.S., Li, S. and Eastman, J.A. (1999). Measuring thermal conductivity of fluids containing oxide nanoparticles. *ASME Journal of Heat Transfer* 121, 280–289.

81. Eastman, J.A., Choi, S.U.S., Li, S., Yu, W. and Thomson, L.J. (2001). Anomalously increased effective thermal conductivities of ethylene glycol based nanofluids containing copper nanoparticles. *Applied Physics Letters* 78, 718–720.

82. Octanicar, T.P., Phelan P.E. and Golden, J.S. (2009). Optical properties of liquids for direct absorption solar thermal energy systems. *Journal of Solar Energy* 83, 969–977.

83. Taylor, R.A., Phelan, P.E., Octanicar, T.P., Adrian, R. and Prasher, R. (2011). Nanofluid optical property characterisation: Towards efficient direct absorption solar collectors. *Nanoscale Research Letters* 6, 225.

84. Drotning, W.D. (1978). Optical properties of solar-absorbing oxide particles suspended in a molten salt heat transfer fluid. *Solar Energy* 20, 313–319.

85. Palambo, N. and Park, K. (2011). Investigation of dynamic near-field radiation b/w quantum dots and plasmonic nanoparticles for effective tailoring of the solar spectrum. In *Proceedings of the ASME 2011 International Mechanical Engineering Congress and Exposition*, Denver, CO, November 11–17, pp. 1–5.

86. Al-Madani, H. (2006). The performance of a cylindrical solar water heater. *Renewable Energy* 31, 1751–1763.

87. Lee, D.W. and Sharma, A. (2007). Thermal performance of the active and passive water heating systems based on annual operation. *Solar Energy* 81, 207–215.

88. Eastop, T.D. and McConkey, A. (1977). *Applied Thermodynamics for Engineering Technologists*. Longman: London, UK.

89. Taylor, R.A., Phelan, P.E., Octanicar, T.P., Walker, C.A., Nguyen, M., Trimble, S. and Prasher, R. (2011). Applicability of nanofluids in high flux solar collectors. *Journal of Renewable and Sustainable Energy* 3, 023104.

90. Li, Y., Xie, H., Yu, W. and Li, J. (2011). Investigation on heat transfer performances of nanofluids in solar collector. *Material Science Forum* 694, 33–36.

91. Yousefi, T., Veysi, F., Shojaeizadeh, E. and Zinadini, S. (2012). An experimental investigation on the effect of Al_2O_3-H_2O nanofluid on the efficiency of flat plate solar collectors. *Renewable Energy* 39, 293–298.

92. Khullar, V., Tyagi, H., Phelan, P.E., Octanicar, T.P. and Singh Taylor, R.A. (2012). Solar energy harvesting using a nanofluids-based concentrating solar collector. In *Proceedings of MNHMT 2012 3rd Micro/Nanoscale Heat & Mass Transfer International Conference*, March 3–6, Atlanta, GA.

93. Octanicar, T.P., Phelan, P.E., Prasher, R.S., Rosengarten, G. and Taylor, R.A. (2012). Nanofluid based direct absorption solar collector. *Journal of Renewable and Sustainable Energy* 2, 033102.

94. Saleh, S., Ali, K., Hamid, N., Omid, M. and Somchai, W. (2015). Experimental investigation on the thermal efficiency and performance characteristics of a flat plate solar collector using SiO_2/EG-water nanofluids. *International Communications in Heat and Mass Transfer* 65, 71–75.

95. Kalogirou, S.A. (2004). Solar thermal collectors and applications. *Process in Energy and Combustion Science* 30(3), 231–295.

96. Tian, Y. and Zhao, C.Y. (2012). A review of solar collector and thermal energy storage in solar thermal applications. *Applied Energy* 104, 538–553.

97. Moghadam A.J., Farzane-Gord, M., Sajadi, M. and Hoseyn-Zadeh, M. (2014). Effects of CuO/water nanofluid on the efficiency of a flat-plate collector. *Experimental Thermal and Fluid Science* 58, 9–14.

98. Sozen, A., Menlik, T. and Unvar S. (2008) Determination of efficiency of flat-plate solar collectors using neural approach. *Expert Systems with Applications* 35, 1533–1539.

99. Sukhatme, S.P. (1993). *Solar Energy*. New York: McGraw-Hill, pp. 83–139.

100. Francia, G. (1961). A new collector of solar radiant energy, In *Proceedings of the United Nations Conference On New Sources of Energy*, paper no. F6. Rome, Italy, p. 572.

101. Slaman, M. and Griessen, R. (2009). Solar collector overheating protection. *Solar Energy* 83, 982–987.

102. Kumar, K.R. and Reddy, K.S. (2009). Thermal analysis of solar parabolic trough with porous disc receiver. *Applied Energy* 86, 1804–1812.

103. Sopian, K., Alghoul, M.A., Alfegi, E.M., Sulaiman, M.Y. and Musa, E.A. (2009). Evaluation of thermal efficiency of double-pass solar collector with porous–nonporous media. *Renewable Energy* 34, 640–645.

104. Yousefi, T., Veysi, F., Shojaeizadeh, E. and Zinadini, S. (2012). An experimental investigation on the effect of $MWCNT$-H_2O nanofluid on the efficiency of flat plate solar collector. *Experimental Thermal and Fluid Science* 39, 207–212.

105. Yousefi, T., Veysi, F., Shojaeizadeh, E. and Zinadini, S. (2012). An experimental investigation on the effect of PH variation of $MWCNT$-H_2O nanofluid on the efficiency of a flat-plate solar collector. *Solar Energy* 86, 771–779.

106. Jamal-Abad, M.T., Zamzamian, A., Imami, E. and Mansouri, M. (2013). Experimental study of the performance of a flat-plate collector using Cu-water nanofluid. *Journal of Thermophysics and Heat Transfer* 27, 756–760.

107. Faizal, M., Saidur, R., Mekhilef, S. and Alim, M.A. (2013). Energy, economic and environmental analysis of metal oxides nanofluid for flat-plate solar collector. *Energy Conversation Management* 76, 162–168.

108. Colangelo, G., Favale, E., de Risi, A. and Laforgia, D. (2013). A new solution for reduced sedimentation flat panel solar thermal collector using nanofluid. *Applied Energy* 111, 80–93.
109. Parvin, S., Nasrin, R. and Alim, M.A. (2014). Heat transfer and entropy generation through nanofluid filled direct absorption solar collector. *International Journal of Heat and Mass Transfer* 50, 117–127.
110. Chaji, H., Ajabshirchi, Y., Esmaeilzadeh, E., Heris, S.Z., Hedayatizadeh, M. and Kahani, M. (2013). Experimental study on thermal efficiency of flat plate solar collector using TiO₂/water nanofluid. *Modern Applied Science* 7(10), 60–69.
111. Zamzamian, A., KeyanpourRad, M., KianiNeyestani, M. and Jamal-Abad, M.T. (2014). An experimental study on the effect of Cu-synthesized/EG nanofluid on the efficiency of flat-plate solar collectors. *Renewable Energy* 71, 658–664.
112. Mohadam, A.J., Farzane-Gord, M., Sajadi, M. and Hoseynzadeh, M. (2014). Effects of CuO/water nanofluid on the efficiency of a flat plate solar collector. *Experimental Thermal and Fluid Science* 58, 9–14.
113. Hendrie, S.D., Raghuraman P. and Cox, C.H. (1981). 3, liquid photovoltaic/thermal collectors for residential applications. *Proceedings of 15th IEEE Photovoltaic Specialists Conference*, Orlando, FL (May).
114. Natarajan, E. and Sathish, R. (2009). Role of nanofluids in solar water heater. *International Journal of Advanced Manufacturing Technology*. doi:10.1007/s00170–008–1876–8.
115. Tyagi, H., Phelan, P. and Prasher, R. (2009). Predicted efficiency of a low-temperature nanofluid based direct absorption solar collection. *Journal of Solar Energy Engineering* 31, 0410041.
116. Octanicar, T. and Golden, J. (2009). Comparative environmental and economic analysis of conventional and nanofluid solar hot water technologies. *Environmental Science and Technology* 43, 6082–6087.
117. Octanicar, T., Phelan, P.E., Prasher, R.S., Rosengarten, G. and Taylor, R.A. (2010). Nanofluid-based direct absorption solar collector. *Journal of Renewable Sustainable Energy* 2, 033102.
118. Sani, E. (2011). Potential of carbon nanohorn-based suspensions for solar thermal collectors. *Solar Energy Materials and Solar Cells* 95, 2994–3000.
119. Mercatalli, L. (2011). Absorption and scattering properties of carbon nanohorn-based nanofluids for direct sunlight absorbers. *Nanoscale Research Letters* 6, 282.
120. Wang, H. and Zeng, S. (2013). Experimental investigation on photo thermal properties of nanofluids for direct absorption solar thermal energy systems. *Energy Conversation and Management* 73, 150–157.
121. Liu, Z.H., Hu, R.H., Zhao, F. and Xiao, H.S. (2013). Thermal performance of an open thermosyphon using nanofluid for evacuated tubular high temperature air solar collector. *Energy Conversation and Management* 20173, 135–143.
122. Javadi, F.S., Saidur, R. and Kamalisarvestani, M. (2013). Investigating performance improvement of solar collectors by using nanofluids. *Renewable Sustainable Energy Reviews* 28, 232–245.
123. Zhang, H., Chen, H.J., Du, X.Z. and Wen, D.S. (2014). Photo thermal conversion characteristics of gold nanoparticle dispersions. *Solar Energy* 100, 141–147.
124. Tyagi, H., Phelan, P. and Prasher, R. (2009). Predicted efficiency of a low-temperature nanofluid-based direct absorption solar collector. *Journal of Solar Energy Engineering* 131, 041004.
125. Sani, E., Barison, S., Pagura, C., Mercatelli, L., Sansoni, P. and Fontani, D. (2010). Carbon nanohorns-based nanofluids as direct sunlight absorbers. *Optics Express* 18, 5179–5187.
126. Han, D., Meng, Z., Wu, D., Zhang, C. and Zhu, H. (2011). Thermal properties of carbon black aqueous nanofluids for solar absorption. *Nanoscale Research Letters* 6, 457.
127. Yousefi, T., Veysi, F., Shojaeizadeh, E. and Zinadini, S. (2012). Experimental investigation on the effect of Al₂O₃-H₂O nanofluid on the efficiency of flat-plate solar collectors. *Renewable Energy* 39, 293–298.
128. Yousefi, T., Veysi, F., Shojaeizadeh, E. and Zinadini, S. (2012). Experimental investigation on the effect PH variation of MWCNT-H₂O nanofluid on the efficiency of a flat-plate collector. *Solar Energy* 86, 771–779.
129. Zambo, L.E. and Del, C.D. (2010). Experimental analysis of thermal performance of flat plate and evacuated tube solar collectors in stationary standard and daily conditions. *Solar Energy* 84, 1382–1396.
130. Ayompe, L.M., Duffy, A., Mc Keever, M., Conlon, M. and McCormack, S.J. (2011). Comparative field performance study of flat plate and heat pipe evacuated tube collectors (ETCs) for domestic water heating systems in a temperature climate. *Energy* 36, 3370–3378.
131. Norton, B. (2014). Flat-plate and evacuated tube collectors. In *Harnessing Solar Heat*, Dordrecht, the Netherlands: Springer, pp. 91–113.

132. Muhammad, M.J., Muhammad, I.A., Sidik, N.A.C. and Muhammad Yazid, M.N.A.W.M. (2016). Thermal performance enhancement of flat-plate evacuated tube solar collectors using nanofluid: A review. *International Communications in Heat and Mass Transfer* 76, 6–15.

133. Bracamonte, J., Parada, J., Dimas, J. and Baritto, M. (2015). Effect of the collector tilt angle on thermal efficiency and stratification of passive water in glass evacuated tube solar water heater. *Applied Energy* 155, 648–659.

134. Zhang, M., Miao, I., Kang, Y.P., Tanemura, S., Fisher, C.A.J. and Xu, G. (2013). Efficient, low-cost solar thermoelectric cogenerators comprising evacuated tubular solar collectors and thermoelectric modules. *Applied Energy* 109, 51–59.

135. Kalogirou, S. (2003). The potential of solar industrial process heat applications. *Applied Energy* 76, 337–361.

136. Muhammad, M.J., Muhammad, I.A., Sidik, N.A.C., Muhammad Yazid, M.N.A.W., Mamat, R. and Najafi, G. (2016). The use of nanofluids for enhancing the thermal performance of stationary solar collectors: A review. *Renewable Sustainable Energy Reviews* 63, 226–236.

137. Chun, W., Kang, Y.H., Kwak, H.Y. and Lee, Y.S. (1999). An experimental study of the utilization of heat pipes for solar water heaters. *Applied Thermal Engineering* 19, 807–817.

138. Riffat, S.B., Zhao, X. and Doherty, P.S. (2005). Developing a theoretical model to investigate thermal performance of a thin membrane heat pipe solar collector. *Applied Thermal Engineering* 25, 899–915.

139. Khandekar, S., Joshi, Y.M. and Mehta, B. (2008). Thermal performance of closed two-phase thermosyphon using nanofluids. *International Journal of Thermal Science* 47, 659–667.

140. Liu, Z.H., Yang, X.F., Wang, G.S. and Guo, G.I. (2010). Influence of carbon nanotube suspension on the thermal performance of a miniature thermosyphon. *International Journal of Heat and Mass Transfer* 53, 1914–1920.

141. Liu, Z.H. and Li, Y.Y. (2012). A new frontier of nanofluids research–application of nanofluids in heat pipes. *International Journal of Heat and Mass Transfer* 55, 6786–6797.

142. Liu, Z.H., Li, Y.Y. and Bao, R. (2011). Application of aqueous nanofluids in a horizontal mesh heat pipe. *Energy Conversation Management* 52, 292–300.

143. Khonkar, H.E.I. and Sayign, A.A.M. (1995). Optimization of the tubular absorber using a compound parabolic concentrator. *Renewable Energy* 6, 17–21.

144. Eames, P.C. and Norton, B. (1995). Thermal and optical cosequences of the introduction of baffles into compound parabolic concentrating solar energy collector cavities. *Solar Energy* 55, 139–150.

145. Tripanagnostopoulos, Y., Yianoulis, P., Papaefthimiou, S., Souliotis, M. and Nousia, T. (1999). Cost effective asymmetric CPC solar collectors. *Renewable Energy* 16, 628–631.

146. Oommen, P. and Jayaraman, S. (2001). Development and performance analysis of compound parabolic solar concentrators with reduced gap losses-oversized reflector. *Energy Conversation Management* 42, 1379–1399.

147. Sandeep, H.M. and Arunachala, U.C. (2017). Solar parabolic trough collectors: A review on heat transfer augmentation techniques. *Renewable Sustainable Energy Reviews* 69, 1218–1231.

148. Giglio, A., Lanzini, A., Leone, P., Rodriguez Garcia, M.M. and Zarza Moya, E. (2017). Direct steam generation in parabolic-trough collectors: A review about the technology and a thermo-economic analysis of a hybrid system. *Renewable Sustainable Energy Reviews* 74, 453–473.

149. Baharoon, D.A., Rahaman, H.A., Omar, W.Z.W. and Fadhl, S.O. (2015). Historical development of concentrating solar power technologies to generate clean electricity efficiently: A review. *Renewable Sustainable Energy Reviews* 41, 996–1027.

150. Salgado Conrado, L., Rodriguez-Pulido, A. and Calderon, G. (2017). Thermal performance of parabolic trough solar collectors. *Renewable Energy Reviews* 67, 1345–1359.

151. Cheng, Z.D., He, Y.L., Du, B.C., Wang, K. and Liang, Q. (2015). Geometric optimization on optical performance of parabolic trough solar collector system using particle swarm optimization algorithm. *Applied Energy* 148, 282–293.

152. Liang, H., You, S. and Zhang, H. (2015). Comparison of different heat transfer models for parabolic trough solar collectors. *Applied Energy* 148, 105–114.

153. Hachicha, A.A., Rodriguez, I. and Oliva, A. (2014). Wind speed effect on the flow field and heat transfer around a parabolic trough solar collector. *Applied Energy* 130, 200–211.

154. Cabrera, F.J., Fernandez-Garcia, A., Silva, R.M.P. and Perez-Garcia, M. (2013). Use of parabolic trough solar collectors for solar refrigeration and air-conditioning applications. *Renewable and Sustainable Energy Reviews* 20, 103–118.

155. Bakos, G.C. (2006). Design and construction of a two-axis sun tracking system for Parabolic Trough Collector (PTC) efficiency improvement. *Renewable Energy* 31, 2411–2421.

156. Mousdzadeh, H., Keyhani, A., Javadi, A., Mobli, H., Abrinia, K. and Sharifi, A. (2009). A review of principle and sun-tracking methods for maximizing solar systems output. *Renewable and Sustainable Energy Reviews* 13, 1800–1818.

157. Kalogirou, S.A. (2002). Parabolic trough collector for industrial process heat in Cyprus. *Energy* 27, 813–830.

158. Price, H., Lupfert, E., Kearney, D., Zarza, E., Cohen, G. and Gee, R. (2002). Advances in parabolic trough solar power technology. *Journal of Solar Energy Engineering Transactions of the ASME* 124, 109–125.

159. Lupfert, E., Zarza-Moya, E., Geyer, M., Nava, P., Lagenkamp, J., Schiel, W., Esteban, A., Osuna, R. and Mandelberg, E. (2003). A. Eurotrough collector qualification complete-performance test results from PSA. In *Proceedings of the ISES Solar World Congress, 2003, Solar Energy for a Sustainable Future*, Goteborg, Sweden.

160. Castaneda, N., Vazquez, J., Domingo, M., Fernandez, A. and Leon, J. (2006). SENER parabolic trough collector design and testing. In *Proceedings of the 13th Solar PACES International Symposium on Concentrated Solar Power and Chemical Energy Technologies*, Seville, Spain.

161. Zarza, E., Valenzuela, L., Leon, J., Hennecke, K., Eck, M. and Weyers, H.D. (2004). Direct steam generation in parabolic troughs: Final results and conclusions of the DISS project. *Energy* 29, 635–644.

162. Collins, T.P. (2000). S.A. Parabolic trough solar water heating: Renewable technology for reducing water heating costs. *Federal Technology Alert. Washington, DC: U.S. Department of Energy (DOE) by the National Renewable Energy Laboratory, a DOE National Laboratory.*

163. Kimura, D.T. (2012). Commercial and industrial thermal applications of Micra-CSP. *NSTI-Nanotechnology*, 451–454.

164. SRCC, The solar rating and certification corporation collector database, 2012, http://www.solar-rating.org.

165. Weiss, W. and Rommel, M. (2008). Process heat collectors. State of art within Task 33/IV. IEA SHC-Task 33 and Solar PACES Task IV: Solar heat for industrial processes. Gleisdorf (Austria). *International Energy Agency. AEE INTEC.*

166. Millioud, A.D. (2008). Novel parabolic trough collector for roof mounted solar cooling applications. In *1st International Congress on Heating, Cooling and Buildings Eurosun*, Lisbon, Portugal.

167. Mwesigye, A., Huan, Z. and Meyer, J.P. (2016). Thermal performance and entropy generation analysis of a high concentration ratio parabolic trough solar collector with Cu-Therminol® VP-1 nanofluid. *Energy Conversation Management* 120, 449–465.

168. Daly, J.C. (1979). Solar concentrator flux distributions using backward ray tracing. *Applied Optics* 18(15), 2696–2699.

169. Riveros, H.G. and Oliva, A.I. (1986). Graphical analysis of sun concentrating collectors. *Solar Energy* 36, 313–322.

170. Shuai, Y., Xia, X.L. and Tan, H.P. (2008). Radiation performance of dish solar concentrator/cavity receiver systems. *Solar Energy* 82, 13–21.

171. He, Y.L., Xiao, J., Cheng, Z.D. and Tao, Y.B. (2010). A MCRT and FVM coupled simulation method for energy conversion process in parabolic trough solar collector. *Intentional Communications in Heat and Mass Transfer* 73, 782–787.

172. Odeh, S.D., Morrison, G.L. and Behnia, M. (1998). Modelling of parabolic trough direct steam generation solar collectors. *Solar Energy* 62, 395–406.

173. Hou, Z., Zheng, D., Jin, H. and Sui, J. (2007). Performance analysis of non-isothermal solar reactors for methanol decomposition. *Solar Energy* 81, 415–423.

174. Martinez, I. and Almaza, R. (2007). Experimental and theoretical analysis of annular two-phase flow regimen in direct steam generation of a low-power system. *Solar Energy* 81, 216–226.

175. Eck, M., Steinman, W.D. and Rheinl, J. (2004). Maximum temperature difference in horizontal and tilted absorber pipes with direct steam generation. *Energy* 29, 665–676.

176. Eck, M. and Steinman, W.D. (2005). Modelling and design of direct solar steam generating collector fields. *Journal of Solar Energy Engineering* 127, 371–380.

177. Hang, Q., Jun, Z., Xiao, Y. and Junkui, C. (2008). Prospect of concentrating solar power in China the sustainable future. *Renewable and Sustainable Energy Reviews* 12, 2505–2514.

178. Li, J. (2009). Scaling up concentrating solar thermal technology in China. *Renewable and Sustainable Energy Reviews* 13, 2051–2060.

179. Fluri, T.P. (2009). The potential of concentrating solar power in South Africa. *Energy Policy* 37, 5075–5080.

180. Karathanassis, I.K., Papanicolaou, E., Belessiotis, V. and Bergeles, G.C. (2017). Design and experimental evaluation of a parabolic-trough concentrating photovoltaic/thermal (CPVT) system with high-efficiency cooling. *Renewable Energy* 101, 467–483.

181. Xu, Z. and Weinstreuer, C. (2014). Concentration photovoltaic-thermal energy co-generation system using nanofluids for cooling and heating. *Energy Conversation Management* 87, 504–512.

182. Nilsson, J., Hakansson, H. and Karlsson, B. (2007). Electrical and thermal characterization of a PV-CPC hybrid. *Solar Energy* 81, 917–928.

183. Assoa, Y.B., Menezo, C., Fraisse, G., Yezou, R. and Brau, J. (2007). Study of a new concept of photovoltaic–thermal hybrid collector. *Solar Energy* 81, 1132–1143.

184. Conventry, J. (2005). Performance of a concentrating photovoltaic/thermal solar collector. *Solar Energy* 78, 211–222.

185. Chaudhari, K.S. and Walke, P.V. (2014). Applications of nanofluid in solar energy: A review. *International Journal of Engineering Research and Technology* 3, 460–463.

186. Elmir, M., Mehdaoui, R. and Mojtabi, A. (2012). Numerical simulation of cooling a solar cell by forced convection in the presence of a cooling a solar cell by forced convection in the presence of a nanofluid. *Energy Procedia* 18, 594–603.

187. Chang, H., Kao, M.J., Chok, C., Chen, S.J., Chu, K.H. and Chen, C.C. (2011). Integration of CuO thin films and dye-sensitized solar cells for thermoelectric generators. *Current Applied Physics* 11, S19–S22.

188. Omer, S.A. and Infield, D.G. (1998). Design optimization of thermoelectric devices for solar power generation. *Solar Energy Materials and Solar Cells* 53, 67–82.

189. Huen, P. and Daound, W.A. (2017). Advances in hybrid solar photovoltaic and thermoelectric generators. *Renewable and Sustainable Energy Reviews* 72, 1295–1302.

190. Zheng, X.F., Liu, C.X., Yan, Y.Y. and Wang, Q. (2014). A review of thermoelectrics research—Recent developments and potentials for sustainable and renewable energy applications. *Renewable and Sustainable Energy Reviews* 32, 486–503.

191. Twaha, S., Zhu, J., Yan, Y. and Li, B. (2016). A comprehensive review of thermoelectric technology: Materials, applications, modelling and performance improvement. *Renewable and Sustainable Energy Reviews* 65, 698–726.

192. Stobart, R., Wijewardane, M.A. and Yang, Z. (2017). Comprehensive analysis of thermoelectric generations system for automotive applications. *Applied Thermal Engineering* 112, 1433–1444.

193. Huang, K., Li, B., Yan, Y., Li, Y., Twaha, S. and Zhu, J. (2017). A comprehensive study on a novel concentric cylindrical thermoelectric power generation system. *Applied Thermal Engineering* 117, 501–510.

194. Sun, X., Zhang, L. and Liao, S. (2017). Performance of thermoelectric solar system integrated with a gravity-assisted heat pipe for cooling electronics. *Applied Thermal Engineering* 116, 433–444.

195. Lee, H., Seshadri, C.R., Han, S.J. and Sampath, S. (2017). TiO_{2-x} based thermoelectric generators enabled by additive and layered manufacturing. *Applied Energy* 192, 24–32.

196. Benday, N.S., Dryden, D.M., Kornbluth, K. and Stroeve, P. (2017). A temperature variant method for performance modelling and economic analysis of thermoelectric generators: Linking material properties to real world conditions. *Applied Energy* 190, 764–771.

197. Ahammed, N., Asirvatham, L.G. and Wongwises, S. (2016). Thermoelectric cooling of electronic devices with nanofluid in a multiport minichannel heat exchanger. *Experimental Thermal Fluid Science* 74, 81–90.

198. Mohammadian, S.K. and Zhang, Y. (2014). Analysis of nanofluid effects on thermo electric cooling by micro-pin-fin heat exchangers. *Applied Thermal Engineering* 70, 282–290.

199. Ahammed, N., Asirvatham, L.G. and Wongwises, S. (2016). Entropy generation analysis of graphene-alumina hybrid nanofluid in a multiport minichannel heat exchanger coupled with thermo electric cooler. *International Journal of Heat and Mass Transfer* 103, 1084–1097.

200. Sharshir, S.W., Peng, G., Wu, L., Yang, N., Essa, FA. and Elsheikh, A.H. (2017). Enhancing the solar still performance using nanofluids and glass cover cooling: Experimental study. *Applied Thermal Engineering* 113, 684–693.

201. Sharshir, S.W., Elsheikh, A.H., Peng, G., Yang, N., EL-Samadony, M.O.A. and Kabeel, A.E. (2017). Thermal performance and exergy analysis of solar stills: A review. *Renewable and Sustainable Energy Reviews* 73, 521–544.

202. Xiao, G., Wang, X., Ni, M., Wang, F., Zhu, W. and Luo, Z. (2013). A review on solar stills for brine desalination. *Applied Energy* 103, 642–652.

203. Pandey, R., Tripathi, R. and Varshney, P.K. (2016). Current status of solar distillation: A review. *International Journal of Research in Applied, Natural and Social Sciences* 4, 37–48.

204. Sharshir, W., Peng, G., Wu, L., Essa, F.A., Kabeel, A.E. and Yang, N. (2017). The effects of flake graphite nanoparticles phase change material and film cooling on the solar still performance. *Applied Energy* 191, 358–366.

205. Kabeel, A.E., Omara, Z.M. and Essa, F.A. (2014). Improving the performance of solar still by using nanofluids and providing vacuum. *Energy Conversation Management* 86, 268–274.

206. Sahota, L. and Tiwari, G.N. (2016). Effects of nanofluids on the performance of passive double slope solar still: A comparative study using characteristic curve. *Desalination* 388, 9–21.

207. Sahota, L. and Tiwari, G.N. (2017). Analytical characteristic equation of nanofluid loaded active double slope solar still coupled with helically coiled heat exchanger. *Energy Conversation Management* 135, 308–326.

208. Mahian, O., Kianifar, A., Heris, S.Z., Wen, D., Sahin, A.Z. and Wongwises, S. (2017). Nanofluids effects on the evaporation rate in a solar still equipped with a heat exchanger. *Nano Energy* 36, 134–155.

209. Elango, T., Kannan, A. and Murugavel, K.K. (2015). Performance study on Single basin single slope solar still with different water nanofluids. *Desalination* 360, 45–51.

210. Nijmeh, S., Odeh, S. and Akash, B. (2005). Experimental and theoretical study of a single basin solar still in Jordan. *International Communications in Heat and Mass Transfer* 32, 565–572.

211. Mahian, O., Kianifar, A., Kalogirou, S.A., Pop, I. and Wongwises, S. (2013). A review of the applications of nanofluids in solar energy. *International Journal of Heat and Mass Transfer* 57, 582–594.

212. Akbarzaden, A., Andrews, J. and Golding, P. (2008). Solar ponds. In *Encylopedia of Life Support Systems* (*EOLSS*), developed under the Auspices of the UNESCO. EOLSS, Oxford, UK.

213. Akbarzaden, A., Andrews, J. and Golding, P. (2005). Integration of solar ponds in salinity mitigation schemes to produce low grade heat for industrial process heating, desalination and power. *ISES-2005 Solar World Congress*, Orlando, FL, August 6–12.

214. Yousefi, T., Veisy, F., Shojaeizadeh, E. and Zinadini, S. (2012). An experimental investigation on the effect of Al_2O_3–H_2O nanofluid on the efficiency of flat-plate solar collectors. *Renewable Energy* 39, 293–298.

215. Kim, H., Kim, J. and Cho, H. (2017). Experimental study on performance improvement of U-tube solar collector depending on nanoparticle size and concentration of Al_2O_3 nanofluid. *Energy* 118, 1304–1312.

216. Chaji, H., Ajabshirchi, Y., Esmaeilzadeh, E., Heris, S.Z., Hedayatizadeh, M. and Kahani, M. (2015). Experimental study on thermal efficiency of the flat plate collectors with nanofluids. *Applied Thermal Engineering* 88, 165–171.

217. He, Q., Zeng, S. and Wang, S. (2015). Experimental investigation on the efficiency of flat-plate solar collectors with nanofluids. *Applied Thermal Engineering* 88, 165–171.

218. Jamal-Abad, M.T., Zamzamian, A., Imani, E. and Mansouri, M. (2013). Experimental study of the performance of a flat plate collector using Cu-Water nanofluid. *Journal of Thermophysics and Heat Transfer* 27, 756–760.

219. Moghadam, A.J., Farzane-Gord, M., Sajadi, M. and Hoseynzadeh, M. (2014). Effects of CuO/Water nanofluid on the efficiency of a flat-plate solar collector. *Experimental Thermal Fluid Science* 58, 9–14.

220. Mahian, O., Kianifar, A., Sahin, A.Z. and Wongwises, S. (2014). Performance analysis of mini-channel based solar collector using different nanofluids. *Energy Conversation Management* 88, 729–738.

221. Tong, Y., Kim, J. and Cho, H. (2015). Effects of thermal performance of enclosed-type evacuated U-tube solar collector with multi-walled carbon nanotube/water nanofluid. *Renewable Energy* 83, 463–473.

222. Sabiha, M.A., Saidur, R., Hassant, S., Said, Z. and Mekhilef, S. (2015). Energy performance of an evacuated tube solar collector using single walled carbon nanotubes nanofluids. *Energy Conversation Management* 105, 1377–1388.

223. Kasaeian, A., Daviran, S., Azarian, R.D. and Rashidi, A. (2015). Performance evaluation and nanofluid using capability study of a solar parabolic trough collector. *Energy Conversation Management* 89, 368–375.

224. Ghasemi, S.E. and Mehdizadeh Ahangar, G.R. (2014). Numerical analysis of performance of solar parabolic trough collector with Cu-water nanofluid. *International Journal of Nano Dimension* 5, 233–240.

225. Menbari, A., Alemrajabi, A.A. and Rezaei, A. (2016). Heat transfer analysis and the effect of CuO/Water nanofluid on direct absorption concentrating collector. *Applied Thermal Engineering* 104, 176–183.

226. Wang, Y., Xu, J., Liu, Q., Chen, Y. and Liu, H. (2016). Performance analysis of a parabolic trough solar collector using Al₂O₃/synthetic oil nanofluid. *Applied Thermal Engineering* 107, 469–478.

227. Khullar, V. and Tyagi, H. (2012). A study on environmental impact of nanofluid-based concentrating solar water heating system. *International Journal of Environmental Studies* 69, 220–232.

228. Sokhansefat, T., Kasaeian, A.B. and Kowsary, F. (2014). Heat transfer enhancement in parabolic trough collector using Al₂O₃/synthetic oil nanofluid. *Renewable and Sustainable Energy Reviews* 33, 636–644.

14

Nanofluids for Solar Steam Generation

Yurong He, Xinzhi Wang, and Yanwei Hu

CONTENTS

14.1 Background

Solar energy has been regarded as an inexhaustible and sustainable energy source that could serve as a promising alternative to fossil fuels [1]. Although currently the solar energy constitutes only a small percentage of the total energy consumption, it has received enthusiastic support from politicians, regulators, technologists, and environmental groups [2]. Solar energy can be used to produce electricity, fuel, or heat. One of the most important issues in the solar energy management and conversion is ensuring that it is used efficiently [3,4]. As a typical process of utilizing solar thermal energy, solar steam generation has been widely used in seawater desalination [5,6], power plants [7], distillation [8], and water purification [9]. In some special situations, such as those involving island navy stations, aboriginals in remote areas, or ocean voyagers, solar energy might be the only energy source available in sufficient abundance to generate clean water [10]. Energy-efficient clean water generation technologies using solar energy play a critical role in augmenting freshwater resources.

In traditional solar steam generation systems, the incident sunlight is first collected via absorption by an absorber painted in black [4]. Subsequently, the collected solar energy is converted into thermal energy, which is transferred to the heat transfer fluid via thermal convection and thermal conduction. After that, the thermal energy is used to heat water to its boiling point to generate steam. In the traditional strategy, highly concentrated solar light is usually required, which may lead to high thermal loss and optical loss during their applications [11]. To eliminate or minimize the energy loss in the steam generation process, volumetric solar steam generation via a working fluid has been suggested as a promising alternative capable of maximizing the efficiency of solar energy harvesting.

Nanofluids, as a kind of functional nanoparticle (NP) dispersion, have attracted enormous attention owing to their excellent thermophysical properties, which could contribute to the enhancement of heat transfer [12,13]. Recently, researchers have also discovered that nanofluids have distinctive spectral absorption and radiation properties, which lead to significant potential applications of nanofluids in the utilization of solar energy [14]. In nanofluid-based volumetric solar collection systems, solar energy could be converted into thermal energy in working fluids, even at low particle concentrations [15]. Therefore, nanofluids have emerged as a promising means of improving the efficiency of solar steam generation owing to their superior thermophysical and solar absorption properties [14]. Using nanofluids as volumetric solar absorbers, it might be possible to take the advantage of directly capturing the latent heat of vaporization that is required for liquid–vapor phase transitions [16]. In this chapter, we comprehensively review the recent investigations on nanofluid-based solar steam generation. Subsequently, we review and explain the mechanism of nanofluid-based solar steam generation. The experimental methods for preparing different types of solar absorption nanofluids and experimental setups for solar steam generation are also described. In addition, the solar steam generation performance with different types of nanofluids is discussed. Finally, the prospect of further applications of nanofluids to high-temperature and high-pressure steam generation is delineated.

14.2 Research Progress

Plasmonic nanomaterials such as doped metal oxides or metal materials have a strong light collection ability owing to their localized surface plasmon resonance arising from the interactions between the free electrons on the NP surface and incident light [17]. Thus, plasmonic NPs exhibit significant potential for use in solar energy harvesting systems. As solar light is absorbed by the NPs, localized heat can be generated in the NPs, which can then be transferred to the surrounding medium through phonon vibrations. Neumann et al. [18,19] experimentally studied the steam generation performance of nanofluids based on plasmonic Au and Au-SiO$_2$ nanoshell NPs. Under 1000 times concentrated solar power (1 sun = 1 kW m^{-2}), the authors achieved a device efficiency of 24%. Guo et al. [20] conducted solar steam generation experiments with Au nanofluids containing differently sized NPs and investigated the effect of NP size on the steam generation performance. Jin et al. [21] proposed that nanofluid-based solar steam generation occurs by the localized boiling and vaporization of water using a concentrated solar light of 220 suns to promote the phase change of Au nanofluids. Liu et al. [22] employed a biomimetic NP-based nanofluid to realize a highly efficient solar steam generation. Ag-based nanofluids were also studied by Wang et al. [23], who investigated the effect of morphologies and diameters of Ag NPs on their solar steam generation performance. Although plasmonic nanofluids are generally based on noble NPs such as Au and Ag NPs, they are usually expensive and might undergo oxidation in air. In order to protect the noble NPs and enhance the solar light absorption, core-shell-structured plasmonic NPs have been designed with noble NPs as the core and metallic oxide/carbon/silica as the shell material. Li et al. [10] studied the solar steam generation performance of Ag@TiO$_2$ nanofluids. Gan et al. [24] prepared nanofluids based on Au-cored carbon nanospheres and found that the conversion efficiency of these nanofluids is much higher than those of nanofluids based on pure carbon spheres or Au NPs.

Apart from noble metal NPs, carbon-based NPs are also promising materials for solar steam generation. Compared to noble metal NPs, carbon-based NPs usually have the advantage of low cost. Ni et al. [25] studied the solar steam generation performance of three different types of carbon-based nanofluids (graphitized carbon black, carbon black, and graphene) and found that nanofluids based on graphitized carbon black could lead to an efficiency of 69% under 10-sun irradiance. The authors also indicated that the mechanism of enhanced steam generation with nanofluids stems from the heating of the bulk nanofluids. Wang et al. [26] conducted solar steam generation experiments with carbon nanotube nanofluids and also obtained a high evaporation efficiency. Reduced graphene oxide (rGO) has recently attracted considerable attention owing to its superior thermal conduction, electric conduction, strength, and optical properties. This material can also work at high temperatures and retain its excellent features in many potential application areas [27,28]. Thus, rGO-based nanofluids are another promising material alternative for volumetric solar steam generation. Liu et al. [29] investigated the solar steam generation using rGO nanofluids experimentally and discussed the effects of light intensity and rGO mass concentration on the performance. This work showed that a hot area emerges near the water-air interface, which is the main factor for rapid water evaporation.

In some special applications of nanofluid-based solar steam generation systems, NPs may cause secondary pollution of water. Therefore, it is imperative to develop strategies for separating the NPs from nanofluids to allow their subsequent reuse. Previously, magnetic NPs have been used successfully for solar steam generation, after use, the magnetic NPs were collected using an appropriate magnetic field [30]. Fe_3O_4 NPs are a good alternative for separating NPs from aqueous solutions owing to their magnetic nature. However, Fe_3O_4 NPs have low photothermal conversion ability. Therefore, Fe_3O_4 NPs are usually composited with light-absorbing NPs to obtain reusable nanofluids. For example, Shi et al. [31] combined Fe_3O_4 with carbon nanotubes (CNTs) to prepare a nanofluid. The Fe_3O_4@CNT nanofluid not only accelerated the evaporation of water, but also enabled high reusability of NPs. Wang et al. [32] designed a new type of composite NPs by decorating graphene oxide (GO) with magnetic NPs. The resulting material absorbed over 95% of solar energy, and the evaporation reached up to 70% even under 1-sun solar light intensity. In addition, the magnetic NP-decorated GO could be recovered from the nanofluid using a magnetic field. Liu et al. [33] designed and developed a novel solar evaporation system with Fe_3O_4 NPs in nanofluids as solar light absorbers. To enhance the solar absorption properties of such nanofluids, graphene oxide was coated onto the magnetic NPs. To further enhance the solar energy harvesting performance, Chen et al. [34] developed a recoverable nanofluid containing Fe_3O_4 NP-modified multi-walled CNTs (MWCNTs). The recoverable nanofluids showed high recyclability and stability.

It is apparent that volumetric solar steam generation has attracted significant attention in the recent years, driven by the demand for more efficient and competitive vapor generation. Researchers are currently trying to develop solutions that would permit the use of the full spectrum of solar energy and to better understand the physical mechanisms of nanofluid-enhanced solar steam generation.

14.3 Mechanism

14.3.1 Generation of Nanobubbles

According to one mechanism, nanoscale evaporation takes place and vapor bubbles are generated around the NPs as they are illuminated with light [35]. The NPs absorb the light energy, allowing photothermal conversion to occur within them, consequently, the heat generation is confined to the close proximity of the particle-liquid interface. There is a non-equilibrium thermal distribution between the NPs and the liquid. The NPs are heated up to a high temperature while the surrounding bulk fluid remains cold, which leads to the local nucleation of a vapor bubble. The generated vapor subsequently separates from the NP and rises up to the water-air interface as a result of the buoyancy lift. The evaporation and bubble generation in the nanoscale to microscale environment around the NPs have been studied by several researchers. For example, Lukianova-Hleb et al. [36] used a short pulse laser to illuminate plasmonic gold NPs and generate transient vapor nanobubbles in their vicinity. The results indicated that a critical laser intensity equivalent to 3×10^8 suns is required to realize plasmonic nanobubbles. Fang et al. [35] applied a continuous laser to illuminate gold NPs on a substrate immersed in water and observed that the initial nanoscale vapor envelopes the NPs, growing into a microscale bubble owing to the coalescence of the expanded vapor envelopes. In this work, a threshold incident laser intensity of 3×10^7 suns was required for nanobubble formation. Lombard et al. [37] simulated the formation of a nanobubble following the application of incident light onto a gold NP. The results indicated that up to $\sim 1 \times 10^{10}$ sun intensity is required to generate a vapor nucleate around the gold NP. For this mechanism, therefore, a very high optical intensity is required to generate nanobubbles around the NPs.

14.3.2 Increasing the Bulk Fluid Temperature for Evaporation

According to the second mechanism, the temperature of the NPs rapidly equilibrates with the surrounding medium [25,38]. Consequently, the fluid temperature increases quickly in the optical concentrated region, resulting in fast evaporation of water. Some experiments have demonstrated that the temperature of the fluid surrounding the NPs could reach as high as the spinodal decomposition temperature of the fluid outside, was it not for the formation of the nanobubbles [39,40]. Jin et al. [21] investigated experimentally and numerically the volumetric solar steam generation of Au nanofluids under a concentrated solar light intensity of 220 suns and found that the steam generation is mainly caused by localized boiling and evaporation in the initial stage. A highly non-uniform temperature distribution was observed in the nanofluids, suggesting that the superheated region is located at the light irradiation spot while the bulk fluid remains sub-cooled. The authors claimed that the hypothesized nanobubble cannot be produced under a normal concentrated solar system. Keblinski et al. [41] studied numerically the heating process of 5000 NPs and reported that the heating of the bulk fluid is on the macrosecond scale, whereas the heating in the vicinity of the NPs is on the nanosecond scale. Hogan et al. [42] simulated light propagation through a gold nanofluid and demonstrated the effect of light scattering by Au NPs on the localized solar absorption and heating of the evaporating surface. By comparing the numerical results with experimental data, the authors concluded that the heat transfer in nanofluids during steam generation can be described by the Fourier-law heat conduction.

14.4 Experimental Work

14.4.1 Preparation of Nanofluids

14.4.1.1 Typical Plasmonic Nanofluids

Au NPs are typical plasmonic materials that have been widely used in nanofluids for solar energy harvesting. A typical method of synthesizing Au NP-based nanofluids involves the reduction of gold (III) chloride hydrate ($HAuCl_4$) [43,44]. First, 950 mL of the $HAuCl_4$ solution (0.189 mg/mL) are heated to boiling under vigorous stirring. After boiling for ~20 minutes, 50 mL of a sodium citrate dihydrate solution (1 mg/mL) are rapidly added, and the mixture is boiled for another ~20 minutes, during which the color of the mixture changes from pale yellow to colorless and eventually to wine red. After further heating for ~20 minutes to ensure the reaction completion, the mixture is cooled to room temperature under continuous magnetic stirring. Subsequently, the Au NPs are separated from the unreacted chemicals via centrifugation. The isolated Au NP precipitate is dispersed in a certain volume of water to obtain a high concentration of the nanofluid. Eventually, Au NP nanofluids with different target concentrations are prepared via additional dilution with water.

The characterization of the thus-prepared Au nanofluids is demonstrated in Figure 14.1 [44]. Photographs of the wine red plasmonic Au nanofluids and the transmission electron micrograph (TEM) of the prepared spherical Au NPs are shown in Figure 14.1a and b, respectively. The diameters of the Au NPs were analyzed using the Nano Measurer 1.2 software (Fudan University, China), and the NP size distribution is shown in Figure 14.1c. The characterization of the Au NPs reveals that the prepared Au nanofluids are stable dispersions and the NP size is relatively uniform.

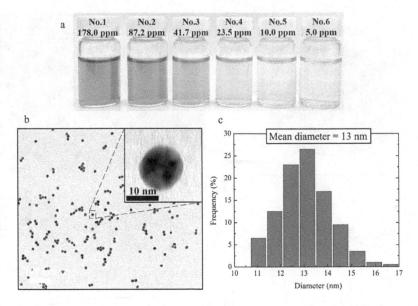

FIGURE 14.1 (a) Photographic images of plasmonic nanofluids with different Au NP concentrations; (b) TEM images of Au NPs; and (c) diameter distributions of Au NPs. (Reprinted from *Sol. Energy*, 157, Wang, X. et al., Investigation of photothermal heating enabled by plasmonic nanofluids for direct solar steam generation, 35–46, Copyright 2017, with permission from Elsevier.)

14.4.1.2 Nanofluids with Core-Shell Nanoparticles

The near-field enhancement of plasmonic Ag NPs is recognized to be 10 times greater than that of similar Au NPs [45]. However, metallic Ag NPs are easily oxidized in air, which limits their practical applications. In this section, the design and synthesis of Ag@TiO_2 core-shell NPs are described [10]. In this material, Ag cores are completely covered with TiO_2 shells, here, Ag NPs act as the solar light absorber and TiO_2 acts as the protector and regulator of absorption. The Ag@TiO_2 core-shell NPs with broadband absorbance are synthesized using a modified procedure based on the reduction of $AgNO_3$ and decomposition reaction of $Ti(OC_4H_9)_4$ (TBT) in C_3H_7NO (DMF) [46].

A typical synthesis involves the following steps [10]: (1) an $AgNO_3$ solution (8.44 mL, 8.3 mM) in isopropanol alcohol (IPA) is mixed with a TBT solution (1.56 mL, 15 mM) in IPA isopropanol alcohol under vigorous stirring; (2) DMF (4.4 mL) and a CTAC solution (0.6 mL, 4.8 mg/mL) in DMF are added to the reactants; (3) after mixing thoroughly, the solution is transferred to an oil bath at 80°C and allowed to react for 70 minutes; and (4) the purified NPs are suspended in pure water for further use. It is noteworthy that it is impossible to prevent the aggregation of the NPs during their isolation and drying. Therefore, the purified Ag@TiO_2 core-shell NPs are dispersed in a certain amount of water under ultrasonication to obtain stable nanofluids without the use of a dispersant. Finally, nanofluids with different concentrations are obtained by diluting the concentrated nanofluid with water.

The characterization of the Ag@TiO_2 NPs is shown in Figure 14.2 [10]. The average diameter of the Ag core and the thickness of the TiO_2 shell are 19.7 ± 1.4 nm and 2.2 ± 0.7 nm, respectively. The TEM image of the prepared Ag@TiO_2 NP is shown in Figure 14.2a. The TEM electron diffraction (ED) patterns (Figure 14.2b) and X-ray diffraction (XRD) patterns (Figure 14.2c) of the prepared Ag@TiO_2 NPs confirm the formation of Ag nanocrystals and the amorphous nature of the TiO_2 shell. Furthermore, the absorption spectra of the Ag@TiO_2 nanofluids (Figure 14.2d) show the broadband light absorption of Ag@TiO_2 core-shell NPs, with strong absorptivity in the visible light range. In accordance with the

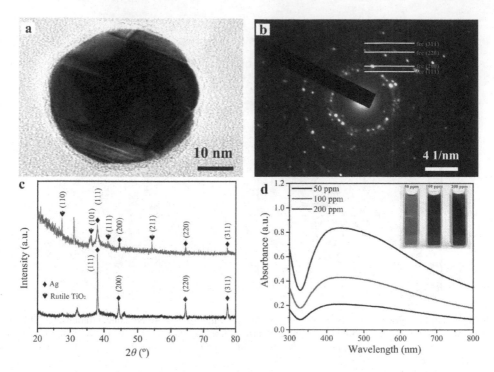

FIGURE 14.2   Characterization of Ag@TiO$_2$ NPs. (a) TEM image; (b) electron diffraction (ED) pattern; (c) XRD patterns of the prepared NPs before (black line) and after (red line) annealing at 800°C for 2 hours; and (d) Absorption spectra of the Ag@TiO$_2$ nanofluid. The inset shows the photographs of the nanofluids with Ag@TiO$_2$ NP at concentrations of 50, 100, and 200 ppm. (Reprinted from *Appl. Therm. Eng.*, 121, Li, H. et al., Synchronous steam generation and heat collection in a broadband Ag@TiO$_2$ core/shell nanoparticle-based receiver, 617–627, Copyright 2017, with permission from Elsevier.)

Beer-Lambert law [47], the absorbance of the Ag@TiO$_2$ nanofluids increases proportionally with the increase in the Ag@TiO$_2$ NP concentration. The broadband absorption of Ag@TiO$_2$ NPs is caused by a combination of direct absorption of the photons and light scattering by the NPs.

14.4.1.3 Carbon-Based Nanofluids

Owing to the high solar light absorption of rGO, rGO-based nanofluids have also been investigated for solar steam generation [29]. For the synthesis of rGO, chemical reagents such as graphite powder, KMnO$_4$, H$_2$SO$_4$, aqueous ammonia, L-ascorbic acid (L-AA), NaNO$_3$, and H$_2$O$_2$ are typically used. First, chemical exfoliation of the graphite powder is carried out by a modification of the Hummers' method [29,48] to obtain GO. Then, GO is reduced with L-AA to rGO. Finally, rGO is dispersed in water to obtain rGO nanofluids with different concentrations. The characterization of the prepared rGO nanofluids is shown in Figure 14.3. The morphology of rGO was characterized by atomic force microscopy (AFM) (Figure 14.3a) and TEM (Figure 14.3c). rGO nanosheets can be observed in the AFM image, and their average thickness was determined to be ~1 nm based on the AFM height profile (Figure 14.3b). In addition, rGO was found to be transparent with a tulle shape. The photograph of the rGO nanofluids in Figure 14.3d shows that the rGO nanofluid gets darker with increasing concentration.

14.4.1.4 Magnetic Nanofluids

The synthesis of the Fe$_3$O$_4$@TiO$_2$ magnetic NPs was carried out as follows [49]: First, an aqueous solution of FeCl$_2$·4H$_2$O and FeCl$_3$·6H$_2$O was stirred vigorously, and the pH of the suspension was adjusted to 10 using a 25-wt.% NH$_3$·H$_2$O solution. Afterwards, the black precipitate was added slowly to a solution

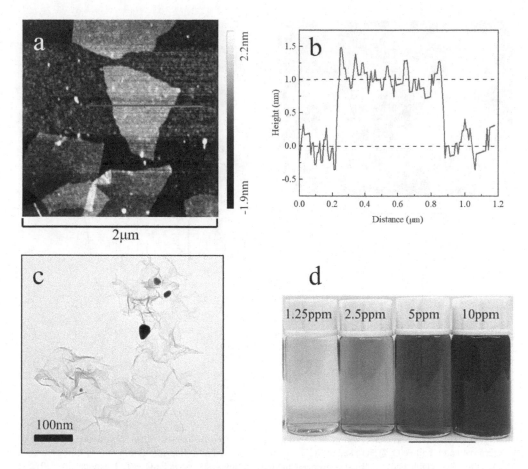

FIGURE 14.3 (a) AFM image; (b) AFM height profile; (c) TEM image of rGO; and (d) photograph of rGO nanofluids at different concentrations. (Reprinted from *Applied Energy.*, 220, Liu, X. et al., Volumetric solar steam generation enhanced by reduced graphene oxide nanofluid, 302–312, Copyright 2018, with permission from Elsevier.)

of TiF$_4$ and transferred to a Teflon-sealed autoclave along with 50 mL of deionized water. The sealed autoclave was loaded into an oven and heated at 180°C for 48 hours. After this time, the autoclave was cooled to room temperature in ambient air and the final product, Fe$_3$O$_4$@TiO$_2$ NPs, was magnetically separated from the reaction solution.

The scanning electron microscope (SEM) and TEM images, XRD pattern, and X-ray photoelectron spectroscopy (XPS) spectra of the prepared Fe$_3$O$_4$@TiO$_2$ NPs are shown in Figure 14.4a–d. The prepared Fe$_3$O$_4$@TiO$_2$ NPs have a uniform morphology and size (diameter of ~280 nm). The chemical composition of the Fe$_3$O$_4$@TiO$_2$ NPs is confirmed by the XRD pattern, which shows peaks characteristic for magnetic Fe$_3$O$_4$ (Figure 14.4c). The XPS spectra indicate the formation of the Fe$_3$O$_4$ phase in the composite. The major peaks in the XPS spectra are assigned to Ti 2p, O 1s, and Fe 2p levels.

14.4.2 Steam Generation

14.4.2.1 Experimental Setup

A schematic diagram illustrating the experimental setup used to investigate nanofluid-based solar steam generation is shown in Figure 14.5 [10]. The set up consists mainly of four parts. The first part is the test section and is made of an acrylic tube into which some thermocouples are inserted. During the solar steam generation test, nanofluids are added to the test tube, followed by the immersion

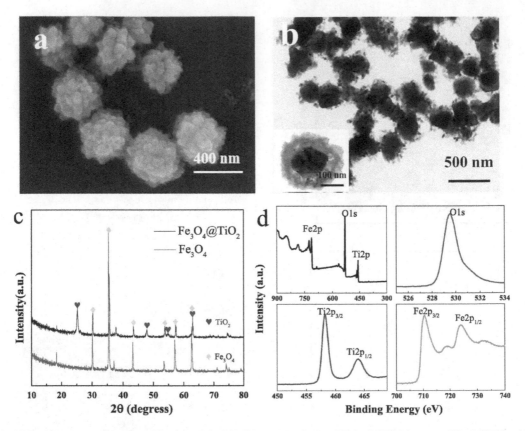

FIGURE 14.4 Structural characterization of $Fe_3O_4@TiO_2$ and Fe_3O_4: (a) SEM; (b) TEM images of $Fe_3O_4@TiO_2$; (c) XRD patterns of Fe_3O_4 and $Fe_3O_4@TiO_2$; and (d) XPS spectra of $Fe_3O_4@TiO_2$. (Reprinted from *Energy Convers. Manag.*, 171, Shi, L. et al., Recyclable photo-thermal conversion and purification systems via $Fe_3O_4@TiO_2$ nanoparticles, 272–278, Copyright 2018, with permission from Elsevier.)

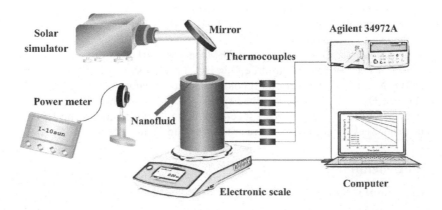

FIGURE 14.5 Schematic diagram illustrating the experimental setup used to investigate nanofluid-based solar steam generation. (Reprinted from *Appl. Therm. Eng.*, 121, Li, H. et al., Synchronous steam generation and heat collection in a broadband $Ag@TiO_2$ core/shell nanoparticle-based receiver, 617–627, Copyright 2017, with permission from Elsevier.)

of the thermocouples. The second section contains a light source, typically, a simulated solar light generator (CEL-HXF300, CEAULIGHT, Beijing, China) is used to generate different intensities of incident energy. The simulated solar light is generated using a xenon lamp, which could produce a light spectrum that matches well that of the sunlight. The third section contains an electric balance to monitor the weight loss of the test section caused by the evaporation of water. The electric balance is connected to a computer, and the weight change is recorded as a function of elapsed test time. The fourth section is dedicated to monitoring the temperature using thermocouples immersed in the nanofluid at different heights. The thermocouples are also connected to a data acquisition system that exports the temperature of the nanofluid during the experiment to a computer. During the experiment, the test tube is placed on an electric balance. As the simulated solar light is turned on, the light irradiates the top surface of the working fluid. The incident light interacts with the NPs through scattering and absorption of light photons. Consequently, solar energy is converted to thermal energy in the nanofluid, resulting in solar steam generation at the water-air interface.

14.4.2.2 Model for Calculating the Efficiency of Steam Generation

To better describe the solar steam generation performance of the nanofluids by volumetric solar energy collection, the evaporation efficiency (η) in the evaporation stage is defined [26]. First, the rate of evaporation per unit area is defined as follows:

$$\dot{m} = \Delta m / (At), \tag{14.1}$$

where $\dot{m}$ represents the steady-state evaporation flux of water (kg m^{-2} h^{-1}), Δm is the weight loss of the testing system caused by the evaporation of water (kg), A is the cross-sectional area of the test tube (m^2), and t is the time for evaporation.

Furthermore, the evaporation efficiency can be calculated by dividing the heat generated by the evaporation during the steam generation with the total light energy illuminated on the working fluids. The efficiency is expressed as follows:

$$\eta = \dot{m} h_{\text{fg}} / I, \tag{14.2}$$

where h_{fg} is the phase-change enthalpy of water (2.257 kJ g^{-1} at 1 atm) and I is the incident light intensity (kW m^{-2}).

In addition, the specific vapor productivity (SVP) is defined to evaluate the capability of the NPs to generate solar steam per unit mass, which is defined as follows:

$$SVP = \dot{V}/m_{\text{NPs}} \tag{14.3}$$

where $\dot{V}$ is the volume of the steam flux (m^3 g^{-1} h^{-1}) with $\dot{V} = \dot{m} A/\rho_{\text{vapor}}$, m_{NPs} is the mass of the NPs dispersed in the tested nanofluid (g), and ρ_{vapor} is the vapor density (0.6 kg m^{-3} for water at 1 atm and 100°C).

14.5 Solar Steam Generation Performance of Nanofluids

14.5.1 Effect of Nanoparticle Concentrations on Solar Steam Generation Performance

The NP concentration in the nanofluid affects its solar steam generation performance significantly, particularly because the NPs are the main light absorption medium. The solar steam generation performances of CNT-based nanofluids with four different concentrations under 10 kW/m^2 solar illumination intensity are shown in Figure 14.6 [26]. The weight change in the nanofluids caused by the evaporation of water is displayed in Figure 14.6a. As the CNT concentration increased, the weight of the generated steam also increased. Compared to the steam generation with pure water under solar light

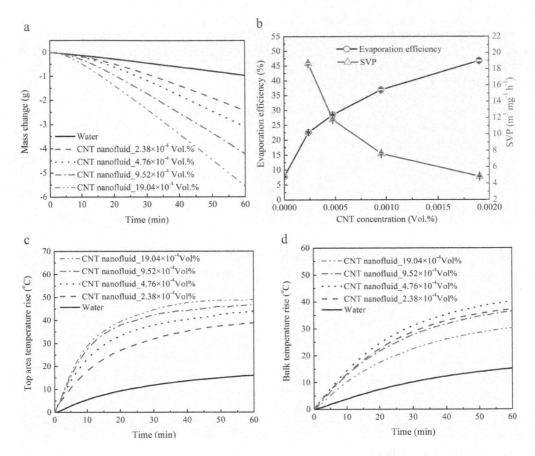

FIGURE 14.6 Evaporation performance of CNT nanofluids with different concentrations (0 vol.%, 2.38×10^{-4} vol.%, 4.76×10^{-4} vol.%, 9.52×10^{-4} vol.%, and 19.04×10^{-4} vol.%) under a constant solar power of 10 suns: (a) evaporation–mass change curves; (b) changes in evaporation efficiency and SVP as a function of CNT concentration; and temperature increase in (c) top area (height = 50 mm), and (d) bulk liquid as a function of time. (Reprinted from *Energy Convers. Manag.*, 130, Wang, X. et al., Direct vapor generation through localized solar heating via carbon-nanotube nanofluid, 176–183, Copyright 2016, with permission from Elsevier.)

irradiation, the evaporation of water from the nanofluid was obviously enhanced. Furthermore, as shown in Figure 14.6b, the evaporation efficiency (η) increased gradually, whereas the SVP decreased gradually with increasing concentration of the CNTs in the nanofluid. As the CNT concentration increased from 0 to 19.04×10^{-4} vol.%, the evaporation efficiency increased from ~7.8% to ~46.8%. However, as the CNT concentration increased from 2.38×10^{-4} vol.% to 19.04×10^{-4} vol.%, the SVP decreased from 18.7 m^{-3} mg^{-1} h^{-1} to 4.8 m^{-3} mg^{-1} h^{-1}. These results demonstrate that high CNT concentrations lead to a high evaporation efficiency as a result of the high optical absorption of the nanofluid. The decrease in the SVP can be ascribed to two factors: (1) the CNTs might have agglomerated at increased CNT concentrations, and (2) the transparency of CNT nanofluids decreases with increasing CNT concentrations. In this case, the incident solar light is mainly concentrated at the top area of the nanofluid, and a large amount of the CNTs in the interior parts of the fluid cannot be effectively used to absorb the solar light.

The temperature changes in the CNT nanofluids with different concentrations under consistent solar light irradiation are shown in Figure 14.7 [26]. It is apparent that as the light illumination time increased, the temperature of the nanofluids increased gradually. In addition, the temperature difference at different heights of the nanofluids increased with increasing CNT concentration. For instance, the temperature difference between the bottom and top parts of the CNT nanofluid with a concentration of 19.04×10^{-4} vol.% reached up to ~30°C. By contrast, almost no temperature difference was observed

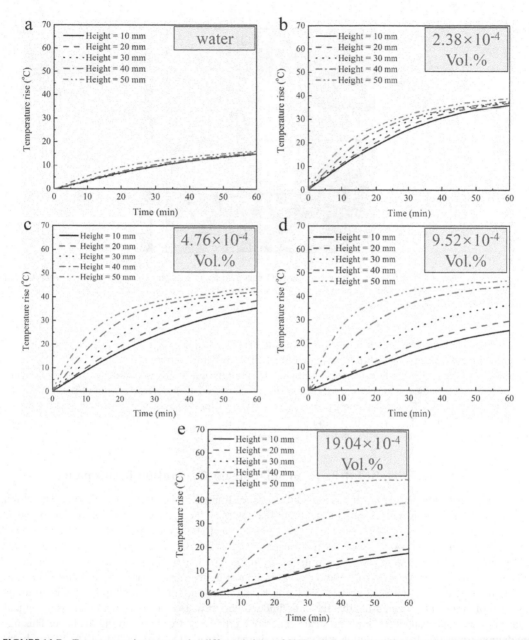

FIGURE 14.7 Temperature rise measured at different heights of CNT nanofluids with different concentrations: (a) 0 vol.%; (b) 2.38×10^{-4} vol.%; (c) 4.76×10^{-4} vol.%; (d) 9.52×10^{-4} vol.%; and (e) 19.04×10^{-4} vol.% under a constant solar power of 10 suns. (Reprinted from *Energy Convers. Manag.*, 130, Wang, X. et al., Direct vapor generation through localized solar heating via carbon-nanotube nanofluid, 176–183, Copyright 2016, with permission from Elsevier.)

in pure water. These results indicate that the upper section of the nanofluid absorbs most of the solar energy and could be attributed to localized solar heating at the nanofluid-air interface when the NP concentration is high. The temperature of the upper section of the nanofluid increased with increasing CNT concentration, evidencing highly localized temperature increase in the systems employing nanofluids (Figure 14.6c). However, although the bulk nanofluid temperature increased gradually at first, it subsequently decreased as the CNT concentration increased (Figure 14.6d). The temperature distributions in the nanofluids of different concentrations and the effects of NP concentration on the steam generation

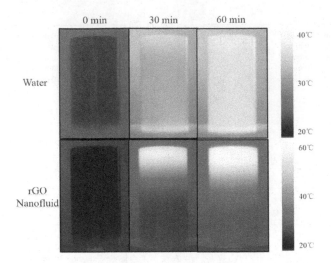

FIGURE 14.8 Infrared thermal images of water and rGO nanofluid as a function of time. (Reprinted from *Applied Energy.*, 220, Liu, X. et al., Volumetric solar steam generation enhanced by reduced graphene oxide nanofluid, 302–312, Copyright 2018, with permission from Elsevier.)

performance demonstrate that high evaporation efficiency and high evaporation rate can be achieved using highly concentrated nanofluids with a relatively low bulk fluid temperature.

Figure 14.8 shows the temperature distributions observed in nanofluids and water during solar steam generation [29]. It is clear that a concentrated hot area was formed at the upper section of the nanofluids, which harvests most of the solar energy and results in fast evaporation of water at the nanofluid-water interface. The localized solar heating can reduce thermal losses to the bulk fluid and ensure that more energy is used for steam generation at the phase interface.

14.5.2 Effect of Solar Light Intensity on Solar Steam Generation Performance

The steam generation performance of nanofluids is rather limited under natural conditions. Therefore, the solar light is usually concentrated to increase the power of the incident energy. In order to understand the effect of the solar light intensity on steam generation efficiency of nanofluids, experiments were conducted under different optical concentrations [10]. The mass change in $Ag@TiO_2$ nanofluids under different optical concentrations is demonstrated in Figure 14.9a. With the increase in the solar irradiation intensity, the mass change in the nanofluid increases, indicating the generation of more steam. The bulk fluid temperature measured as a function of the illumination time under different optical concentrations is shown in Figure 14.9b, revealing that the bulk fluid temperature also increases with increasing optical concentration. The evaporation rate and steam generation efficiency are shown in Figure 14.9c. This figure demonstrates that the evaporation rate could be increased from 0.86 kg/(m²·h) to 10.67 kg/(m²·h) by increasing the solar light intensity from 1 to 10 suns. However, the evaporation efficiency first decreased from 53.6% to 45.9% on going from 1 sun to 3 suns illumination and subsequently increased to 66.9% when the illumination was increased to 10 suns.

This seemingly contradictory thermodynamic optical relationship has been reported by various researchers [11,50]. This changing relationship is influenced by many factors, including various heat losses mentioned above and the resistance to mass transfer from the location of bubble formation to the surface of the nanofluid. A possible explanation for this relationship is that when the light intensity is only 1 sun, the light propagation distance is very short and most of the energy is used to heat the surface water, causing higher evaporation efficiency. As the light intensity increases, the distance of light propagation increases. Therefore, although some solar energy is directly absorbed by the surface water, a portion heats the hot zone, resulting in decreased evaporation efficiency. According to the Bouguer's law, the light intensity in nanofluids increases logarithmically with the increase in the long

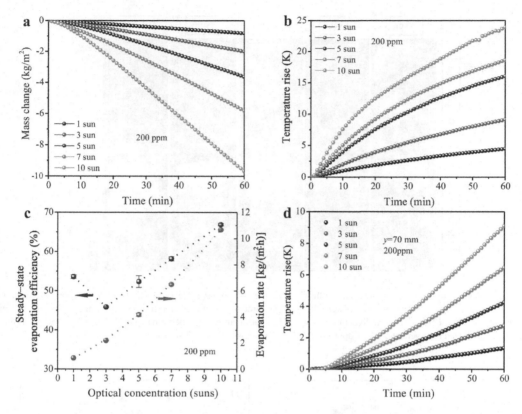

FIGURE 14.9 Effect of optical concentration on the photothermal performance of a nanofluid with Ag@TiO$_2$ NPs at a concentration of 200 ppm. (a) Mass change; (b) temperature increase in the bulk of the nanofluid observed as a function of illumination time and intensity; (c) evaporation rate and steady-state evaporation efficiency measured as a function of optical concentration; and (d) temperature increase measured at the bottom area of the nanofluid as a function of illumination time and intensity. (Reprinted from *Appl. Therm. Eng.*, 121, Li, H. et al., Synchronous steam generation and heat collection in a broadband Ag@TiO$_2$ core/shell nanoparticle-based receiver, 617–627, Copyright 2017, with permission from Elsevier.)

wave distance. In addition, from the temperature increase observed in the lower bottom part of the plot shown in Figure 14.9d and the digital images shown in Figure 14.10, it is apparent that even under 10-sun illumination, no light passes through the collector. Therefore, as the intensity of light continues to increase, the growth rate of the light propagation distance decreases, and the temperature in the hot zone increases rapidly (Figure 14.10, infrared images), thereby accelerating the evaporation of water and directly causing higher evaporation efficiency for intensities between 3 and 10 suns.

14.5.3 Reusability and Stability of Nanofluids during Solar Steam Generation

As a heat transfer medium, the nanofluids used in practical engineering must have good reusability and long-term stability. In order to demonstrate the reusability of nanofluids, the evaporation performance of nanofluids was previously continuously evaluated by recording the mass change during the evaporation process and repeating the experiment 30 times under the same conditions [10]. The cyclic experiment was performed as follows [10]: the nanofluid with a concentration of 200 ppm was irradiated for 60 minutes under 10-sun irradiation intensity. After 60 minutes of irradiation, the nanofluid was cooled to room temperature and a supplementary amount of water equivalent to the amount of evaporated water was added to the heat collector and completely mixed with the rest of the nanofluid. Subsequently, the next cycle of heat generation was performed. Figure 14.11a shows the maximum evaporation, minimum evaporation, and average evaporation of water during the steam generation process, and Figure 14.11b shows the evaporation capacity measured at the end of each cycle. The evaporation capacity of the nanofluid with

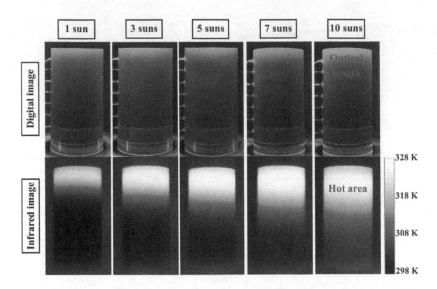

FIGURE 14.10   Digital images (top) of a nanofluid containing 200 ppm Ag@TiO$_2$ NPs taken under different optical concentrations, and the respective infrared images (bottom) captured at an illumination time of 60 minutes. (Reprinted from *Appl. Therm. Eng.*, 121, Li, H. et al., Synchronous steam generation and heat collection in a broadband Ag@TiO$_2$ core/shell nanoparticle-based receiver, 617–627, Copyright 2017, with permission from Elsevier.)

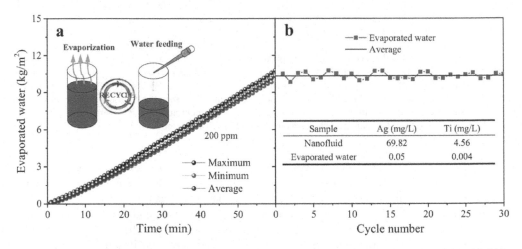

FIGURE 14.11   (a) The amount of evaporated water observed as a function of time using a nanofluid with 200 ppm Ag@TiO$_2$ NPs under 10-sun illumination; and (b) the total amount of evaporated water in each experiment cycle; the inset table shows the mass concentration of Ag and Ti elements in the nanofluid and the evaporated water. (Reprinted from *Appl. Therm. Eng.*, 121, Li, H. et al., Synchronous steam generation and heat collection in a broadband Ag@TiO$_2$ core/shell nanoparticle-based receiver, 617–627, Copyright 2017, with permission from Elsevier.)

Ag@TiO$_2$ NPs is 10.33 kg/m^2. After 30 cycles, the deviation between the evaporation capacity in the 30th cycle and the first cycle is very small, confirming the good reusability of the nanofluid. In addition, to prove that the NPs do not escape during the water evaporation, the vapor was collected and condensed to obtain 10 mL of condensed water. The mass concentrations of Ag and Ti in the condensed water were determined to be 0.05 mg/L and 0.004 mg/L, respectively, by inductively coupled plasma-atomic emission spectrometry (Optima 7300 DV PerkinElmer). These concentrations are significantly lower than the concentrations of these two elements in the nanofluid (69.82 mg/L and 4.56 mg/L), and, consequently, the number of NPs escaping during the evaporation is considered to be negligible.

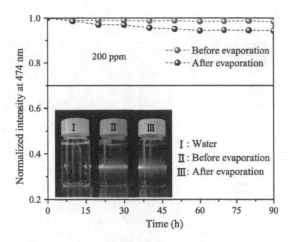

FIGURE 14.12 Normalized absorbance of a nanofluid containing Ag@TiO₂ NPs (200 ppm) before (red-dotted line) and after (blue-dotted line) evaporation. The inset shows the Tyndall effect of the nanofluid. (Reprinted from *Appl. Therm. Eng.*, 121, Li, H. et al., Synchronous steam generation and heat collection in a broadband Ag@TiO₂ core/shell nanoparticle-based receiver, 617–627, Copyright 2017, with permission from Elsevier.)

The stability of a nanofluid can also be verified by measuring the changes in its absorbance with time at a fixed wavelength [10]. The absorbance of a nanofluid is directly proportional to its concentration, and, therefore, if the absorbance remains unchanged after leaving the nanofluid to stand for a given period of time, it is considered to be stable within this time range. In addition, the nanofluid, being a colloidal solution, can exhibit the Tyndall effect [51]. The Ag@TiO₂ nanofluid with the highest mass concentration (200 ppm) was selected for testing. The absorbance of the nanofluid at 474 nm before and after evaporation was measured for verification. Figure 14.12 shows the relationship between the absorbance of the nanofluid before and after evaporation and the intervening time. The inset shows a picture of the Tyndall effect of the nanofluid. Before evaporation, the nanofluid can still maintain a concentration of ~98% after standing for 90 hours, which demonstrates that the prepared nanofluid has good stability, which can be attributed to the small size of the NPs and their good hydrophilicity. In addition, the NPs synthesized by the solvothermal method were used directly after several cycles of washing, thus preventing the agglomeration of the particles during drying, and ensuring their dispersibility. After the evaporation, the concentration of the nanofluid decreased to ~94% after standing for 90 hours, and this decrease can be attributed to the fact that the stability of the nanofluid is related to the temperature, that is, as a result of the higher temperature experienced during evaporation, some of the NPs agglomerate and settle down at an accelerated rate. Note that in order to eliminate the influence of external factors, this section directly demonstrated that the enhancement of solar steam generation is caused by the action of the NPs. Thus, no dispersant or surfactant was added to the nanofluid. It should be noted, however, that a small amount of dispersant would not cause a large change in the heat exchange performance of a nanofluid. Therefore, if a nanofluid is to be used for an extended time under non-flowing hydrodynamic conditions, an appropriate amount of dispersant can be added to the nanofluid.

14.5.4 Magnetic Separation Performance of Nanofluids in Solar Steam Generation

As shown in Figure 14.13a, a magnetic block can be used to apply magnetic field to separate Fe₃O₄ and Fe₃O₄@CNT NPs from the nanofluid [31]. Compared to Fe₃O₄ NPs, the Fe₃O₄@CNT NPs could be separated from the nanofluid faster at the same NP concentration. A possible reason for this result is that the density of the Fe₃O₄@CNT NPs is lower than that of Fe₃O₄ NPs. Furthermore, as the Fe₃O₄ NPs decorated the surface of CNTs in the Fe₃O₄@CNT structures, Fe₃O₄ and the CNT NPs could be separated together. Fe₃O₄@CNT nanofluids with three different mass concentration ratios of Fe₃O₄ and CNT (Fe₃O₄: CNT = 1:1, 2:1, and 4:1) were prepared, and their magnetic properties are shown in

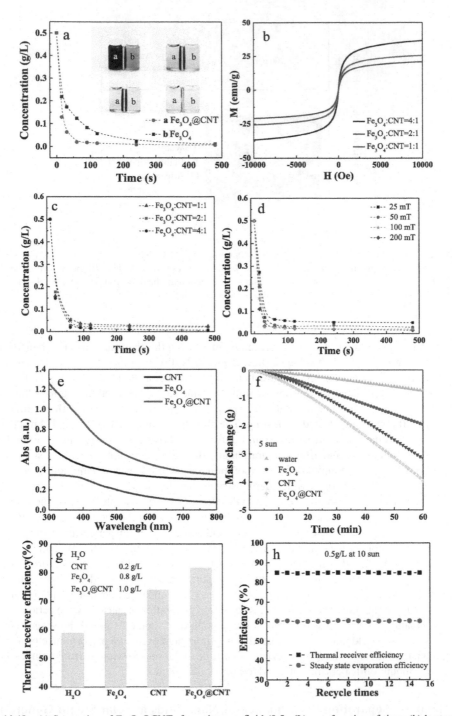

FIGURE 14.13 (a) Separation of Fe_3O_4@CNTs from the nanofluid (0.5 g/L) as a function of time; (b) hysteresis loops of Fe_3O_4@CNT nanofluids prepared at different Fe_3O_4:CNT ratios; (c) separation performance of Fe_3O_4@CNT nanofluids prepared with different Fe_3O_4:CNT ratios at a magnetic field of 100 mT; (d) separation performance of Fe_3O_4@CNT nanofluid as a function of the magnetic field strength; (e) UV-Vis absorbance spectra of Fe_3O_4, CNTs, and Fe_3O_4@CNTs; (f) solar steam generation performances of different nanofluids; (g) thermal receiver efficiencies of water and various nanofluids; and (h) thermal receiver efficiency and steady-state evaporation efficiency of Fe_3O_4@CNTs nanofluid over 15 cycles. (Reprinted from *Energy Convers. Manag.*, 149, Shi, L. et al., Recyclable Fe_3O_4@CNT nanoparticles for high-efficiency solar vapor generation, 401–408, Copyright 2017, with permission from Elsevier.)

Figure 14.13b. In addition, the possibility of separating them was evaluated in a 100-mT magnetic field (Figure 14.13c). The speed and efficiency of the separation process increased as the concentration of Fe_3O_4 increased. The effect of the magnetic field intensity on the separation was also characterized (Figure 14.13d), showing that an increase in the magnetic field intensity can speed up the separation process. Compared to Fe_3O_4 and CNT NPs, Fe_3O_4@CNT NPs exhibited a better light absorption capability (Figure 14.13e). Finally, the solar steam generation performances of water, Fe_3O_4 nanofluid, CNT nanofluid, and Fe_3O_4@CNT nanofluid were studied (Figure 14.13f). The Fe_3O_4@CNT nanofluid displayed the best vapor generation characteristics, reaching a thermal receiver efficiency of up to 81.7%. After separation, the Fe_3O_4@CNT NPs could be re-dispersed in water. Figure 14.13g shows the steady-state evaporation of the Fe_3O_4@CNT nanofluid over 15 cycles of their separation and re-dispersion, thus demonstrating their good reusability.

14.5.5 Water Purification of Nanofluids during Solar Steam Generation

Apart from the solar steam generation, nanofluid-based volumetric solar collection could be applied to simultaneous purification of contaminated water [49]. Organic dyes are one of the most common contaminants in water—they are both highly toxic and slow to degrade. Photocatalysis using nanofluids could be applied to treat organic pollutants in water. Previously, for example, Rhodamine B (RhB)-contaminated water was treated using Fe_3O_4@TiO_2 nanofluid under 1-sun illumination [49]. The absorption spectra of the RhB-contaminated water measured after different treatment times are shown in Figure 14.14a, revealing that the absorption peaks characteristic for RhB decreased as the processing time increased. As shown in Figure 14.14b, the concentration of RhB decreased as a function of time. The degradation of RhB could be increased from 85% to 94% after 60 minutes of solar treatment by increasing the solar light intensity from 1 to 10 suns. The enhanced photodegradation might be attributed to the larger temperature increase under 10 suns. The pore size distribution, pore diameter distribution, and N_2 adsorption-desorption isotherms of the utilized Fe_3O_4@TiO_2 NPs are shown in Figure 14.14c. The Fe_3O_4@TiO_2 NPs have large pore diameters and a high specific surface area, which improve their photocatalytic capacity. In addition, the thermal receiver efficiency and degradation efficiency of the Fe_3O_4@TiO_2 nanofluid are shown in Figure 14.14d. This chart shows that as the solar light intensity increased, the degradation efficiency increased, while the thermal receiver efficiency decreased, the latter could be caused by increased thermal losses under the high light intensity as a result of the high temperature in the top area.

14.5.6 Numerical Simulation of Heat Localization of Nanoparticles

Under low-concentration solar irradiation, no nanobubbles are formed around the NPs [16,25]. Therefore, heat diffusion between the NPs and the surrounding media occurs in the absence of heat convection and phase change. The corresponding energy balance equation is expressed as follows [52,53]:

$$\rho(\mathbf{r})c(\mathbf{r})\frac{\partial T(\mathbf{r},t)}{\partial t} = \nabla k(\mathbf{r})\nabla T(\mathbf{r},t) + Q(\mathbf{r},t) \tag{14.4}$$

where $\rho(\mathbf{r})$ is the mass density, $c(\mathbf{r})$ is the specific heat capacity, $k(\mathbf{r})$ is the thermal conductivity, $\partial T(\mathbf{r},t)$ is the temperature of the fluids as a function of the coordinate r and time t, and $Q(\mathbf{r},t)$ is the heat generated by the photothermal conversion process in NPs.

For a plasmonic Au NP, the heat generation could occur from the Joule effect [52,53]:

$$Q(\mathbf{r},t) = \mathbf{j}(\mathbf{r},t) \cdot \mathbf{E}(\mathbf{r},t) \tag{14.5}$$

where $\mathbf{j}(\mathbf{r},t)$ is the electronic current density in NPs, and $\mathbf{E}(\mathbf{r},t)$ is the complex amplitude of the electric field in the NPs.

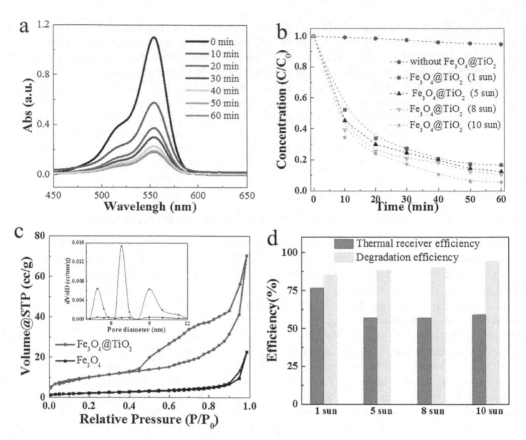

FIGURE 14.14   (a) Changes in the UV-Vis spectrum of a Rhodamine B solution under 1-sun illumination; (b) photocatalytic degradation of $Fe_3O_4@TiO_2$ as a function of time under different solar illuminations; (c) N_2 adsorption-desorption isotherms and pore size distributions of Fe_3O_4 and $Fe_3O_4@TiO_2$ NPs; and (d) thermal receiver efficiency and degradation efficiency of the $Fe_3O_4@TiO_2$ nanofluid at different solar power intensities. (Reprinted from *Energy Convers. Manag.*, 171, Shi, L. et al., Recyclable photo-thermal conversion and purification systems via $Fe_3O_4@TiO_2$ nanoparticles, 272–278, Copyright 2018, with permission from Elsevier.)

A time-averaged heat during the constant photothermal conversion could be defined as follows:

$$Q(\mathbf{r}) = \left\langle \mathbf{j}(\mathbf{r},t) \cdot \mathbf{E}(\mathbf{r},t) \right\rangle_t = \frac{1}{2} \mathrm{Re}\left[\mathbf{j}^*(\mathbf{r}) \cdot \mathbf{E}(\mathbf{r}) \right] \tag{14.6}$$

where $\mathbf{j}(\mathbf{r},t) = i\omega \mathbf{D}(\mathbf{r}) = \varepsilon_0 \varepsilon(\omega) \mathbf{E}(\mathbf{r})$, $\varepsilon(\omega)$ is the dielectric constant, $\mathbf{D}(\mathbf{r})$ is the electric displacement vector, $\varepsilon(\omega)$ is the permittivity of Au NPs, and ω is the angular frequency of light.

Furthermore, the generated heat in NPs can be expressed as follows:

$$Q(\mathbf{r}) = \frac{1}{2} \varepsilon_0 \omega \mathrm{Im}\left[\varepsilon(\omega) \right] \left| \mathbf{E}(\mathbf{r}) \right|^2 \tag{14.7}$$

The photothermal conversion in the NPs usually happens in picoseconds. The generated heat is subsequently transferred to the surrounding fluid through phonon-phonon interactions and a thermal equilibrium is reached in nanoseconds. Therefore, the heat transfer process can be regarded as a steady-state thermal equilibrium process, and the temperature profile around the Au NPs can be calculated as follows:

$$\Delta T(\mathbf{r}) = \frac{V_{NP}Q}{4\pi k_0} \frac{1}{r} \quad (r > R_{NP}) \tag{14.8}$$

where V_{NP} is the volume of the Au NPs, k_0 is the thermal conductivity of water, and r is the distance from the center of the Au NPs.

In this section, the photothermal conversion in Au NPs and the heat transfer between the NPs and surrounding water is investigated numerically [44]. Specifically, irradiation of an Au NP with 518 nm incident light at a power of 10 kW m^{-2}, propagated from the positive z-direction, was simulated. The photothermal conversion was evaluated using finite-difference time-domain (FDTD) simulations. The heat transfer between the hot NP and the surrounding water was determined by finite element analysis. The generated heat power volume density in the Au NP is illustrated in Figure 14.15a. A periodic structure in the nanoparticle dispersion is assumed, as shown in Figure 14.15b. To simulate the Au NP nanofluid with a concentration of 128 ppm, Au NPs with a diameter of 13 nm were arranged in a cubic box with a length of 225 nm. The NP was set as the heat source, and the boundary water box was set to be symmetric. Owing to the small length scale, only transient heat conduction was considered [25]. The initial and environmental temperature was set as 300 K. The photoconversion was simulated over 1 second.

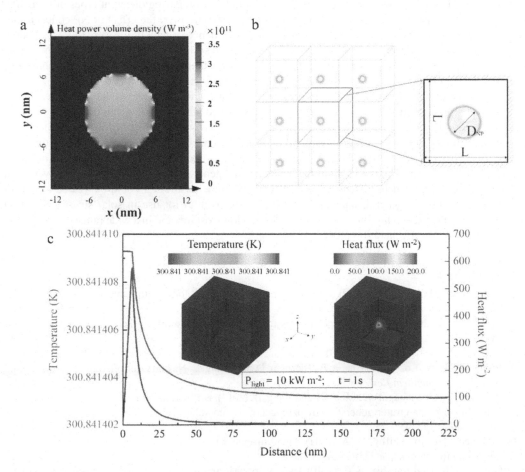

FIGURE 14.15 (a) Heat power volume density of an Au NP under light irradiation ($\lambda = 518$ nm) with a power of 10 kW m^{-2}; (b) schematic diagram illustrating the lattice of the nanoparticles with a lattice constant; and (c) temperature and heat flux distribution from the center of the nanoparticle to the lattice boundary along the x-axis under light illumination ($\lambda = 518$ nm; power = 10 kW m^{-2}) after 1 second. (Reprinted from *Sol. Energy*, 157, Wang, X. et al., Investigation of photothermal heating enabled by plasmonic nanofluids for direct solar steam generation, 35–46, Copyright 2017, with permission from Elsevier.)

The temperature and heat flux distribution from the center of the Au NP to the boundary of the water box is shown in Figure 14.15c. Compared to the temperature of the initial state, the temperature of both the NP and water increased. The temperature decreased on going from the center of the NP to the boundary of the water box. However, the temperature difference between the NP and surrounding water was simulated to be less than 0.001 K, which can be considered negligible. Put together, therefore, the results indicate fast thermal equilibrium between the base fluid water and the NP heated by light.

14.6 Summary and Prospects

In this chapter, nanofluid-based volumetric solar steam generation was introduced and some key recent developments in this area were summarized. Moreover, the mechanisms of nanofluid-based solar steam generation were described. Usually, steam generation mediated by nanobubble formation around the NPs is realized under high-intensity laser light. Under commonly used concentrated solar irradiation conditions, by contrast, steam generation is mainly caused by the increase in the temperature of the bulk fluid. In this case, the nanofluids localize the solar light absorption near the water-air interface, thereby causing fast and efficient evaporation of water. The methods for preparing different types of solar light absorption nanofluids were also introduced. Moreover, investigations on the effect of NP concentration and solar light intensity on the solar steam generation performance of nanofluids were discussed, along with the reusability and stability of nanofluids in the examined systems. In addition to the steam generation process, it might also be possible to utilize nanofluids for water purification owing to the photocatalytic properties of the NPs.

In the future, it might be interesting to develop systems that can further increase the pressure and temperature of the generated steam. High-temperature pressurized steam is an essential element of power generation, consequently, it is expected that this area of research will intensify in the upcoming years, with primary focus on investigating the possibility of generating superheated steam by harvesting solar energy through nanofluids. Another area of prime interest is and will be the development of water-purification systems that rely on nanofluid-based solar steam generation. In such systems, pure water can be obtained from the condensed steam, while at the same time wastewater could be treated to remove harmful ingredients. In general, nanofluids exhibit great potential for use in solar steam generation, which can be exploited as a highly efficient, scalable, and low-cost method in a wide range of distillation, sterilization, and power generation processes and applications.

REFERENCES

1. Chu, S., and Majumdar, A. (2012). Opportunities and challenges for a sustainable energy future. *Nature* 488, 294–303.
2. Lewis, N. S. (2016). Research opportunities to advance solar energy utilization. *Science* 351(6271), aad1920.
3. Lewis, N. S. (2007). Toward cost-effective solar energy use. *Science* 315(5813), 798–802.
4. Weinstein, L. A., Loomis, J., Bhatia, B., Bierman, D. M., Wang, E. N., and Chen, G. (2015). Concentrating solar power. *Chemical Reviews* 115(23), 12797–12838.
5. Huang, J., He, Y., Wang, L., Huang, Y., and Jiang, B. (2017). Bifunctional Au@TiO$_2$ core–shell nanoparticle films for clean water generation by photocatalysis and solar evaporation. *Energy Conversion and Management* 132, 452–459.
6. Elimelech, M., and Phillip, W. A. (2011). The future of seawater desalination: Energy, technology, and the environment. *Science* 333(6043), 712–718.
7. Gupta, M. K., and Kaushik, S. C. (2010). Exergy analysis and investigation for various feed water heaters of direct steam generation solar-thermal power plant. *Renewable Energy* 35(6), 1228–1235.
8. Sharshir, S. W., Peng, G., Wu, L., Essa, F. A., Kabeel, A. E., and Yang N. (2017). The effects of flake graphite nanoparticles, phase change material, and film cooling on the solar still performance. *Applied Energy* 191, 358–366.

9. Shannon, M. A., Bohn, P. W., Elimelech, M., Georgiadis, J. G., Mariñas, B. J., and Mayes, A. M. (2010). Science and technology for water purification in the coming decades. In *Nanoscience and Technology: A Collection of Reviews from Nature Journals*, Macmillan, London, pp. 337–346.

10. Li, H., He, Y., Liu, Z., Huang, Y., and Jiang, B. (2017). Synchronous steam generation and heat collection in a broadband Ag@TiO$_2$ core/shell nanoparticle-based receiver. *Applied Thermal Engineering* 121, 617–627.

11. Ghasemi, H., Ni, G., Marconnet, A. M., Loomis, J., Yerci, S., Miljkovic, N., and Chen, G. (2014). Solar steam generation by heat localization. *Nature Communications* 5, 4449.

12. Choi, S. U. S., and Eastman, J. A. (1995). Enhancing thermal conductivity of fluids with nanoparticles. *ASME-Publications-Fed* 231, 99–106.

13. Wen, D., and Ding, Y. (2004). Experimental investigation into convective heat transfer of nanofluids at the entrance region under laminar flow conditions. *International Journal of Heat and Mass Transfer* 47(24), 5181–5188.

14. Otanicar, T. P., Phelan, P. E., Prasher, R. S., Rosengarten, G., and Taylor, R. A. (2010). Nanofluid-based direct absorption solar collector. *Journal of Renewable and Sustainable Energy* 2(3), 033102.

15. Zhang, H., Chen, H. J., Du, X., and Wen, D. (2014). Photothermal conversion characteristics of gold nanoparticle dispersions. *Solar Energy* 100, 141–147.

16. Jin, H., Lin, G., Bai, L., Zeiny, A., and Wen, D. (2016). Steam generation in a nanoparticle-based solar receiver. *Nano Energy* 28, 397–406.

17. Brongersma, M. L., Halas, N. J., and Nordlander, P. (2015). Plasmon-induced hot carrier science and technology. *Nature Nanotechnology* 10(1), 25.

18. Neumann, O., Neumann, A. D., Silva, E., Ayala-Orozco, C., Tian, S., Nordlander, P., and Halas, N. J. (2015). Nanoparticle-mediated, light-induced phase separations. *Nano Letters* 15(12), 7880–7885.

19. Neumann, O., Urban, A. S., Day, J., Lal, S., Nordlander, P., and Halas, N. J. (2013). Solar vapor generation enabled by nanoparticles. *ACS Nano* 7(1), 42–49.

20. Guo, A., Fu Y., Wang, G., and Wang, X. (2017). Diameter effect of gold nanoparticles on photothermal conversion for solar steam generation. *RSC Advances* 7(8), 4815–4824.

21. Jin, H., Lin, G., Bai, L., Amjad, M., Bandarra Filho, E. P., and Wen, D. (2016). Photothermal conversion efficiency of nanofluids: An experimental and numerical study. *Solar Energy* 139, 278–289.

22. Liu, C., Huang, J., Hsiung, C. E., Tian, Y., Wang, J., Han, Y, and Fratalocchi, A. (2017). High-performance large-scale solar steam generation with nanolayers of reusable biomimetic nanoparticles. *Advanced Sustainable Systems* 1(1–2), 1600013.

23. Wang, H., Miao, L., and Tanemura, S. (2017). Morphology control of Ag polyhedron nanoparticles for cost-effective and fast solar steam generation. *Solar RRL* 1(3–4), 1600023.

24. Gan, Z., Chen, Z., Liu, L., Zhang, L., Tu, W., and Liu, Y. (2017). Highly efficient solar-driven photothermal performance in Au-carbon core-shell nanospheres. *Solar RRL* 1(3–4), 1600032.

25. Ni, G., Miljkovic, N., Ghasemi, H., Huang, X., Boriskina, S. V., Lin, C. T., Wang, J. et al. (2015). Volumetric solar heating of nanofluids for direct vapor generation. *Nano Energy* 17, 290–301.

26. Wang, X., He, Y., Cheng, G., Shi, L., Liu, X., and Zhu, J. (2016). Direct vapor generation through localized solar heating via carbon-nanotube nanofluid. *Energy Conversion and Management* 130, 176–183.

27. Cui, P., Lee, J., Hwang, E., and Lee, H. (2011). One-pot reduction of graphene oxide at subzero temperatures. *Chemical Communications* 47(45), 12370–12372.

28. Liu, J., Wang, F., Zhang, L., Fang, X., and Zhang, Z. (2014). Thermodynamic properties and thermal stability of ionic liquid-based nanofluids containing graphene as advanced heat transfer fluids for medium-to-high-temperature applications. *Renewable Energy* 63, 519–523.

29. Liu, X., Wang, X., Huang, J., Cheng, G., and He, Y. (2018). Volumetric solar steam generation enhanced by reduced graphene oxide nanofluid. *Applied Energy* 220, 302–312.

30. Shahsavar, A., Saghafian, M., Salimpour, M. R., and Shafii, M. B. (2016). Experimental investigation on laminar forced convective heat transfer of ferrofluid loaded with carbon nanotubes under constant and alternating magnetic fields. *Experimental Thermal and Fluid Science* 76, 1–11.

31. Shi, L., He, Y., Huang, Y., and Jiang, B. (2017). Recyclable Fe$_3$O$_4$@CNT nanoparticles for high-efficiency solar vapor generation. *Energy Conversion and Management* 149, 401–408.

32. Wang, X., Ou, G., Wang, N., and Wu, H. (2016). Graphene-based recyclable photo-absorbers for high-efficiency seawater desalination. *ACS Applied Materials & Interfaces* 8(14), 9194–9199.

33. Liu, Y., Wang, X., and Wu, H. (2017). High-performance wastewater treatment based on reusable functional photo-absorbers. *Chemical Engineering Journal* 309, 787–794.

34. Chen, W., Zou, C., Li, X., and Liang, H. (2019). Application of recoverable carbon nanotube nanofluids in solar desalination system: An experimental investigation. *Desalination* 451, 92–101.

35. Fang, Z., Zhen, Y. R., Neumann, O., Polman, A., García de Abajo, F. J., Nordlander, P., and Halas, N. J. (2013). Evolution of light-induced vapor generation at a liquid-immersed metallic nanoparticle. *Nano Letters* 13(4), 1736–1742.

36. Lukianova-Hleb, E., Hu, Y., Latterini, L., Tarpani, L., Lee, S., Drezek, R. A., Hafner, J. H., and Lapotko, D. O. (2010). Plasmonic nanobubbles as transient vapor nanobubbles generated around plasmonic nanoparticles. *ACS Nano* 4(4), 2109–2123.

37. Lombard, J., Biben, T., and Merabia, S. (2014). Kinetics of nanobubble generation around overheated nanoparticles. *Physical Review Letters* 112(10), 105701.

38. Jin, H., Lin, G., Bai, L., Zeiny, A., and Wen, D. (2016). Steam generation in a nanoparticle-based solar receiver. *Nano Energy* 28, 397–406.

39. Carlson, M. T., Green, A. J., and Richardson, H. H. (2012). Superheating water by CW excitation of gold nanodots. *Nano Letters* 12, 1534–1537.

40. Baral, S., Green, A. J., Livshits, M., Govorov, A. O., Richardson, H. H., and States, U. (2014). Comparison of vapor formation of water at the solid/water interface to colloidal solutions using optically excited gold nanostructures. *ACS Nano* 8(2), 1439–1448.

41. Keblinski, P., Cahill, D. G., Bodapati, A., Sullivan, C. R., and Taton, T. A. (2006). Limits of localized heating by electromagnetically excited nanoparticles. *Journal of Applied Physics* 100(5), 054305.

42. Hogan, N. J., Urban, A. S., Ayala-Orozco, C., Pimpinelli, A., Nordlander, P., and Halas, N. J. (2014). Nanoparticles heat through light localization. *Nano Letters* 14, 4640–4645.

43. Ji, X., Song, X., Li, J., Bai, Y., Yang, W., and Peng, X. (2007). Size control of gold nanocrystals in citrate reduction: The third role of citrate. *Journal of the American Chemical Society* 129(45), 13939–13948.

44. Wang, X., He, Y., Liu, X., Shi, L., and Zhu, J. (2017). Investigation of photothermal heating enabled by plasmonic nanofluids for direct solar steam generation. *Solar Energy* 157, 35–46.

45. Cobley, C. M., Skrabalak, S. E., Campbell, D. J., and Xia, Y. (2009). Shape-controlled synthesis of silver nanoparticles for plasmonic and sensing applications. *Plasmonics* 4, 171–179.

46. Li, H., He, Y., Liu, Z., Jiang, B., and Huang, Y. (2017). Rapid synthesis of broadband Ag@TiO$_2$ core–shell nanoparticles for solar energy conversion. *Solar Energy Materials and Solar Cells* 166, 52–60.

47. Swinehart, D. F. (1962). The Beer-Lambert law. *Journal of Chemical Education* 39(7), 333.

48. Guo, P., Chen, P., and Liu, M. (2013). One-dimensional porphyrin nanoassemblies assisted via graphene oxide: Sheetlike functional surfactant and enhanced photocatalytic behaviors. *ACS Applied Materials & Interfaces* 5(11), 5336–5345.

49. Shi, L., He, Y., Wang, X., and Hu, Y. (2018). Recyclable photo-thermal conversion and purification systems via Fe$_3$O$_4$@TiO$_2$ nanoparticles. *Energy Conversion and Management* 171, 272–278.

50. Bae, K., Kang, G., Cho, S. K., Park, W., Kim, K., and Padilla, W. J. (2015). Flexible thin-film black gold membranes with ultrabroadband plasmonic nanofocusing for efficient solar vapour generation. *Nature Communications* 6, 10103.

51. Blees, M. H. (2002). Foundations of colloid science. *Colloids and Surfaces A: Physicochemical and Engineering Aspects* 210, 125.

52. Baffou, G., Girard, C., and Quidant, R. (2010). Mapping heat origin in plasmonic structures. *Physical Review Letters* 104(13), 136805.

53. Govorov, A. O., Zhang, W., Skeini, T., Richardson, H., Lee, J., and Kotov, N. A. (2006). Gold nanoparticle ensembles as heaters and actuators: Melting and collective plasmon resonances. *Nanoscale Research Letters* 1(1), 84–90.

15

Supersteam Production Using Magnetic Nanofluids

K.R.V. Subramanian*, Avinash Balakrishnan, Tubati Nageswara Rao, and A. Radhakrishna

CONTENTS

15.1 Introduction

Magnetic ferrofluids or nanofluids have been the subject of study mainly for their unique magnetic (superparamagnetic) and thermal conductivity enhancements (Nkurikiyimfura et al. 2013, Parekh 2014). The ferrofluids consist of colloidal magnetic nanocrystals dispersed in a medium of water or organic mediums like ethylene glycol. The nanocrystals are stabilized by surfactants to prevent agglomeration due to van der Waal's forces and exchange interaction. Also, the constant motion of the nanocrystals take place due to Brownian motion forces. Brownian motion has been proven to aid thermal conductivity and heat transfer due to diffusion and micro-convection currents as per reference above. The thermal conductivity enhancements yield good values of Nusselt numbers in the range of 5–50 (Bhimani et al. 2013), which are far greater than typical values for conventional heat transfer fluids.

Nanofluids (non-magnetic) have been used in an organic rankine cycle to improve overall efficiencies (Quoilin et al. 2011). The organic rankine cycle is usually coupled with solar concentrator system to capture solar heat and convert to useful output like electricity. It has also been reported (J. Appl. Phys, 105, 07B324 2009) that a magnetic nanofluid when subjected to an alternating current magnetic field produces heat energy due to the fact that alternating current (AC) magnetic fields cause the magnetization field vector to repeatedly relax and to produce heat. Brown and Neel relaxations are dominant means of power dissipation.

There have been previous successful attempts to couple solar heat (using parabolic concentrators) to run a solar sanitation system (or solar toilet) (http://www.colorado.edu/news/releases/2014/03/12/innovative-solar-powered-toilet-developed-cu-boulder-ready-india-unveiling, accessed on 14 April 2014).

The self-contained, waterless toilet, designed and built using a $777,000 grant from the Bill & Melinda Gates Foundation, has the capability of heating human waste to a high enough temperature to sterilize human waste and create biochar, a highly porous charcoal. The Colorado University (CU)-Boulder invention consists of eight parabolic mirrors that focus concentrated sunlight to a spot no larger than a postage stamp on a quartz-glass rod connected to eight bundles of fiber-optic cables, each consisting

---

* This author has given maximum contribution to this chapter.

of thousands of intertwined, fused fibers. The energy generated by the sun and transferred to the fiber-optic cable system—similar in some ways to a data transmission line—can heat up the reaction chamber to over 600°F to treat the waste material, disinfect pathogens in both feces and urine, and produce char. Tests have shown that each of the eight fiber-optic cables can produce between 80 and 90 W of energy, meaning the whole system can deliver up to 700 W of energy into the reaction chamber. Tests at CU-Boulder showed the solar energy directed into the reaction chamber could easily boil water and effectively carbonize solid waste.

Other attempts have also been made by Caltech (http://www.caltech.edu/news/caltech-wins-toilet-challenge-23635). Michael Hoffmann, the James Irvine Professor of Environmental Science at Caltech, and his team were awarded a $400,000 grant to create a toilet that can safely dispose of human waste for just 5 cents per user per day. The lavatory can't use a septic system or an outside water source or produce pollutants. Hoffmann's proposal—which won one of the eight grants given—was to build a toilet that uses the sun to power an electrochemical reactor. The reactor breaks down water and human waste into fertilizer and hydrogen, which can be stored in hydrogen fuel cells as energy. The treated water can then be reused to flush the toilet or for irrigation. It starts with a photovoltaic or solar panel, which converts the sun's rays into enough energy to power an electrochemical reactor that Hoffmann designed to break down water and human waste material into hydrogen gas. The hydrogen gas can then be stored in hydrogen fuel cells to provide a backup energy source for nighttime operation or for use under low-sunlight conditions. Hoffmann also envisions equipping the units with self-cleaning toilets that would also be powered by the energy from the sun and fuel cells.

So far, there have been no attempts nationally (or worldwide) to build a system using magnetic nanofluids to capture and process solar heat and subsequently use for the purposes of solar sanitation/de-salination.

15.2 Design and Process

15.2.1 Background and Societal Needs

- In our country, a large proportion of people do not have access to adequate sanitation and toilet facilities. Even if toilets are provided, they are not clean
- Need for an effective methodology for sanitation of toilets (solar toilets) to effectively kill the microbes and germs and maintain a sanitized environment

15.2.2 Challenges Faced

- Natural disasters
- Refugee camps
- Rural and remote areas
- Urban slums
- To meet the sanitation requirements of above
- Solar energy latrines as a solution to the challenges

15.2.3 Solar Energy Green Toilets

- Challenges faced by the present human society during the emergency camps set up for the people taking refuge from events such as wars and natural disasters. These conditions are worse for conflict/disaster hit areas where a large human population is suddenly gathered, in such refugee camps
- To meet the sanitation requirements for emergency camps
- Propose solar energy latrines as a solution to the challenges faced by the present human society during the emergency camps

FIGURE 15.1 A typical solar toilet unit.

- The solar powered latrine unit is perfect for those events where quality and safety is a must, where power is not available
- The solar latrine unit features (Figure 15.1):
  - A cottage interior with a granite counter top,
  - Hardwood simulated floors and
  - A stainless steel sink (as an example) or any suitable sink
- Current waterless toilets—dry pit latrines, ventilated improved pit (VIP) latrines, and composting toilets

15.2.4 Current Toilet Technologies

- Dry pit latrines and VIPs are plagued with difficulties associated with the removal of contents from full pits, and the need to treat the waste offsite once it is removed poses additional risks to public health and the environment
- Composting toilets require an external power source or intensive user input in order to properly maintain the compost, are often associated with undesirable odors, and require extended

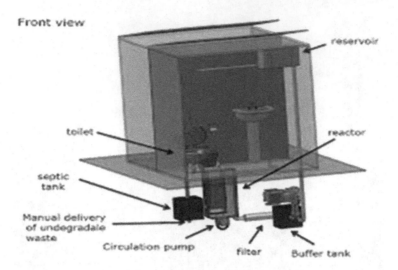

FIGURE 15.2   An internal view of the solar toilet unit. The magnetic nanofluid is processed in the reactor chamber.

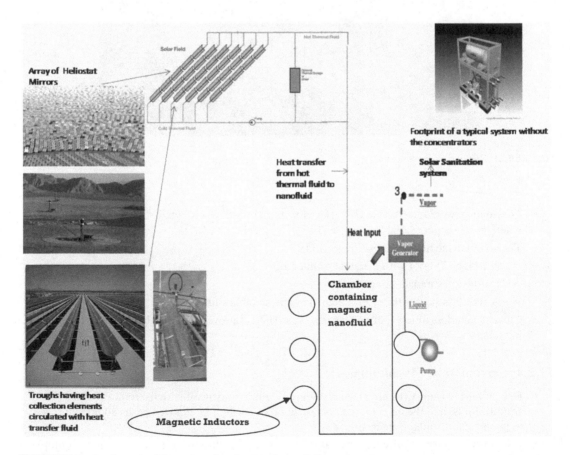

FIGURE 15.3   A schematic of the process flow in our solar sanitation system.

lengths of time before the waste is rendered safe and suitable for use as a soil amendment in agriculture
- The present proposed solar latrines can also overcome these difficulties as they can be integrated to a waste treatment circuit to develop sanitation solutions that are affordable and desirable to use, to render fecal waste harmless within a short time-span
- The proposed solar latrines are self-contained without the need for flush water or electricity and can also produce valuable end products such as manure and biogas, which further can be used as energy source for the emergency camps cooking needs and the manure for to grow the vegetables in the near agriculture fields (Figures 15.2 and 15.3)

1. Development of the magnetic nanofluids consisting of simple Fe_3O_4 and characterize for stability, magnetic properties, and thermal conductivity properties:

    Magnetic ferrofluids or nanofluids have been the subject of study mainly for their unique magnetic (superparamagnetic) and thermal conductivity enhancements (Nkurikiyimfura et al. 2013). The ferrofluids consist of colloidal magnetic nanocrystals dispersed in a medium of water or organic mediums like ethylene glycol. The nanocrystals are stabilized by surfactants to prevent agglomeration due to van der Waal's forces and exchange interaction. Also, the constant motion of the nanocrystals take place due to Brownian motion forces. Brownian motion has been proven to aid thermal conductivity and heat transfer due to diffusion and micro-convection currents.

    It is proposed to synthesize colloidal magnetite nanocrystals by sol-gel methods using organometallic precursors and stabilize them using suitable surfactants. Using a similar approach, the researchers have prior experience (Subramanian, 2006) working on and successfully synthesizing yttrium-iron-garnet nanomagnetic systems with their property characterization.

2. Design of a parabolic dish for capture of solar heat with the primary heat transfer fluid system (Figure 15.4):

    Important parameters:

    - Vertex V
    - Focus F
    - Point of incidence of light (sun) on the parabolic dish R
    - Point on directrix line D
    - Focal length f
    - Rim angle ψ
    - Height of dish h
    - Diameter of dish opening d
    - Parabolic radius p

    Why parabolic shape is used?

    - It has the property that, for any line parallel to the axis of the parabola, the angle p between it and the surface normal is equal to the angle between the normal and a line to the focal point
    - Solar radiation arrives at the earth in essentially parallel rays
    - By Snell's law, the angle of reflection equals the angle of incidence, all radiation parallel to the axis of the parabola will be reflected to a single point F, which is the focus

    Useful equations for design:

    - Height of parabola

$$h = d^2/16f$$

- Rim angle

$$\tan \psi = 1/((d/8h)-(2h/d))$$

Paraboloid (Figure 15.5):

Useful equations for design:

- Full surface area of a paraboloid

$$A = (8\pi f^2/3) \{[(d/4f)^2 + 1]^{3/2} - 1\}$$

- Concentrator aperture area (Figure 15.6)

$$A = \pi d^2/4$$

Total radiant flux:

- $dA = I.ds = I.p. \ d\psi/(\cos (\psi/2))$
- $d\phi$, radiant flux $= dA.I.\cos (\psi/2) = I.p.I. \ d\psi$
- $p = 2f/(1 + \cos \psi)$, parabolic radius
- Radiant flux from a parabolic dish

$$d\phi = 8 \ \pi If^2 \sin \psi \ d \ \psi/(1 + \cos \psi)^2$$

I is intensity of incident solar radiation

Solution of integral for flux:

- Let $1 + \cos \psi = t$

$$-\sin \psi \ d \ \psi = dt$$

Solving the integral $8 \ \pi If^2 \int dt/t^2$

Between rim angle limits 15° and 150°, it becomes:

$1333 \ \pi If^2$

Contribution of errors and losses:

- Treatment is based on the incident light rays being parallel. This is not true for solar applications. Because of the finite angular size of the sun's disc (about 33 minutes of arc or 9.6 mrad), the sun's rays reaching the concentrator are not parallel
- Rather than all the incident rays being reflected to the focal point (or line in the case of a trough), the reflected rays form an "image" of finite size centered about the focus
- Total effective error = 7 mrad

This will cause about 5% losses going by 2 sigma limits

Important figures of merit for the solar toilet (by putting values):

- Diameter of dish opening $d = 2.5$ m (assuming dish is mounted on top of a single toilet unit of similar span)
- Height of dish $h = 1.5$ m
- Focal length $f = 0.26$ m
- Rim angle $\psi = 135°$
- Full surface area of paraboloid $A = 0.022$ m^2
- Concentrator aperture area = 4.9 m^2
- Total radiant solar flux = 0.28 MW/m^2 (assuming intensity of sunlight at earth's surface is 985 W/m^2)

- Assuming 5% losses, flux converted to heat energy in 3 hours daytime irradiation is: $2.9 * 10^9$ Joules/m^2.

Primary heat transfer fluid (HTF):

Heat transfer fluid	Peak flux (MW/m^2)
Liquid sodium	1.5
Molten nitrate salt	0.8
Steam vapor	0.4
Air	0.22

Preferred HTF's are NaK and molten salt

- NaK

  Operating temperature: 785°C, 1100°C (pressurized)

  Specific heat capacity: 0.94 kJ/kg. K

  Thermal conductivity: 26 W/m. K

- Molten salt (sodium nitrate 60% + potassium nitrate 40%)

  Operating temperature: 567°C

  Specific heat capacity: 1.56 kJ/kg. K

  Thermal conductivity: 0.4 W/m. K

3. Design of heat transfer to the magnetic nanofluid together with AC magnetization heating of the magnetic fluid

Main components of evaporator chamber containing magnetic nanofluid:

- The primary heat transfer fluid (NaK or molten salt) is enabled to transfer solar captured heat to this evaporator chamber through a HTF pump.
- The evaporator chamber is also surrounded by a current carrying coil which produces a magnetic field of suitable strength when current flow is active and can be de-magnetized when the current flow is decreased to zero, i.e., current flow is tunable.
- The magnetic ferrofluid can be additionally subjected to an AC magnetic field. It has been studied that such a process produces heating effects and has the ability to convert the fer-rofluid to high temperature vapor (or steam).
- The evaporator chamber contains a secondary HTF composed of stabilized magnetic nanocrystals suspended in water/ethylene glycol.

Some design variables of evaporator unit:

- Considering the heat flux from the parabolic dish as 0.23 MW/m^2 (refer earlier section), the evaporator unit is designed as a 20 kW (small scale unit: 10^8 Joules/m^2) high tem-perature system with magnetic nanofluid as working fluid and evaporating temperature of ~150°C–170°C and solar insolation as 800 W/m^2.
- Excess heat from the solar radiant flux (MW $\rightarrow$ kW) will be stored in a thermal storage unit and then drawn into the evaporator unit when necessary.
- Corrugated plate heat exchanger is used, hot fluid is heated, magnetic nanofluid fed from the accessory chamber. Cold fluid is fluid like pentane.

15.2.5 Design: Evaporator

- Design by mean temperature difference method for counter flow heat exchangers
- Turbulent heat transfer:

  $Q = 2\pi k (\Delta T)$; heat input into evaporator can be taken as enthalpy of the magnetic nanofluid; k is heat transfer coefficient.

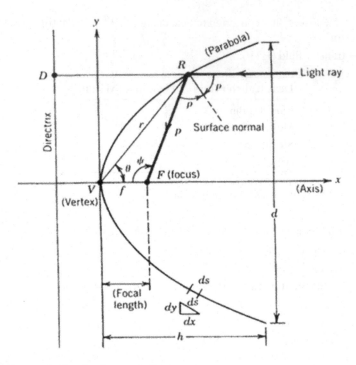

FIGURE 15.4 Ray diagram for focusing in a parabolic concentrator.

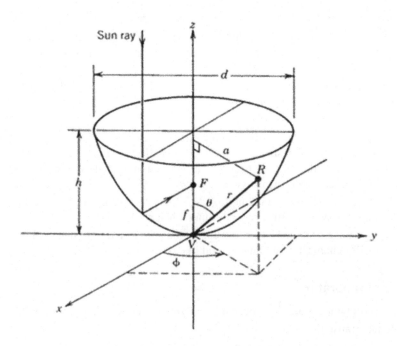

FIGURE 15.5 Ray diagram and parametric representation for a paraboloid.

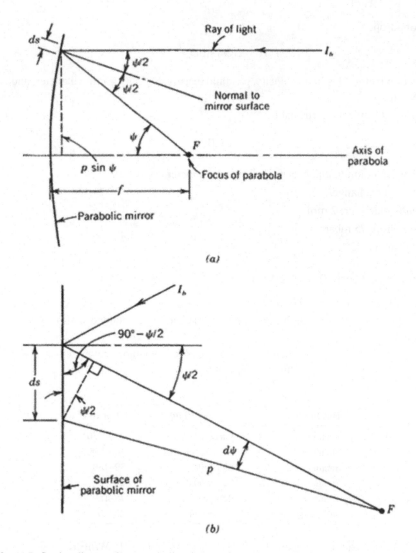

FIGURE 15.6 (a) Reflection diagram for a parabolic mirror surface; (b) Angular representations for the same.

- Calculate evaporating temperature from ΔT (this can be taken as "super steam" temperature)
- Heat exchanger divided into three moving boundary zones, each of them characterized by heat transfer area A and heat transfer coefficient U; $1/U = (1/h_1) + (1/h_2)$
- Total heat transfer area of the heat exchanger:

$$(Np\text{–}2).\ L.\ W;$$

$$L = A_{tot}/W$$

where Np is number of plates, L is plate length, and W is plate width.
- Forced convection heat transfer coefficient h_1, Nusselt number:
  $CRe^m Pr^n$, where Re is Reynold's number and Pr is Prandtl number

- Pressure drop:

$$2fG^2L/(\rho D_h),$$

f is friction factor, G is mass velocity, ρ is fluid density, D_h is hydraulic diameter, and L is plate length.
- Boiling heat transfer coefficient h_2:

$$Ch_1B_o^{0.5},$$

B_o is boiling number, and h_1 is no-boiling heat transfer coefficient.
- Size of heat exchanger:
  Hydraulic diameter: 2 mm
  Pressure drop: 75 mbar
  Chevron angle: 45°

15.2.6 Figures of Merit (Putting Values)

- Heat input to evaporator: 20,000 W
- Nusselt number: 40 (considering a nominal value for, e.g., obtained with magnetic nanofluid in water)
- Boiling Heat Transfer Coefficient (HTC): 80 (please refer table below) considering output as steam with light oils as cold fluid in evaporator
- Evaporators

Hot Fluid	Cold Fluid	Overall U
Steam	Water	350–750
Steam	Organic solvents	100–200
Steam	Light oils	80–180
Steam	Heavy oils (vacuum)	25–75
Water	Refrigerant	75–150
Organic solvents	Refrigerant	30–100

- Overall heat transfer coefficient: $(1/40) + (1/80) = (1/U)$; $U = 25$ W/m^2.K
- Number of plates = 10; length = 0.5 m; width = 0.2 m; and area = 0.9 m^2
- Assuming ambient temperature of fluid as 298 K,

$$20000 = 2\pi * 25 * 0.9 (T - 298)$$

- Solving, evaporator temperature T = 439 K (166°C); super steam/vapor temperature is 165°C

It has also been reported (J. Appl. Phys, 105, 07B324 2009) that a magnetic nanofluid when subjected to an alternating current magnetic field produces heat energy due to the fact that AC magnetic fields cause the magnetization field vector to repeatedly relax and to produce heat. Brown and Neel relaxations are dominant means of power dissipation.

4. Conversion of magnetic nanofluid to superheated vapor/steam: A design estimation of the expected superheat of the vapor/steam has been completed
5. Demonstration of production of biochar from human fecal waste by contact with the superheated vapor/steam

15.3 Conclusions

This work is positioned as a proof-of-concept and if successfully executed can form the basis of development of solar toilets and solar de-salination systems which are based on elegant and technologically superior designs and sub-systems and result in higher process efficiencies with lower costs and footprints.

REFERENCES

Bhimani, V. L., Rathod, P. P., and Sorathiya, A. S. (2013) Experimental study of heat transfer enhancement using water based nanofluids as a new coolant for car radiators. *The International Journal of Emerging Technology and Advanced Engineering* 3(6), 295–302.

Nkurikiyimfura, I., Wamg, Y., and Pan, Z. (2013). Heat transfer enhancement by magnetic nanofluids—A review. *Renewable and Sustainable Energy Reviews* 1, 548–561.

Parekh K. J. (2014). Thermo-magnetic properties of ternary polydispersed $Mn_{0.5}Zn_{0.5}Fe_2O_4$ ferrite magnetic fluid. *Solid State Communications* 187, 33–37.

Quoilin, S., Orosz, M., Hemond, H., and Lemort, V. (2011). Performance and design optimization of a low-cost solar organic Rankine cycle for remote power generation. *Solar Energy* 85(5), 955–966.

Subramanian, K. R. V. (2006). PhD thesis, Spin-coatable oxide resists for electron beam nanolithography, Cambridge University.

16

Enhanced Thermophysical Properties of the Heat Transfer Fluids for Concentrating Solar Power by Using Metal-Based Nanofluids: An Experimental and Theoretical Overview

Antonio Sánchez-Coronilla, Javier Navas, Elisa I. Martín, Roberto Gómez-Villarejo, Teresa Aguilar, Juan Jesús Gallardo, Miriam Teruel, Rodrigo Alcántara, and Concha Fernández-Lorenzo

CONTENTS

16.1 Introduction

To find solutions for the growing demand for energy while minimizing the impact that energy sources have on the environment is one of the main goals of the society of twenty-first century [1]. Solar energy is one of the options generating most interest [2], and transforming the sun energy by using concentrating solar power (CSP) is one of the most interesting alternatives for transforming thermal energy from the sun into electricity [3]. One of the most common radiation collectors designed for this technology is the parabolic cylinder collector [3], where the radiation is targeted onto a collector in which a heat transfer fluid (HTF) circulates, and it is heated. The heat is stored and transferred to a heat exchanger to generate steam, which moves a turbine [4]. Thus, improving the thermal properties of the HTF used in CSP plants improves the efficiency of the plants. The use of nanofluid (colloidal suspensions of nanometric systems in a base fluid) seems to be an interesting alternative to lead to improvements in the heat transfer processes taking place by improving the thermal properties of the base fluids [5–7]. Improvements in properties of nanofluids such as isobaric specific heat [6,8–12] or thermal conductivity have been reported [6,9,13–15]. However, there are few studies in the literature about nanofluids using the typical HTFs used in CSP as a base fluid, being very different in nature to water or ethylene glycol, which are the fluids that have been studied most for use as the base fluid for nanofluids.

In turn, the analysis of nanofluids based on the base fluids that are used as HTFs in CSP plants is of interest. Therefore, in this chapter, the preparation of Cu, Ag, and Au nanofluids based on the HTF used in CSP plants, which is the eutectic mixture of biphenyl ($C_{12}H_{10}$) and diphenyl oxide ($C_{12}H_{10}O$) is presented. Metallic nanoparticles were used to prepare the nanofluid because they are known to increase the thermal conductivity of base fluids [6,12,14–16]. The stability of the nanofluids is presented by using particle size and ζ potential measurements and UV-V is spectroscopy. Properties related to the performance of the nanofluids such as density, viscosity, isobaric specific heat, and thermal conductivity were also analyzed. In addition, those metallic nanofluids are rationalized through a molecular level perspective by using molecular dynamic calculations to gain an insight into the molecular behavior of this kind of nanofluid system. The isobaric specific heat values were obtained and compared with the experimental values following a good qualitative correlation. The structural properties of the metallic nanofluid systems were obtained by analyzing their spatial distribution function (SDF). Theoretical results show diphenyl oxide molecules occupy a first layer of molecules around the metal playing a crucial role in enhancing the thermal properties.

16.2 Methods

16.2.1 Preparation of the Cu, Ag, and Au-Nanofluids

The nanofluids were prepared by means of the two-step method that consists of the dispersion of nanoparticles into the base fluid. The nanoparticles, or any other kind of nanomaterial such as nanofibers or nanotubes, are synthesized (first step) and then dispersed into the base fluid (second step). The base fluid used was supplied by The Dow Chemical Company©, model Dowtherm A, and it is a eutectic mixture of biphenyl ($C_{12}H_{10}$, 26.5%) and diphenyl oxide ($C_{12}H_{10}O$, 73.5%). The nanoparticles used were supplied by Sigma-Aldrich©. According to the supplier, the Cu nanoparticles (purity ≥99.5, density 8940 kg m^{-3} at 298 K) show a particle size of 40 nm–60 nm. To prepare the nanofluids, an initial nanofluid with a nanoparticle mass concentration of 0.01 wt% was prepared. For this, 100 mL of the base fluid, the adequate amount of the nanoparticles of Cu, and the same proportion of polyethylene glycol (PEG, MW: 400, Sigma-Aldrich©) were mixed and kept under sonication for 3 hours (~50 W output power) using a Sonics Vibra Cell VCX 750 sonicator. Immediately, portions of these nanofluids were used to prepare nanofluids with a nanoparticle mass concentration of 0.5×10^{-4} wt%, 1.0×10^{-4} wt%, and 5.0×10^{-4} wt%. Next, the required amount of the base fluid to obtain 100 mL of the nanofluid with these nanoparticle mass concentrations was added, and the mixture was kept under sonication again for 3 hours (~50 W output power). In the case of silver nanofluid, we have used commercial nanoparticles (purity ≥99%, density 10.490 kg m^{-3} at 298 K, Sigma-Aldrich©) with a particle size below 100 nm. Gold nanoparticles were synthesized by the citrate reduction method ($C_6H_5Na_3O_7$ · 2H$_2$O, purity >99.0%, supplied by PanReac AppliChem©) using tetrachloroauric acid (HAuCl$_4$, assay 99.9%, Sigma-Aldrich©) as gold precursor agent. This reduction method constitutes the first step in the nanofluid preparation. The second step consists on dispersing the nanomaterial in the base fluid. Three nanofluids of each metal, with a mass nanoparticle concentration of 0.5×10^{-4} wt%, 1.0×10^{-4} wt%, and 5.0×10^{-4} wt% were prepared by dilution of other nanofluid with 0.01 wt% and the same amount of polyethylene glycol (PEG, average molecular: 5000–7000, supplied by PanReac AppliChem©). All nanofluids were prepared using 100 mL of the base fluid. A sonication method was used in all of them to obtain the dispersion of nanoparticles. It was applied for 3 hours (~50 W output power) using a Sonics Vibra Cell VCX 750 sonicator, controlling the temperature at 293 K in a thermal bath. However, for gold nanofluid, one-step method with the presence of a surfactant has been reported based on the synthesis of gold nanoparticles in the base fluid [17]. An initial nanofluid with a nanoparticle mass concentration of 0.01 wt% was prepared through a one-step method which involves dissolving a certain amount of tetraoctylammonium bromide ($(C_8H_{17})_4NBr$, TOAB, purity >98%, supplied by Sigma-Aldrich©) in 100 mL of the base fluid.

16.2.2 Characterization of Nanofluids

Stability is of importance in nanofluids because their thermal properties depend heavily on whether they are stable or unstable. UV-vis spectroscopy (UV-vis) can provide a measurable characterization of stability by evaluating the absorbance of a suspension [18]. To this end, UV-vis spectra were recorded using a halogen lamp DH-2000-Bal supplied by Ocean Optics© and a monochromator USB-2000 + supplied by Ocean Optics© operating in the range of wavelength of 400 nm–880 nm, using a glass cuvette at room temperature. Also, ζ potential measurements were performed using Zetasizer Nano ZS supplied by Malvern Instruments Ltd. This system uses the principle of electrophoresis light scattering (ELS), in which charged particles suspended in a fluid are attracted toward the oppositely charged electrode when an electric field is applied. The ζ potential measurements were performed applying a potential of 120 V at 313 K. Particle size and size distribution were also measured by Zetasizer Nano ZS, using the principle of dynamic light scattering (DLS) as a simple method for analyzing suspension stability and particle size measurements in solution [19,20].

To characterize the nanofluids performance, density, viscosity, isobaric specific heat, and thermal conductivity were determined. Density was estimated using a pycnometer and a thermal bath supplied by Select© to control the temperature of the measurements. The density values of each nanofluid were determined five times to obtain the most accurate value. Dynamic viscosity was measured using a SV-10 viscometer supplied by Malvern Instruments Ltd. A water-circulating bath supplied by VWR© with heating and cooling features was used to keep the nanofluids at room temperature during the experiment. The measurements were performed five times to calculate the average values. The isobaric specific heat measurements were performed using a temperature modulated differential scanning calorimeter (TMDSC), supplied by TA Instruments©, model Q20. To perform the measurements, a program was created which can be summarized as: the temperature was equilibrated at 341 K to remove contaminants and kept isothermal for 10 minutes, then the samples were equilibrated at 301 K and then ramped to 391 K at 1 K/min. A modulation was programmed around the studied temperatures with an amplitude of ±1 K a period of 120 seconds. Finally, cooling was performed at 1 K/min. The isobaric specific heat of the base fluid was measured to test the method used with regard to the values reported by the supplier. Finally, the thermal conductivity of the nanofluids was measured at several temperatures using the laser flash technique (LFA 1600 equipment, supplied by Linseis Thermal Analysis©). This technique really measures thermal diffusivity, which is the thermophysical property that defines the speed of heat propagation by conduction during changes of temperature. According to Standard ASTM E1461-01, the relationship between both properties is given by the equation:

$$k(T) = D(T) \cdot C_P(T) \cdot \rho(T) \tag{16.1}$$

where k is the thermal conductivity, D the thermal diffusivity, C_P is the isobaric specific heat, and ρ is the density. All the thermal measurements were performed in triplicate.

16.2.3 Computational Framework

The transferable potentials for phase equilibria (TraPPE)-Explicit Hydrogen (EH) force field [21,22] was used to describe the intra and intermolecular interactions of the HTF fluid (diphenyl oxide/biphenyl blend). The TraPPE-EH force field treats aromatic rings and the directly connected atoms as rigid entities. The phenyl rings were treated as rigid, but were allowed to rotate with regard to each other around the C1–C1′ bond of the biphenyls. The parameters which have been used in our simulations to describe the metal nanoparticle were adapted to a non-bonded dummy model of six particles, referred to as "dummy atoms", placed around a central metal particle in an octahedral geometry. The geometry of the dummy complex itself is kept rigid by the imposition of large force constants on the metal-dummy bonds. However, as there are no bonds between the dummy complex and the surrounding ligands, overall rotation of the six-center frame about the nucleus is allowed, and no internal forces are associated with such rotation. Therefore, the coordination geometry is not constrained to the geometry of the dummy model used, but rather, the system is free to exchange ligands.

The TraPPE and the non-bonded dummy model for the metals use Lennard-Jones (LJ) and Coulomb potentials to represent the non-bonded interactions:

$$u\left(r_{ij}\right) = 4\varepsilon_{ij}\left[\left(\frac{\sigma_{ij}}{r_{ij}}\right)^{12} - \left(\frac{\sigma_{ij}}{r_{ij}}\right)^{6}\right] + \frac{q_i q_j}{4\pi\varepsilon_0 r_{ij}} \tag{16.2}$$

where r_{ij}, ε_{ij}, σ_{ij}, q_i, q_j, and ε_0 are the distance between interaction sites i and j, the LJ well depth, the LJ diameter, the partial charges on interaction sites i and j, and the permittivity of vacuum, respectively. The Lorentz-Berthelot combining rules were used to determine LJ parameters for unlike interactions.

Molecular dynamics simulations were performed with the DL POLY code [23] in the canonical ensemble (NVT) using a Nose-Hoover thermostat and periodic boundary conditions. The initial configurations were built with the PACKMOL code [24] providing cubic boxes in which the length of the box sides are chosen to keep the density of the experimental HTF at 298 K (1059 kg m^{-3}). A time step of 0.5 fs was employed and simulation runs lasted for 1 ns. For the trajectory analysis, structures were saved every 100 time steps. A cut-off distance of 9 Å was applied in all the cases and the Ewald sum methodology [25] applied to account for the electrostatic interactions.

16.3 Results and Discussion

16.3.1 Nanofluid Stability

For considering to be stable, the nanofluids concentration or particle size should remain constant [26]. After preparation, the nanoparticles tend to agglomerate due to attraction between them, resulting in some clusters with a high weight that tends to precipitate and sediment [27,28].

UV-vis spectroscopy is one of the techniques used to analyze the stability of the nanofluids and the sedimentation process that can occur. By using Vis-near-infrared (NIR) spectroscopy, it is possible to determine whether the addition of nanoparticles produced chemical modifications in the base fluid that affect negatively to its properties. Vis-NIR spectra were recorded in a range between 400 and 1800 nm. As an example, Figure 16.1 shows the spectra recorded for the Cu nanofluid in the range between 400 and 1800 nm. It shows that the spectra of the nanofluid have a higher absorbance than the base fluid due to the dispersion processes

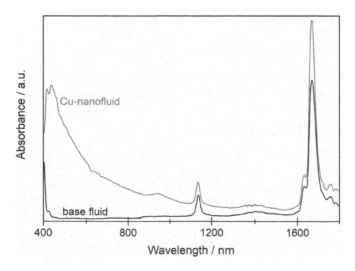

FIGURE 16.1   Vis-NIR spectra for the Cu nanofluid and base fluid (included for comparison purposes).

that are usually produced in colloidal suspensions. A wide absorption band assignable to the presence of nanoparticles can also be seen centered between 400 and 450 nm, which is in agreement with other studies in the literature and with the coloring of the nanofluids prepared. Meanwhile, the spectrum of the base fluid showed two clearly differentiated bands centered at approximately 1136 and 1668 nm, while the spectra obtained for the nanofluids also showed these bands and correlate well with those of the base fluid. Thus, the incorporation of Cu nanoparticles into the fluid does not cause any chemical changes in it.

The chemical stability of the Ag and Au nanofluids were also analyzed. Vis-NIR spectra shows that the nanofluids have a wide band up to approximately 700 nm, which is typical for colloidal suspension systems. Scattering process generated by the nanoparticles in the colloidal suspension produces absorbance around 500 nm–600 nm, which is higher at lower wavelengths, as reported previously [29,30]. Furthermore, this process is known to mainly take place at short wavelengths. All the spectra show two bands at approximately 1135 and 1679 nm. It is well-known that the highest absorbance of Au nanoparticles occurs at $\lambda = 520$ nm [17,31,32]. These bands are observed in the spectrum for the base fluid, so they can be assigned to the fluid itself, the addition of nanoparticles having no effect. No other bands are observed in the spectra. So, as expounded above for Cu nanofluid, that result suggests that no chemical changes take place in the base fluid. Bands at 630 and 870 nm appear both in Au nanofluids and the base fluid, and they can be assigned to the fluid itself. In general, there is a linear relationship between the absorbance and concentration of nanoparticles in the base fluid [26]. In the case of Ag nanofluids, it appears only as a soft peak at 870 nm that corresponds also of the base fluid. Therefore, in this case, there is no chemical modification in the base fluid by the incorporation of silver nanoparticles.

The physical stability of the nanofluids is of interest for the applications of these systems in concentrating solar power. DLS technique was used for the analysis of the physical stability of the nanofluids by measuring the particle sizes. These measurements were performed for approximately 1 week, taking several measurements each day and in triplicate. The results for silver and gold nanofluids at 5.0 · wt% are shown in Figure 16.2 as examples of less stable nanofluids. With the DLS technique, the particle size obtained corresponds to the hydrodynamic diameter, so values recorded were higher than the nominal values of the nanoparticles [33], and the values obtained using this technique always overestimate the size of the particle. The hydrodynamic diameter is determined as the sum of the particle diameter and the Debye length. The Debye length is the thickness of the diffuse layer, a layer of species between the surface of the nanoparticle and the slipping plane that moves with the nanoparticle within the base fluid [33]. For the nanofluids at lower concentration, the particle size increases, but then remains practically stable. However, in the case of the nanofluid with the highest concentration (Figure 16.2), it starts with a higher value at time zero and a steadier increase is produced suggesting those Ag and Au nanofluids are less stable. Thus, except for the fluid with the highest concentration (Figure 16.2), the nanoparticles added to the base fluid agglomerate after a certain time and then remains table [19,34,35]. The ζ potential is the surface charge at the "slipping plane", based on the electrophoresis theory. A colloidal-nanoparticle

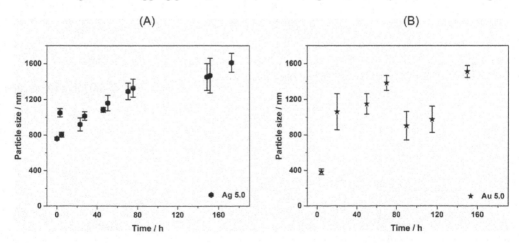

FIGURE 16.2 Particle size measurements using the dynamic light scattering technique for the Ag (a) and Au nanofluids (b).

solution in terms of the ζ potential of ± 30 mV is considered stable as reported elsewhere [36]. The $|\zeta|$ values are always higher for the nanofluid with the lowest nanoparticle concentration, therefore, it tends to agglomerate more slowly than the other two nanofluids that are more concentrated, which is coherent with UV-vis and the particle size measurements shown above.

16.3.2 Density

Density values measurements are of interest for being used as heat transfer fluids because they affect its efficiency, in this sense, high density materials are generally beneficial to the heat transfer performance [13]. The density of the Cu, Ag, and Au nanofluids were determined to establish the relationship between their values and the possible enhancement of their thermal properties. The density of the base fluid was also estimated for comparison purposes to assess the method of measurement and check the measured values with those provided by the manufacturer. The density obtained for the base fluid at room temperature was 1056.0 kg m^{-3}, in good agreement with the value of 1055.7 kg m^{-3} provided by the supplier at 298 K. The evolution of the density of the Cu, Ag, and Au nanofluids versus the nanoparticle mass concentration is shown in Figure 16.3. It shows a slight increase in the density values at higher nanoparticle concentrations when a certain amount of nanoparticles is added, with an increase in density around 0.2% for the nanofluids with the highest concentration of nanoparticles. Of note is that there is not a linear increase in the density when the concentration of nanoparticles increased, which could be explained by interactions being generated that can lead to the system being rearranged [37]. It appreciates an increase in the density of up to 1.3% for the nanofluid with the high nanoparticle content in the case of Au nanofluids and an increase up to 0.2% for Ag nanofluid.

16.3.3 Viscosity

The viscosity values measurements of a heat transfer fluid are of interest in heat transfer applications because they may affect pumping power or possible drops in pressure important parameters for CSP. Particles in suspension increase the viscosity of the base fluid being necessary to control that increase [38–40]. Figure 16.4 shows the dynamic viscosity values of the Cu, Ag, and Au nanofluids. The viscosity

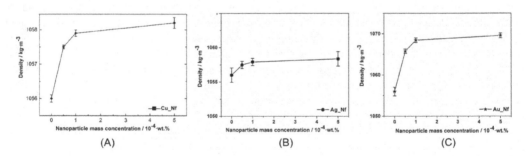

FIGURE 16.3 Density values versus nanoparticle mass concentration of Cu (a), Ag (b), and Au (c) nanofluids.

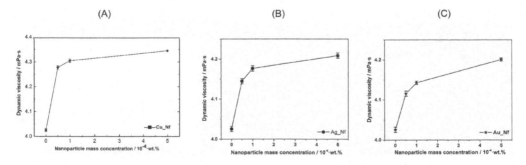

FIGURE 16.4 Dynamic viscosity values for the Cu (a), Ag (b), and Au (c) nanofluids.

values increase with mass concentration. The Einstein model [41] is a typical model to predict the effective viscosity of colloidal suspensions for dynamic viscosity for suspensions with low nanoparticle concentration. This model considers spherical particles and not interaction between nanoparticles. The Einstein model can be expressed as $\mu_{nf} = \mu_{bf}(1 + 2.5\phi_{eff})$, where μ is the dynamic viscosity of the nanofluid and the base fluid and ϕ_{eff} the effective volume fraction. Brinkman [42] generalized the Einstein model for more concentrated systems. The results of viscosity are coherent with the tendency of increase viscosity values when nanoparticles are added. The number of interactions between the nanoparticles inside the fluid considerably increasing, producing highest viscosity values, the tendencies are not linear, as the Einstein model predicts. The viscosity seems to deviate from those models, suggesting the existence of particle-particle interactions, which is not considered in the Einstein viscosity model.

16.3.4 Isobaric Specific Heat

The isobaric specific heat is a thermal property of relevance in the study of heat transport by means of heat transfer fluids. Thus, it is of interest the comparison of the isobaric specific heat values of the nanofluids with those of the base fluid. The isobaric specific heat of the nanofluids and base fluid were measured at several temperatures between room temperature and 373 K, approximately. Although the isobaric specific heat of nanofluids may be lower than that of the base fluids [13,43,44], the opposite behavior has been reported on occasions whereby the isobaric specific heat in the nanofluids increases [8,10,11]. It has been suggested that the specific interaction between the nanoparticle and the fluid base forms a kind of internal structure in the heart of the nanofluid generated that explains that behavior [11,45]. Figure 16.5 shows the values obtained for the Cu, Ag, and Au nanofluids with the base fluid that was included for comparison purposes. Figure 16.5 shows that the Cu nanofluid presents isobaric specific heat values around 15% higher that the base fluid. Moreover, Ag nanofluids improve the isobaric specific heat values regarding to base fluid, and of note is the Ag nanofluid with the highest concentration of nanoparticles that shows an improvement around 7%. However, in the case of Au nanofluids, the values of the isobaric specific heat do not increase, and this value decreases when the nanoparticle mass concentration increases. This behavior may be because the specific heat of solids is generally lower than that of liquids, so the specific heat of nanofluids would be expected to be lower than that of base fluids, as reported previously. So, the highest values are for Cu nanofluid, followed by Ag nanofluid, the base fluid, and finally the gold nanofluid (prepared following the two-step method). This behavior is due to the formation of a certain internal structure in the heart of the nanofluid generated by the specific interaction between the nanoparticle and the base fluid in each case, with a lower internal structure of base fluid molecules around the gold nanofluid as compared to the silver and copper nanofluid, as shown below by using theoretical calculations of molecular dynamics.

16.3.5 Thermal Conductivity

Thermal conductivity of Cu, Ag, and Au nanofluids and base fluid was measured at different temperatures, and the measurements were performed using the laser flash technique. Laser flash technique

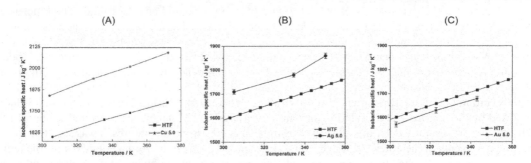

FIGURE 16.5   Isobaric specific heat values for the Cu (a), Ag (b), and Au (c) nanofluids.

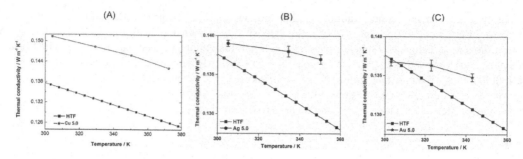

FIGURE 16.6 Thermal conductivity values measured for the Cu (a), Ag (b), and Au (c) nanofluids.

measures thermal diffusivity and estimates the conductivity from the density and isobaric specific heat values according to Eq. (16.1). The suspension of nanoparticles in a heat transfer fluid leads to an increase in its thermal conductivity [9,15], and high thermal conductivity values for a material improve its heat transfer efficiency [13,14]. The thermal conductivity values for Cu, Ag, and Au nanofluids and for the base fluid are shown in Figure 16.6. These values show that there is an increase in thermal conductivity for the nanofluids prepared with Cu, Ag, and Au nanoparticles. Moreover, the Cu nanofluid showed the higher thermal conductivity values, followed by Ag and Au nanofluids that show an enhancement of thermal conductivity regarding to the base fluid. Upon increasing mass concentration, this effect is mainly observed up to a nanoparticle mass concentration of $1.0 \cdot 10^{-4}$ wt% with a reduced effect at higher concentrations. This result suggests a saturation process of the base fluid. Thermal conductivity for Cu, Ag, and Au nanofluids decreases as the temperature increased, which is coherent with the tendency shown by the base fluid. For the case of Au nanofluid, the improvement of thermal conductivity appears significantly at high temperature, giving an enhancement up to 3% for the nanofluid with highest nanoparticle concentration.

To determine the efficiency of the nanofluids in heat transfer, applications may be assessed by different *figures of merit (FoM)*. In Figure 16.7, it is shown the ratio of the heat transfer coefficients for Ag and Au nanofluids as an example, according to the Dittus-Boelter correlation. So finally, different *FoM* can be used to assess the improvement in the heat transfer processes of the nanofluid with regard to the base fluid. Although two *FoMs* can be analyzed regarding laminar and turbulent flow conditions, the most interesting case for concentrating solar power applications is the turbulent flows. Under laminar flow conditions, the working conditions are improved by the nanofluid when the dynamic viscosity increase is less than four times the thermal conductivity enhancement [46]. But, under turbulent flow conditions, for the case we present here, it is possible to analyze the ratio of the heat transfer coefficients. According to the Dittus-Boelter correlation, this heat transfer coefficient ratio is given by [47]:

$$\frac{h_{nf}}{h_{bf}} = \left(\frac{\rho_{nf}}{\rho_{bf}} \right)^{0.8} \left(\frac{k_{nf}}{\rho k_{bf}} \right)^{0.6} \left(\frac{C_{P,nf}}{C_{P,bf}} \right)^{0.4} \left(\frac{\mu_{nf}}{\mu_{bf}} \right)^{-0.4}, \tag{16.3}$$

where k_p is the thermal conductivity of the nanoparticles, ρ_p is the density, C_P is the isobaric specific heat, and μ the dynamic viscosity, and *nf* and *bf* mean nanofluid and base fluid. The efficiency of the heat transfer process is usual considered to have improved when this ratio is greater than 1. Other *figures of merit* can be used to assess the improvement in the case of turbulent flow, such as the ratio of Mouromtseff number, which provides similar results. Thus, the *FoMs* were estimated for the nanofluids prepared at the temperatures at which the thermal conductivity and isobaric specific heat were measured. Values for this ratio above 1 ($h_{nf}/h_{bf} > 1$) imply enhanced efficiency of the nanofluid with regard to base fluid. Density and viscosity values were estimated using the evolution of the values reported by the supplier and considering the values measured for the nanofluids at around temperature. The dynamic viscosity and density values were estimated at the same temperatures, it was seen that there are improvements in the efficiency of the heat transfer process for all the Ag nanofluids at all the temperatures assessed (Figure 16.7a). For Ag-nanofluids, best results were obtained for the nanofluid with the highest nanoparticle concentration that shows an enhancement around 6%, while for gold nanofluid,

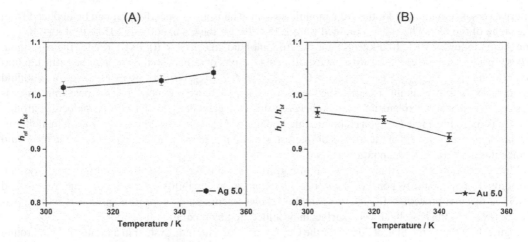

FIGURE 16.7   Values of the ratio of the heat transfer coefficients obtained for Ag (a) and Au-nanofluids (b).

there is a significant improvement in no case. Attending to experimental results, Cu and Ag nanofluids could be a promising alternative as heat transfer systems in concentrating solar power applications. However, gold nanofluids prepared by this two-step method do not present an improvement in the heat transfer coefficient, consequently, their application in concentrating solar power would not be logical. However, it has been reported a greatest enhancement of heat transfer coefficient when Au nanofluids are stabilizers by a surfactant, by one-step synthesis [17]. In any way, presence of surfactant allows an effective stabilization of gold nanoparticles that provides an improvement in their thermal properties. As a result, the role of surfactant is very important in the stabilization of nanofluids, and it opens the possibility to use gold nanofluids in concentrating solar power systems.

16.3.6 Theoretical Analysis by Molecular Dynamics

Based on the results of the experimental characterization of the Cu, Ag, and Au nanofluids, we have performed theoretical calculations to understand the experimental behavior of the nanofluids. Molecular dynamic calculations were performed with the highest experimental nanoparticle mass concentration for Cu, Ag, and Au taking into account the computational cost of this kind of calculations.

To calculate the isobaric specific heat, molecular dynamic calculations were performed at different temperatures ranging from ca. 50 to 500 K. As an example, Figure 16.8 shows the plot of the total

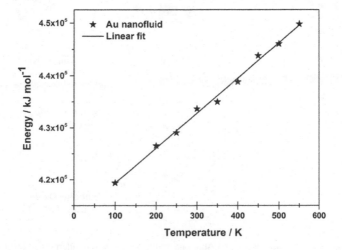

FIGURE 16.8   Plot of the energy versus temperature for the Au-nanofluid system.

energy versus temperature for the Au nanofluid system. The isobaric specific heat can be deduced from the slope of the plot. The values obtained were 2.25×10^3 for the Cu nanofluid, 2.15×10^3 J kg^{-1} K^{-1} for the Ag-nanofluid, 2.01×10^3 J kg^{-1} K^{-1} for the Au-nanofluid, and 1.94×10^3 J kg^{-1} K^{-1} for the base fluid. These results are in accordance with the experimental results obtained, so they follow the same qualitative tendency, $C_{p(\text{Cu nanofluid})} > C_{p(\text{Ag nanofluid})} > C_{p(\text{Au nanofluid})} > C_{p(\text{base fluid})}$. With those results, the Cu nanofluid system has the most suitable isobaric specific heat value to achieve the most efficient heat transfer processes. A reasonable explanation may be because the arrangement of the base fluid molecules around the Cu is more favorable (can you elaborate this arrangement) as compared to the Ag and Au nanofluids. In this sense, the analysis of the SDF is of interest because it is possible to determine how the base fluid molecules are arranged around the metals.

Moreover, most recent studies show the importance of the use of surfactants in this type of system, not only from a stability point of view, but also from their capability to reorganize the system and improve thermal processes. In fact, Au-nanofluid shows a remarkable improvement on thermal properties in regard to the base fluid when surfactant is added in the preparation process.

Figure 16.9 shows the SDF at 300 K for the Cu, Ag, and Au nanofluid system in a radius of 11 Å around the metal. It is of interest the first layer of molecules of the fluid base around the metals adopts an arrangement with the base of the benzene ring of the biphenyl, the metals in the center, and three, two, and one diphenyl oxide molecules around the Cu, Ag, and Au, respectively. The oxygen atoms are oriented toward the metal, in this sense, the radial distribution function for the pair metal-oxygen shows an intense peak that indicates strong orientation binding [48]. A second outer layer occupied by a diphenyl oxide molecule can be seen for Au nanofluid in Figure 16.9, that is depicted in Figure 16.9d. This structure in Figure 16.9d clarifies the important role that diphenyl oxide molecules play in the enhancement of the thermal properties as isobaric specific heat, with the highest value for Cu nanofluid with three diphenyl molecules around the metal. The structural arrangement may generate a directionality of movement that involves effective heat transport, leading to increased thermal properties compared with the base fluid [49].

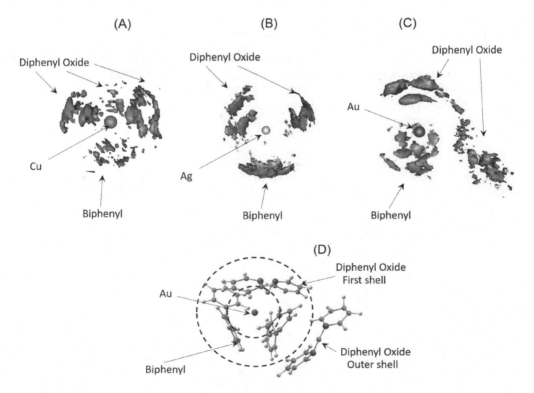

FIGURE 16.9 SDFs from the Cu (a), Ag (b), and Au (c) nanofluid at different temperatures. Structure around the metal (d) in the Au nanofluid system.

16.4 Conclusions

This chapter presents the preparation, characterization, and theoretical analysis of Cu, Ag, and Au nanofluids on a non-polar heat transfer fluid normally used in CSP plants (the eutectic mixture of biphenyl and diphenyl oxide).

The addition of metallic nanoparticles did not cause chemical changes in the base fluid. An increase in both density and viscosity was observed in all the nanofluids prepared. Regarding to the thermal properties, Cu nanofluids showed the highest value in both isobaric specific heat and thermal conductivity, followed by Ag and Au nanofluid. In fact, for Cu nanofluid, the thermal conductivity increased up to 13%. For Ag nanofluid an improvement by up to 6% of heat transfer coefficient was obtained with regard to the base fluid according to an *FoM* based on the Dittus-Boelter correlation. Consequently, the nanofluids based on Cu and Ag enhance heat transfer efficiency, and its application in CSP plants, in which this kind of fluid is often used as a HTF, would be interesting. While Au nanofluids prepared following the two-step method show limited improvements in the thermal properties cited above, so its application such as HTF would be dismissed. However, for Au nanofluid synthesized by one-step method reveals the importance of the use of surfactants in this type of systems, playing a role not only from a stability point of view, but also from its capability to reorganize the system and improve thermal processes. By the study of the SDF, the structural analysis around the metals showed the existence of a first-layer structure where the presence of diphenyl oxide molecules is related with the enhancement of the thermal properties. Thus, the Cu nanofluid with three diphenyl oxide molecules around the metal presents the highest isobaric specific heat, followed by the Ag nanofluid with two diphenyl oxide molecules, with the lowest value for the Au nanofluid with only one molecule in the first shell and a second molecule outer. So the rearrangement of base fluid molecules around the metals is related to the increase of thermal properties of those systems.

REFERENCES

1. Allen, M. P., and D. J. Tildesley. *Computer Simulation of Liquids: Clarendon*, Oxford, UK; 1989.
2. Andreu-Cabedo, P., R. Mondragon, L. Hernandez, R. Martinez-Cuenca, L. Cabedo, and J. E. Julia. (2014). Increment of specific heat capacity of solar salt with SiO2 nanoparticles. *Nanoscale Research Letters* 9, 582.
3. Angayarkanni, S. A., and J. Philip. (2015). Review on thermal properties of nanofluids: Recent developments. *Advances in Colloid and Interface Science* 225, 146–176.
4. Annapureddy, H. V. R., S. K. Nune, R. K. Motkuri, B. P. McGrail, and L. E. X. Dang. (2015). A combined experimental and computational study on the stability of nanofluids containing metal organic frameworks. *Journal of Physical Chemistry B* 119, 8992–8999.
5. Brinkman, H. C. (1952). The viscosity of concentrated suspensions and solutions. *Journal of Chemical Physics* 20, 571–571.
6. Cabaleiro, D., C. Gracia-Fernandez, J. L. Legido, and L. Lugo. (2015). Specific heat of metal oxide nanofluids at high concentrations for heat transfer. *International Journal of Heat and Mass Transfer* 88, 872–879.
7. Chakraborty, S., I. Sarkar, K. Haldar, S. K. Pal, and S. Chakraborty. (2015). Synthesis of Cu-Al layered double hydroxide nanofluid and characterization of its thermal properties. *Applied Clay Science* 107, 98–108.
8. Chandrasekar, M., S. Suresh, and T. Senthilkumar. (2012). Mechanisms proposed through experimental investigations on thermophysical properties and forced convective heat transfer characteristics of various nanofluids—A review. *Renewable & Sustainable Energy Reviews* 16, 3917–3938.
9. Chen, H. Y., S. Witharana, Y. Jin, Y. L. Ding, and C. Kim. (2008). Predicting the thermal conductivity of nanofluids based on suspension viscosity. *Paper read at 4th International Conference on Information and Automation for Sustainability*, IEES publishing, Sri Lanka.
10. Das, S. K., S. U. S. Choi, W. Yu, and T. Pradeep. (2008). *Nanofluids: Science and Technology*. Hoboken, NJ: John Wiley & Sons.

11. Desideri, U., F. Zepparelli, V. Morettini, and E. Garroni. (2013). Comparative analysis of concentrating solar power and photovoltaic technologies: Technical and environmental evaluations. *Applied Energy* 102, 765–784.
12. Devabhaktuni, V., M. Alam, S. S. S. R. Depuru, R. C. Green, D. Nims, and C. Near. (2013). Solar energy: Trends and enabling technologies. *Renewable & Sustainable Energy Reviews* 19, 555–564.
13. Dittus, F. W., and L. M. K. Boelter. (1930). Pioneers in heat transfer—Heat transfer in automobile radiators of the tubular type. *University California Publications in Engineering* 2, 443–461.
14. Eastman, J. A., S. U. S. Choi, S. Li, W. Yu, and L. J. Thompson. (2001). Anomalously increased effective thermal conductivities of ethylene glycol-based nanofluids containing copper nanoparticles. *Applied Physics Letters* 78, 718–720.
15. Einstein, A. (1906). A new determination of the molecular dimensions. *Annalen Der Physik* 19, 289–306.
16. Fedele, L., L. Colla, S. Bobbo, S. Barison, and F. Agresti. (2011). Experimental stability analysis of different water-based nanofluids. *Nanoscale Research Letters* 6, 300.
17. Fernandez-Garcia, A., E. Zarza, L. Valenzuela, and M. Perez. (2010). Parabolic-trough solar collectors and their applications. *Renewable & Sustainable Energy Reviews* 14, 1695–1721.
18. Goodwin, J. W., and R. W. Hughes. (2008). *Rheology for Chemists: An Introduction*. 2nd ed. Cambridge, UK: The Royal Society of Chemistry.
19. Hwang, Y., J. K. Lee, C. H. Lee, Y. M. Jung, S. I. Cheong, C. G. Lee, B. C. Ku, and S. P. Jang. (2007). Stability and thermal conductivity characteristics of nanofluids. *Thermochimica Acta* 455, 70–74.
20. Kambli, P., and V. Kelkar-Mane. (2016). Nanosized Fe3O4 an efficient PCR yield enhancer comparative study with Au, Ag nanoparticles. *Colloids and Surfaces B-Biointerfaces* 141, 546–552.
21. Khan, J., and M. H. Arsalan. (2016). Solar power technologies for sustainable electricity generation—A review. *Renewable & Sustainable Energy Reviews* 55, 414–425.
22. Lee, S., S. U. S. Choi, S. Li, and J. A. Eastman. (1999). Measuring thermal conductivity of fluids containing oxide nanoparticles. *Journal of Heat Transfer-Transactions of the Asme* 121, 280–289.
23. Martinez, J. M., and L. Martinez. (2003). Packing optimization for automated generation of complex system's initial configurations for molecular dynamics and docking. *Journal of Computational Chemistry* 24, 819–825.
24. Mousavi, N. S. S., and S. Kumar. (2014). Effective heat capacity of ferrofluids—Analytical approach. *International Journal of Thermal Sciences* 84, 267–274.
25. Murdock, R. C., L. Braydich-Stolle, A. M. Schrand, J. J. Schlager, and S. M. Hussain. (2008). Characterization of nanomaterial dispersion in solution prior to in vitro exposure using dynamic light scattering technique. *Toxicological Sciences* 101, 239–253.
26. Mwesigye, A., Z. J. Huan, and J. P. Meyer. (2015). Thermodynamic optimisation of the performance of a parabolic trough receiver using synthetic oil-Al2O3 nanofluid. *Applied Energy* 156, 398–412.
27. Navas, J., A. Sánchez-Coronilla, E. I. Martín, M. Teruel, J. J. Gallardo, T. Aguilar, R. Gómez-Villarejo, R. Alcántara, C. Fernández-Lorenzo, J. C. Piñero, and J. Martín-Calleja. (2016). On the enhancement of heat transfer fluid for concentrating solar power using Cu and Ni nanofluids: An experimental and molecular dynamics study. *Nano Energy* 27, 213–224.
28. Nieh, H. M., T. P. Teng, and C. C. Yu. (2014). Enhanced heat dissipation of a radiator using oxide nanocoolant. *International Journal of Thermal Sciences* 77, 252–261.
29. Pastoriza-Gallego, M. J., C. Casanova, R. Paramo, B. Barbes, J. L. Legido, and M. M. Pineiro. (2009). A study on stability and thermophysical properties (density and viscosity) of Al2O3 in water nanofluid. *Journal of Applied Physics* 106.
30. Pati, S. S., V. Mahendran, and J. Philip. (2013). A simple approach to produce stable ferrofluids without surfactants and with high temperature stability. *Journal of Nanofluids* 2, 94–103.
31. Philip, J., P. D. Shima, and B. Raj. (2007). Enhancement of thermal conductivity in magnetite based nanofluid due to chainlike structures. *Applied Physics Letters* 91, 203108.
32. Prasher, R., D. Song, J. L. Wang, and P. Phelan. (2006). Measurements of nanofluid viscosity and its implications for thermal applications. *Applied Physics Letters* 89, 133108.
33. Rai, N., and J. I. Siepmann. (2007). Transferable potentials for phase equilibria. 9. Explicit hydrogen description of benzene and five-membered and six-membered heterocyclic aromatic compounds. *Journal of Physical Chemistry B* 111, 10790–10799.
34. Rai, N., and J. I. Siepmann. (2013). Transferable potentials for phase equilibria. 10. Explicit-hydrogen description of substituted benzenes and polycyclic aromatic compounds. *Journal of Physical Chemistry B* 117, 273–288.

35. Sarsam, W. S., A. Amiri, M. N. M. Zubir, H. Yarmand, S. N. Kazi, and A. Badarudin. (2016). Stability and thermophysical properties of water-based nanofluids containing triethanolamine-treated graphene nanoplatelets with different specific surface areas. *Colloids and Surfaces A-Physicochemical and Engineering Aspects* 500, 17–31.

36. Shin, D., and D. Banerjee. (2013). Enhanced specific heat capacity of nanomaterials synthesized by dispersing silica nanoparticles in eutectic mixtures. *Journal of Heat Transfer-Transactions of the Asme* 135, 032801.

37. Shin, D., and D. Banerjee. (2014). Specific heat of nanofluids synthesized by dispersing alumina nanoparticles in alkali salt eutectic. *International Journal of Heat and Mass Transfer* 74, 210–214.

38. Singh, D., E. V. Timofeeva, M. R. Moravek, S. Cingarapu, W. H. Yu, T. Fischer, and S. Mathur. (2014). Use of metallic nanoparticles to improve the thermophysical properties of organic heat transfer fluids used in concentrated solar power. *Solar Energy* 105, 468–478.

39. Smith, W., and T. R. Forester. (1996). DL_POLY_2.0: A general-purpose parallel molecular dynamics simulation package. *Journal of Molecular Graphics* 14, 136–141.

40. Song, Y. Y., H. K. D. H. Bhadeshia, and D. W. Suh. (2015). Stability of stainless-steel nanoparticle and water mixtures. *Powder Technology* 272, 34–44.

41. Suganthi, K. S., V. L. Vinodhan, and K. S. Rajan. (2014). Heat transfer performance and transport properties of ZnO-ethylene glycol and ZnO-ethylene glycol-water nanofluid coolants. *Applied Energy* 135, 548–559.

42. Timofeeva, E. V., W. H. Yu, D. M. France, D. Singh, and J. L. Routbort. (2011). Nanofluids for heat transfer: An engineering approach. *Nanoscale Research Letters* 6, 182.

43. Wang, B. G., X. B. Wang, W. J. Lou, and J. C. Hao. (2011). Ionic liquid-based stable nanofluids containing gold nanoparticles. *Journal of Colloid and Interface Science* 362, 5–14.

44. Wang, C. X., J. Yang, and Y. L. Ding. (2013). Phase transfer based synthesis and thermophysical properties of Au/Therminol VP-1 nanofluids. *Progress in Natural Science-Materials International* 23, 338–342.

45. Witharana, S., H. S. Chen, and Y. L. Ding. (2011). Stability of nanofluids in quiescent and shear flow fields. *Nanoscale Research Letters* 6, 231.

46. Yoo, D. H., K. S. Hong, and H. S. Yang. (2007). Study of thermal conductivity of nanofluids for the application of heat transfer fluids. *Thermochimica Acta* 455, 66–69.

47. Yu, W., and H. Q. Xie. (2012). A review on nanofluids: Preparation, stability mechanisms, and applications. *Journal of Nanomaterials* 1.

48. Zhao, Y., D. Yang, H. C. Hu, L. Chen, Y. Xu, L. L. Qu, P. P. Yang, and Q. Zhang. (2016). A simple approach to the synthesis of eccentric Au@SiO2 Janus nanostructures and their catalytic applications. *Surface Science* 648, 313–318.

49. Zhu, H. T., Y. S. Lin, and Y. S. Yin. (2004). A novel one-step chemical method for preparation of copper nanofluids. *Journal of Colloid and Interface Science* 277, 100–103.

17

Potential Utilization of Nanofluids for Concentrating Solar Power (CSP)

Amine Allouhi and Mahmut Sami Buker

CONTENTS

17.1 Introduction

Civilization continues to be changed completely by our endeavor to utilize energy beyond any conventional means. A sequence of industrial and agricultural advances has granted many people worldwide easy accessibility to various needs. All of this development is sustained by our capability to discover, extract, and utilize power with ever-increasing skill and grace. Extensive research in energy is progressing toward more sustainable societies based on cleaner power generation technologies [1]. Among the solutions to global energy problems, the utilization of solar energy is, without a doubt, one of the most promising ecological avenues [2]. In general, solar energy benefits from low carbon emission, having virtually no fossil fuel requirement, abundant solar resources, and plunging life-cycle costs [3].

The exploitation of solar energy is achieved through solar collectors, which are devices that absorb incident solar irradiation and partially convert it to electricity, useful heat, or both at the same time [4]. At present, various types of solar collectors are in practical use based on purpose of use that can be power generation, water heating, or solar cooling [5].

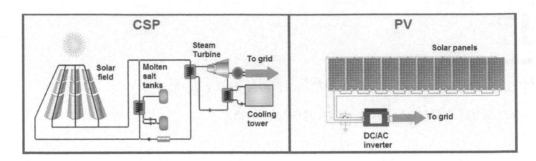

FIGURE 17.1   CSP and PV technologies [7].

Solar energy conversion into electricity can be rendered by either using photovoltaics (PV) or concentrating solar power (CSP) [6]. Figure 17.1 envisions the most common PV and CSP technologies.

Since invention, photovoltaic cells have constituted a significant portion of the worldwide installation and now totaling around 400 GW of global cumulative installed capacity [8]. These cells are made of semi-conducting materials to convert solar energy directly into electricity using photoelectric and electrochemical processes. The most efficient commercially available solar panels have efficiency ratings as high as 22%, whereas majority of the panels on the market today range from 15% to 17% efficiency ratings. Based on the efficiency ratings, PV technology can be subdivided into monocrystalline, polycrystalline, and thin film solar PV modules. Generally, monocrystalline silicon solar PV is the best technology to deliver maximum efficiency, which is typically around 15%, but this efficiency can come with cost due to the complicated manufacturing process. Due to the impurities in crystal, polycrystalline silicon is less efficient comparing to the monocrystalline, which is around 12%. However, simpler manufacturing process consumes less energy and materials, giving it a significant cost advantage over monocrystalline silicon. Thin-film solar PV technology is an ideal option for the applications with lightweight and portability, but lesser power requirements. Thin film cells can be constructed from a variety of materials including amorphous silicon (a-Si), cadmium telluride (CdTe), and copper indium gallium selenide (CIS-CIGS). As an emerging technology, thin-film panels are less efficient around 6%, but it tends to be easier and cheaper to produce [9]. Figure 17.2 illustrates the efficiency chart for various PV cell technologies.

Besides PV incorporated systems, CSP or solar thermal electricity (STE) is a competent technology for utility scale electricity generation. While doing so, CSPs offer firm capacity and dispatchable power on demand by integrating thermal energy storage or hybrid operation [11,12].

It is projected that, with high level of energy efficiency and technological advancement, CSPs could meet up to 6% of global energy demand by 2030 and 12% by 2050 [13]. Although dating back to 1970s, nearly all commercial CSP plants have been launched in the last decade [14]. Currently, although Spain and the United States are the two largest markets for the CSP technologies, Dubai has newly approved CSP project that will generate 1000 MW by 2020 and a whopping 500 MW by 2030 and Morocco's Ouarzazate will provide around 580 MW of power by 2020. Apart from these giant projects, construction has started in several new markets including China, South Africa, and the Middle East [15].

CSP technologies rely on a simple operating principle: sunlight is reflected through mirrors (heliostats) onto a receiver, where the heat is accumulated by a thermal energy carrier (primary circuit) called heat transfer fluid (HTF). The HTF, then, can be used directly to drive a turbine to produce power or, more commonly, be integrated with heat exchanger to generate steam via a secondary cycle [16]. Generally, major components in a CSP plant can be sorted as solar concentrators, receiver, steam turbine, and electrical generator. At present, there are four kinds of CSP technologies available as parabolic trough collectors (PTCs), solar power tower (SPT), linear Fresnel reflectors (LFR) and solar parabolic dishes (SPD) from most prominent to least [17]. Figure 17.3 illustrates the most common CSP types and concentration modes.

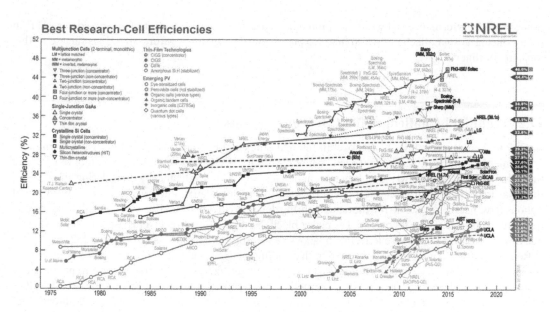

FIGURE 17.2 Efficiency chart for various cell technologies [10].

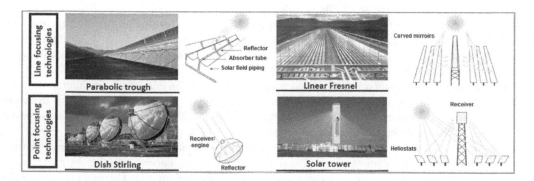

FIGURE 17.3 The most common CSP types (adapted from [142]).

Currently, several technical challenges are being faced by CSP technologies to improve their economic viability. Among these challenges, finding innovative working fluids with enhanced properties leading to higher thermal performances is a critical aspect that has gained particular attention recently.

Numerous research works have discussed the properties of nanofluids mainly used in solar systems [18–21] and their potential application in various solar energy technologies including flat plate [22], evacuated tube [23], solar still [24], and more [25,26].

Newly applied CSP systems require a high operating temperature, and nanofluids could be a way to advance them and improve their thermal storage capacities [27,28]. In this direction, many researchers have recently studied the effects of nanofluids on the enhancement of heat transfer in CSP systems, both experimentally and theoretically [29,30].

This chapter highlights one of the most promising and less discussed subjects entitled applications of nanofluids particularly in CSP systems. Overview of the past and most recent studies from various literature is presented. On top of that, mechanisms of heat transfer improvement in such systems along with the most used assessment criteria to measure the performance of nanofluid based systems are introduced. Future directions, research opportunities, perspectives, and challenges are discussed as well.

The remainder of this chapter is organized as follows: first, key challenges related to CSP developments are discussed with a focus on the limitations of conventional HTFs. The potential of nanofluids in such systems is reported after reviewing the main categories and physical properties of nanofluids.

Some useful performance indexes that commonly characterize the potential of nanofluids as HTF in solar concentrating collectors (SCCs) are also presented. Finally, main future directions related to this topic are outlined.

17.2 Technical Challenges for CSP Power Plants

In CSP technologies, increasing the overall thermal efficiency is crucial challenge to make them more economically competitive. The levelized cost of electricity (LCOE) is often used as a tool to compare the competitiveness of power generation technologies. LCOE is evaluated by considering the ratio of the total investment paid over the project lifespan to the cumulative energy generation [31]. Several options have been proposed in the literature to bring down the LCOE of CSP technologies. These include improvements of the solar loop as well as power block. Next generation receivers based on more efficient and durable coatings can cause a decline of about 12% of the plant LCOE as indicated in [32].

Reducing water consumption by adopting dry cooling has the potential of saving more than 90% of water quantity [33]. Also, hybrid cooling system allows the utilization of several systems in parallel: dry and wet cooling depending on the external ambient temperature. As revealed by [34], this can result in economizing up to 80% of the water requirement in cooling towers. Moreover, due to accumulation of dust in receivers, the reflectance of mirrors and transmittance of solar absorbers diminish which causes degradation on the optical efficiency of the concentrating technology. As indicated by [35], 1% decrease in reflectance engenders 1% rise in LCOE.

The second contributor in the water consumption in CSP plants is cleaning processes of the mirrors. The rarity of water in deserts incurs huge expenses. Some procedures have been suggested to clean receiver mirrors while minimizing the quantity of water used. One promising alternative is the use of super-hydrophobic coatings, which cause lotus effect inducing self-cleaning [36].

CSP-desalination cogeneration is another innovative method that can enhance the economic viability of future CSP plants. In such a configuration, it is possible to produce power and fresh water at the same time. The main configurations discussed in the literature are Multi-Effect Distillation, Reverse Osmosis, and Multi-stage flash coupled to parabolic trough collectors [37–39].

The intermittent nature of renewable energies is a pivotal issue, which constitutes a real barrier against their development and wide deployment. CSP technologies are not exceptions, and thereby energy storage systems are required to ensure continuous electricity generation [40,41]. Thermal energy storage allows many benefits over other mechanical storage categories such as reduced investment costs and good efficiencies. These systems store thermal energy under sensible, latent, or thermochemical form. Up to now, the first storage category (sensible) is dominant in the CSP market, whereas other categories remain under development. Sensible thermal energy storage (TES) systems are financially and technologically viable, but require huge storage volumes. Using latent storage systems have larger energy densities, but they are less efficient compared to sensible systems as a result of reduced thermal conductivity of phase changes materials (PCM). Moreover, latent storage configurations present certain complexities in their designs, therefore it is difficult to have suitable materials that resist to excessive temperature environments as required in CSP plants [42,43]. Currently, most exploratory research is directed into searching new materials and processes that can achieve high thermal efficiency for storage. Many new tendencies abound for improvement of TES utilization in CSP plants, the major development areas concern the utilization of coupled TES technologies to achieve higher temperature levels, cascade design with modularized units, and mechanical circulation of granular particles [44].

HTF could be a liquid or gas, used to transport heat from one part to other. Therefore, it is one of the most important components of CSP systems in terms of improving the thermoelectric conversion efficiency and overall performance.

As CSP systems are immense and a huge quantity of HTF is needed to drive the CSP plant, it is vital to keep the costs of HTF at a minimum while retaining the competing performance. In addition to moving heat from receiver to steam generator, hot HTF can be accumulated for power generation when the solar radiation is unavailable. Hence, HTFs should possess a number of favorable characteristics including high boiling and low melting points, thermal stability, low vapor pressure at high temperature, low corrosion with metal alloys used to contain the HTF, low viscosity and high thermal conductivity, high heat capacity for energy storage, non-toxicity, low environmental impacts, and low cost [45]. Since the emerging CSP systems operate in critical condition (550°C–900°C) and supercritical condition (700°C–1200°C), the high temperature HTF is supposed to endure higher temperature limits [46]. Currently, a wide range of commercialized HTFs exist including thermal oils [47], water/steam [48], molten salts [49], pressurized gases like air, and N_2 and CO_2 [50]. Nevertheless, each HTF category exhibits some limitations [51].

Specific heat of water is high with respect to others due to which it can absorb vast amounts of heat. Water is preferred usually in low temperature applications. Whereas the major drawback of water as HTF is its low boiling point (100°C) and high freezing points (0°C). As it is known, water loses its neutrality due to the contamination so preventing corrosion hazard and eventually reducing its heat transfer capability. Recently, salts are the most promising options with upper temperature limits, excellent thermal stability, large heat capacity, and low cost. Molten salts are recommended for improving heat storage capacity and financial feasibility of CSP plant [52,53]. Moreover, Hitec salts were considered to replace solar salts due to their low freezing point (142°C). It is a combination of wt. 40% sodium nitrate ($NaNO_2$), 53% potassium nitrate (KNO_3), and 7% sodium nitrate ($NaNO_3$). Hitec salt has thermal stability at temperatures up to 454°C and can be used at temperatures up to 530°C for a short period [54].

Nitrates are commonly utilized HTFs in central receiver and parabolic trough type CSP plants. However, they can dissolve at relatively high temperatures above 550°C [55]. Dowtherm A is a combination of very stable elements, namely, biphenyl ($C_{12}H_{10}$) and diphenyl oxide ($C_{12}H_{10}O$). These pair compounds have practically same vapor pressure, and their mixture can compose a distinct stable compound. This HTF can be utilized in either liquid or vapor phase. It was found that low purity biphenyl and diphenyl oxide-based fluids might be low cost initially, but this may cause high degradation over the life span, which may eventually increase the cost of the plant [56].

United States Department of Energy has proposed halide-based salt on CSPs in order to obtain the desired properties of low melting point, low vapor pressure, and great stability at temperatures at least 800°C. Brayton cycle can be achieved with this type of HTF, which can concurrently improve the efficiency of heat transformation to the power and may substitute the nitrous and other types of salt-based HTFs [57].

The imperfection of solar thermal systems, if not thermal engineering devices, is the low thermal conductivity of conventional heat transfer fluids such as water, ethylene glycol, or oil. Improving heat transfer practices in solar energy systems can be accomplished by utilizing heat transfer fluids with remarkable thermophysical properties. One of the competent methods is to utilize nanofluids to improve the heat transfer process.

Nanofluids are kind of embryonic fluids that show exceptional thermal properties greater than other fluids. Structurally, nanofluids are formed by colloidal suspension of nanometric particles in a base fluid (smaller than 100 nm) [58,59]. Lately, great quanta of attempts have been made to utilize the nanofluids in diverse solar engineering applications, taking profit from their remarkably improved characteristics:

- Thermal conductivity of the nanofluids is substantially higher in comparison to the base fluid because of the presence of nanoparticles
- Favorable stability under a wide range of temperature changes
- Optical features of nanofluids are improved (good absorption and extinction coefficients)
- Utilization of nanofluids eliminates clogging, sedimentation, and fouling of pipes and pumps due as a result of their nano size against micro or millimeter size particles
- Nanofluids require smaller heat transfer area, therefore diminish expenses of solar equipment
- Nanofluids usually acquire large densities and high convective heat transfer coefficients with relatively low specific heats leading to improved efficiency of the solar devices [60]

17.3 Nanofluids Types and Thermal/Optical Properties

Nanotechnology studies matter at atomic or molecular levels. Over other materials, solid metals possess higher heat transfer effectiveness. Therefore, incorporating solid particles in liquids advances the heat transfer substantially. Yet, there are restraints, and hence various conditions must be optimized in order to enhance the heat transfer while at the same time avoid choking and clogging [61]. Table 17.1 shows the thermophysical properties of some commonly used nanofluids. It is obvious that nanofluids exhibit improved thermal conductivity over base fluids. Moreover, metallic nanofluids have superior diffusivity, thermal conductivity, convective heat transfer coefficient, viscosity, emissivity, and optical absorption.

17.3.1 Nanofluids/Nanomaterial Types

Nanomaterials can be generally categorized into four types [62]:

1. 0-dimensional: This type of nanomaterial covers all nano dimensions, typically larger than 100 nm
2. 1-dimensional: with one dimension out of the nanoscale range, e.g., nanorods, nanotubes, and nanowires
3. 2-dimensional: Under the form of plate-like shape, this category is outside the nanoscale range, e.g., nano coatings, nanolayers, and nanofilms
4. 3-dimensional: This final class is not limited to nanoscale in any dimension. Arbitrary 3-D microstructures area over 100 nm, e.g., bundles of nanowires, dispersion of nanoparticles and nanotubes

TABLE 17.1

Thermophysical Properties of Nanofluids [62]

Material	Crystalline Structure	Thermal Conductivity W/(m-K)	Viscosity (Cp) with 5.0 vol% (30°C)	Density (g/cm^3)	Thermal Conductivity Enhancement of Nanofluids (%) with 5.0 vol%
MgO	Cubic	48.4	17.4	2.9	40.6
TiO$_2$	Anatase	8.4	31.2	4.1	27.2
ZnO	Wurtzite	13.0	129.2	5.6	26.8
Al$_2$O$_3$	γ	36.0	28.2	3.6	28.2
SiO$_2$	Noncrystalline	10.4	31.5	2.6	25.3

Material	Specifications	Thermal Conductivity W/(m-K)
Non-metallic solids	Silicon	148
	Alumina	40
	CNT	2000
Non-metallic liquids	Water	0.613
	EG (ethylene glycol)	0.253
Metallic solids	Copper	401
	Aluminum	237
	Silver	429

17.3.2 Thermal and Optical Properties of Nanofluids

Nanofluids possess enhanced thermal characteristics, like high thermal conductivity, lower specific heat, and good heat transfer coefficients when compared to ordinary fluids. Also, they present remarkable optical properties. More specifically, they do not engender clogging in the pipes if applied as heat transfer fluids. Additionally, among the important outcomes of nanofluids, one can refer their larger surface area, good stability of the colloidal suspension, decreased pumping power required to achieve the same heat transfer, lower particle clogging compared to ordinary colloids, good adjustment of the thermophysical and transport properties by varying the particle concentration, size, and shape demonstrating greater heat transfer capabilities [63]. Utilization of HTFs in CSP applications requires having excellent thermophysical properties, unique optical properties, and most importantly, stability under various operating conditions.

New research trends in CSP applications deal with seeking for HTFs that can yield to better thermodynamic efficiencies thus reduced LCOE and equipment sizes. In this sense, nanofluids have been considered for these applications. In what follows, some specific features and main optical and thermal properties of nanofluids are presented.

17.3.2.1 Optical Properties

Specifying optical properties of nanofluids includes the evaluation of absorption [64], extinction [65], and scattering coefficients [66], dielectric constant and strength [67], and refractive index [68].

Optical properties of nanofluids are strongly dependent on particle size and shape and the optical properties of the base fluid. A minor change on these factors can cause important deviations in scattering and absorption. As working fluid, selecting appropriate concentrations of nanoparticles is a challenging task. In fact, the majority of input thermal energy is absorbed in a thin surface layer with high concentrations of nanoparticles, meanwhile weak absorption occurs in the opposite case.

The models describing absorption and scattering of nanoparticles in a host medium are influenced by various parameters such as nanoparticle type, size, shape, base fluid, and nanofluid volume fraction. Theoretical models are used to characterize the optical properties of nanofluids, including Rayleigh scattering approach, Maxwell-Garnett theory, and Mie scattering theory and discrete dipole approximation. An overview of these models can be found in [69].

17.3.2.1.1 Absorption

Absorption identifies the intensity loss of the incident light when penetrating through the working medium. The absorption, for the case of nanofluids, characterizes both the base fluid and nanoparticles, which is not simply the summation of both terms due to particular behavior of nanoparticles when dispersed into the base fluid [70].

17.3.2.1.2 Transmittance

The transmittance of nanofluids changes with path depth, nature of base fluid, particle size, and fractional volume of nanoparticles. Nanofluids manifest a behavior of decline in transmittance when concentration of nanoparticle augments. Also, it diminishes with decreased path lengths due the decrease in the distance that light takes to transmit through nanofluids [20]. In an experimental investigation [71], the optical and thermal properties of Cu-Water nanofluids are studied. It was shown that the transmittance of nanofluid and water were around zero in an interval of 1370 nm–2500 nm. This indicates that both fluids have excellent absorption of light in the infrared zone. Overall, it was reported that the transmittance of Cu-Water nanofluids are much less than that of deionized water in full range of solar spectrum. Maximum temperature of nanofluids with 0.1 wt% was 25.3% higher than base fluid. The high absorption of Cu-Water nanofluid for solar energy proved that it is a promising candidate for direct absorption thermal collectors.

17.3.2.1.3 Scattering

Scattering results from the obstacles created by the presence of nanoparticles in the base fluid. These obstacles arouse a deviation of light from its straight path [72]. As reported by [73,74], scattering is marginal when the particles radius is below 10 nm or when the volume fraction is <0.6% However, when the particle size increases, the effect of scattering becomes more influential. Also, a good dispersion of nanoparticles into the base fluid helps to reduce the scattering effects.

17.3.2.1.4 Extinction Coefficient

In addition to scattering and absorption, extinction coefficient is also one of the most crucial optical parameters of nanofluids. It presents the imaginary component of complex refractive index. This parameter is influenced by nature of the particle and its size, dielectric constant of the medium, temperature, and dispersed quantity of nanoparticles. The extinction coefficient of various nanofluids becomes higher with increasing particle volume fraction and size. In this context, the extinction coefficient of nanofluids based on different metals was examined by altering type of base fluid, particle size, and volume fraction. For instance [75], the extinction coefficient of water was calculated to be 0.01 cm^{-1} in the visible range (400 nm–800 nm). By adding single wall carbon nanotubes (SWCNTs) nanoparticles, a substantial increase occurred in the nanofluid (a maximum value of 9.48 cm^{-1} was reached by using of 0.05 vol.% of SWCNTs nanofluid). This proves the optical properties of the nanofluids that were enhanced greatly by dispersing SWCNTs nanoparticles. Generally, the extinction coefficient follows a linear trend with respect to added nanoparticle volume.

17.3.2.2 Thermal Properties

To assess the efficiency of any thermal mechanism employing nanofluids, the accurate estimation of thermophysical properties is mandatory. These properties are highly impacted by the quantity of nanoparticles dispersed in the base fluid. The quantity of nanoparticles needed to obtain the nanofluid at a specific volume fraction can be calculated using the next expression [76]:

$$\Phi = \frac{\left[\dfrac{m_{np}}{\rho_{np}} \right]}{\left[\dfrac{m_{np}}{\rho_{np}} + \dfrac{m_{bf}}{\rho_{bf}} \right]} \tag{17.1}$$

with Φ is the volume fraction of nanoparticles in the base fluid. m and V are, respectively the mass and volume, the subscripts np and bf denote for nanoparticle and base fluid, respectively.

In order to estimate the thermal properties of nanofluid, analytical and experimental methods are applied by supposing that the nanoparticles are uniformly dispersed in the base fluid. Many investigators developed useful correlations to evaluate the thermal properties of nanofluid. The assessment of these properties depends on various considerations such as nanoparticle type, size and shape, base fluid, and temperature. The developed correlations, if appropriately applied, give a good matching with experimental studies [77]. The main thermal properties of nanofluids are presented in the followings sub-sections along with the most used correlation for their estimation.

17.3.2.2.1 Density

Density of nanofluids is an extremely important physical property. Density strongly influences the Reynolds number and friction coefficient and thereby the pumping power. The density of nanofluids augments following the increase of percentage of nanoparticles in the base fluid. This evidently occurs as the density of nanoparticles surpasses those of the base fluid. Research on nanofluids density is until now too scarce and presents controversial results [78]. The density of nanofluids can be determined by applying the mixture model as follows:

$$\rho_{nf} = \left(\frac{m}{V}\right)_{nf} = \frac{m_{bf} + m_{np}}{V_{bf} + V_{np}} = \frac{\rho_{bf}V_{bf} + \rho_{np}V_{np}}{V_{bf} + V_{np}} = \Phi\rho_{np} + (1-\Phi)\rho_{bf} \tag{17.2}$$

where m and V are, respectively, the mass and volume, the subscripts nf, np, and bf denote for nanofluid, nanoparticle, and base fluid, respectively. Φ is the volume fraction of nanoparticles in the base fluid.

17.3.2.2.2 Thermal Conductivity

Thermal conductivity is among the crucial properties of nanofluids inducing an improvement of the heat transfer coefficient. It is recognized that the thermal conductivity rises in the case of nanofluids due to existence of nanoparticle dispersions. The thermal conductivity of nanofluids is influenced by various parameters such as particle type and size, volume fraction, pH, operating temperature, particle size, as well and preparation procedure [79]. Experimentally, this property can be determined by using several methods, among which the transient hot-wire method is the widely adopted [80,81].

Theoretically, Maxwell formulation is mostly considered as the first equation to compute the thermal conductivity of nanofluids. This formulation is constructed for suspension comprising a low concentration of homogenous spherical particles and considering non-interactions between particles. The thermal conductivity of nanofluids, noted k, according to the previous model is given as [82]:

$$k_{nf} = k_{bf} \frac{k_{np} + 2k_{bf} + 2(k_{np} - k_{bf})\Phi}{k_{np} + 2k_{bf} - (k_{np} - k_{bf})\Phi} \tag{17.3}$$

It is interesting to state that other similar expressions were proposed in [83,84], however, a main limitation is that these expressions underestimate nanofluids thermal conductivity in the case of nanoparticles having high thermal conductivities [82].

17.3.2.2.3 Specific Heat Capacity

The specific heat capacity of nanofluids is reduced proportionally with the increasing volume fraction of nanoparticles [85]. This is due to the fact that nanoparticles specific heat capacities are lower than the base fluid. Assuming thermal equilibrium between particles and the liquid phase [86], the effective specific heat capacity of nanofluids can be expressed as:

$$c_{p,nf} = \frac{\rho_{bf}(1-\Phi)}{\rho_{nf}} c_{p,bf} + \frac{\rho_{np}\Phi}{\rho_{nf}} c_{p,np} \tag{17.4}$$

ρ_{bf} and ρ_{nf} are the densities of base fluid and nanofluid, respectively. $c_{p,bf}$ and $c_{p,np}$ are the specific heat capacities of base fluid and nanoparticle, respectively. This is the most accurate model used to evaluate this property according to Khanafer and Vafai [87].

17.3.2.2.4 Viscosity

Viscosity controls the flow properties of nanofluids, which plays a fundamental role in their viability in many engineering applications. A net increase in dynamic viscosity takes place, which could be a real issue, due to the important pressure drop rise for applications using forced convection [88].

Determination of nanofluids viscosity is therefore fundamental to decide if their utilization is to bring a benefit compared to the base liquid. Many correlations were introduced for calculating the dynamic viscosity of nanofluids. Among these equations, Batchelor correlation [89] that considers the Brownian motion of nanoparticles is widely used in simulation studies as indicated in [90]:

$$\mu_{nf} = \mu_{bf}\left(1 + 2.5\Phi + 6.5\Phi^2\right) \tag{17.5}$$

17.3.3 Heat Transfer and Pressure Drop Correlations of Nanofluids

As stated earlier, nanofluids possess unusual features, which make them potential options in a wide range of engineering systems. Increasing numerical and experimental research have been performed recently to take benefit from hydrothermal properties of nanofluids [91].

The performance enhancement obtained by the use of nanofluids has been interpreted by multiple theoretical approaches indicating that thermal improvement originates from various microscale combined phenomenon covering particles' Brownian motion, rotation, migration, clustering, fluid layers encapsulating the nanoparticles, and boundary layer disturbance [92–95].

Similarly, to thermal properties, the ordinary correlations are not successful in estimating thermal features of nanofluids. As such, many investigators introduced various expressions for nanofluids as a result of their experimental or theoretical studies. In addition, it was highlighted that the improvement of heat transfer coefficient is accompanied by penalty in pressure drop due to presence of nanoparticles. In this sense, new correlations to predict the friction coefficient were developed for nanofluids, as well.

A comprehensive overview of some correlations for determination of the Nusselt number and friction factor are listed per operating condition. At this stage, it is important to emphasize that the most of developed expressions are suitable for forced convective heat transfer of nanofluids and limited expressions exist for the natural and mixed convection heat transfer. Depending on the nature of nanofluids and their specific applications, several correlations have been introduced. Here, widely applied correlations to characterize heat exchange of nanofluids in plain tube without inserts are quoted [91]:

A basic Nusselt number correlation was developed very early by [96]:

$$Nu_{nf} = 0.023 Re_{nf}^{0.8} Pr_{nf}^{0.4} \tag{17.6}$$

Pak and Cho [76] performed an experimental investigation using Al_2O_3/water and TiO_2/water nanofluids for $0 \leq \Phi \leq 3\%$, $10^4 \leq Re_{nf} \leq 10^5$, and $6.54 \leq Pr_{nf} \leq 12.33$. This study allowed developing the following Nusselt number correlation:

$$Nu_{nf} = 0.021 Re_{nf}^{0.8} Pr_{nf}^{0.5} \tag{17.7}$$

Based also on an experimental study, the Gnielinski's Nusselt number correlation is given as [97]:

$$Nu_{nf} = \frac{f/8\left(Re_{nf} - 1000\right)Pr_{nf}}{1 + 12.7\left(f/8\right)^{0.5}\left(Pr_{nf}^{2/3} - 1\right)} \tag{17.8}$$

for $3 \times 10^3 \leq Re_{nf} \leq 5 \times 10^6$ and $0.5 \leq Pr_{nf} \leq 2000$.

For the nanofluids Al_2O_3/water, CuO/water, and SiO_2/water, the experimental findings in [98] led to the following Nusselt number expression:

$$Nu_{nf} = 0.065\left(Re_{nf}^{0.65} - 60.22\right)\left(1 + 0.0169\Phi^{0.15}\right)Pr_{nf}^{0.542}. \tag{17.9}$$

The above equation was derived for $1 \leq \Phi \leq 10\%$ and $3 \times 10^3 \leq Re_{nf} \leq 1.7 \times 10^4$.

In [99,100], some other correlations for Nusselt number were obtained in the case of laminar flow:

$$Nu_{nf} = 0.4328\left(1.0 + 11.285\Phi^{0.754} Pe_d^{0.218}\right) Re_{nf}^{0.333} Pr_{nf}^{0.4} \tag{17.10}$$

and turbulent flow

$$Nu_{nf} = 0.0059\left(1.0 + 7.6286\Phi^{0.6886} Pe_d^{0.001}\right) Re_{nf}^{0.9238} Pr_{nf}^{0.4} \tag{17.11}$$

This correlation is obtained experimentally for Cu/water with $0.3 \leq \Phi \leq 2\%$ and $800 \leq Re_{nf} \leq 25,000$.

Several other Nusselt number correlations were obtained numerically using single phase or mixture models. For instance:

The correlation developed by [101]:

$$Nu_{nf} = 0.085 Re_{nf}^{0.71} Pr_{nf}^{0.35} \tag{17.12}$$

using Al_2O_3/water with $0 \leq \Phi \leq 10\%$ and $10^4 \leq Re_{nf} \leq 5 \times 10^5$.

The Nusselt number correlation for Al_2O_3/water given by [102]:

$$Nu_{nf} = 0.716 Re_{nf}^{0.314} Pr_{nf}^{0.6} \Phi^{0.3} \tag{17.13}$$

In the case when $1 \leq \Phi \leq 4\%$ and $250 \leq Re_{nf} \leq 1050$.

The Nusselt number correlation established by [103]:

$$Nu_{nf} = 0.02178 Re_{nf}^{0.811} Pr_{nf}^{0.266} \left(1 + \Phi\right)^{1.140} \tag{17.14}$$

This expression was derived for Al_2O_3/water, and $0.1 \leq \Phi \leq 2\%$, and $3 \times 10^3 \leq Re_{nf} \leq 1.4 \times 10^4$.

Concerning friction factors correlations, the first one was developed for over a century by [104]:

$$f_{nf} = 0.3164 Re_{nf}^{-0.25} \tag{17.15}$$

The well-known Colebrook correlation [105] was also derived very early for turbulent flow, it is given as:

$$\frac{1}{\sqrt{f_{nf}}} = -2 Log_{10}\left(\frac{\varepsilon / D}{3.7} + \frac{2.51}{Re_{nf}\sqrt{f_{nf}}}\right) \tag{17.16}$$

where D is inner diameter of pipe, and ε is the roughness of pipe. For more recent friction factors correlations used for forced convective heat transfer of nanofluids in a smooth tube, the correlation was established by [106]:

$$f_{nf} = 26.44 Re_{nf}^{-0.8737} (1 + \Phi)^{156.23} \tag{17.17}$$

17.3.4 Performance Assessment Indexes for Nanofluids-Based SCC

For CSP technologies which are characterized by high concentration ratios, efficient heat transfer mechanism is fundamental and has the potential to improve the overall plant efficiency and avoid excessive absorber failures due the high temperature gradients. Also, since the absorber tube temperatures are extremely high, thermodynamic irreversibilities augment remarkably. Dispersing nanoparticles into the base fluid induces improvements in thermal properties of the given fluid, which enhance the heat transfer process as well.

The figure of merit, referring to the ratio between the convective heat transfer coefficients with respect to the base fluid, is widely used as preliminary performance criterion. It is given as [107]:

$$FoM = \frac{h_f \mid_{nf}}{h_f \mid_{bf}} \tag{17.19}$$

The enhancement in the convective heat transfer coefficient will provoke maximizing the useful energy captured by the HTF. This will stimulate an increase in the annual thermal energy, E, generated by the solar field when nanofluids are used. The relative gain can be obtained by [108]:

$$\Delta E\% = 100 \frac{E_{nf} - E_{bf}}{E_{bf}} \tag{17.20}$$

The widely used evaluation criteria of adopting nanofluids as HTF are the energy and exergy efficiencies. The energy efficiency of a given CSP describes to the ratio between the useful energy rates gained by the HTF to the solar power received on the reflector. It is expressed as:

$$\eta_{th} = \frac{\dot{q}_u}{\dot{q}_S} = \frac{\dot{m}c_p \left(T_{out} - T_{in} \right)}{I_b A_c} \tag{17.21}$$

where I_b and A_c are the solar beam irradiation per unit area and the collector area, respectively.

Similarly, the exergetic efficiency presents the ratio of useful produced exergy rate to the available solar radiation exergy. It is given as [109]

$$\eta_{ex} = \frac{\dot{e}_u}{\dot{e}_S} = \frac{\dot{q}_u - \dot{m}c_p T_{am} ln \left(\dfrac{T_{out}}{T_{in}} \right)}{\dot{q}_s \left[1 - \dfrac{4}{3} \left(\dfrac{T_{am}}{T_{sun}} \right) + \dfrac{1}{3} \left(\dfrac{T_{am}}{T_{sun}} \right)^4 \right]} = \frac{\dot{m}c_p \left(T_{out} - T_{in} \right) - \dot{m}c_p T_{am} ln \left(\dfrac{T_{out}}{T_{in}} \right)}{I_b A_c \left[1 - \dfrac{4}{3} \left(\dfrac{T_{am}}{T_{sun}} \right) + \dfrac{1}{3} \left(\dfrac{T_{am}}{T_{sun}} \right)^4 \right]} \tag{17.22}$$

T_{am} is the ambient temperature, while T_{sun} is the sun's apparent temperature (equal to 6000 K according to [110]).

The amelioration of overall thermal performance via nanofluids is not free. Due to the high viscosity of nanofluids, pressure drops are increased and in turn higher pumping power will be required. In this sense, it is mandatory to introduce some performance indexes which quantify both aspects to decide properly about the superiority of nanofluids compared to ordinary fluids.

The commonly used index is the thermal enhancement index (TEI) which is assessed under the same conditions of pumping work. This index measures the improvement in the heat transfer, but also the increase in the friction coefficient, f. It is expressed as [111]:

$$TEI = \frac{h_{nf}}{h_{bf}} \mid_w = \frac{\left(Nu_{nf} / Nu_{bf} \right)}{\left(f_{nf} / f_{bf} \right)^{1/3}} \tag{17.23}$$

f denotes the friction factor that can be computed using the equations previously described.

Moreover, the formulas used for the thermal and exergy efficiencies can be modified by taking into account the penalty in the pumping power. These formulas become, respectively, [112,113]:

$$\eta_{th} = \frac{\dot{q}_u - \dot{W}_p / \eta_{el}}{I_b A_c} = \frac{\dot{m}c_p \left(T_{out} - T_{in} \right) - \dot{V}\Delta P / \eta_{el}}{I_b A_c} \tag{17.24}$$

and

$$\eta_{ex} = \frac{\dot{m}c_p\left[\left(T_{out} - T_{in}\right) - T_{am}ln\left(\frac{T_{out}}{T_{in}}\right)\right] - \dot{V}\Delta P}{I_b A_c\left[1 - \frac{4}{3}\left(\frac{T_{am}}{T_{sun}}\right) + \frac{1}{3}\left(\frac{T_{am}}{T_{sun}}\right)^4\right]} \tag{17.25}$$

η_{el} is the electrical efficiency used to transform the pumping work to useful heat energy, $\dot{V}$ is volumetric flow rate, and ΔP is the pressure drops occurring due to the circulation of nanofluids in the receiver. It can be given as:

$$\Delta P = \frac{1}{2}\rho_{nf}u_m^2\frac{L}{d_{ri}}f \tag{17.26}$$

with ρ_{nf} is the density of nanofluid, u_m the mean velocity of nanofluids, L the pipe length, and d_{ri} is the internal diameter.

17.4 Research Trends of Using Nanofluids in Solar Concentrating Collectors

As stated in the introduction section, utilization of nanofluids in stationary solar collectors has been widely discussed in the literature. Substantial enhancement in the thermal efficiency of this collectors' category has been reported. This improvement can be more crucial in CSPs, which are characterized by high operating temperatures [114]. Employing nanofluids in such systems for various heat generation applications is a topic of high interest that attracted many researchers around the world. By examining a wide range of research works, it was found that PTC is the extensively studied technology for which the use of nanofluids was assessed. Latest developments of nanofluid utilization in SCC are reported in the next sub-sections.

17.4.1 Parabolic Trough Collectors

The PTC type has been subjected to many improvements during the last years. In fact, this technology is the most used in CSP plants and has been proved to be commercially and technologically viable option. Some intriguing studies have focused on making geometrical changes in the flow, which can increase the flow turbulence and as a result enhance the heat transfer coefficient. Practically, flow inserts and the modified absorbers are widely studied in the literature [115,116].

Using nanofluids as working fluids in PTC has been discussed both theoretically and experimentally. Various combinations of base fluids and nanoparticles were considered for the energetic, exergetic, and economic analyses.

A solar transparent PTC operating with gas-based nanofluid has been suggested by de Risi et al. [117]. A proper mixture of CuO and Ni nanoparticles was performed to permit a full absorption of solar energy through the receiver. For a volume concentration of 0.3%, the achieved thermal efficiency and outlet temperature were estimated to be 62.5% and 650°C, respectively.

Sokhansefat et al. [118] presented a 3-D computational fluid dynamics (CFD) model of a PTC using Al_2O_3/synthetic oil as a working fluid. Some parametric analyses were investigated such as operating temperature and volume fraction of nanoparticles. Authors reported that the heat transfer coefficient was augmented by 6% for the nanoparticle volumetric concentration of 5% at a working temperature of 500 K.

Kasaeian et al. [119] have built a prototype in which they experimentally tested the use of nanofluids in PTCs. The global efficiency of PTC is improved by up to 7%, when 0.3% multi-walled carbon nanotubes (MWCNTs)/mineral oil nanofluid is employed comparing to the base fluid. The same research team [120] conducted a hybrid optimization process comprising genetic algorithm and sequential quadratic programming. The objective was to assess the Nusselt number and pressure drops when using Al_2O_3/ synthetic oil as a working fluid in a PTC. It was found that heat transfer improved with the higher nanoparticle concentration ratios, but decreased with higher operating temperatures. Since the year 2016, the number of investigations related to the utilization of nanofluids in PTC has increased remarkably.

Kaloudis et al. [121] proposed a simulation work based on a two-phase model testing Syltherm 800/ Al_2O_3 nanofluid as a potential HTF in PTCs. An improvement of 10% on the collector efficiency was achieved for nanoparticles concentration of 4%.

Based on CFD study, Ghasemi and Ranjbar [122] demonstrated that the use of copper and alumina nanoparticles improved heat transfer coefficient by 28% and 35%, respectively, compared to base fluid (water) for a volume fraction of 3%. Moreover, adding Al_2O_3 to synthetic oil (Syltherm 800) was investigated using CFD as well by Wang et al. [123]. The particularity of the study is the use of an optic-thermal-stress coupling model to analyze the overall PTC performance. It was evidenced that used nanofluid generates a net enhancement of the PTC thermal efficiency, ensures lower thermal gradients, and, consequently, reduces the thermal stress and deformation of the receiver.

Bellos et al. [124] numerically studied the modifications of the absorber geometry and benefits resulting from using nanofluids. The simulation work was undertaken using the Flow Simulation Studio of SOLIDWORKS and concerned the commercial Industrial Solar Technology Corp. (IST)-PTC collector. Both methods permitted a thermal efficiency improvement in the range of 4% compared to the initial configuration. For low-enthalpy applications, Coccia et al. [108] performed annual simulations for six water-based nanofluids (i.e., Al_2O_3, ZnO, Fe_2O_3, TiO_2, Au, and SiO_2) and concluded that the improvement in thermal performance is marginal for the studied range of temperatures and volume concentrations. They also recommended developing further experimental studies for higher temperature levels. Since the dynamic viscosity will decrease while the thermal conductivity will be improved for high operating temperatures. Later, Potenza et al. [125] experimentally investigated a transparent PTC using a mixture of CuO nano-powder and air as HTF. Average daily thermal efficiency was found to be 65%, whereas the fluid temperature was 180°C. Gómez-Villarejo et al. [126] dispersed Ag nanoparticles in a base fluid (Dowtherm A) to employ in CSP applications and confirmed via a series of experiments that the heat transfer coefficient can be enhanced by about 6% compared to the base fluid.

Bellos et al. [127] studied the possible improvement in the PTC's energetic performances by using Al_2O_3, CuO, TiO_2, and Cu in Syltherm 800. The studied PTC was designed to an organic Rankine cycle (ORC). The overall assessment indicated that studied system can generate a net electric power of 167.05 kW with an efficiency of 20.11% when CuO is used as nanoparticle.

Mwesigye and Meyer [128] presented a three-dimensional, steady-state, and turbulent model to analyze the potential of using nanofluids as working fluids in a PTC. It was concluded that Ag-Therminol VP-1 nanofluid with enhanced thermal properties achieved a thermal efficiency improvement of about 13.9% compared to base fluid. Allouhi et al. [107] presented energy and exergy analyses based on 1-D transient thermal model of a PTC operating at high temperature levels. Four nanoparticle types were investigated and parametric studies were performed considering Therminol VP-1 as base fluid. CuO-nanofluid led to a maximum exergetic efficiency of 9.05%. The maximum daily relative benefit in the produced thermal energy was assessed to be 1.46% by dispersing Al_2O_3 in the base fluid.

The use of multi-wall CNT (MW)CNT nanoparticles in thermal oil as working fluid in PTC was investigated energetically and financially by [129]. An increase of 15% in the heat transfer coefficient was predicted while the payback period was reduced from 3.98 to 3.87 years by utilizing the nanofluid.

Subramani et al. [130] managed to improve the efficiency of PTC by 8.66% by using TiO_2/DI-H_2O (De-Ionized water) nanofluids. It was also reported that the convective heat transfer coefficient was increased up to 22.76% by using the same nanofluids. Consequently, 9.5% greater absorbed energy parameter was attained as well.

Bellos and Tzivanidis [131] theoretically studied the potential of using hybrid nanofluids in the LS-2 PTC. Parametric studies in terms of operating temperature and flow rates were performed for various combinations of Al_2O_3 and TiO_2 in the Syltherm 800 HTF. It was revealed that hybrid nanofluids achieve thermal improvement two times higher than mono nanofluids. Platinum nanoparticles at a volume concentration of 0.005% were synthesized and dispersed via pulsed ultrasonication in a working fluid commonly utilized in CSP plants [132]. It was proved experimentally that the thermal conductivity and heat transfer coefficient increased by around 37% and 20%, respectively.

Recently, Okonkwo et el. [133] introduced experimentally water-based green synthesized nanofluids to be used in PTCs. The innovative aspect in that study is that nanoparticles were obtained from natural product (olive leaf extract and barley husk). Better thermal improvement at reduced cost compared to Al_2O_3 nanofluid was achieved.

17.4.2 Parabolic Dish

In comparison to the parabolic trough collectors based on technological viewpoint, the parabolic dish type CSP systems occupy less area and suit for small and medium scale applications with high temperature output [134]. Loni et al. [135] considered a dish concentrator with a cylindrical cavity receiver and investigated four different nanofluids including Al_2O_3, Cu, SiO_2, and TiO_2 as the nanoparticles. The results indicated that the thermal efficiency of the system decreased with increasing nanoparticle volume concentration. In addition, it was concluded that Cu/thermal oil type nanofluids have shown the best exergy manner among others. Abid et al. [136] performed a study using nanofluids and molten salts for energy and exergy analysis of two types of CSPs including parabolic dish and parabolic trough with the steam power plant. The heat transfer fluids used were aluminum oxide (Al_2O_3)- and ferric oxide (Fe_2O_3)-based nanofluids, LiCl-RbCl, and $NaNO_3$-KNO_3 molten salts. The results obtained disclosed that the outlet temperature of parabolic dish was higher than the outlet temperature of parabolic trough solar collector under identical operating conditions. It was also observed that the nanofluids offer higher energetic and exergetic efficiencies as compared to molten salts. Moreover, the overall performance of parabolic dish solar collector was found to be higher through using nanofluids as HTFs. Pavlovic et al. [137] dispersed Al_2O_3, Cu, CuO, and TiO_2 on thermal oil and investigated the performance characteristics of a solar dish collector by using these nanofluids. According to the results, the utilization of oil-based nanofluids resulted in higher exergetic performance, while the use of Cu nanoparticle was the most suitable exegetically among the examined cases. The maximum exergetic efficiency is 12.19% which corresponds to an improvement of about 2.5% compared to the scenario with corrugated tube and pure oil.

17.4.3 Solar Tower

A solar power tower employs several mirrors that track the sun's trajectory across the sky. The nanofluids are used in this category of collectors to further reduce heat losses and increase performance at high temperatures and thermal detoxification [61]. Taylor et al. [138] experimentally investigated graphite/Therminol VP-1 nanofluids for 10 MW–100 MW solar power tower and reported possible efficiency enhancement. In theory, 10% efficiency improvement could be achieved under solar concentration ratio between 10 and 1000. However, experimental results revealed that the efficiency increase around 5%–10% could be obtained while utilizing nanofluids in the receiver part. The authors additionally found that $3.5 million per year revenue could be achieved by convenient employment. Kandasamy et al. [139] stated that power tower type CSPs could benefit from the potential efficiency improvements that arise from utilization of nanofluids as working fluids.

17.4.4 Linear Fresnel

LFR is a relatively immature, but promising low cost CSP technology with high concentration ratios and relatively low land utilization factor. Bellos and Tzivanidis [140] investigated the utilization of nanofluids as thermal enhancement method. CuO nanoparticle dispersed on Syltherm 800 (6% volumetric concentration) was used on a LFR collector with the concentration ratio of 58.36. The operation with nanofluids was compared against pure thermal oil for inlet temperatures ranging from 350 K to 650 K. The results illustrated that the maximum thermal efficiency enhancement with the nanofluids was attained to be 0.8%, while pumping work demand was spiked by 50%. In consideration of various criteria including exergy efficiency, overall efficiency, and entropy generation, the operation with nanofluids was found to be beneficial, especially in high temperature levels. Bellos et al. [141], this time, conducted a study to explore thermal efficiency methods in a linear Fresnel reflector with evacuated tube receiver with Syltherm/CuO nanofluids. The use of internal fins with the combination of nanofluids as a means of thermal enhancement methods resulted in 4% enhancement in the overall thermal efficiency.

17.5 Challenges of Nanofluids as HTF in CSP

The literature review truly reflects that nanofluids can potentially improve the thermal performance of SCC, however, many issues should be addressed in future investigations.

1. Nanoparticle cost is extremely high which makes the economic viability of the utilization of nanofluids in SCC questionable. This cost must be huge especially when employed in large-scale CSP plants with large collecting areas. One should note that examining both the energetic, exergetic, and economic (3E) analyses to further investigation of the feasibility of nanofluids in CSP is an important research gap that needs to be undertaken.

2. Excessive pumping power due to the presence of nanoparticles is a technical problem that limits the full utilization of nanofluids in CSP plants. Accurate estimation of additional expenditures related to extra pumping power should be made.

3. For operational and long-term usage of nanofluids in solar energy applications and processes, it is essential to guarantee a high level of stability of suspensions. To face this challenge, surfactants or additives are included to act on the surface chemistry, thus enhancing the suspension of nanoparticles. Nevertheless, these elements cause a degradation of the thermophysical properties of the nanofluids, especially at high operating temperatures. To ensure better stability while preserving thermophysical properties of the prepared nanofluid, one solution consists on coating metallic nanoparticles with ceramic layers.

4. There is currently a lack of experimental investigations related to the utilization of nanofluids in high temperature applications, i.e., CSP technologies. Moreover, most of the experimental works concern water as a base fluid. In this sense, further studies are required to proper characterization of thermal and hydraulic performance of nanofluids as base fluids.

17.6 Concluding Remarks

The IEA's 2050 target has stimulated significant development in various CSP technologies including solar power tower, parabolic dish, parabolic trough, and linear Fresnel reflector systems. The central theme of these emerging technologies is to exploit and transfer solar thermal energy through heat transfer fluids. Currently, solar power tower technology can achieve higher temperatures from 300°C to 656°C. This could be as high as 800°C, expected to be anticipated in the near future. In this context, specially engineered HTFs with unprecedentedly high boiling points and thermal stability are required for the emerging CSP technologies. Therefore, this chapter has provided a comprehensive review on the nanofluids and their practical use in CSP systems. The details of various nanofluids including types, thermal and optical properties, along with performance assessment indices were specifically evaluated under sub-sections. Several drawbacks associated with nanofluids use in CSP systems were also discussed besides the methods to tackle with these setbacks. Likewise, particularly the advancement in recent years and future research trends were provided as well.

Eventually, it has been stated that the utilization of nanofluids in CSP systems are promising. This concept is not only offering performance wise amelioration, but also many environmental and economic benefits including potential reduction in CO_2 emissions through efficiency enhancements. However, ideal nanofluids for CSP application do not exist yet. Therefore, more research is needed to identify the fundamental methods to engage in some of the challenges inherently existed in nano-fluids. In addition, the selection of appropriate nanofluids type must be optimized considering the parameters such as thermal conductivity, particle size, pressure drop, agglomeration, viscosity, pH, and stability under high temperatures and pressures. Clearly, further experimental data are required to verify the performance characteristics via imparting nanofluids. Potential research directions could include:

- Surfactant degradation under high temperature applications
- Functionalized nanoparticles for long-term stability
- Enhancing specific heat of nanofluids for harnessing more thermal energy
- Effects of nanofluids on the thermal efficiency of various CSP systems
- Hybrid nanofluids in CSP application
- Further studies on the optical properties of nanofluids and their effects on the performance of CSP systems
- Energy, exergy and economic analyses of nanofluids use in CSP systems.

Undoubtedly, research on the utilization of nanofluids in CSP systems is at its fundamental level. Using nanotechnologies in solar systems has enormous potential and is under global focus to generate clean and green power.

NOMENCLATURE

ΔP	pressure drops [Pa]
A_c	collector area [m²]
d_{ri}	internal diameter [m]
E	annual thermal energy [MJ]
f	friction coefficient [-]
m	mass [kg]
T	Temperature [K]
TEI	thermal enhancement index [-]
u_m	mean velocity of nanofluids [m/s]
V	Volume [m³]
$\dot{V}$	volumetric flow rate [m³/s]

ABBREVIATIONS

a-Si	amorphous silicon
CdTe	cadmium telluride
CIS-CIGS	copper indium gallium selenide
CPV	concentrating photovoltaic
CSP	concentrating solar power
LCOE	levelized cost of electricity
LFR	linear Fresnel reflectors
MWCNT	multi-wall carbon nanotubes

PTC	parabolic trough collectors
PV	photovoltaic
PV/T	photovoltaic/thermal
SCC	solar concentrating collectors
SPD	solar parabolic dishes
SPT	solar power tower
STE	solar thermal electricity
SWCNT	single wall carbon nanotubes

SUBSCRIPTS

am	ambient
bf	base fluid
nf	nanofluid
np	nanoparticle
sun	sun

GREEK LETTERS

ρ_{nf}	density of nanofluid [kg/s]
Φ	volume fraction of nanoparticles [-]
ε	roughness of pipe [-]
η_{el}	electrical efficiency [-]
μ	dynamic viscosity [Pas]

REFERENCES

1. Chu, S., Cui, Y., and Liu, N. (2017). The path towards sustainable energy. *Nature materials* 16(1), 16–22.
2. Allouhi, A., Kousksou, T., Jamil, A., El Rhafiki, T., Mourad, Y., and Zeraouli, Y. (2015). Economic and environmental assessment of solar air-conditioning systems in Morocco. *Renewable and Sustainable Energy Reviews* 50, 770–781.
3. Buker, M. S., and Riffat, S. B. (2015). Building integrated solar thermal collectors—A review. *Renewable and Sustainable Energy Reviews* 51, 327–346.
4. Allouhi, A., Agrouaz, Y., Amine, M. B., Rehman, S., Buker, M. S., Kousksou, T., Jamil, A., and Benbassou, A. (2017). Design optimization of a multi-temperature solar thermal heating system for an industrial process. *Applied Energy* 206, 382–392.
5. Allouhi, A., Kousksou, T., Jamil, A., Bruel, P., Mourad, Y., and Zeraouli, Y. (2015). Solar driven cooling systems: An updated review. *Renewable and Sustainable Energy Reviews* 44, 159–181.
6. Allouhi, A., Saadani, R., Kousksou, T., Saidur, R., Jamil, A., and Rahmoune, M. (2016). Grid-connected PV systems installed on institutional buildings: Technology comparison, energy analysis and economic performance. *Energy and Buildings* 130, 188–201.
7. Renewable Energy World—News, Resources, Companies, Jobs and more. (n.d.). Retrieved July 29, 2018, from https://www.renewableenergyworld.com/index.html.
8. Feldman, D. J., Margolis, R. M., and Hoskins, J. (2018). *Q4 2017/Q1 2018 Solar Industry Update.* National Renewable Energy Lab (NREL), Golden, CO.
9. Parida, B., Iniyan, S., and Goic, R. (2011). A review of solar photovoltaic technologies. *Renewable and Sustainable Energy Reviews* 15(3), 1625–1636.

10. Photovoltaic Research|NREL. (n.d.). Retrieved July 29, 2018, from https://www.nrel.gov/pv/.

11. Kousksou, T., Allouhi, A., Belattar, M., Jamil, A., El Rhafiki, T., Arid, A., and Zeraouli, Y. (2015). Renewable energy potential and national policy directions for sustainable development in Morocco. *Renewable and Sustainable Energy Reviews* 47, 46–57.

12. Islam, M. T., Huda, N., Abdullah, A. B., and Saidur, R. (2018). A comprehensive review of state-of-the-art concentrating solar power (CSP) technologies: Current status and research trends. *Renewable and Sustainable Energy Reviews* 91, 987–1018.

13. Teske, S., Leung, J., Crespo, L., Bial, M., Dufour, E., Richter, C., and Rochon, E. (2016). Solar thermal electricity: Global outlook 2016. *European Solar Thermal Electricity Association.*

14. Vignarooban, K., Xu, X., Arvay, A., Hsu, K., and Kannan, A. M. (2015). Heat transfer fluids for concentrating solar power systems—A review. *Applied Energy* 146, 383–396.

15. Kost, C., Engelken, M., and Schlegl, T. (2012). Value generation of future CSP projects in North Africa. *Energy Policy* 46, 88–99.

16. Zhang, H. L., Baeyens, J., Degrève, J., and Cacères, G. (2013). Concentrated solar power plants: Review and design methodology. *Renewable and Sustainable Energy Reviews* 22, 466–481.

17. Allouhi, A., Benzakour Amine, M., Kousksou, T., Jamil, A., and Lahrech, K. (2018). Yearly performance of low-enthalpy parabolic trough collectors in MENA region according to different sun-tracking strategies. *Applied Thermal Engineering* 128, 1404–1419.

18. Chen, W., Zou, C., Li, X., and Li, L. (2017). Experimental investigation of SiC nanofluids for solar distillation system: Stability, optical properties and thermal conductivity with saline water-based fluid. *International Journal of Heat and Mass Transfer* 107, 264–270.

19. Sundar, L. S., Sharma, K. V, Singh, M. K., and Sousa, A. C. M. (2017). Hybrid nanofluids preparation, thermal properties, heat transfer and friction factor—A review. *Renewable and Sustainable Energy Reviews* 68, 185–198.

20. Ahmad, S. H. A., Saidur, R., Mahbubul, I. M., and Al-Sulaiman, F. A. (2017). Optical properties of various nanofluids used in solar collector: A review. *Renewable and Sustainable Energy Reviews* 73, 1014–1030.

21. Gupta, M., Singh, V., Kumar, R., and Said, Z. (2017). A review on thermophysical properties of nanofluids and heat transfer applications. *Renewable and Sustainable Energy Reviews* 74, 638–670.

22. Sundar, L. S., Singh, M. K., Punnaiah, V., and Sousa, A. C. M. (2018). Experimental investigation of Al2O3/water nanofluids on the effectiveness of solar flat-plate collectors with and without twisted tape inserts. *Renewable Energy* 119, 820–833.

23. Iranmanesh, S., Ong, H. C., Ang, B. C., Sadeghinezhad, E., Esmaeilzadeh, A., and Mehrali, M. (2017). Thermal performance enhancement of an evacuated tube solar collector using graphene nanoplatelets nanofluid. *Journal of Cleaner Production* 162, 121–129.

24. Rashidi, S., Akar, S., Bovand, M., and Ellahi, R. (2018). Volume of fluid model to simulate the nanofluid flow and entropy generation in a single slope solar still. *Renewable Energy* 115, 400–410.

25. Khodabandeh, E., Safaei, M. R., Akbari, S., Akbari, O. A., and Alrashed, A. A. (2018). Application of nanofluid to improve the thermal performance of horizontal spiral coil utilized in solar ponds: geometric study. *Renewable Energy* 122, 1–16.

26. Wang, X., He, Y., Chen, M., and Hu, Y. (2018). ZnO-Au composite hierarchical particles dispersed oil-based nanofluids for direct absorption solar collectors. *Solar Energy Materials and Solar Cells* 179, 185–193.

27. Martín, E. I., Sánchez-Coronilla, A., Navas, J., Gómez-Villarejo, R., Gallardo, J. J., Alcántara, R., and Fernández-Lorenzo, C. (2018). Unraveling the role of the base fluid arrangement in metal-nanofluids used to enhance heat transfer in concentrating solar power plants. *Journal of Molecular Liquids* 252, 271–278.

28. Chieruzzi, M., Cerritelli, G. F., Miliozzi, A., Kenny, J. M., and Torre, L. (2017). Heat capacity of nanofluids for solar energy storage produced by dispersing oxide nanoparticles in nitrate salt mixture directly at high temperature. *Solar Energy Materials and Solar Cells* 167, 60–69.

29. Zhang, Z., Yuan, Y., Ouyang, L., Sun, Q., Cao, X., and Alelyani, S. (2017). Enhanced thermal properties of Li 2 CO 3–Na 2 CO 3–K 2 CO 3 nanofluids with nanoalumina for heat transfer in high-temperature CSP systems. *Journal of Thermal Analysis and Calorimetry* 128(3), 1783–1792.

30. Wu, Y., Li, J., Wang, M., Wang, H., Zhong, Y., Zhao, Y., Wei, M., and Li, Y. (2018). Solar salt doped by MWCNTs as a promising high thermal conductivity material for CSP. *RSC Advances* 8(34), 19251–19260.

31. Tran, T. T. D., and Smith, A. D. (2018). Incorporating performance-based global sensitivity and uncertainty analysis into LCOE calculations for emerging renewable energy technologies. *Applied Energy* 216, 157–171.

32. Boubault, A., Ho, C. K., Hall, A., Lambert, T. N., and Ambrosini, A. (2016). Levelized cost of energy (LCOE) metric to characterize solar absorber coatings for the CSP industry. *Renewable Energy* 85, 472–483.

33. Liqreina, A., and Qoaider, L. (2014). Dry cooling of concentrating solar power (CSP) plants, an economic competitive option for the desert regions of the MENA region. *Solar Energy* 103, 417–424.

34. Poullikkas, A. (2009). Economic analysis of power generation from parabolic trough solar thermal plants for the Mediterranean region—A case study for the Island of Cyprus. *Renewable and Sustainable Energy Reviews* 13(9), 2474–2484.

35. Hunter, S. R., Smith, D. B., Polizos, G., Schaeffer, D. A., Lee, D. F., and Datskos, P. G. (2014). Low cost anti-soiling coatings for CSP collector mirrors and heliostats. In *High and Low Concentrator Systems for Solar Energy Applications IX* (Vol. 9175, p. 91750J). International Society for Optics and Photonics.

36. Ennaceri, H., El Alami, H., Brik, H., Mokssit, O., and Khaldoun, A. (2014). Lotus effect and super-hydrophobic coatings for concentrated solar power systems (CSP). In *Composite Materials and Renewable Energy Applications (ICCMREA), 2014 International Conference on* (pp. 1–4). IEEE.

37. Palenzuela, P., Zaragoza, G., Alarcón-Padilla, D. C., Guillén, E., Ibarra, M., and Blanco, J. (2011). Assessment of different configurations for combined parabolic-trough (PT) solar power and desalination plants in arid regions. *Energy* 36(8), 4950–4958.

38. Al-Othman, A., Tawalbeh, M., Assad, M. E. H., Alkayyali, T., and Eisa, A. (2018). Novel multi-stage flash (MSF) desalination plant driven by parabolic trough collectors and a solar pond: A simulation study in UAE. *Desalination* 443, 237–244.

39. Wellmann, J., Meyer-Kahlen, B., and Morosuk, T. (2018). Exergoeconomic evaluation of a CSP plant in combination with a desalination unit. *Renewable Energy* 128, 586–602.

40. Calderón, A., Palacios, A., Barreneche, C., Segarra, M., Prieto, C., Rodriguez-Sanchez, A., and Fernández, A. I. (2018). High temperature systems using solid particles as TES and HTF material: A review. *Applied Energy* 213, 100–111.

41. Liu, M., Tay, N. H. S., Bell, S., Belusko, M., Jacob, R., Will, G., Saman, W., and Bruno, F. (2016). Review on concentrating solar power plants and new developments in high temperature thermal energy storage technologies. *Renewable and Sustainable Energy Reviews* 53, 1411–1432.

42. Chirino, H., Xu, B., and Xu, X. (2018). Parametric study of cascade latent heat thermal energy storage (CLHTES) system in Concentrated Solar Power (CSP) plants. *Journal of the Energy Institute.* In press.

43. Almendros-Ibáñez, J. A., Fernández-Torrijos, M., Díaz-Heras, M., Belmonte, J. F., and Sobrino, C. (2018). A review of solar thermal energy storage in beds of particles: Packed and fluidized beds. *Solar Energy.* In press.

44. Pelay, U., Luo, L., Fan, Y., Stitou, D., and Rood, M. (2017). Thermal energy storage systems for concentrated solar power plants. *Renewable and Sustainable Energy Reviews* 79, 82–100.

45. Pacio, J., and Wetzel, T. (2013). Assessment of liquid metal technology status and research paths for their use as efficient heat transfer fluids in solar central receiver systems. *Solar Energy* 93, 11–22.

46. Zhang, Z., Yuan, Y., Zhang, N., Sun, Q., Cao, X., and Sun, L. (2017). Thermal properties enforcement of carbonate ternary via lithium fluoride: A heat transfer fluid for concentrating solar power systems. *Renewable Energy* 111, 523–531.

47. Ouagued, M., Khellaf, A., and Loukarfi, L. (2013). Estimation of the temperature, heat gain and heat loss by solar parabolic trough collector under Algerian climate using different thermal oils. *Energy Conversion and management* 75, 191–201.

48. Moya, E. Z. (2017). Innovative working fluids for parabolic trough collectors. In *Advances in Concentrating Solar Thermal Research and Technology.* Woodhead Publishing, pp. 75–106.

49. Grosu, Y., Bondarchuk, O., and Faik, A. (2018). The effect of humidity, impurities and initial state on the corrosion of carbon and stainless steels in molten HitecXL salt for CSP application. *Solar Energy Materials and Solar Cells* 174, 34–41.

50. Cabeza, L. F., de Gracia, A., Fernández, A. I., and Farid, M. M. (2017). Supercritical CO2 as heat transfer fluid: A review. *Applied Thermal Engineering* 125, 799–810.

51. Xu, X., Vignarooban, K., Xu, B., Hsu, K., and Kannan, A. M. (2016). Prospects and problems of concentrating solar power technologies for power generation in the desert regions. *Renewable and Sustainable Energy Reviews* 53, 1106–1131.

52. Galione, P. A., Pérez-Segarra, C. D., Rodríguez, I., Torras, S., and Rigola, J. (2015). Multi-layered solid-PCM thermocline thermal storage for CSP. Numerical evaluation of its application in a 50 MWe plant. *Solar Energy* 119, 134–150.

53. Li, C.-J., Li, P., Wang, K., and Molina, E. E. (2014). Survey of properties of key single and mixture halide salts for potential application as high temperature heat transfer fluids for concentrated solar thermal power systems. *AIMS Energy* 2(2), 133–157.

54. Bradshaw, R. W., Cordaro, J. G., and Siegel, N. P. (2009). Molten nitrate salt development for thermal energy storage in parabolic trough solar power systems. In *ASME 2009 3rd International Conference on Energy Sustainability collocated with the Heat Transfer and InterPACK09 Conferences* (pp. 615–624). American Society of Mechanical Engineers.

55. Fernández, A. G., Ushak, S., Galleguillos, H., and Pérez, F. J. (2014). Development of new molten salts with LiNO3 and Ca (NO3) 2 for energy storage in CSP plants. *Applied Energy* 119, 131–140.

56. Lang, C., and Lee, B. (2015). Heat transfer fluid life time analysis of diphenyl oxide/biphenyl grades for concentrated solar power plants. *Energy Procedia* 69, 672–680.

57. Vignarooban, K., Pugazhendhi, P., Tucker, C., Gervasio, D., and Kannan, A. M. (2014). Corrosion resistance of Hastelloys in molten metal-chloride heat-transfer fluids for concentrating solar power applications. *Solar Energy* 103, 62–69.

58. Kasaeian, A., Eshghi, A. T., and Sameti, M. (2015). A review on the applications of nanofluids in solar energy systems. *Renewable and Sustainable Energy Reviews* 43, 584–598.

59. Mahian, O., Kianifar, A., Kalogirou, S. A., Pop, I., and Wongwises, S. (2013). A review of the applications of nanofluids in solar energy. *International Journal of Heat and Mass Transfer* 57(2), 582–594.

60. Elsheikh, A. H., Sharshir, S. W., Mostafa, M. E., Essa, F. A., and Ali, M. K. A. (2017). Applications of nanofluids in solar energy: A review of recent advances. *Renewable and Sustainable Energy Reviews* 82, 3483–3502.

61. Nagarajan, P. K., Subramani, J., Suyambazhahan, S., and Sathyamurthy, R. (2014). Nanofluids for solar collector applications: A review. *Energy Procedia* 61, 2416–2434.

62. Singh, T., Hussien, M. A. A., Al-Ansari, T., Saoud, K., and McKay, G. (2018). Critical review of solar thermal resources in GCC and application of nanofluids for development of efficient and cost effective CSP technologies. *Renewable and Sustainable Energy Reviews* 91, 708–719.

63. Azmi, W. H., Sharif, M. Z., Yusof, T. M., Mamat, R., and Redhwan, A. A. M. (2017). Potential of nanorefrigerant and nanolubricant on energy saving in refrigeration system—A review. *Renewable and Sustainable Energy Reviews* 69, 415–428.

64. Rativa, D., and Gómez-Malagón, L. A. (2015). Solar radiation absorption of nanofluids containing metallic nanoellipsoids. *Solar Energy* 118, 419–425.

65. Lee, S.-H., and Jang, S. P. (2013). Extinction coefficient of aqueous nanofluids containing multi-walled carbon nanotubes. *International Journal of Heat and Mass Transfer* 67, 930–935.

66. Tan, J., Xie, Y., Wang, F., Jing, L., and Ma, L. (2017). Investigation of optical properties and radiative transfer of TiO2 nanofluids with the consideration of scattering effects. *International Journal of Heat and Mass Transfer* 115, 1103–1112.

67. Aberoumand, S., and Jafarimoghaddam, A. (2018). Tungsten (III) oxide (WO3)–Silver/transformer oil hybrid nanofluid: Preparation, stability, thermal conductivity and dielectric strength. *Alexandria Engineering Journal* 57, 169–174.

68. Kim, J. B., Lee, S., Lee, K., Lee, I., and Lee, B. J. (2018). Determination of absorption coefficient of nanofluids with unknown refractive index from reflection and transmission spectra. *Journal of Quantitative Spectroscopy and Radiative Transfer* 213, 107–112.

69. Hossain, M. S., Saidur, R., Sabri, M. F. M., Said, Z., and Hassani, S. (2015). Spotlight on available optical properties and models of nanofluids: A review. *Renewable and Sustainable Energy Reviews* 43, 750–762.

70. Leong, K. Y., Ong, H. C., Amer, N. H., Norazrina, M. J., Risby, M. S., and Ahmad, K. Z. K. (2016). An overview on current application of nanofluids in solar thermal collector and its challenges. *Renewable and Sustainable Energy Reviews* 53, 1092–1105.

71. He, Q., Wang, S., Zeng, S., and Zheng, Z. (2013). Experimental investigation on photothermal properties of nanofluids for direct absorption solar thermal energy systems. *Energy Conversion and Management* 73, 150–157.

72. Gorji, T. B., and Ranjbar, A. A. (2017). A review on optical properties and application of nanofluids in direct absorption solar collectors (DASCs). *Renewable and Sustainable Energy Reviews* 72, 10–32.

73. Noguez, C. (2005). Optical properties of isolated and supported metal nanoparticles. *Optical Materials* 27(7), 1204–1211.

74. Taylor, R. A., Phelan, P. E., Otanicar, T. P., Adrian, R., and Prasher, R. (2011). Nanofluid optical property characterization: Towards efficient direct absorption solar collectors. *Nanoscale Research Letters* 6(1), 225.

75. Said, Z. (2016). Thermophysical and optical properties of SWCNTs nanofluids. *International Communications in Heat and Mass Transfer* 78, 207–213.

76. Pak, B. C., and Cho, Y. I. (1998). Hydrodynamic and heat transfer study of dispersed fluids with submicron metallic oxide particles. *Experimental Heat Transfer an International Journal* 11(2), 151–170.

77. Babu, J. A. R., Kumar, K. K., and Rao, S. S. (2017). State-of-art review on hybrid nanofluids. *Renewable and Sustainable Energy Review* 77, 551–565.

78. Said, Z., Kamyar, A., and Saidur, R. (2013). Experimental investigation on the stability and density of TiO2, Al2O3, SiO2 and TiSiO4. In *IOP Conference Series: Earth and Environmental Science* (Vol. 16, p. 12002). IOP Publishing.

79. Philip, J., and Shima, P. D. (2012). Thermal properties of nanofluids. *Advances in Colloid and Interface Science* 183, 30–45.

80. Xie, H., Wang, J., Xi, T., and Liu, Y. (2002). Thermal conductivity of suspensions containing nanosized SiC particles. *International Journal of Thermophysics* 23(2), 571–580.

81. Murshed, S. M. S., Leong, K. C., and Yang, C. (2005). Enhanced thermal conductivity of TiO2—Water based nanofluids. *International Journal of Thermal Sciences* 44(4), 367–373.

82. Yang, L., Xu, J., Du, K., and Zhang, X. (2017). Recent developments on viscosity and thermal conductivity of nanofluids. *Powder Technology* 317, 348–369.

83. Bruggeman, D. A. G. (1935). Dielectric constant and conductivity of mixtures of isotropic materials. *Annales de Physique* (Leipzig) 24, 636–679.

84. Timofeeva, E. V, Gavrilov, A. N., McCloskey, J. M., Tolmachev, Y. V, Sprunt, S., Lopatina, L. M., and Selinger, J. V. (2007). Thermal conductivity and particle agglomeration in alumina nanofluids: Experiment and theory. *Physical Review E* 76(6), 61203.

85. Murshed, S. M. S. (2011). Determination of effective specific heat of nanofluids. *Journal of Experimental Nanoscience* 6(5), 539–546.

86. Xuan, Y., and Roetzel, W. (2000). Conceptions for heat transfer correlation of nanofluids. *International Journal of Heat and Mass Transfer* 43(19), 3701–3707.

87. Khanafer, K., and Vafai, K. (2011). A critical synthesis of thermophysical characteristics of nanofluids. *International Journal of Heat and Mass Transfer* 54(19–20), 4410–4428.

88. Corcione, M. (2011). Empirical correlating equations for predicting the effective thermal conductivity and dynamic viscosity of nanofluids. *Energy Conversion and Management* 52(1), 789–793.

89. Batchelor, G. K. (1977). The effect of Brownian motion on the bulk stress in a suspension of spherical particles. *Journal of Fluid Mechanics* 83(1), 97–117.

90. Bellos, E., Tzivanidis, C., and Tsimpoukis, D. (2018). Enhancing the performance of parabolic trough collectors using nanofluids and turbulators. *Renewable and Sustainable Energy Reviews* 91, 358–375.

91. Ambreen, T., and Kim, M.-H. (2018). Heat transfer and pressure drop correlations of nanofluids: A state of art review. *Renewable and Sustainable Energy Reviews* 91, 564–583.

92. Ding, Y., and Wen, D. (2005). Particle migration in a flow of nanoparticle suspensions. *Powder Technology* 149(2–3), 84–92.

93. Keblinski, P., Phillpot, S. R., Choi, S. U. S., and Eastman, J. A. (2002). Mechanisms of heat flow in suspensions of nano-sized particles (nanofluids). *International Journal of Heat and Mass Transfer* 45(4), 855–863.

94. Jang, S. P., and Choi, S. U. S. (2004). Role of Brownian motion in the enhanced thermal conductivity of nanofluids. *Applied Physics Letters* 84(21), 4316–4318.

95. Ghofrani, A., Dibaei, M. H., Sima, A. H., and Shafii, M. B. (2013). Experimental investigation on laminar forced convection heat transfer of ferrofluids under an alternating magnetic field. *Experimental Thermal and Fluid Science* 49, 193–200.

96. Dittus, F. W., and Boelter, L. M. K. (1985). Heat transfer in automobile radiators of the tubular type. *International Communications in Heat and Mass Transfer* 12(1), 3–22.

97. Yunus, C. A., and Afshin, J. G. *Heat and Mass Transfer: Fundamentals and Applications.* New Delhi, India: Tata McGraw-Hill; 2011.

98. Vajjha, R. S., Das, D. K., and Kulkarni, D. P. (2010). Development of new correlations for convective heat transfer and friction factor in turbulent regime for nanofluids. *International Journal of Heat and Mass Transfer* 53(21–22), 4607–4618.

99. Xuan, Y., and Li, Q. (2003). Investigation on convective heat transfer and flow features of nanofluids. *Journal of Heat Transfer* 125(1), 151–155.

100. Li, Q., and Xuan, Y. (2002). Convective heat transfer and flow characteristics of Cu-water nanofluid. *Science in China Series E: Technological Science* 45(4), 408–416.

101. El Bécaye Maïga, S., Tam Nguyen, C., Galanis, N., Roy, G., Maré, T., and Coqueux, M. (2006). Heat transfer enhancement in turbulent tube flow using Al2O3 nanoparticle suspension. *International Journal of Numerical Methods for Heat and Fluid Flow* 16(3), 275–292.

102. Moraveji, M. K., and Esmaeili, E. (2012). Comparison between single-phase and two-phases CFD modeling of laminar forced convection flow of nanofluids in a circular tube under constant heat flux. *International Communications in Heat and Mass Transfer* 39(8), 1297–1302.

103. Hejazian, M., Moraveji, M. K., and Beheshti, A. (2014). Comparative study of Euler and mixture models for turbulent flow of Al2O3 nanofluid inside a horizontal tube. *International Communications in Heat and Mass Transfer*, 52, 152–158.

104. White, F. M. *Fluid Mechanics*, Boston, MA: WCB. Ed McGraw-Hill; 1999.

105. Colebrook, C. F., Blench, T., Chatley, H., Essex, E. H., Finniecome, J. R., Lacey, G. et al. (1939). Correspondence. Turbulent flow in pipes, with particular reference to the transition region between the smooth and rough pipe laws (Includes plates). *Journal of the Institution of Civil Engineers* 12(8), 393–422.

106. Suresh, S., Venkitaraj, K. P., Selvakumar, P., and Chandrasekar, M. (2012). Effect of Al2O3–Cu/water hybrid nanofluid in heat transfer. *Experimental Thermal and Fluid Science* 38, 54–60.

107. Allouhi, A., Benzakour Amine, M., Saidur, R., Kousksou, T., and Jamil, A. (2018). Energy and exergy analyses of a parabolic trough collector operated with nanofluids for medium and high temperature applications. *Energy Conversion and Management* 155, 201–217.

108. Coccia, G., Di Nicola, G., Colla, L., Fedele, L., and Scattolini, M. (2016). Adoption of nanofluids in low-enthalpy parabolic trough solar collectors: Numerical simulation of the yearly yield. *Energy Conversion and Management* 118, 306–319.

109. Mwesigye, A., Huan, Z., and Meyer, J. P. (2016). Thermal performance and entropy generation analysis of a high concentration ratio parabolic trough solar collector with Cu-Therminol® VP-1 nanofluid. *Energy Conversion and Management* 120, 449–465.

110. Petela, R. (2003). Exergy of undiluted thermal radiation. *Solar Energy* 74(6), 469–488.

111. Hasanpour, A., Farhadi, M., and Sedighi, K. (2014). A review study on twisted tape inserts on turbulent flow heat exchangers: The overall enhancement ratio criteria. *International Communications in Heat and Mass Transfer* 55, 53–62.

112. Wirz, M., Petit, J., Haselbacher, A., and Steinfeld, A. (2014). Potential improvements in the optical and thermal efficiencies of parabolic trough concentrators. *Solar Energy* 107, 398–414.

113. Madadi, V., Tavakoli, T., and Rahimi, A. (2014). First and second thermodynamic law analyses applied to a solar dish collector. *Journal of Non-Equilibrium Thermodynamics* 39(4), 183–197.

114. Bellos, E., Said, Z., and Tzivanidis, C. (2018). The use of nanofluids in solar concentrating technologies: A comprehensive review. *Journal of Cleaner Production* 196, 84–99.

115. Bellos, E., and Tzivanidis, C. (2018). Investigation of a star flow insert in a parabolic trough solar collector. *Applied Energy* 224, 86–102.

116. Huang, Z., Li, Z.-Y., Yu, G.-L., and Tao, W.-Q. (2017). Numerical investigations on fully-developed mixed turbulent convection in dimpled parabolic trough receiver tubes. *Applied Thermal Engineering* 114, 1287–1299.

117. De Risi, A., Milanese, M., and Laforgia, D. (2013). Modelling and optimization of transparent parabolic trough collector based on gas-phase nanofluids. *Renewable Energy* 58, 134–139.
118. Sokhansefat, T., Kasaeian, A. B., and Kowsary, F. (2014). Heat transfer enhancement in parabolic trough collector tube using Al2O3/synthetic oil nanofluid. *Renewable and Sustainable Energy Reviews* 33, 636–644.
119. Kasaeian, A., Daviran, S., Azarian, R. D., and Rashidi, A. (2015). Performance evaluation and nanofluid using capability study of a solar parabolic trough collector. *Energy Conversion and Management* 89, 368–375.
120. Zadeh, P. M., Sokhansefat, T., Kasaeian, A. B., Kowsary, F., and Akbarzadeh, A. (2015). Hybrid optimization algorithm for thermal analysis in a solar parabolic trough collector based on nanofluid. *Energy* 82, 857–864.
121. Kaloudis, E., Papanicolaou, E., and Belessiotis, V. (2016). Numerical simulations of a parabolic trough solar collector with nanofluid using a two-phase model. *Renewable Energy* 97, 218–229.
122. Ghasemi, S. E., and Ranjbar, A. A. (2016). Thermal performance analysis of solar parabolic trough collector using nanofluid as working fluid: A CFD modelling study. *Journal of Molecular Liquids* 222, 159–166.
123. Wang, Y., Xu, J., Liu, Q., Chen, Y., and Liu, H. (2016). Performance analysis of a parabolic trough solar collector using Al2O3/synthetic oil nanofluid. *Applied Thermal Engineering* 107, 469–478.
124. Bellos, E., Tzivanidis, C., Antonopoulos, K. A., and Gkinis, G. (2016). Thermal enhancement of solar parabolic trough collectors by using nanofluids and converging-diverging absorber tube. *Renewable Energy* 94, 213–222.
125. Potenza, M., Milanese, M., Colangelo, G., and de Risi, A. (2017). Experimental investigation of transparent parabolic trough collector based on gas-phase nanofluid. *Applied Energy* 203, 560–570.
126. Gómez-Villarejo, R., Martín, E. I., Navas, J., Sánchez-Coronilla, A., Aguilar, T., Gallardo, J. J., Alcántara, R., De los Santos, D., Carrillo-Berdugo, I., and Fernández-Lorenzo, C. (2017). Ag-based nanofluidic system to enhance heat transfer fluids for concentrating solar power: Nano-level insights. *Applied Energy* 194, 19–29.
127. Bellos, E., and Tzivanidis, C. (2017). Parametric analysis and optimization of an organic Rankine cycle with nanofluid based solar parabolic trough collectors. *Renewable Energy* 114, 1376–1393.
128. Mwesigye, A., and Meyer, J. P. (2017). Optimal thermal and thermodynamic performance of a solar parabolic trough receiver with different nanofluids and at different concentration ratios. *Applied Energy* 193, 393–413.
129. Kasaiean, A., Sameti, M., Daneshazarian, R., Noori, Z., Adamian, A., and Ming, T. (2018). Heat transfer network for a parabolic trough collector as a heat collecting element using nanofluid. *Renewable Energy* 123, 439–449.
130. Subramani, J., Nagarajan, P. K., Mahian, O., and Sathyamurthy, R. (2018). Efficiency and heat transfer improvements in a parabolic trough solar collector using TiO2 nanofluids under turbulent flow regime. *Renewable Energy* 119, 19–31.
131. Bellos, E., and Tzivanidis, C. (2018). Thermal analysis of parabolic trough collector operating with mono and hybrid nanofluids. *Sustainable Energy Technologies and Assessments* 26, 105–115.
132. Gómez-Villarejo, R., Martín, E. I., Sánchez-Coronilla, A., Aguilar, T., Gallardo, J. J., Martínez-Merino, P., Carrillo-Berdugo, I., Alcántara, R., Fernández-Lorenzo, C., and Navas, J. (2018). Towards the improvement of the global efficiency of concentrating solar power plants by using Pt-based nanofluids: The internal molecular structure effect. *Applied Energy* 228, 2262–2274.
133. Okonkwo, E. C., Essien, E. A., Akhayere, E., Abid, M., Kavaz, D., and Ratlamwala, T. A. H. (2018). Thermal performance analysis of a parabolic trough collector using water-based green-synthesized nanofluids. *Solar Energy* 170, 658–670.
134. Rajendran, D. R., Sundaram, E. G., and Jawahar, P. (2017). Experimental studies on the thermal performance of a parabolic dish solar receiver with the heat transfer fluids SiC + water nano fluid and water. *Journal of Thermal Science* 26(3), 263–272.
135. Loni, R., Askari Asli-ardeh, E., Ghobadian, B., Kasaeian, A. B., and Gorjian, S. (2017). Thermodynamic analysis of a solar dish receiver using different nanofluids. *Energy* 133, 749–760.
136. Abid, M., Ratlamwala, T. A. H., and Atikol, U. (2015). Performance assessment of parabolic dish and parabolic trough solar thermal power plant using nanofluids and molten salts. *International Journal of Energy Research* 40(4), 550–563.

137. Pavlovic, S., Bellos, E., and Loni, R. (2018). Exergetic investigation of a solar dish collector with smooth and corrugated spiral absorber operating with various nanofluids. *Journal of Cleaner Production* 174, 1147–1160.

138. Taylor, R. A., Phelan, P. E., Otanicar, T. P., Walker, C. A., Nguyen, M., Trimble, S., and Prasher, R. (2011). Applicability of nanofluids in high flux solar collectors. *Journal of Renewable and Sustainable Energy* 3(2), 023104–023115.

139. Kandasamy, R., Muhaimin, I., and Rosmila, A. K. (2014). The performance evaluation of unsteady MHD non-Darcy nanofluid flow over a porous wedge due to renewable (solar) energy. *Renewable Energy* 64, 1–9.

140. Bellos, E., and Tzivanidis, C. (2018). Multi-criteria evaluation of a nanofluid-based linear Fresnel solar collector. *Solar Energy* 163, 200–214.

141. Bellos, E., Tzivanidis, C., and Papadopoulos, A. (2018). Enhancing the performance of a linear Fresnel reflector using nanofluids and internal finned absorber. *Journal of Thermal Analysis and Calorimetry* 1–9.

18

Influence of the Base Fluid and Surfactant Arrangement on the Enhancement of Heat Transfer in Metal–Nanofluids Used in Concentrating Solar Power Plants: A Molecular Level Perspective

Elisa I. Martín, Antonio Sánchez-Coronilla, Javier Navas, Roberto Gómez-Villarejo, Rodrigo Alcántara, and Concha Fernández-Lorenzo

CONTENTS

18.1 Introduction

The society requires increasing amounts of energy, and one of the challenges facing society today is to find solutions for that demand while minimizing the environmental impact on the planet [1]. Given the possible imminent end to the age of fossil fuels for electricity generation, renewable energy sources have grown in importance due to their capacity to generate sustainable energy to cater for the world's electricity requirements. The energy from the sun, solar energy [2], is an important option and being the most available energy source on the planet can be used to convert thermal energy into electricity. In this sense, concentrating solar power (CSP) is one of the most interesting resources for transforming thermal energy from the sun into electricity [3]. To this end, high temperatures (near 500°C) must be reached by concentrating solar radiation onto special receivers, the parabolic-trough collector (PTC). This collector is one of the most common for commercial interest [3]. The PTC is integrated indirectly into a steam turbine plant in a heat exchanger and thermal oil is heating for generating steam. This technology is known as heat transfer fluid (HTF) technology [3–5]. By improving the thermal properties of the heat transfer fluid used in CSP plants, this should improve the efficiency of the plants. In this sense, the use of nanofluids (colloidal suspensions in a base fluid) seems to be an alternative for improving the thermal properties of the base fluids [5–7]. Thus, improvements in properties such as isobaric specific heat [6,8–11] or thermal conductivity [6,9,12–14] have been reported on metal based nanofluids. A typical heat transfer fluid used in CSP is an eutectic mixture of biphenyl ($C_{12}H_{10}$) and diphenyl oxide ($C_{12}H_{10}O$) [15,16].

The use of gold nanofluids in CSP plants could be a promising alternative on the enhancement of heat transfer, as stated elsewhere [17,18]. Experimental results with gold nanofluids show the interest in using a phase transfer surfactant to provide a greater stability to the nanofluid [17,18]. Therefore, a consequence of the experimental preparation of the Au nanofluid itself is that part of the surfactant is transferred to the nanofluid, forming part of the nanofluid system [17,18]. In this sense, exploring the behaviour and effect of the surfactant on the properties of the system is of interest. The influence of the base fluid arrangement around the metal nanoparticles should be related in the enhancement of heat transfer properties of metal nanofluid.

Thus, in this chapter, molecular dynamic calculations were performed on metal nanofluid (Cu, Ag, Ni, and Au with and without surfactant) using an eutectic mixture of biphenyl ($C_{12}H_{10}$, 26.5%) and diphenyl oxide ($C_{12}H_{10}O$, 73.5%) as the base fluid. The surfactant selected was tetraoctylammonium bromide, $(C_8H_{17})_4NBr$ (TOAB). This surfactant was chosen because it was the experimental surfactant employed in the experimental synthesis of Au nanofluids [17,18], and it leads to a good stability for the system. The isobaric specific heat, diffusivity, and thermal conductivity values were obtained for Cu, Ag, Au, and Ni nanofluid systems, and for Au nanofluid, it was studied with and without surfactant and compared with the values obtained for the base fluid. The structural properties were obtained by analysing their radial distribution functions (RDFs) and spatial distribution functions (SDFs). The role of the base fluid and the surfactant, and more concrete, the role of diphenyl oxide molecules around the metal were analysed. Our theoretical results showed, in all cases, diphenyl oxide molecules play a key role in the enhancement of the thermal properties of the nanofluids. Moreover, in the case of Au nanofluid, the competition between the base fluid-surfactant and base fluid-gold nanoparticle interactions was studied. In this case, it was detected an exchange of surfactant and diphenyl oxide molecules in the inner layer of molecules around the metal taking place. Such structural reorganisation is related to the enhancement of the thermal properties of the systems.

18.2 Computational Framework

The intra- and intermolecular interactions of the HTF fluid (diphenyl oxide/biphenyl blend) were described by using the transferable potentials for phase equilibria-explicit hydrogen force field (TraPPE-EH) [19,20]. The phenyl rings and the directly connected atoms are treated as rigid entities, but were allowed to rotate with regard to each other around the carbon bond.

The transferable potentials for phase equilibria-united atom force field (TraPPE-UA) [21–23] was used to describe the intra- and intermolecular interactions of the TOAB surfactant. This force field treats the alkyl groups as a fully flexible model based on single interaction sites (pseudo-atoms). A pseudo-atom represents a carbon atom together with all of its bonded hydrogen atoms (CH_3 and CH_2) and treats the polar nitrogen atom and the carbon atoms bonded to it as explicit interaction sites in a rigid unit.

The metal nanoparticles of Cu, Ag, Ni were described using a non-bonded dummy model [6,24]. That model consists of six particles, referred to as 'dummy atoms', placed around a central metal particle in an octahedral geometry. The geometry of the dummy complex itself is kept rigid by the imposition of large force constants on the metal-dummy bonds. However, as there are no bonds between the dummy complex and the surrounding ligands, overall rotation of the six-center frame about the nucleus is allowed, and no internal forces are associated with such rotation. Therefore, the coordination geometry is not constrained to the geometry of the dummy model used, but rather, the system is free to exchange ligands.

In the case of Au nanoparticle, a rigid unit cell of 14 gold atoms with the point group Fm-3m was used to perform the simulation of the metal nanoparticle. The parameters used in previous studies [25] were adopted to describe the non-bonded force field of the metal nanoparticle.

The TraPPE-EH and TraPPE-UA force fields and the non-bonded force field of the metal nanoparticle use Lennard-Jones (LJ) and Coulomb potentials to represent the non-bonded interactions:

$$u\left(r_{ij}\right) = 4\varepsilon_{ij}\left[\left(\frac{\sigma_{ij}}{r_{ij}}\right)^{12} - \left(\frac{\sigma_{ij}}{r_{ij}}\right)^{6}\right] + \frac{q_i q_j}{4\pi\varepsilon_0 r_{ij}} \tag{18.1}$$

where r_{ij}, ε_{ij}, σ_{ij}, q_i, q_j, and ε_0 are the distances between interaction sites i and j, the LJ well depth, the LJ diameter, the partial charges on interaction sites i and j, and the permittivity of vacuum, respectively. The Lorentz-Berthelot combining rules were used to determine LJ parameters for unlike interactions.

The DLPOLY code [26] was used to carry out the molecular dynamics simulations. The canonical ensemble (N: number of particles, V: volume, T: temperature [NVT]) was applied in all the simulations performed by using periodic boundary conditions and a Nose-Hoover thermostat. A cubic box providing the initial configuration was built with the PACKMOL code [27]. The length of the box sides were chosen to keep the density of the experimental HTF at 298 K (1059 kg m⁻³). The Ewald sum methodology [28] was applied to account for electrostatic interactions using a cut-off distance of 9 Å in all the cases. The simulations run lasted for 1 ns employing a time step of 0.5 fs, and the structures were saved every 100 time steps for analysing the trajectory.

18.3 Theoretical Analysis by Molecular Dynamics

18.3.1 Thermophysical Properties

18.3.1.1 Isobaric Specific Heat

Molecular dynamic simulations were performed to discover a predictive model of the experimental behaviour of the nanofluids. To calculate the isobaric specific heat, molecular dynamic simulations were performed at different temperatures ranging from ca. 50–500 K. As an example, Figure 18.1 shows a linear tendency for the plot of the total energy of the system versus temperature for the base fluid. The isobaric specific heat can be obtained from the slope of this plot (Figure 18.1). The values of 1940 J kg⁻¹ K⁻¹ for the base fluid and 3700, 2250, 2150, 1990, and 1830 J kg⁻¹ K⁻¹ for the Au with surfactant, Cu, Ag, Au without surfactant, and Ni nanofluid, respectively, were obtained. These values, although slightly higher, qualitatively coincide quite well with the experimental values and show the same experimental tendency, $Cp_{(Au\ nanofluid\ with\ surfactant)} > Cp_{(Cu\ nanofluid)} > Cp_{(Ag\ nanofluid)} > Cp_{(base\ fluid)} > Cp_{(Ni\ nanofluid)}$. The comparison of the isobaric specific heat values obtained for the base fluid and the nanofluids simulated would suggest that the Au with surfactant and Cu nanofluid systems present the most suitable isobaric specific heat values to achieve efficient heat transfer processes followed by the Ag nanofluid system. Those results are in agreement with experimental studies that show that metal nanoparticles enhance the thermal properties of the base fluid [6,12,14,17,18]. But, for the Ni nanofluid system, a decrease in the value for isobaric specific heat was observed. These behaviours are related to the different arrangement of the base

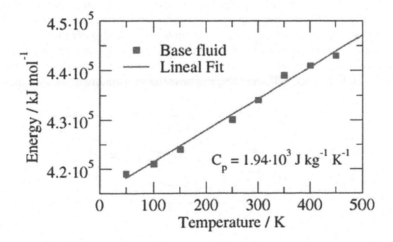

FIGURE 18.1 Plot of the energy versus temperature for the base fluid.

fluid around of the metal, and in the case of Au nanofluid system with surfactant, the reorganisation of the surfactant with the base fluid around the Au happens as will be demonstrated below in the study of the structural properties of these systems.

18.3.1.2 Diffusivity and Thermal Conductivity

The translational diffusion coefficients were computed according to the Einstein relation by computing the mean square displacement (MSD). This translational diffusion coefficient is the thermal diffusivity used typically in experimental studies [6,24]. The diffusion coefficients are obtained by the following equation:

$$D_i = \lim_{t \to \infty} \frac{\left| \vec{r}_i(t) - \vec{r}_i(0) \right|^2}{6t}, \tag{18.2}$$

where $\left| \vec{r}_i(t) - \vec{r}_i(0) \right|^2$ is the MSD.

As an example, Figure 18.2 shows the plot of the MSD versus time for the base fluid at 300 K. After approximately 3 ps the mean square displacement varies in line with time. The diffusion coefficient is obtained from the slope of this plot and by applying equation (18.2). The thermal conductivity values were obtained using equation (18.3) from the values for the diffusion coefficient, the density, and the isobaric specific heat.

$$k(T) = D(T) \cdot C_P(T) \cdot \rho(T), \tag{18.3}$$

where k is the thermal conductivity, D the thermal diffusivity, C_P is the isobaric specific heat, and ρ is the density.

The values obtained at 300 K were 0.107 W m^{-1} K^{-1} for the base fluid and around 0.600 W m^{-1} K^{-1} for the Au with and without surfactant, 0.147, 0.125, and 0.097 W m^{-1} K^{-1} for the Cu, Ag, and Ni-nanofluid, respectively. Comparing these values with those obtained experimentally [6,18,24], it can be said to follow the same experimental tendency. High thermal conductivity values are important in heat transfer fluids, the highest value corresponds with the thermal conductivity of the Au with and without surfactant and Cu nanofluid, followed by the Ag nanofluid, the base fluid, and the Ni nanofluid.

However, special attention deserves to the gold system studied at different temperatures. In the case of the gold nanofluid system with and without surfactant, the diffusion coefficient and the thermal conductivity values at different temperatures fit a sigmoid curve, while an exponential increase is produced for the base fluid (Figures 18.3 and 18.4). As Figure 18.3 shows, the greatest changes are detected between 300 and 600 K. The diffusion values for the system without surfactant being much greater than that of the one without that suggest greater movement of the particles in the nanofluid system without surfactant. In the heart of the nanofluid, the surfactant must impede to a certain extent the freedom of movement of the rest of the particles, decreasing the number of effective interparticle collisions in the Au nanofluid system with surfactant. Consequently, it is needed more energy to raise the temperature of the system, which explains

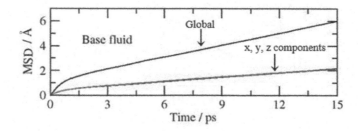

FIGURE 18.2   Plot of the MSD versus time for base fluid at 300 K.

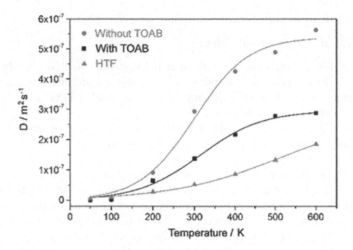

FIGURE 18.3 Plot of the diffusion coefficient versus temperature for gold nanofluid system with and without surfactant and the heat transfer fluid.

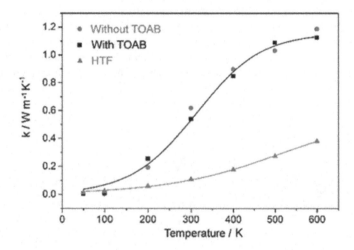

FIGURE 18.4 Plot of the thermal conductivity versus temperature for gold nanofluid system with and without surfactant and the heat transfer fluid.

the high isobaric specific heat value obtained as compared with the system without surfactant. Thus, the surfactant seems to act as a kind of net that cushions the interparticle collisions, as will be shown below.

Figure 18.4 shows a positive trend with temperature when the system is studied in the range of 50–600 K. The influence of temperature and surfactant on the enhancement of the thermal conductivity depends on the system nature [29]. In the interval between 300 and 600 K, it is observed the greatest differences in the thermal conductivity values of the system with and without surfactant compared with the base fluid (Figure 18.4).

The highest values obtained for the thermal conductivity of gold nanofluid have a reasonable explanation. Thermal conductivity values are obtained by applying equation (18.3) with the isobaric specific heat, density, and diffusion coefficient values. Thus, in the present case, high isobaric specific heat values should lead to high thermal conductivity values. These values follow the same experimental trend as reported elsewhere [17,18] with the thermal conductivity of the gold nanofluid system greater than that of the base fluid. Similar values obtained for the system with and without surfactant come from the higher value of the isobaric specific heat and lower diffusion coefficient for the system with surfactant as compared to that without surfactant. Therefore, the presence of the surfactant exerts an effect at molecular level in the enhancement of the thermal properties of the nanofluid.

18.3.2 Structural Properties

18.3.2.1 Base Fluid Arrangement Effect

In order to reveal at molecular-level insight of the nanofluid systems, the RDF and SDF were analysed.

The analysis of the RDFs of the X-O pair (X = Cu, Ag, and Ni) at different temperatures showed very little variation, and for the discussion of RDFs and SDFs corresponding to the Cu, Ag, and Ni nanofluid systems, the temperature of 300 K was chosen.

Figure 18.5 shows the RDFs of the X-O and X-CO pairs for the three metal nanofluids studied. Regarding to the atomic types, CO represents the carbon atoms linked to the oxygen of the diphenyl oxide molecule.

The analysis of the RDF of the X-O pair shows an intense peak centred around 2.0 Å for Cu and Ni nanofluid and 2.2 Å for Ag nanofluid (Figure 18.5). The highest intensity of this peak is for the Cu nanofluid system indicating a more strong orientation binding. The number of integration of this peak for the Cu nanofluid is of three oxygen atoms which belong to three diphenyl oxide molecules, while for the Ag and Ni nanofluid systems, the peak integrates to two and one oxygen atoms, respectively. Those atoms belong to two and one diphenyl oxide molecules, for Ag and Ni nanofluids, respectively.

In the inset of Figure 18.5, it is included the X-CO pairs. The RDF of the Cu-CO, Ag-CO, and Ni-CO pair shows peak at 3.2 Å attributed to six carbons (two carbons for each diphenyl oxide), four carbons and a split wide peak that integrates to two C atoms, respectively.

The analysis of the SDF allows us to determine how the molecules of the fluid base are arranged around the metal nanoparticle. Figure 18.6 shows the SDF for the Cu, Ag, and Ni nanofluids with the metal in the centre. In all cases, the radius of the SDF was chosen to enable a greater clarity of the image of the first layer around the metal. Figure 18.6 shows that for the Cu nanofluid, there are three O atoms

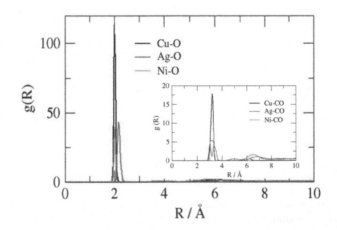

FIGURE 18.5 RDFs for the Cu-O, Ag-O, and Ni-O pairs. Inset: RDFs for Cu-CO, Ag-CO, and Ni-CO pairs.

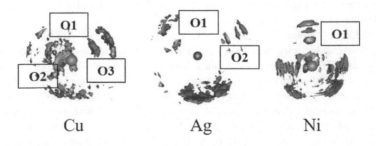

FIGURE 18.6 SDFs from the Cu-nanofluid, Ag-nanofluid, and Ni-nanofluid systems.

(indicated by the corresponding number) from the diphenyl oxide molecules and two O atoms for the Ag nanofluid from two diphenyl oxide molecules. However, for the Ni nanofluid system, only one oxygen atom belongs to a diphenyl oxide molecule is detected. In all cases, the O atoms are oriented towards the metal in the centre. Thus, the Cu, Ag, and Ni nanofluid system present a first layer with three, two, and one molecules of diphenyl oxide around the metal, respectively, with biphenyl molecule placed as base of the systems.

Thus, depending on the kind of metal incorporated into the fluid, a different arrangement of the base fluid molecules around the metal nanoparticle is found. The different arrangement of the benzene rings and the number of O atoms from the diphenyl oxide around the metal nanoparticle agree with reported analyses of complex metals with benzene rings. Due to a stable configuration of d^{10} for Cu atom, no charge transfer is produced, and it shows a quite weak interaction with the benzene ring. In turn, the Cu-O interaction is more stable, and the weak interaction of Cu with the two phenyl groups of the diphenyl oxide allows other diphenyl oxide molecules to approach. The opposite is found for Ni atom that transfers charge to the benzene rings. Consequently, the two phenyl groups from the diphenyl oxide approach the metal nanoparticle, hindering the approximation of more diphenyl oxide molecules. For the Ni nanoparticle, it is found it has preference for the interaction with the benzene rings of both diphenyl oxide and biphenyl, then it interacts only with a diphenyl oxide molecule. For the Ag atom, a behaviour between the Cu and Ni nanoparticle is found. For Ag nanoparticle, interactions similar to those of Cu are generated, although Ag shows less weak interactions with the benzene rings due to its more expanded d^{10} configuration, forming a stable structure with two diphenyl oxide molecules and a ring of biphenyl.

Therefore, the preference for the metal-oxygen interaction favour the rearrangement of the base fluid around Cu and Ag nanoparticles leading to an increase in the isobaric specific heat and thermal conductivity. Whereas in the case of Ni nanoparticles, based on the preference for the interaction with the benzene rings, it is found a different arrangement of the base fluid around the metal. This seems to be the reason why for Ni nanofluid there is no enhancement of the thermal properties as compared with the base fluid.

According to those results, it would seem reasonable to think that the arrangement of the first layer of base fluid molecules around the metal is responsible for the enhanced thermal properties of the Cu and Ag nanofluid. And the directionality of movement that generates the structural arrangement in the first layer should involve effective heat transport, leading to increased thermal conductivity compared with the base fluid.

For testing the directionality of movement, from the simulation trajectory, translational distribution functions of the metal nanoparticle with respect to its average position in the system were calculated. Figure 18.7 shows the translational distribution functions of Cu, Ag, and Ni metals, $P_{Cu}(r)$, $P_{Ag}(r)$, and $P_{Ni}(r)$, respectively, in the x, y, and z directions.

In Figure 18.7, the decrease in intensity of the translational distribution function of Ag nanoparticle, $P_{Ag}(r)$, comparing with the translational distribution functions of Cu and Ni nanoparticles, $P_{Cu}(r)$ and $P_{Ni}(r)$ draws the attention. The loss of intensity of the translational distribution function of Ag reveals a greater amplitude of movement of the Ag nanoparticle. This greater amplitude of movement is corroborated when the value of the function at long distance is observed, as shown in the right part of Figure 18.7. For Cu and Ni nanoparticles, the value of $P_{Cu}(r)$ and $P_{Ni}(r)$ decreases practically to zero at 3 Å and 4 Å, respectively, but $P_{Ag}(r)$ is kept without decreasing to zero until a distance up to 7 Å (Figure 18.7, right). Related with the movement described for the metal nanoparticle in the x, y, and z directions, the behaviour of Cu nanoparticle is different from Ag and Ni nanoparticle. For Cu nanoparticle, a homogeneous translation in the x, y, and z directions that describes a Brownian movement is observed. However, an anisotropic movement in the y and z direction, repectively, is observed for the Ag and Ni nanoparticles. For Ag and Ni nanofluid, it is possible to observe that the movement is slightly longer in y direction for Ag nanoparticle and in x direction for Ni nanoparticle. This anisotropic translation is unexpected and reflects the internal movement of the base fluid molecules belonging to the first-layer of molecules around the metals.

The analysis of the orientational distribution function of the angle formed by each oxygen atom of the first-layer cluster and the benzene ring with regard to the metal and the x, y, and z axes of the box

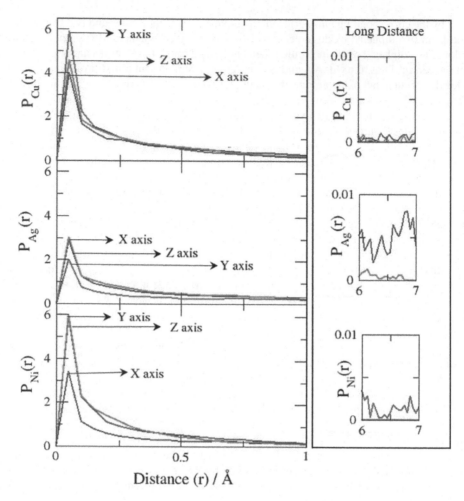

FIGURE 18.7 Translational distribution functions of Cu nanoparticle, $P_{Cu}(r)$, Ag nanoparticle, $P_{Ag}(r)$, and Ni nanoparticle, $P_{Ni}(r)$, in the x, y, and z directions.

is shown in the right part of Figure 18.8, what completes the view of the internal movement of the system. In Figure 18.8, it is possible to observe a symmetric Gaussian form of all orientational distribution functions, for the Cu nanofluid system, showing an amplitude of 45° for the three components, $P_{X\,axis}(\theta)$, $P_{Y\,axis}(\theta)$, and $P_{Z\,axis}(\theta)$.

In the case of Ni nanofluid, a slightly asymmetric Gaussian form is shown for $P_{X\,axis}(\theta)$, the x component of the orientational distribution functions corresponding to the oxygen atom and the benzene (Figure 18.9, right). Therefore, a decrease of the intensity by increasing the amplitude of the function up to 90° with regard to x axis is produced. But a more symmetric with a smaller amplitude, 60°, is observed for $P_{Y\,axis}(\theta)$ and $P_{Z\,axis}(\theta)$, the functions in the y and z axes, that explains the observed asymmetry in the translational movement of the nanoparticle shown in Figure 18.7.

Regarding to the Ag nanofluid, as it is shown in Figure 18.9, $P_{X\,axis}(\theta)$, $P_{Y\,axis}(\theta)$, and $P_{Z\,axis}(\theta)$, the three components of the orientational distribution functions corresponding to the oxygen atom and the benzene show a clear asymmetric Gaussian form with the amplitude of the function increasing up to 90° (Figure 18.9, left). Of notice are the different forms of the functions for the y axis, $P_{Y\,axis}(\theta)$ for the two oxygen atoms, suggesting that the Ag nanoparticle swing during the trajectory due to the asymmetric internal movement of the two diphenyl oxide molecules, describing an asymmetric displacement.

The above results suggest the base fluid molecules in the first layer around of the metal nanoparticle adopt a disposition that encourages the directionality of the movement in the heart of the nanofluid.

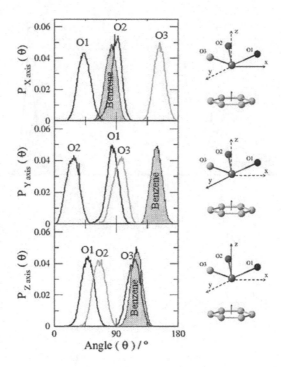

FIGURE 18.8 Orientational distribution functions of the three components of the angle formed by each oxygen atom of the first-layer cluster and the benzene ring with regard to the metal and the x, y, and z axes of the box, $P_{X\,axis}(\theta)$, $P_{Y\,axis}(\theta)$, and $P_{Z\,axis}(\theta)$, for Cu nanofluid.

In the case of Cu nanofluid, the Cu nanoparticle presents a full symmetric internal translational and rotational movement in the heart of the nanofluid, typical of Brownian motion, that ensures that Cu nanofluid reaches the maximum effective heat transport and thermal conductivity. In the case of the Ag nanoparticle, the internal rotation shows a marked asymmetry that causes an asymmetric translational movement, preferably in one direction. This movement can favour effective heat transport and an increase in the thermal conductivity of the silver nanofluid with regard to the base fluid. However, in the Ni nanofluid, the first-layer cluster structure stability prevents the internal rotation that disrupts slightly the symmetry of the translational movement, hence decreasing for Ni nanofluid the thermal conductivity of the nanofluid with regard to the base fluid.

18.3.2.2 Surfactant Arrangement Effect

Experimentally, Au nanofluids were prepared in presence of a surfactant, and it is of interest to understand its role in the system. So, to understand the role and behaviour of the surfactant in the Au nanofluid system, the comparative analysis of the RDFs and SDFs for the Au-nanofluid system with surfactant with regard to the system without was performed. Unless stated otherwise, the RDFs and SDFs have been recorded with regard to the centre of mass of the Au unit cell. To test the effect of temperature in the systems, the temperatures of 100, 300, and 500 K were chosen as representative of the variation in the systems with and without surfactant with temperature.

The analysis of the RDFs for the Au nanofluid system with surfactant is shown below. The surfactant was tetraoctylammonium bromide, $(C_8H_{17})_4NBr$ (TOAB). Thus, for this analysis, the N atom of the surfactant was taken as the reference atom of the interaction between the surfactant and the gold particles. The number of N atoms will indicate the number of surfactant molecules around the central atom. By analysing the RDFs corresponding to the Au-N pair at the three temperatures (100, 300, and 500 K), it is possible to understand the interaction between the Au and the surfactant molecule (Figure 18.10). The RDF for Au-N pair at 100 K shows two peaks centred around 7.3 and 7.9 Å that integer to one N atom

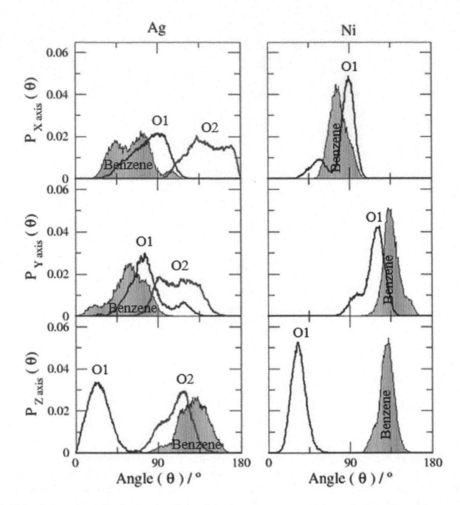

FIGURE 18.9 Orientational distribution functions of the three components of the angle formed by each oxygen atom of the first-layer cluster and the benzene ring with regard to the metal and the x, y, and z axes of the box, $P_{X\,axis}(\theta)$, $P_{Y\,axis}(\theta)$ and $P_{Z\,axis}(\theta)$, for Ag and Ni nanofluid.

each peak (Figure 18.10). With the increasing of the temperature up to 300 K, it is observed a shift of both peaks towards the Au, so, the RDF shows a wide peak centred around 6.7 Å that integrates two N atoms. Those atoms belong to two surfactant molecules at 300 K (Figure 18.10). However, it is of note that at higher temperature only a wide peak centred around 8.5 Å that integrates one N atom it is shown in Figure 18.10 for 500 K. Therefore, two surfactant molecules approach the gold cell occupying a first layer around the gold nanoparticle at low temperature (300 K), while these two surfactant molecules move away from gold unit cell at high temperature (500 K) with only one molecule remaining in the second layer.

Therefore a structural reorganisation of the remaining base fluid molecules around the metal is involved by the movement of surfactant molecules from an inner to outer layer further from the gold unit cell. In this sense, to understand how the presence of the surfactant affects the interaction of the gold with the remaining base fluid molecules, it is analysed the RDF of the interaction between the Au and the diphenyl oxide and biphenyl molecules (Figure 18.11). Figure 18.11 shows the RDF analysis of the Au-O pair from the interaction between the gold and the oxygen of the diphenyl oxide molecule. At 100 K, it is observed an intense peak around 5.0 Å for the Au-O pair that integrates to one O atom. This O atom belong to a diphenyl oxide molecule, and the high intensity of that peak indicates a strong orientation binding [6,30]. At longer distance (from 6.5 to 9.5 Å), there are other peaks that integrate to one O atom from one diphenyl oxide molecule. According to that distance, this diphenyl

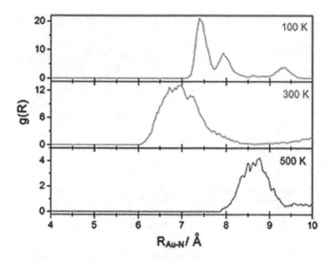

FIGURE 18.10   RDFs for Au-N pair at 100, 300, and 500 K in the Au nanofluid system with presence of TOAB as a surfactant.

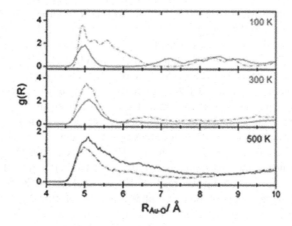

FIGURE 18.11   RDFs for Au-O pair at 100, 300, and 500 K in the Au nanofluid system with and without the presence of a surfactant.

oxide molecule is placed in a second layer around the gold particle. At higher temperature, as the temperature increases, these peaks converge showing an approaching of diphenyl oxide molecules to the gold cell. In fact, the peak centred at 5.2 Å at 300 K integrates two O atoms from two diphenyl oxide molecules. But at higher temperature (500 K), this peak remains and widens, increasing its number of integration to four O atoms.

With those results, it is observed that there is an exchange of diphenyl oxide molecules from the second to the first layer around the gold nanoparticles in which the surfactant is involved. So to understand the role played by the surfactant in this exchange of molecules, the structure of the base fluid around the closest surfactant molecule to the gold nanoparticle was analysed at different temperatures. As commented before, as the surfactant chosen was TOAB, the N atom was taken as reference for this analysis. And more concretely, the variation in the N-O pair with temperature (100, 300, and 500 K) was studied with the N atom from the surfactant molecule closest to the gold unit cell taken as a reference (Figure 18.12).

As the temperature increases, what first draws the attention is the decrease in intensity of the first peaks for the N-O pair at 100 K (Figure 18.12). This result is different to what is observed in Figure 18.11 for the RDFs of the Au nanoparticle, since it is possible to find a structure of the base fluid around the nanoparticle even at high temperatures. However, the structure of the base fluid around the surfactant

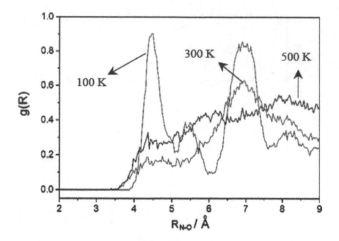

FIGURE 18.12 RDFs for N-O pair at 100, 300, and 500 K in the Au-nanofluid system with presence of a surfactant.

shows a loss of definition upon increases in the temperature up to 500 K. This result is of interest because it reveals the greater ability of the interaction of the surfactant molecules with the base fluid. The first intense N-O peak at 5.0 Å for 100 K integrates to an oxygen atom that belongs to a diphenyl oxide molecule. The intensity of that peak decreases considerably upon increasing the temperature from 100 to 500 K, although without decreasing to zero (Figure 18.12). For 300 and 500 K, the integration of the RDF until 6.0 Å reveals the presence of an O atom. So this average radial distance must take place with the exchange of a diphenyl oxide molecule. At 100 K, there is a second intense peak of the N-O pair centred at 7.0 Å that integrates to two oxygen atoms. So the integration of this peak indicates the presence of two diphenyl oxide molecules (Figure 18.12). Remarkably, this peak remains when the temperature increases to 300 K, but it widens and decreases slightly in intensity, with a total loss of definition at 500 K. Thus, a greater mobility of the two diphenyl oxide molecules should occur. Those molecules must move and increase their radial distance to the surfactant, as will be discussed below with the analysis of the histograms of these molecules. At higher temperatures (500 K), the total loss of definition of this second peak would suggest a continuous exchange of the two diphenyl oxide molecules.

Although it is observed a loss of structure of the base fluid around the surfactant (Figure 18.12), it contrasts with the stabilisation of the base fluid structure around the gold nanoparticle upon increasing temperature (Figure 18.11). In fact, an exchange takes place of the diphenyl oxide molecule closest to the surfactant, and the other two diphenyl oxide molecules shift further away from 100 to 300 K (Figure 18.12). Moreover, as it is shown in Figure 18.11, one diphenyl oxide molecule moves from the outer to the inner layer around the gold nanoparticle. So it can be concluded that the interaction of the surfactant with the base fluid is much more labile than that of the gold particle with the base fluid. In this regard, the observed behaviour is closely related with the high isobaric specific heat value obtained for the system with the surfactant compared with the one without.

A different result is observed from the RDFs for the system without surfactant. The analysis of the RDFs shows a different arrangement of the base fluid molecules around the metal as compared with the system with surfactant. For the Au-O pair, even at low temperatures, 100 K, the analysis of the RDF shows a wide peak with a maximum around 5.0 Å and a shoulder around 5.5 Å that integrates to three and two O atoms, respectively (Figure 18.11). The distant peaks of the RDF in Figure 18.11 gradually shift with the temperature rises until they converge into a wider peak around 5.0 Å that remains unchanged with a longer tail that does not decay as the temperature rises up to 500 K. At 300 and 500 K, that peak integrates to five O atoms of diphenyl oxide molecules. This result is indicative of a system in which there is an exchange of diphenyl oxide molecules from an outer to an inner layer close to the gold unit cell. At high temperatures the system without surfactant reaches an equilibrium. In this system an inner layer is formed, and it consists of five diphenyl oxide molecules around the gold nanoparticle.

To understand the inner behaviour of the surfactant in the system, an analysis was performed of the distance of the three diphenyl oxide molecules around the surfactant with regard to the centre of mass of the gold cell (Figure 18.13). This analysis was performed at 300 K as representative whether there is any relationship between the changes observed in the arrangement of the surfactant around the gold nanoparticle.

The histogram of the distance of the N atom of the surfactant with regard to the centre of the Au cell at 300 K is shown in Figure 18.14. The analysis of the histogram shows the approaching of the surfactant towards the gold nanoparticle, reducing its interaction distance by about 0.5 Å. This result is consistent with the increase in the amplitude of the peak observed for the Au-N RDF at 300 K shown in Figure 18.10.

Figure 18.15 shows the histograms with the distance of the O atom, N atom, and Au cell centre of mass of the structure depicted in Figure 18.13. In Figure 18.15 the distance of the oxygen atoms from the three diphenyl oxide molecules with regard to the N atom of the surfactant and to the centre of mass of the Au cell are shown following the nomenclature indicated in Figure 18.13. It is of note the shape

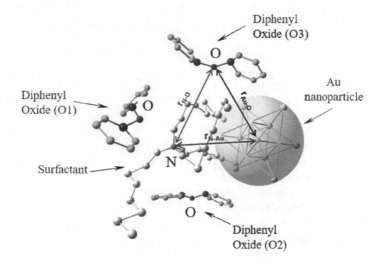

FIGURE 18.13   Structure around the surfactant molecule closest to the gold nanoparticle at 300 K.

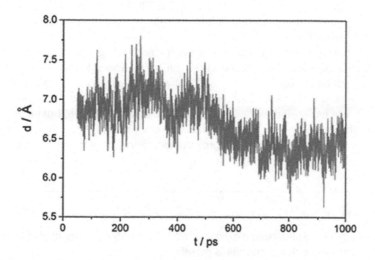

FIGURE 18.14   Histogram of the distance between the N atom of the surfactant molecule and the centre of the Au cell at 300 K.

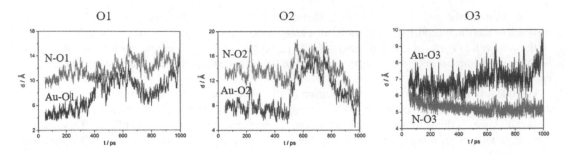

FIGURE 18.15   Histograms of the distance between the O atom of the three diphenyl oxide molecules with regard to the N atom of the surfactant molecule and the centre of the Au cell at 300 K.

of the function described by the N-O and Au-O histograms (Figure 18.15). In Figure 18.15, the same functional shape for the N-O and Au-O distances is observed for the diphenyl oxide molecule from the inner layer (O1) and the one in the outer layer furthest from the Au cell (O2), and in both cases they move more with regard to the surfactant molecule. The molecule from the inner layer (O1) moves away from both the surfactant and the Au cell. This diphenyl oxide molecule does not return to this region throughout its trajectory (Figure 18.15). This result confirms the exchange at a short radial distance from the surfactant of diphenyl oxide molecules previously discussed in the analysis of the RDFs of the N-O pairs (Figure 18.12). Meanwhile, a spring-like movement lengthening and shortening its radial distance is observed for the diphenyl oxide molecule (O2) from the outer layer (Figure 18.15), that corroborates a higher mobility and the lability of this interaction that was previously assessed by the widening of the second peak of the RDF of the N-O pair shown in Figure 18.12.

But the N-O and Au-O histograms from O3, the diphenyl oxide molecule from the outer layer and closest to the Au cell have shown an inverted functional shape (Figure 18.15). In this case, the diphenyl oxide molecule moves away from the surfactant and closer to the gold nanoparticle. This result helps us to understand the increase in the intensity and amplitude of the peak of the inner layer of the RDF of the Au-O pair observed at 300 K in Figure 18.11. In this case, it increases from one to two as regards the number of diphenyl oxide molecules around the Au cell. Therefore, there should be a competition between the base fluid-surfactant and base fluid-gold nanoparticle interactions. This explains the more labile interaction of the diphenyl oxide molecule with the surfactant than with the Au cell. Moreover, this result indicates a correlation between the changes observed in the base fluid structure around the Au cell and the surfactant with the rise of temperature. This result explains the movement of the surfactant molecule from the inner to the outer layer of the Au cell at 500 K. Thus, there is an increase in the gold nanoparticle-diphenyl oxide ratio to a value of 1:4 that leads to an exchange of the three diphenyl oxide molecules around the surfactant.

Finally, the surfactant at high temperatures imposes a structural reorganisation contributing to more efficient heat transfer processes. This effect occurs even from a more distant outer layer from the gold cell. There is a competition between surfactant and diphenyl oxide molecules to interact with the gold in a way that goes beyond a mere steric hindrance due to its four alkyl chains of the TOAB molecule. The surfactants act at a molecular level. They impede the entry of diphenyl oxide molecules into the inner layer around the gold unit cell. Moreover, the surfactant acts at high temperatures even from an outer layer further from the gold unit cell stopping the incorporation of a fifth diphenyl oxide molecule to the inner layer of the experimental system.

18.4 Conclusions

In this chapter, a comparative molecular dynamics study of metal based nanofluids (Au, Ag, Cu, and Ni) with and without the presence of a surfactant is presented.

The isobaric specific heat and thermal conductivity values obtained by molecular dynamics calculations show the same tendency as the experimental results for the base fluid and the Cu, Ag, and Ni nanofluids and Au nanofluid with surfactant. The enhancement produced in the thermal properties of the Cu and Ag nanofluid is related to the rearrangement and the internal movement (rotational and translational) of the base fluid molecules around the metal. In this sense, there is a first-layer structure composed of three, two, and one diphenyl oxide molecules for the Cu, Ag, and Ni nanofluid, respectively, leading to a directionality of the movement. The Brownian motion in the Cu nanofluid system leads to effective heat transport and thermal conductivity.

For the Au nanofluid system with and without the surfactant, the thermal conductivity and isobaric specific heat are higher than those of the base fluid. In the system with surfactant, an exchange takes place between molecules of the surfactant and diphenyl oxide molecules within an inner layer around the gold unit cell at higher temperatures. In this rearrangement, four diphenyl oxide molecules in the inner layer molecules with a surfactant molecule are situated in an outer layer further from the unit cell. So, in the system without surfactant, five diphenyl oxide molecules are detected around the gold cell at high temperatures. Therefore, at higher temperatures, a greater movement of the particles in the nanofluid system without surfactant is produced, explaining the greater diffusion values obtained for gold nanofluid system without surfactant as compared with the system with it. Therefore, there is a competition between the surfactant and diphenyl oxide molecules to interact with the gold. The theoretical results indicate the more labile interaction of the surfactant with the base fluid than that of the gold particle with the base fluid. The surfactant impedes the entry of diphenyl oxide molecules into the inner layer around the gold unit cell at a molecular level. Consequently, the high isobaric specific heat value number of the system with surfactant should be produced by a decreasing number of effective interparticle collisions. Finally, due to the presence of the surfactant (even from an outer layer), the incorporation of the fifth diphenyl oxide molecule to the inner layer of the experimental system is prevented at high temperatures.

REFERENCES

1. Khan, J., and M. H. Arsalan. (2016). Solar power technologies for sustainable electricity generation—A review. *Renewable & Sustainable Energy Reviews* 55, 414–425.
2. Devabhaktuni, V., M. Alam, S. S. S. R. Depuru, R. C. Green, D. Nims, and C. Near. (2013). Solar energy: Trends and enabling technologies. *Renewable & Sustainable Energy Reviews* 19, 555–564.
3. Desideri, U., F. Zepparelli, V. Morettini, and E. Garroni. (2013). Comparative analysis of concentrating solar power and photovoltaic technologies: Technical and environmental evaluations. *Applied Energy* 102, 765–784.
4. Fernandez-Garcia, A., E. Zarza, L. Valenzuela, and M. Perez. (2010). Parabolic-trough solar collectors and their applications. *Renewable & Sustainable Energy Reviews* 14, 1695–1721.
5. Mwesigye, A., Z. J. Huan, and J. P. Meyer. (2015). Thermodynamic optimisation of the performance of a parabolic trough receiver using synthetic oil-Al_2O_3 nanofluid. *Applied Energy* 156, 398–412.
6. Navas, J., A. Sánchez-Coronilla, E. I. Martín, M. Teruel, J. J. Gallardo, T. Aguilar, R. Gómez-Villarejo, R. Alcántara, C. Fernández-Lorenzo, J. C. Piñero, and J. Martín-Calleja. (2016). On the enhancement of heat transfer fluid for concentrating solar power using Cu and Ni nanofluids: An experimental and molecular dynamics study. *Nano Energy* 27, 213–224.
7. Suganthi, K. S., V. L. Vinodhan, and K. S. Rajan. (2014). Heat transfer performance and transport properties of ZnO-ethylene glycol and ZnO-ethylene glycol-water nanofluid coolants. *Applied Energy* 135, 548–559.
8. Andreu-Cabedo, P., R. Mondragon, L. Hernandez, R. Martinez-Cuenca, L. Cabedo, and J. E. Julia. (2014). Increment of specific heat capacity of solar salt with SiO2 nanoparticles. *Nanoscale Research Letters* 9, 582.
9. Lee, S., S. U. S. Choi, S. Li, and J. A. Eastman. (1999). Measuring thermal conductivity of fluids containing oxide nanoparticles. *Journal of Heat Transfer-Transactions of the Asme* 121, 280–289.
10. Shin, D., and D. Banerjee. (2014). Specific heat of nanofluids synthesized by dispersing alumina nanoparticles in alkali salt eutectic. *International Journal of Heat and Mass Transfer* 74, 210–214.
11. Singh, D., E. V. Timofeeva, M. R. Moravek, S. Cingarapu, W. H. Yu, T. Fischer, and S. Mathur. (2014). Use of metallic nanoparticles to improve the thermophysical properties of organic heat transfer fluids used in concentrated solar power. *Solar Energy* 105, 468–478.

12. Chandrasekar, M., S. Suresh, and T. Senthilkumar. (2012). Mechanisms proposed through experimental investigations on thermophysical properties and forced convective heat transfer characteristics of various nanofluids—A review. *Renewable & Sustainable Energy Reviews* 16, 3917–3938.

13. Timofeeva, E. V., W. H. Yu, D. M. France, D. Singh, and J. L. Routbort. (2011). Nanofluids for heat transfer: An engineering approach. *Nanoscale Research Letters* 6, 182.

14. Yoo, D. H., K. S. Hong, and H. S. Yang. (2007). Study of thermal conductivity of nanofluids for the application of heat transfer fluids. *Thermochimica Acta* 455, 66–69.

15. Coco-Enriquez, L., J. Munoz-Anton, and J. M. Martinez-Val. (2017). Dual loop line-focusing solar power plants with supercritical Brayton power cycles. *International Journal of Hydrogen Energy* 42, 17664–17680.

16. Erdogan, A., C. O. Colpan, and D. M. Cakici. (2017). Thermal design and analysis of a shell and tube heat exchanger integrating a geothermal based organic Rankine cycle and parabolic trough solar collectors. *Renewable Energy* 109, 372–391.

17. Wang, C. X., J. Yang, and Y. L. Ding. (2013). Phase transfer based synthesis and thermophysical properties of Au/Therminol VP-1 nanofluids. *Progress in Natural Science-Materials International* 23, 338–342.

18. Gomez-Villarejo, R., J. Navas, E. I. Martin, A. Sanchez-Coronilla, T. Aguilar, J. J. Gallardo, D. De los Santos, R. Alcantara, C. Fernandez-Lorenzo, and J. Martin-Calleja. (2017). Preparation of Au nanoparticles in a non-polar medium: Obtaining high-efficiency nanofluids for concentrating solar power. An experimental and theoretical perspective. *Journal of Materials Chemistry A* 5, 12483–12497.

19. Rai, N., and J. I. Siepmann. (2007). Transferable potentials for phase equilibria. 9. Explicit hydrogen description of benzene and five-membered and six-membered heterocyclic aromatic compounds. *Journal of Physical Chemistry B* 111, 10790–10799.

20. Rai, N., and J. I. Siepmann. (2013). Transferable potentials for phase equilibria. 10. Explicit-hydrogen description of substituted benzenes and polycyclic aromatic compounds. *Journal of Physical Chemistry B* 117, 273–288.

21. Martin, M. G., and J. Ilja Siepmann. (1998). Transferable potentials for phase equilibria. 1. United-atom description of n-Alkanes. *Journal of Physical Chemistry B* 102, 2569–2577.

22. Martin, M. G., and J. Ilja Siepmann. (1999). Novel configurational-bias monte carlo method for branched molecules. Transferable potentials for phase equilibria. 2. United-atom description of branched Alkanes. *Journal of Physical Chemistry B* 103, 4508–4517.

23. Wick, C. D., J. M. Stubbs, N. Rai, and J. Ilja Siepmann. (2005). Transferable potentials for phase equilibria. 7. Primary, secondary, and tertiary amines, nitroalkanes and nitrobenzene, nitriles, amides, pyridine, and pyrimidine. *Journal of Physical Chemistry B* 109, 18974–18982.

24. Gomez-Villarejo, R., E. I. Martín, J. Navas, A. Sánchez-Coronilla, T. Aguilar, J. J. Gallardo, R. Alcántara, D. M. De los Santos, I. Carrillo-Berdugo, and C. Fernández-Lorenzo. (2017). Ag-based nanofluidic system to enhance heat transfer fluids for concentrating solar power: Nano-level insights. *Applied Energy* 194, 19–29.

25. Neek-Amal, M., R. Asgari, and M. R. Rahimi Tabar. (2009). The formation of atomic nanoclusters on graphene sheets. *Nanotechnology* 20, 135602.

26. Smith, W., and T. R. Forester. (1996). DL_POLY_2.0: A general-purpose parallel molecular dynamics simulation package. *Journal of Molecular Graphics* 14, 136–141.

27. Martinez, J. M., and L. Martinez. (2003). Packing optimization for automated generation of complex system's initial configurations for molecular dynamics and docking. *Journal of Computational Chemistry* 24, 819–825.

28. Allen, M. P., and D. J. Tildesley. *Computer Simulation of Liquids*, Clarendon Press, Oxford, UK; 1989.

29. Yang, L., J. Y. Xu, K. Du, and X. S. Zhang. (2017). Recent developments on viscosity and thermal conductivity of nanofluids. *Powder Technology* 317, 348–369.

30. Annapureddy, H. V. R., S. K. Nune, R. K. Motkuri, B. P. McGrail, and L. E. X. Dang. (2015). A combined experimental and computational study on the stability of nanofluids containing metal organic frameworks. *Journal of Physical Chemistry B* 119, 8992–8999.

19

Applications of Nanofluids in Direct Absorption Solar Collectors

M.E. Zayed, S.W. Sharshir, J. Shaibo, F.A. Hammad, Mohamed Kamal Ahmed Ali, Sajjad Sargana, K.K. Salman, Elbager M.A. Edreis, Jun Zhao, Chun Du, and Ammar H. Elsheikh

CONTENTS

19.1 Introduction

Conventional thermal fluids such as water, ethylene/propylene glycol, and oil play a vital role in numerous engineering sectors, including chemical production, electronic applications, power generation, heating and cooling processes, air conditioning, microelectronics, space and defence, and nuclear reactors cooling. Those thermal liquids have intrinsic weak thermal properties compared to solids [1]. To enhance the heat transfer performance of these fluids, micro channels and fins with extended-surface, injection/suction of fluids, electrical/magnetic fields, and the vibration have been utilised, however, all these attempts have reached a dead end. Started from two decades ago, new technologies named nanotechnology and nanoscience with the prospective to improve the thermophysical properties of conventional fluids become a central issue in heat transfer science [2].

Recently, nanotechnology provides modified fabrication procedures and novel materials with sizes in the nanometer range that can be suspended in the traditional fluids. Furthermore, with an increasing focus on economic development, the energy sector plays an unavoidable role in any functional society [3]. With the unprecedented increase in energy demand, the gap between demand and supply widens [4]. This leads to a continuous consumption of fossil fuel, which has limited stocking amount and long formation period. Moreover, excessive exploitation of fossil fuels leads to environmental pollution and global warming, which stand out as the global concern nowadays [5]. All above mentioned factors have prompted researchers all over the world to further use environmentally friendly renewable energy technologies [6], such as solar energy [7], marine energy [8], hydrogen energy [9], geothermal energy [10], and bioenergy [11]. Solar technology has emerged as an important routine to convert solar energy into either electrical energy or thermal energy using devices of solar panel or solar collector in an eco-friendly and inexhaustible way [12]. Nevertheless, the approval of solar energy as an alternative source is not very clear due to its high operation cost and low efficiency [13], which motivate many researchers to optimise and enhance the performance of such devices [14]. Solar air and water collectors are representative existing devices that can efficiently convert solar energy into thermal energy using a heat transfer fluid as an absorption fluid [15]. Flat plate solar collector is the most commonly used collector due to its simple construction and low fabrication cost [16]. The effective way to increase the performance of the solar collectors is to use nanofluids as an absorbent liquids instead of conventional liquids [17].

Before the design of a thermal energy system such as solar collectors (SCs) in which a nanofluid is a working fluid, it is important to fully understand the thermophysical properties of nanofluids, including the thermal conductivity, density, heat capacity, and viscosity. Among all thermophysical properties of nanofluids, thermal conductivity and viscosity are the most important ones as the former refers to heat transfer coefficient, while the latter refers to liquid resistance. When the viscosity increases, the power required for pumping and mixing increases with it. Pumping power and pressure drop are two key factors that depend on viscosity [18]. The nanofluid thermophysical properties are influenced by numerous parameters, such as base fluid type, preparation method, temperature, volume concentration, shape and particle size, acidity (pH value), particle aggregation, shear rate, and surfactants [19]. On the other hand, optical properties, such as absorption, transmittance, scattering, and extinction coefficient, play considerable roles in heat absorbance in nanofluids. Therefore, it is of importance to study the optical properties of nanofluids that are utilised in solar thermal applications. These optical properties are related to many parameters such as particle size and shape, volume fraction, and path length.

This chapter focuses on the application of the nanofluids in direct absorption solar collectors (DASCs) in which the working fluid (nanofluid) is immediately exposed to the incident solar radiation. Moreover, we investigate the influence of various thermophysical properties (such as thermal conductivity, specific heat, and thermal diffusivity) and optical properties (such as absorption, transmittance, and scattering) as well as nanofluid properties (such as particles geometry, particles size, and volume fraction) on the performance of the low-flux and high-flux DASCs. Ultimately, challenges and future perspectives of nanofluids are also discussed.

19.2 Nanofluids for Solar Energy Applications

From a few years ago, nanofluids have exhibited their importance and merits in multiple industrial and energy applications such as heat exchangers [20], automotive industries [21], water desalination [22], solar steam generation [23], and solar power generation [24]. According to nanoparticles suspended in the base fluid, nanofluids can be classified into three major categories, namely, non-metallic, metallic,

TABLE 19.1

Classification of Nanoparticles and Base Liquids

Types	Example
Oxides	TiO_2, Al_2O_3, CuO, ZnO, SiC
Nitride	AIN, SiN
Carbide	SiC, TiC
Metals solid	Ag, Fe, Cu, Al
Composite materials	$Al_{70}Cu_{30}$
Carbon	Single-wall carbon nanotubes (SWCNT), double-wall carbon nanotubes (DWCNT), and multi-wall carbon nanotubes (MWCNT), carbon black, graphite, graphene
Base fluids	Water, oil, ethylene glycol

and carbon nanofluids. Base fluids mostly used are oil, ethylene glycol, and water [25]. The commonly used nanoparticles as well as base fluids types are summarised in the following Table 19.1.

Dispersing nanoparticles in the fluid, water, oil, etc., produces colloidal suspensions. Thermal, physical, chemical, and optical properties of the colloidal depend on the nanoparticles concentration and types of nanoparticles as well as the working fluids applied. Nanoparticles must be completely homogenised with base fluid to avoid unlike features. Nanoparticles with sizes smaller than 100 nm have been considered as the ideal choice, because large size of nanoparticles leads to some undesired consequences [26] as following:

1. Unstable mixture and sedimentation occur
2. Excess usage of power causes increased operation cost
3. Large size particles wear out exchanger and channels wall of devices, while increasing the pressure drop
4. Viscosity and thermal conductivity might be impaired in case of large size particles when compared with smaller ones.

19.3 Nanofluids Preparation

There are two main methods to prepare nanofluids, including single-step and two-step methods as shown in Figure 19.1.

19.3.1 One-Step Method

In this method, nanoparticles are synthesised and dispersed in the base fluid simultaneously. This method represents the best choice for preparing nanofluids with metallic nanoparticles, because of its ability to avoid the oxidation of the nanoparticles. However, it suffers from relatively high cost and low production rate, which hinder its commercial use for large scale production.

19.3.2 Two-Step Method

In this method, two steps are applied to prepare nanofluids. Firstly, nanoparticles are synthesised in powder form either by a chemical or physical process. Then, the produced nanoparticles are dispersed in the base fluid using ultrasonic agitation, intensive magnetic force agitation, homogenising, high-shear mixing, mechanical stirrer, or ball milling. The two-step method is the commonly used method due to its large scalability with cost effectiveness.

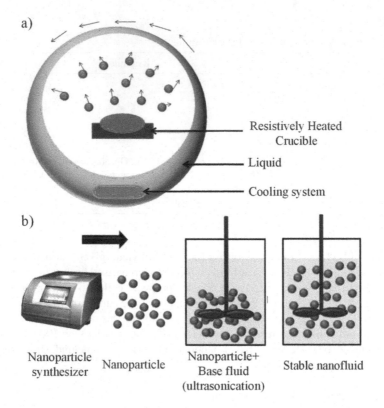

FIGURE 19.1 Nanofluids preparation methods: (a) schematic of the one-step nanofluid preparation method, nanoparticles are simultaneously produced and dispersed into a low vapour-pressure liquid and (b) schematic of the two-step nanofluid preparation method, nanoparticles are produced by microwave synthesiser and dispersed in base fluid by mechanical stirrer. (Reproduced with permission from Angayarkanni, S.A. and Philip, J., *Adv. Colloid Interface Sci.*, 225, 146–176, 2015.)

19.4 Nanofluids Properties

Dispersing nanoparticles in the base fluid has a direct effect on the thermophysical properties such as thermal conductivity, viscosity, density, and specific heat. Meanwhile, the optical properties such as absorption, transmittance, scattering, and extinction coefficient that affect the convective heat transfer are influenced. Previous research findings into the effect of different factors on nanofluids thermophysical properties have been inconsistent and contradictory. Herein, both of thermophysical and optical properties of different types of nanofluids have been briefly discussed.

19.4.1 Thermophysical Properties

Nanomaterials usually have large surface area to volume ratio and enhanced quantum effects, which result in the enhancement of their thermal, optical, physical, mechanical, chemical, electrical, and magnetic properties. Utilising these nanomaterials in nanofluids have shown enhanced thermal properties, minimal clogging in flow passages, long-term stability, higher energy savings, lower pumping power, reduction in friction, and homogeneity compared to base fluids. Numerous studies have been carried out to investigate the effect of different parameters on the thermophysical properties of nanofluids. However, there is inconsistency between the published results even for the same nanofluids under the same test conditions [24]. This inconsistency may arise from the use of different nanoparticles purities and distinct measurement equipment. The published results regarding thermophysical properties of nanofluids can be summarised as follows.

19.4.1.1 Thermal Conductivity

Enhancing the thermal conductivity of nanofluids is of crucial importance in augmentation of their convective heat transfer characteristics [28–32]. Generally speaking, nanofluids thermal conductivity is affected by many factors such as nanoparticles type, nanoparticles shape, base fluid type, potential of hydrogen, particle size, solid volume fraction, temperature, sonication, and surfactants. However, like other thermophysical properties of nanofluids, there are contradictory and inconsistency in the results regarding the effect of these factors on the nanofluids thermal conductivity. According to the findings of Keblinski et al. [33], there are four possible reasons could be used to explain the effect of nanoparticles on enhancing the base fluid thermal conductivity, namely:

1. Brownian motion of the particles: a key mechanism controlling the thermal conductivity of nanofluids in which nanoparticles move through the base fluid and direct solid–solid heat transport occurs due to particle collision, which results in the increase of thermal conductivity.

2. Liquid layering at liquid/particle interface: the liquid molecules adjacent to nanoparticle surface form layered structures, which are significantly better ordered than that of bulk liquid. As a result, it acts as a thermal bridge between nanoparticles and a bulk liquid and enhance the thermal conductivity.

3. The nature of heat transport in the nanoparticles: different from crystalline solids in which heat is transferred by phonons, phonons cannot diffuse through particles in case of nanoparticles. Instead, the heat transferred through ballistic phonon.

4. The effects of nanoparticle clustering: nanoparticles clustering may result in the enhancement of thermal conductivity as localised 'particle-rich' zone with lesser resistance to heat flow is formed compared to a 'particle-free' zone. However, the formed clusters with large masses have larger opportunity to settle and result in 'particle-free' zones, which may decrease the thermal conductivity. Therefore, nanoparticles clustering may result in either enhancing or impairing the nanofluid thermal conductivity.

A considerable amount of literature has been published to investigate the effect of different factors on nanofluids thermal conductivity as follows:

1. Effect of particle size: It has been reported that particle size is a dominant control factor of the thermal conductivity of nanofluids. A considerable enhancement in nanofluids thermal conductivity has been observed when smaller particle size is employed [34]. That is because of the augmentation of Brownian motion of nanoparticles and liquid layering at liquid/particle interface will occur when particles size is decreased [35–40]. However, other researchers claimed that the thermal conductivity increases with increase nanoparticles size, which is attributed to the severe clustering of nanoparticles in case of small particle size [41–44].

2. Effect of solid volume fraction: The solid volume fraction or concentration of nanoparticles in the nanofluid plays a vital role in controlling the thermal conductivity of nanofluids. It has been observed that thermal conductivity is enhanced with increasing particle concentration [2,45–49]. However, some researchers conclude that low particles concentrations result in enhancing the thermal conductivity of nanofluids [50–52]. This is due to the alignment, agglomeration, and clustering of nanoparticle at high concentrations.

3. Effect of particle shape: Cylindrical nanoparticles showed higher thermal conductivity compared with that of spherical nanoparticles. This is due to the larger surface area of cylindrical particles which, consequently, enhance the heat transfer process from the particles to the base fluid and vice versa [53]. In one typical study, Kim et al. [54] reported an obvious enhancement in thermal conductivity when using brick-shape nanoparticles followed by platelet and blade shaped nanoparticles for SiO_2–H_2O nanofluid.

4. Effect of base fluid: Thermophysical properties of the base fluid have a considerable effect on the heat transfer mechanism of the nanofluid. The base fluid viscosity affects the Brownian

motion of nanoparticles and, consequently, the thermal conductivity is also affected [55]. The most commonly used base fluid reported in the literature is water. The thermal conductivity of water-based nanofluids is linearly proportional to temperature and nanoparticle concentration, while for ethylene glycol-based nanofluids, the nanoparticle concentration has a higher effect on the thermal conductivity compared with that of temperature [56]. Therefore, it is recommended to use water-based nanofluids in the applications characterised by prominent increase in temperature. Meanwhile, ethylene glycol-based nanofluids are recommended for high concentrations nanofluids. Moreover, it has been reported that the enhancement in nanofluid thermal conductivity is inversely proportional to the base fluid thermal conductivity [57]. To summarise, water shows the lowest enhancement of thermal conductivity as it has the highest thermal conductivity compared with other base fluids.

5. Effect of nanoparticles type: The thermal conductivity of nanofluids increases with the increase in the nanoparticles thermal conductivity [58]. Therefore, multi-walled carbon nanotubes (MWCNTs)-based nanofluids show the highest thermal conductivity due to the extremely high thermal conductivity of MWCNTs (approximately 3000 $Wm^{-1}K^{-1}$) compared with other nanoparticles [59].

6. Effect of potential of hydrogen (pH): The effect of pH on the thermal conductivity of nanofluids has received relatively less attention from researchers. It has been reported that the increase in the pH value leads to enhance the nanofluids thermal conductivity [60].

7. Effect of temperature: The thermal conductivity of nanofluid increases with the increase of temperature [61]. This is because of two main reasons: the first is that the thermal conductivities of base fluid and nanoparticles increase with temperature increase and the second is that the Brownian motion as well as nanoparticles clustering is obviously affected by temperature [62].

19.4.1.2 Specific Heat

Specific heat is an important property used to characterise any thermal fluid as it plays a vital role in the heat transfer and heat storage processes. It is also very important in any theoretical analysis performed on thermal systems such as heat flow, energy, and exergy analyses [63–65]. Some researchers claimed that specific heat of nanofluids decreases with the increase of volume fraction [66–68]. However, other researchers claimed that specific heat increases with the increase of volume fraction [69–71]. The high surface area of nanoparticles results in the increase of the interfacial thermal resistance, which acts as additional thermal storage due to the interaction of vibration energies between nanoparticles and their surrounding fluid molecules [72]. Consequently, the specific heat of nanofluids increases. Most researchers claimed that the specific heat of nanofluids increases when the temperature rises. However, others researchers claimed that the specific heat of nanofluids dropped at high temperature. Therefore, more investigations are required to determine the optimal operating temperature range as well as particles concentration of specified types of nanoparticles and base fluids.

19.4.1.3 Viscosity

The mass and heat flow of nanofluids are affected by their viscosity. In other words, the pumping power and convective heat transfer have a significant relationship with the nanofluid viscosity. The viscosity of nanofluids increases with the increase in the concentration of nanoparticle. To achieve better heat transfer performance of nanofluids compared with their base fluids, the relative increase in the viscosity should be more than four times of the relative increase in thermal conductivity [44]. The viscosity of nanofluids increases when smaller size of particles are utilised [73]. This is because of higher interface resistance between the nanoparticles and the surrounding fluid molecules, due to the higher surface area of smaller particles compared with bigger ones. Nanofluids viscosity decreased when the temperature rises, which is desirable for high temperature application [74]. This is because of the decrease in intermolecular attraction between the nanoparticles and their base fluids at high temperature. However, some others claim that the temperature rise results in the increase of nanofluids viscosity [75,76]. This inconsistency in results may be attributed to the use of different nanoparticles and base fluids, more information regarding the effect of different parameters on the viscosity of nanofluids could be found in the previous reports [77–79].

19.4.1.4 Density

Density has a significant effect on the heat transfer performances of nanofluids by showing the influence on the Nusselt and Reynolds numbers, pressure loss, and friction factor. The density increases with the increase of nanoparticles concentration.

19.4.2 Optical Properties

Optical properties including absorption, scattering, transmittance, and extinction coefficient have significant effects on heat absorbance of nanofluids. These properties have a strong relationship with particle shape and size, volume fraction, and path length. Sunlight at ground level is composed (by total energy) of about 53% infrared light, 44% visible light, and 3% ultraviolet light [80]. The main goal is to enhance the solar absorption efficiency by adequate utilisation of thermal energy with a broad wavelength. That can be realised by tuning the optical properties to absorb the energy in both infrared and visible regions. One representative work is done by Saidur et al. [81], where they analysed the effect of using aluminum/water nanofluid as working fluid on a DASC. Water was chosen as a base fluid due to its excellent absorption ability at longer wavelength of about 2.6 μm. Another advantage is its high transparency to visible light. Aluminum nanoparticles have a high extinction coefficient for shorter wavelengths at a peak at 0.3 μm. Therefore, aluminum/water nanofluids could be used to enhance the absorption ability of light in both long and short wavelengths. This section provides a brief review of the optical properties of grapheme-, metals-, metals oxide-, and carbon nanotube-based nanofluids including their absorption, scattering, transmittance, and extinction coefficients.

19.4.2.1 Absorption

The light absorption is a key parameter that affects the ability of nanofluids to capture energy for a certain application. It depends on many factors such as particle type, particle concentration, particle size, base fluid type, and path length. In general, nanofluids have good absorptivity due to the trapping of the light for many times on the nanoparticles surfaces [82]. For instance, Rativa et al. investigated the optical absorption of an aqueous colloid containing ellipsoidal and spherical nanoparticles of silver and gold materials [83]. The Maxwell-Garnett model was used to describe the optical properties of nanofluids considering the size of nanoparticles is smaller than the incident wavelength. The size of nanoparticles has a significant influence on the effective optical absorption of nanofluids. The absorption peak for small nanoparticles is higher than that for larger ones. The low volume fraction has also played a better role compared with high volume fraction in tuning the absorption characteristics to match solar irradiance spectrum for both gold and silver nanoparticles. When the aspect ratio increases, two different oscillation modes occur, namely transversal oscillation mode and longitudinal oscillation mode. The first one is spectrally coincident with the peak of plasmon resonance of nanoparticles, while the position of latter one is shifted according to aspect ratio. The tuning of absorption resonance peak can be achieved by controlling the geometry of nanoparticles so that the optical absorption of the nanofluid matches the solar radiation spectrum. For example, Kameya and Hanamura [84] have investigated the absorption characteristics of a Ni/water nanofluid by spectroscopic transmission measurement. The absorption coefficient of the investigated nanofluid was increased in both visible light and near-infrared regions, while it remains the same as that of the base fluid for the infrared region. Jana et al. examined nanofluids containing gold nanoparticles, copper nanoparticles, carbon nanotubes, as well as their hybrids [85]. The absorption of light, at some specific wavelength, depends upon the concentration of nanoparticles in the fluid. The absorbance of carbon nanotubes-based nanofluids is higher than that of copper nanoparticles-based nanofluids, due to the higher exposed surface area of the former compared with the latter one.

A linear relationship between absorption of nanofluids and their concentration has been observed. Moreover, hybrid nanofluids showed better absorption characteristics compared with other nanofluids.

Besides, Said et al. have investigated the optical properties of two nanofluids types (alumina (Al_2O_3) and titania (TiO_2) dispersed in distilled water) with different volume concentrations [86]. It has been observed that nanoparticles concentration has some effects on the absorption coefficient for both nanofluids. Extinction coefficients of TiO_2/water nanofluid were found higher than Al_2O_3/water nanofluid for a wide range of wavelengths. However, TiO_2/water nanofluid is less stable than Al_2O_3/water nanofluid. The TiO_2/water nanofluid showed high absorption trend in visible and ultraviolet spectrum, while the absorption is vanished in infrared region. Vijayaraghavan et al. experimentally assessed the thermal performance of $CuSO_4$ nanofluid as a spectrally selective working fluid in direct absorption collector prototype under concentrated light [87]. Increasing the concentration of nanofluid has little effect on the absorbed solar spectrum, giving the fact that absorption increased by only 2% when the concentration is increased by 50%. Moreover, when the concentration is set to more than 60 g/L, it has a negligible effect on the collector efficiency. Karami et al. investigated the optical properties of nanofluid with CuO nanoparticles dispersed in a blending mixture of distilled water and ethylene glycol [88]. The investigated nanofluid has higher absorption coefficient in ultraviolet, visible, and infrared bands compared with the base fluid even at low nanoparticles concentration. The absorption of the nanofluid (100 ppm and 1 cm depth) was four times larger than that of the base fluid. The absorption peak was found at 1200 nm. Moreover, for a constant collector depth, the absorbed energy fraction is increased with the increase in the volume fraction. Xuan et al. investigated the solar energy absorption of TiO_2, Ag, and TiO_2/Ag composite nanoparticles dispersed in distilled water [89]. While TiO_2 nanofluid absorbs the ultraviolet light, Ag nanofluid absorbs both of ultraviolet and visible light. For TiO_2/Ag composite nanoparticles-based nanofluid, the absorption spectrum is further extended to longer wavelengths. TiO_2/Ag composite nanoparticle showed the best absorption characteristics due to the existence of Ag nanoparticles which localised plasmon resonance effect excited on their surfaces. The obtained results revealed that enhancing the absorption performance of plasmonic nanofluids in both the near- and mid-infrared wavelength ranges may be easily achieved by using different sizes of TiO_2 and Ag with proper volume fractions. Moreover, for most studied nanofluids, the absorption increased with increasing volume fraction. However, this increase is negligible for concentrations more than 0.015. Karami et al. investigated the optical properties of carbon nanotubes/water nanofluid at low-temperature [90]. The nanofluid was prepared by distributing carbon nanotubes into alkaline media. A linear relation between the absorbance and volume fraction has been established. Long-term stability of the investigated nanofluid (approximately 1 month after the preparation) has been observed, as absorption of the investigated nanofluid showed very slight change with increasing sediment time. The absorption peak was found at 1200 nm. Hordy et al. investigated the optical properties of MWCNTs-based nanofluids [91]. Four different base fluids were investigated, namely, water, ethylene glycol, propylene glycol, and Therminol® VP-1. MWCNTs dispersed in ethylene glycol and propylene glycol showed good long-term stability and absorbed the majority of the solar spectrum. Mercatelli et al. evaluated the spectral absorption properties of water-based and glycol-based single-walled carbon nanohorns suspensions [92]. The investigated nanofluids showed good stability when the test temperature is less than 120°C and 150°C for water and glycol, respectively. The absorption of both types of nanofluids increased with increasing path length. The absorption reached approximately 100% with 0.05 g/l concentration and 1 cm sunlight path length. Sani et al. investigated the optical properties of single wall carbon nanohorns suspended in water [93]. The absorption coefficient is affected by the particles sizes for small particle sizes, while this effect is diminished for large particle sizes. This is due to significant scattering amount for the large particle size compared to smaller ones. The absorption of light is also affected by particle concentration when the concentration is changed from 0.001 to 0.05 g/l. The extinction coefficient is enhanced by more than 300%. He et al. experimentally studied the optical properties of a single-walled carbon nanotube dispersed into polystyrene sulfonate solution, copper sulfide nanoparticles dispersed into polystyrene sulfonate solution, and a hybrid nanofluid consists of both of them [94]. Copper sulfide nanoparticles-based solution absorbs much light compared with water in the range of 530–930 nm. Hybrid nanofluid showed the maximum absorbance followed by single-walled carbon nanotube-based solution. The experimental results revealed that increasing carbon nanotube concentration and/or dilution of the copper sulfide in the hybrid nanofluid result in the reduction of the light absorption. Therefore, the new hybrid nanofluid is a better choice for direct solar absorption. Han et al. investigated the optical properties of carbon black distilled water nanofluid [95]. The investigated nanofluid optical properties were increased with increasing volume fraction. Besides, it has good absorption in the entire wavelengths ranging from 200 to 2500 nm. Gorji et al. investigated optical properties of carbon nanotube/deionised water nanofluid for wide range of temperatures from 25°C to 90°C [96]. The nanofluid absorption increased from 2 to 6 when the concentration enhances from 10 to 60 mg/l. However, the absorption dropped by the point that the concentration reached a certain limit. This is because that when the concentration reached beyond this limit, partial settling of nanoparticles happened. Moreover, beyond this limit, the distance between particles decreased and hence van der Waals forces caused particles to agglomerate. Lee and Jang [97] studied the extinction coefficients of water-based MWCNTs. The obtained results revealed that the investigated nanofluid can completely absorb solar energy with very low concentrations (0.0005 Vol.%) at approximately 10 cm penetration depth. Moreover, the absorbed energy is linearly proportional to penetration depth for a given concentration in the range from 0.0005 Vol.% to 0.005 Vol.%. When the penetration depth is increased from 0.1 to 10 cm, the absorptivity increases from 0.6 to 1. Mehrali et al. investigated the properties of nitrogen-doped graphene nanofluids using water as a base fluid and Triton X-100 as surfactant [98]. The concentration of nanoparticles has a direct effect on the absorbance of the nanofluid. The investigated nanofluid showed good stability over a time span of 6 months as the concentration reduced by about 20%. Luo et al. investigated the performance of a DASC using graphite nanofluid as a working fluid [99]. It has been established that increasing particle size can improve the absorption of nanofluids. Thereby, the agglomeration of nanoparticles may be a good factor to increase the nanofluid absorptivity as the equivalent diameter of nanoparticles after agglomeration is greater than primary particle size before agglomeration. Ladjevardi et al. investigated the applicability of graphite nanofluids in DASCs [100]. Different particle sizes and concentrations of graphite nanoparticles are investigated to figure out their effects on absorption of solar energy. The increase in the nanoparticles concentration and size results in increasing the extinction coefficient of nanofluids in ultraviolet and visible ranges. Jia et al. investigated the optical properties of Graphene/water nanofluids using carboxyl methyl cellulose as a surfactant [101]. It was observed that electromagnetic field has a positive effect on absorbance of the nanofluid as it hinders agglomeration. Additionally, the employment of surfactant has better effect on stability and light absorptivity of the nanofluid. Wang et al. investigated Chinese ink (the carbon black particles coated with bone glue dissolved in water) for DASCs applications [102]. The absorbance of Chinese ink was approximately constant for 24 hours, while it dramatically decreased for Cu and CuO water-based nanofluids.

19.4.2.2 Extinction Coefficient

The extinction coefficient is affected by nanoparticles as well as base fluid properties. The concentration and size of nanoparticles have the largest effect among other properties on the extinction coefficient, because it increases with increasing particle concentration and size. Zhang et al. investigated the radiative properties of a nanofluid containing Ni, Cu, and carbon-coated Ni (Ni/C) nanoparticles [103]. The extinction coefficient of carbon-coated Ni (Ni/C)-based nanofluid is higher than that of Ni-based nanofluid and Cu-based nanofluid with similar average nanoparticle size. Taylor et al. investigated the optical properties of five different nanofluids, with Ag, Au, Cu, Al, and graphite nanoparticles dispersed in water [104]. They revealed that the extinction coefficient increases with increasing particle size especially at long wavelengths. He et al. found that the extinction coefficients of Cu-H$_2$O nanofluid with distinct concentrations decrease at the very beginning and then increase along with the wavelength [105]. Wu et al. investigated the optical and thermal radiative properties of non-metallic nanoparticles (Si and SiC) with metallic nanoshells (Au, Ag, Cu, and Al) dispersed in water [106]. The use of this type of nanoparticles enhances the efficiency of solar absorption due to localised surface plasmon resonance effect.

The obtained results revealed that the wavelengths at the maximum extinction coefficient of Si as a core of nanoparticles shift to the long wavelength region. Cu shell has the maximum extinction coefficient followed by Al shell, and both of them are larger than that of Ag or Au shell. Moreover, the peaks of metal plasma resonance absorption for composite nanoparticles are slightly shifted to red compared with pure metal nanoparticle. Zhang et al. investigated the optical properties of many types of nanofluids consists of Au, Si, Fe_3O_4, Al_2O_3, and diamond and their hybrids dispersed in water [107]. For all studied nanofluids, the extinction coefficient was generally increased with the addition of nanoparticles. Fe_3O_4-Au hybrid nanofluid possessed a higher efficiency compared with that of pure Au. Shende and Sundara investigated the optical properties of nitrogen doped hybrid structure of reduced graphene oxide and multi-walled carbon nanotubes dispersed in water and ethylene glycol [108]. For both types of nanofluid, extinction coefficient significantly increased with increasing concentration.

19.4.2.3 Scattering

The scattering behaviours of nanoparticles dispersed in nanofluid depend on many aspects such as nanoparticle type, base fluid type, nanoparticle shape, nanoparticle size, and concentration. In general, light scattering increases with increasing size and/or concentration of nanoparticles. However, the scattering coefficient's effect is negligible compared with that of absorption coefficient in most areas of the wavelengths specially for small particle sizes [81]. Moreover, particle agglomeration results in significant amount of light scattering [104]. The scattering coefficient increases in near ultraviolet region and vanishes in the visible region of the light [92].

19.4.2.4 Transmittance

The transmittance of nanofluids, like other optical properties, depends on many elements such as nanoparticles type, base fluid type, path depth, particle size, and concentration. In general, the transmittance of nanofluids decreases with increasing nanoparticles concentrations and/or path length. The transmissivity of pure water is 100% at the visible light region, which reveals the ineffectiveness of water as light absorber in the visible light region. Then, the transmissivity decreases when light wave goes into the near infrared region with a dip at 1.9 μm. After 1.9 μm, transmissivity of the pure water increases again until the wavelength of 2.60 μm, at which the transmissivity suddenly drops. The transmittance of light is reduced throughout the visible light region by mixing the nanoparticles into water [81]. The nanofluid becomes opaque to light from the values of near ultraviolet to about 0.5 μm wavelengths at a path length equal to 0.1 mm. Moreover, increasing the nanoparticles volume fraction up to 1% may result in losing transmissivity and the passed light may be totally absorbed. Zhang et al. investigated the radiative properties of two different nanofluids with carbon-coated Ni, Ni, or Cu nanoparticles dispersed in ionic liquid [103]. A decrease in nanofluid transmittance has been observed after adding a little amount of nanoparticles especially for large path lengths. The nanofluid containing Ni/C nanoparticles showed the lowest transmission compared with Ni or Cu nanofluids with the same volume fraction. The transmissivity of Ni/C ionic fluid is approximately completely vanished at concentration 40 ppm and path length of 1 cm. Sani et al. investigated the optical properties of nanofluids consisting of aqueous suspensions of single wall carbon nanohorns [93]. The transmissivity of the investigated nanofluid is significantly decreased even at low volume fractions. That is because of the direct capture of photons as well as scattering by carbon nanohorns.

19.5 Solar Collectors

SC is a thermal device which used to utilise the solar energy via collecting and concentrating solar radiation. The collected thermal energy is transferred via a flowing fluid that can be used in various thermal applications, such as water heating, building heating, and other industrial applications. SCs are generally classified into two main categories, namely, non-tracking and tracking SCs as shown in Figure 19.2. The non-tracking SCs are kept stationary at a prespecified orientation according to the location. The most common used

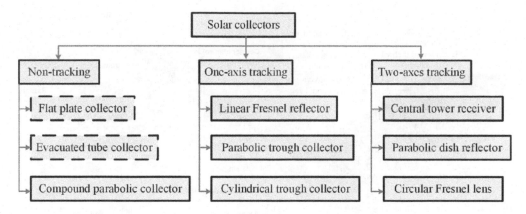

FIGURE 19.2 Solar collectors' classifications.

non-tracking SCs are flat plate, evacuated tube, and compound parabolic collectors. The tracking collectors have the capability to track sun to maximise the received solar radiation. The tracking SCs are classified into one axis tracking and two axes tracking collectors. The most common used types of the one axis tracking SCs are linear Fresnel reflector, parabolic trough collector, and cylindrical trough collector, while the most common used types of the two axes tracking collectors are central tower receiver, parabolic dish reflector, and circular Fresnel lens. It is worthy mentioned, among all aforementioned types of SCs, flat plate collector and evacuated tube collector are non-concentrated collectors (dashed boxes in Figure 19.2.), while all remaining types are concentrating collectors. Herein, the most common used solar collectors will be briefly discussed.

19.5.1 Flat Plate Collectors

A flat plate collector, shown in Figure 19.3a, consists of a transparent glass cover, copper tubes, an absorber plate, and aluminum back plate. Copper tubes are designed according to the working fluid type as well as operating conditions. To minimise the heat losses, the side and bottom walls of the collector are insulated. To minimise the heat losses from the upper surface of the absorber plate, a glass cover is placed over it. However, for long operating period, dust settles on the upper surface of the glass cover which impairs the collector performance. Therefore, cleaning the glass cover is required which may be achieved using many methods such as water jet washing, air jet blowing, use of super-hydrophilic nano-film, and use of automatic wiper [109].

19.5.2 Evacuated Tube Collectors

An evacuated tube collector, shown in Figure 19.3b, is a thermal device used to collect solar energy via capturing the solar insolation through a fluid such as methanol, ethanol, and water which flows inside a heat pipe enclosed in a glass enclosure to minimise the heat loss. This fluid absorbs solar radiation which heats it up and then it is undergoing evaporation-condensation cycles. These cycles occur due to the heating of the fluid by the solar insolation which results in phase change from liquid to vapour. This generated vapour rises in the upper direction under the effect of buoyancy force. Then the vapour reject heat to another working fluid using a heat exchanger, which results in change its phase from vapour to liquid and flows back to the bottom direction of heat pipe under the effect of gravity force. The evacuated tube SC may consist of a single tube or multiple integrated tubes.

19.5.3 Compound Parabolic Collectors

The compound parabolic collector consists of two parabolic reflecting surfaces, an absorber tube, and a glass cover, as shown in Figure 19.3c. The focal point of the first parabolic reflecting surface is lying on

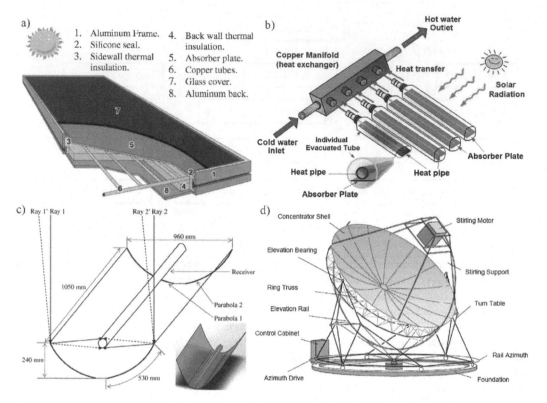

FIGURE 19.3 Common used SCs types: (a) Flat plate. (Reproduced with permission from Sarsam, W.S. et al. *Sol. Energy.*, 122, 1245–1265, 2015); (b) evacuated tube. (Reproduced with permission from Iranmanesh, S. et al. *J. Clean. Prod.*, 162, 121–129, 2017); (c) compound parabolic. (Reproduced with permission from Gudekar, A.S. et al., *Sol. Energy.*, 90, 43–50, 2013); and (d) parabolic dish reflectors. (Reproduced with permission from Hafez, A.Z. et al., *Energy Convers. Manag.*, 126, 60–75, 2016).

the second parabolic and vice versa. The parts of the parabolic surfaces below the focal line (line joins the two focal points) are truncated as they do not contribute towards collecting the solar radiation. The absorber tube is located up to the two focal points at the mid plane between them. To avoid the reflection of the solar radiation to the ambience through the upper aperture, the angle of incidence must be less than half of the acceptance angle. As there is no tracking system in such collectors, the orientation of the parabolic reflectors should be adjusted so that the absorber tube is along east-west direction.

19.5.4 Parabolic Dish Reflectors

The parabolic dish collector, shown in Figure 19.3d, contains a parabolic dish coupled with two-axes tracking system. The sun rays are reflected and concentrated at the focus of the parabolic dish where the receiver is placed. A heat exchanger is placed in the receiver to transfer the collected heat to the working fluid. Throughout the day, the solar parabolic dish automatically follows the sun movements using tracking control system.

19.6 Classification of Direct Absorption Solar Collectors

DASCs are a new generation of SCs in which solar radiation is directly absorbed via a transport medium. DASCs are classified into low-flux DASCs and high-flux DASCs. Initially, conventional low-flux DASCs operate without any solar concentrating devices that commonly used water and ethylene glycol as a heat transfer fluid (HTF). Most investigations emphasising on low-flux collectors adopt the incident fluxes more or less than 1367 W/m², which is basically the same as that of solar constant [114]. Many research efforts have been done to study the influence of wide variety of nanofluids on the performance of low-flux DASCs. Moreover, high-flux DASCs are promising technologies for solar energy harvesting, in which concentrating and absorbing surfaces are used to efficiently transfer solar energy into heat energy that could be applicable for high temperature applications [115]. It frequently uses liquid metals, molten salts, and heat transfer oils (e.g., Therminol and Paratherm) as working fluids. The low overall thermal efficiency is the main drawback of high-flux DASC systems, due to its high radiative losses that result from the large temperature difference at the interface between the HTF and the absorber surface [116].

19.6.1 Developments of Low-Flux DASCs Using Nanofluids

Tyagi et al. presented the basic framework for the theoretical simulation of a nanofluid-based DASC [117]. In this study, the model presumed independent scattering conditions and the incident solar intensity was estimated by employing the black body radiation provided at 5800 K. Their theoretical findings revealed that device with Al-water nanofluid showed 10% improvement in the energy efficiency compared with using water as an HTF, at an incident solar radiation of 1000 W/m². It has been also seen that the collector efficiency was established to enhance abruptly with increasing the nanofluid volume fraction, and it reached up to 80% at volume fraction of 2%. The schematic drawing of the nano-based DASC is demonstrated in Figure 19.4a.

Low-flux DASC with three different nanofluids is numerically and experimentally studied by Otanicar et al. [118]. They used CNT/water (20 nm), graphite/water (30 nm), and silver/water (40 nm) as nanofluids. Their experiments indicated that CNT-water nanofluid gave the best performance using 0.5 wt.% CNT. The thermal efficiency of the collector was 5% higher than the corresponding ones using pure water as a base fluid at solar flux of 1000 W/m².

Gupta et al. investigated experimentally an outdoor DASC-based Al_2O_3/H_2O nanofluid [119]. The collector is tested with different low nanofluid concentrations varied from 0.001% to 0.05%. Their results indicated that the energy efficiency of the DASC improved up to 39.5%, when Al_2O_3/H_2O nanofluid with weight fraction 0.005% is used. Moreover, instantaneous efficiency against temperature graphs showed that enhancing the volume fraction beyond 0.005 vol.% had a negative impact on efficiency.

Parvin et al. numerically analysed the effect of three different nanofluids types, namely, Cu/H_2O, Ag/H_2O, and Al_2O_3/H_2O on the energy efficiency and entropy generation of a DASC under forced convection mode and heat flux of 1000 W/m² [120]. They developed a numerical model using Galerkin finite element method in order to solve momentum, energy, and radioactive heat transfer equations. They observed that, for all proposed nanofluids, with increase concentration to 3%, the DASC efficiency increased. After that, any further increase had reverse effect on the collector performance. Also, they concluded that 40% improvement in the energy efficiency of the DASC has been obtained, as $Cu-H_2O$ and $Ag-H_2O$

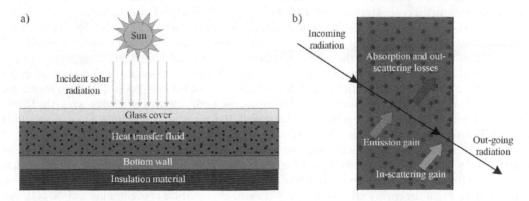

FIGURE 19.4 Direct absorption solar collector: (a) schematic drawing of a direct absorption solar collector and (b) a representative diagram of radiative transport equation.

nanofluids were used at 3% weight fraction. An empirical formula was suggested to calculate the energy efficiency (η_{th}) of the DASC in terms of nanoparticles concentration (ϕ) and Reynolds number (Re) that was given by:

$$\eta_{th} = \left[0.327\phi + 2.48\right] Re^{0.4684}, \tag{19.1}$$

when $100 \leq Re \leq 1000$, $0\% \leq \phi \leq 3.0\%$.

Kundan and Verma experimentally investigated an outdoor DASC using Al_2O_3/water nanofluid as a HTF [121]. Two nanofluid volume fractions of 0.05% and 0.005% were investigated at three mass flow rates of 100, 80, and 60 mL/h. They concluded that, when using 0.05 wt.% Al_2O_3-water nanofluid, the collector efficiency enhanced by 15% compared with using water at the mass flux of 60 mL/h. Kundan and Sharma carried out an experiment in order to investigate the effect of utilising CuO/water as nanofluid on the efficiency of DASC at the same mentioned previous volume fractions and nanofluid mass flow rates [122]. They found that, the improvement in the efficiency of DASC was 6% when CuO/water nanofluid was used with 0.05% volume fraction and mass flow rate of 60 mL/h.

Karimi et al. experimentally studied the impact of CuO nanoparticles that dispersed in water-ethylene glycol mixture on the performance of low temperature outdoor DASC [123]. In this study, different operating variables of the DASC, namely, the inlet temperature of the nanofluid, nanoparticles weight fraction, mass flow rate, and bottom wall internal surface emissivity of the collector were studied, and their effects on the DASC were investigated. It was unveiled that a black bottom wall surface improved the efficiency by 11.4% compared with the reflective internal surface. Moreover, the enhancement of the energy efficiency varied between 9% and 17% compared with water, at nanoparticles concentration ranged from 25 to 100 ppm. Delfani et al. [124] tested the experimental apparatus of Karimi et al. [123] to study the impact of employing MWCNT/water-ethylene glycol nanofluid on the efficiency of the DASC. They found that application of MWCNT-water-ethylene glycol nanofluid can improve the energy efficiency by 10% when compared with CuO-water-ethylene glycol. With the same sequence, Vakili et al. [125] also used the same test setup of Karami et al. [123] to evaluate the performance of the DASC when graphene nanoplatelet nanofluid operating as an absorbing fluid. They studied nanoparticles weight fractions of 0.0005% and 0.005%, whereas the nanofluid mass flow rates changed from 0.0075 to 0.0225 kg/s. They observed that the energy efficiency was enhanced by 9.35% when graphene nanoplatelet/water nanofluid was used with particle fraction of 0.005 wt.% and flow rate of 0.015 kg/s. It was noticed that as the volume friction increased, the collector efficiency increased. However, increment of mass flow rate from 0.015 to 0.0225 kg/s, the energy efficiency surprisingly decreased. They attributed it to the low exposure time and, consequently, less solar absorption rate interfaced at higher flow rates. The influence of the change of both nanoparticle size and concentration of graphite-water nanofluid is numerically conducted by Ladjevardi et al. [100], in order to optimise the performance of low-flux DASC. They analysed their model under a large range of graphite nanoparticles concentrations (0.00001 wt.%–0.5 wt.%) and particles diameters (50–300 nm). It was observed that 0.00025 wt.% of graphite nanoparticles is sufficient to absorb 50% of the solar irradiance that strikes the collector. Whereas the conventional DASC that used water as a HTF could absorb 27% of the incident flux at the same conditions.

Saidur et al. developed a theoretical model to evaluate the impact of concentration and size of Al nanoparticles on the extinction coefficient and thermal efficiency of low-flux DASC [81]. They found that 1.0 wt.% of Al nanoparticles achieved moderate enhancement to the thermal efficiency, while the DASC efficiency increased slightly as the particle size increased. It was recommended that Al-water nanofluid was an effective candidate for enhancing the performance of the DASC.

Chen et al. numerically investigated the potentiality of silver/water nanofluid-based DASC under forced convection mode [126]. Various parameters were analysed, such as the nanoparticle concentration, the irradiance time, the height of collector, and nanofluid heat gain temperature. Their simulation results indicated that the energy efficiency increased as the nanoparticle concentration and the collector height increased. The performance of the simulated DASC was optimised, when the nanoparticle concentration and collector height were 0.03 wt.% and 10 mm, respectively, the collector efficiency improved up to 90% under this condition. They also reported that the efficiency decreased with the increase of the irradiance time due to the increased thermal losses. They also carried out an experimental investigation to verify the numerical modeling, by acquiring the nanofluid mean temperature profiles, which demonstrated a good verification between the numerical and experimental results.

Filho et al. carried out an experimental study to investigate the influence of silver-water nanofluid-based DASC, under realistic conditions for time of 10 hours and wide range of silver nanoparticles concentrations (0.0001625%, 0.000325%, 0.00065%, 0.0065%, and 0.065%), on the performance of DASC [127]. Their results indicated that the stored thermal energy was enhanced by 144%, 93%, 52% at nanoparticles weight fraction of 0.00065%, 0.000325%, and 0.0001625%, respectively. They introduced the parameter of specific absorption rate (SAR) which describes the nanoparticle's ability in absorbing energy per unit mass. They indicated that SAR was improved from 0.01 to 0.6 kw/g, when the concentration of nanofluid decreased from 0.00065% to 0.0001625%.

Khullar et al. [128] numerically investigated the heat transfer behaviour of a volumetric DASC applying two types of nanofluids, namely, MWCNTs nanoparticles suspended in deionised water and carbon amorphous dispersed in ethylene glycol, into an absorber surface involving copper-coated substrate as a selective coating under low-flux irradiance. In this study, they observed that higher stagnation temperatures were attained by the DASC in which the experienced flux permitted to directly interplay with the HTF bulk. In addition, MWCNT-deionised water nanofluid obtained higher stagnation temperatures in comparison to the amorphous carbon-ethylene glycol nanofluid. However, it was shown that the performance of the DASC is extremely delicate to the nanoparticle concentration and provided an optimal value of nanoparticle concentration at which stagnation temperature is the maximum.

The effect of Al-water nanofluid on the thermal efficiency and temperature distribution of an inclined DASC was evaluated numerically by Cergan and Myers [129] by applying Planck's black body approximation (5800 k) as a heat flux source. They carried out a comprehensive analytical-numerical model using radiative transport equation (RTE) and power low function. A representative diagram of RTE is shown in Figure 19.4b. Their analyses demonstrated that the collector performance relied significantly on the nanoparticle weight fraction, however, there is a threshold beyond which it is useless to add additional particle. Besides, the collector efficiency increased with the increasing of fluid depth, and the nanoparticle size had no considerable impact on it.

Turkyilmazoglu developed a theoretical approximation to describe the heat transfer performance of a DASC applying Al-water nanofluid under irradiance of 1000 W/m^2[130]. He completed the earlier analytical work of Cregan and Myres [129] to estimate the impact of non-zero flux and explained the isothermal conditions at the bottom of wall instead of adiabatic bottom wall assumptions which were considered in earlier simulations. It was found that the existence of higher nanoparticle weight fraction improves the outlet average temperature and augments the DASC performance. Energy efficiency of 85.63% was obtained in the case of isothermal bottom wall and concentration of 0.006.

Another interesting study was done by Gorji and Ranjbar [131], they conducted a numerical optimisation investigation of a DASC-based nanofluid. They determined the impact of its design dimensions on its thermal performance by resolving the RTE of specific medium and joining convection and conduction heat transfer equations. Response surface methodology (RSM) was employed on the numerical findings to know the impact of varying dimensions on the collector performance. Positive effect of ascending collector height on its overall thermal performance was observed, because higher amount of energy is achieved as the collector height increased. Also, as the collector width increased, the efficiency decreased, whereas the increment in the length had a mere effect on the thermal efficiency.

The effect of three different kinds of nanofluids, i.e., silver-water, graphite-water, and magnetite-water as HTFs on the thermal performance of a low-flux DASC under halide lamps illumination was experimentally and numerically investigated by Gorji and Ranjbar [132]. In this study, the impacts of changing the solar flux (600–1000 W/m^2), nanoparticle weight fraction (5–40 ppm), and nanofluid flow rate (2.5–10 mL/min) on the collector performance were analysed. Their numerical results revealed that more diversified temperature profile was attained at lower nanoparticle weight fractions. Their experiments unveiled that magnetite nanofluid gave the best performance in which the collector efficiency was improved by 57.4% relative to water as a HTF, at 40 ppm nanoparticles concentration, 10 mL/min flow rate, and flux of 1000 W/m^2.

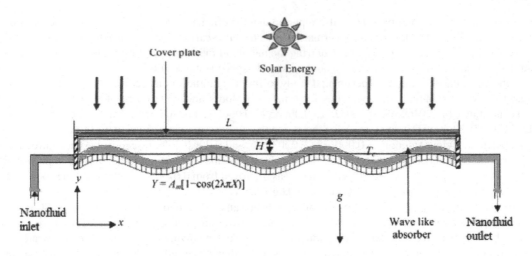

FIGURE 19.5 A layout of the wavy DASC. (Reproduced with permission from Hatami, M. and Jing, D., *J. Mol. Liq.*, 229, 203–211, 2016.)

Lee and Jang [133] conducted an analytical modeling to find out the vital parameters affecting the thermal efficiency and temperature distribution of a DASC using MWCNT-water nanofluid under low solar flux. Their work revealed that the theoretical efficiency of the DASC was proportional to the nanoparticle weight fraction and Peclet number. Whereas the heat loss Nusselt number and the collector aspect ratio have adverse effect on the efficiency. Hatami and Jing [134] analysed numerically the heat transfer behaviour of a wavy DASC with various nanofluids types, i.e., Al_2O_3/H_2O, TiO_2/H_2O and CuO/H_2O, as shown in Figure 19.5. The performance of the collector was optimised by comparing various boundary conditions (insulated, adiabatic, and isothermal) and several wall types (flat top, wavy bottom, and vertical side) in order to get the optimal status. Their results showed that the maximum Nusselt numbers were achieved by wavy bottom collector with TiO_2/H_2O nanofluid compared to other types, at low nanoparticles concentrations.

19.6.2 Developments of High-Flux DASCs Using Nanofluids

Various recent investigations have proven that high-flux DASCs-based nanofluids as HTFs are effective devices for harvesting solar energy, in which the concentrated solar flux is efficiently absorbed with more uniform distribution and less radiative losses within the HTF.

Recently, Li et al. studied numerically and experimentally the impact of $MWCNT/H_2O$ nanofluid on the efficiency of high-flux DASC under different absorber surfaces [135], the volumetric absorber was consisted of a glass pipe containing MWCNTs in Therminol oil or water (depending on the working temperatures which ranged from 80°C to 250°C). The conventional absorber surface composed of a black chrome-coated copper tube with Therminol oil or water as an HTF. Three-dimensional computational fluid dynamics simulation was used to solve the governing equations numerically. The results indicated that the vacuum-packaged volumetric receiver had an efficiency of 54% and 26% at operating temperature of 80°C and 200°C, respectively. While for vacuum-packaged black chrome-coated receiver, it had an efficiency of 68% and 47%, respectively, at the same operating temperatures.

Taylor et al. [136] carried out an experimental investigation of dish receiver as a high-flux DASC using graphite/Therminol VP-1 oil nanofluid at two nanoparticle concentrations of 0.25 and 0.125 wt.%. They found that at 0.125 wt.% graphite nanoparticles, the thermal efficiency of the dish receiver improved by 11% compared with Therminol oil as a base fluid. While as the nanofluid weight fraction increased to 0.25 wt.%, the thermal efficiency did not enhance, as at high weight fractions all the incident flux was absorbed in a thin layer near the receiver surface, rather than volumetrically. A quantitative model was also put forward which demonstrated the overestimated efficiency of the receiver at large volume fractions and the instability of the nanofluid.

The outlet temperature, energy efficiency, and photo thermal efficiency of high-flux DASCs in terms of solar flux and inlet velocity of nanofluid under various types of nanofluids was experimentally and numerically investigated by Luo et al. [99]. Seven different types of nanoparticles, namely, Cu, Ag, CNT, Al$_2$O$_3$, TiO$_2$, SiO$_2$, and graphite were added to Texatherm oil to produce a stable nanofluid which were used as HTFs. The flow rate of nanofluid varied from 9.5 to 47.5 mL/min. They found that graphite suspension gave the best performance, whereas the least performance was obtained by TiO$_2$ nanofluid. It also reported that the outlet temperature and the energy efficiency were improved by 30°C–100°C and 3%–25%, respectively, relative to Texatherm oil. The thermal-photo efficiency decreased with increasing the incident solar flux.

Liu et al. [137] studied the performance of a high-flux DASC using graphene/ionic liquid nanofluids as a working fluid. The nanoparticle concentration varied from 0.0005 wt.% to 0.01 wt.%. One-dimensional transient heat transfer simulation was adopted to evaluate the DASC efficiency and temperature at weight fraction of 0.001 wt.%–0.0005 wt.% and DASC's height of 7.5 and 3.8 cm. Their experimental findings showed a good agreement with the numerical results. With the increase in graphene concentration, the solar radiation absorption can reach 100%. The DASC's efficiency was found to be increased with solar flux and DASC's height, but was found to be decreased with increasing graphene concentration. The DASC efficiency was found to be decreased from 0.72 to 0.23 at 550 K as concentration was varied from 0.0005 wt.% to 0.01 wt.%.

Hector and Singh [138] developed a theoretical model of a direct absorption receiver system (DARS) based on finite different method. Aluminum and graphene nanospheres and therminol oil nanofluids were tested at different weight fractions to determine the optimum geometry of the DARS. Their findings unveiled that DARS-based nanofluid has a high outlet temperature approximately 265°C with a receiver tube diameter of 5 mm, opposed to commercially available 70 mm diameter metallic absorbers.

To find out the impact of nanoparticle concentration, receiver geometry, and thickness of HTFs on the DASC efficiency and temperature profiles, Lenert et al. [139] developed a 1-D unsteady heat transfer numerical model for a stationary high-flux DASC. Their quantitative findings presented as the temperature profile developed over time, and it was observed that the average HTF temperatures were higher than the base temperature. In a comparison involving selective surfaces, it is unleashed that volumetric DASC can attain higher solar efficiencies (greater than 55%) at higher nanoparticle fractions (>100 ppm) and larger receiver heights (>10 cm). They also developed a cylindrical receiver design with average incident radiative flux of 2620 W/m². Temperature profiles obtained experimentally were found to be in good agreement with the quantitative model. Lenert and Wang expanded their previous investigation by taking into account the impacts of the solar concentration, nanofluid height, and optical thickness to optimise the performance of volumetric DASC in terms of receiver efficiency and receiver side net system efficiency [140]. It was demonstrated that if the optical thickness is too small, the receiver was not able to absorb all of the incident solar radiation. While, by increasing the optical thickness above a specific value, the photo penetration depth was reduced and volumetric absorption more like surface absorption. The optimum collector efficiency enhanced by augmenting solar concentration level, and it was determined at larger temperatures greater than 800 K. Also, the nanofluid height had a constructive implication on the system efficiency because of temperature reversal. It was also anticipated that receiver-side efficiency for optimised nanofluid volumetric DASC reached up to 35%. Veeraragavan et al. [141] theoretically investigated the impact of nanoparticle weight fraction, heat losses, solar concentration, and channel height on the efficiency of a volumetric DASC with graphite-Therminol VP-1 nanofluid as an HTF. They indicated that the nanofluid temperature profile revealed that at downstream places of the inlet, surface temperature turns less as compared to the bulk temperature, which reveals the advantage of the volumetric absorption system. They concluded that the maximum efficiency of DASC was 35%, with DASC channel of 1 cm depth, having solar concentration of 10 and 0.01 wt.% graphite nanoparticle.

19.7 Current Challenges

Despite of the growing interest in recent years with nanofluids applications in solar collectors, there are some weakness and difficulties that need to be addressed when optimising the application of the nanofluids in the DASCs. Herein, the primary difficulties and challenges of utilising nanofluids in DASCs are introduced as follows:

1. The low stability of nanofluids represents the main problem, especially when nanoparticles concentrations are high. The nanoparticles have to conglomerate over time, due to their high surface/volume ratio and high surface tension [142]. After that, the nanoparticles instability leads to a reduction in thermal conductivity with time [143,144]. Several methods have been performed and examined to improve the stability and homogeneity of the nanofluids [145]. Nonetheless, the most cost-effective method is by adding surfactants, also known as 'non-covalent functionalisation' process, which should be used to produce more homogenous and stable nanofluids [146].

2. Complexity in the nanoparticles manufacturing process, two-step method and single-step method are the two common methods for producing nanofluids [147]. In both methods, nanoparticles are manufactured in processes that incorporate reactants in the form of ion exchange or reduction reactions, which in turn impact the nanofluids performance used in solar collectors.

3. It takes a long time to stabilise nanofluids during the preparation process when compared with the conventional fluid.

4. Nanofluids are relatively expensive, due to the difficulty involved in the synthesis of nanoparticles. This represents one of their main defects, which impedes the use of nanofluids in engineering applications [148,149]. For that, the lowest weight fraction of nanoparticles that possesses relatively high thermal conductivity should be employed [150].

5. The high viscosity of the nanofluids over that of the conventional fluids results in an undesirable increase in the pumping power [151].

19.8 Conclusions and Future Opportunities

The current chapter represents an attempt to encompass the latest advances and research in the field of the nanofluids and their application in direct-absorption solar thermal collectors. As normally employed heat transfer fluids are weak absorber over the ultra-violet and visible ranges of the solar spectrum. From the literature study, significant conclusions are summarised as below:

1. The concentration of the nanoparticle must be selected with great care as the performance of DASCs is highly sensitive to the nanofluid concentration. In most of the situations, raising the nanoparticle volume fraction over an optimum value, the incident radiation is absorbed by a thin top layer of the nanofluid rather than by the fluid medium volumetrically. Subsequently, the system achieves the surface-based absorber case.

2. More detailed studies should be carried to optimise the nanoparticle types, base fluid types, nanoparticle size, and operating temperature.

3. Significant improvement in performance of the DASCs have been obtained by using carbon nanostructure-based nanofluids, specifically MWCNT, single wall carbon nanotubes (SWCNT), carbon nanoparticles (CNP), and graphene compared to metal- and metal oxides-based nanofluids, but this is true only at low nanoparticles weight fractions.

4. The use of hybrid nanofluids (combination of two or more nanoparticles) on the thermal performance of DASCs should be investigated.

Furthermore, some future trends for optimising the use of nanofluids-based DASCs are also suggested as follows:

1. Investigating the commercial viability of the nanofluid-based DASC system as compared to surface absorbers. Future trends showed tendency towards increasing nanofluids long-term and high temperature stability by maintain simple and cost-effective surface functionalisation process.

2. More studies are needed to predict the effect of blended plasmonic nanofluids with controllable and broad-band absorption properties over the entire solar spectrum. It also should be promising requirement for harvesting solar energy, increasing optical properties of nanofluid, and optimise the performance of DASCs.

3. Future experimental and theoretical studies should be directed to the using of nanofluids-based DASCs built-in phase change materials as latent heat storage to assess the combined effect on the collector performance, due to the lack of research published on this topic.

4. More studies are needed to predict the economic benefits and environmental aspects through the use of nanofluids in DASCs.

REFERENCES

1. Pizzarelli M. The status of the research on the heat transfer deterioration in supercritical fluids: A review. *International Communications in Heat and Mass Transfer.* 2018;95:132–138.
2. Duangthongsuk W, Wongwises S. An experimental study on the heat transfer performance and pressure drop of TiO2-water nanofluids flowing under a turbulent flow regime. *International Journal of Heat and Mass Transfer.* 2010;53:334–344.
3. Ozturk I, Aslan A, Kalyoncu H. Energy consumption and economic growth relationship: Evidence from panel data for low and middle income countries. *Energy Policy.* 2010;38:4422–4428.
4. Liu Z. *Global Energy Interconnection.* Academic Press; 2015.
5. Akhmat G, Zaman K, Shukui T, Sajjad F, Khan MA, Khan MZ. The challenges of reducing greenhouse gas emissions and air pollution through energy sources: Evidence from a panel of developed countries. *Environmental Science and Pollution Research.* 2014;21:7425–7435.
6. Alper A, Oguz O. The role of renewable energy consumption in economic growth: Evidence from asymmetric causality. *Renewable and Sustainable Energy Reviews.* 2016;60:953–959.
7. Al-Karaghouli A, Renne D, Kazmerski LL. Solar and wind opportunities for water desalination in the Arab regions. *Renewable and Sustainable Energy Reviews.* 2009;13:2397–2407.
8. Zanuttigh B, Martinelli L, Lamberti A. Wave overtopping and piling-up at permeable low crested structures. *Coastal Engineering.* 2008;55:484–498.
9. Ali D, Gazey R, Aklil D. Developing a thermally compensated electrolyser model coupled with pressurised hydrogen storage for modelling the energy efficiency of hydrogen energy storage systems and identifying their operation performance issues. *Renewable and Sustainable Energy Reviews.* 2016;66:27–37.
10. AlZaharani AA, Dincer I, Naterer G. Performance evaluation of a geothermal based integrated system for power, hydrogen and heat generation. *International Journal of Hydrogen Energy.* 2013;38:14505–14511.
11. Liu H, Jiaqiang E, Deng Y, Xie C, Zhu H. Experimental study on pyrolysis characteristics of the tobacco stem based on microwave heating method. *Applied Thermal Engineering.* 2016;106:473–479.
12. Devabhaktuni V, Alam M, Depuru SSSR, Green II RC, Nims D, Near C. Solar energy: Trends and enabling technologies. *Renewable and Sustainable Energy Reviews.* 2013;19:555–564.
13. Liu S-Y, Perng Y-H, Ho Y-F. The effect of renewable energy application on Taiwan buildings: What are the challenges and strategies for solar energy exploitation? *Renewable and Sustainable Energy Reviews.* 2013;28:92–106.
14. Elsheikh AH, Abd Elaziz M. Review on applications of particle swarm optimization in solar energy systems. *International Journal of Environmental Science and Technology.* 2018; 1–2.
15. Tian Y, Zhao C-Y. A review of solar collectors and thermal energy storage in solar thermal applications. *Applied Energy.* 2013;104:538–553.

16. Duffie JA, Beckman WA. *Solar Engineering of Thermal Processes.* John Wiley & Sons; 2013.

17. Das SK, Choi SU, Yu W, Pradeep T. *Nanofluids: Science and Technology.* John Wiley & Sons; 2007.

18. Mahbubul I, Saidur R, Amalina M. Investigation of viscosity of R123-TiO 2 nanorefrigerant. *International Journal of Mechanical and Materials Engineering.* 2012;7:146–151.

19. Mahbubul I, Saidur R, Amalina M. Latest developments on the viscosity of nanofluids. *International Journal of Heat and Mass Transfer.* 2012;55:874–885.

20. Huminic G, Huminic A. Application of nanofluids in heat exchangers: A review. *Renewable and Sustainable Energy Reviews.* 2012;16:5625–5638.

21. Ali MKA, Xianjun H, Abdelkareem MAA, Gulzar M, Elsheikh AH. Novel approach of the graphene nanolubricant for energy saving via anti-friction/wear in automobile engines. *Tribology International.* 2018;124:209–229.

22. Sharshir SW, Peng G, Wu L, Yang N, Essa FA, Elsheikh AH et al. Enhancing the solar still performance using nanofluids and glass cover cooling: Experimental study. *Applied Thermal Engineering.* 2017;113:684–693.

23. Elsheikh AH, Sharshir SW, Ahmed Ali MK, Shaibo J, Edreis EMA, Abdelhamid T et al. Thin film technology for solar steam generation: A new dawn. *Solar Energy.* 2019;177:561–575.

24. Elsheikh AH, Sharshir SW, Mostafa ME, Essa FA, Ahmed Ali MK. Applications of nanofluids in solar energy: A review of recent advances. *Renewable and Sustainable Energy Reviews.* 2018;82:3483–3502.

25. Khan I, Saeed K, Khan I. Nanoparticles: Properties, applications and toxicities. *Arabian Journal of Chemistry.* 2017.

26. Hussein AK. Applications of nanotechnology in renewable energies—A comprehensive overview and understanding. *Renewable and Sustainable Energy Reviews.* 2015;42:460–476.

27. Angayarkanni SA, Philip J. Review on thermal properties of nanofluids: Recent developments. *Advances in Colloid and Interface Science.* 2015;225:146–176.

28. Murshed SMS, Leong KC, Yang C. A combined model for the effective thermal conductivity of nanofluids. *Applied Thermal Engineering.* 2009;29:2477–2483.

29. Das K. Flow and heat transfer characteristics of nanofluids in a rotating frame. *Alexandria Engineering Journal.* 2014;53:757–766.

30. Ganvir RB, Walke PV, Kriplani VM. Heat transfer characteristics in nanofluid—A review. *Renewable and Sustainable Energy Reviews.* 2017;75:451–460.

31. Das PK. A review based on the effect and mechanism of thermal conductivity of normal nanofluids and hybrid nanofluids. *Journal of Molecular Liquids.* 2017;240:420–446.

32. Chandrasekar M, Suresh S. A review on the mechanisms of heat transport in nanofluids. *Heat Transfer Engineering.* 2009;30:1136–1150.

33. Keblinski P, Phillpot SR, Choi SUS, Eastman JA. Mechanisms of heat flow in suspensions of nano-sized particles (nanofluids). *International Journal of Heat and Mass Transfer.* 2002;45:855–863.

34. Keblinski P, Eastman JA, Cahill DG. Nanofluids for thermal transport. *Materials Today.* 2005;8:36–44.

35. Teng T-P, Hung Y-H, Teng T-C, Mo H-E, Hsu H-G. The effect of alumina/water nanofluid particle size on thermal conductivity. *Applied Thermal Engineering.* 2010;30:2213–2218.

36. Nemade K, Waghuley S. A novel approach for enhancement of thermal conductivity of CuO/H_2O based nanofluids. *Applied Thermal Engineering.* 2016;95:271–274.

37. Chopkar M, Kumar S, Bhandari DR, Das PK, Manna I. Development and characterization of Al_2Cu and Ag_2Al nanoparticle dispersed water and ethylene glycol based nanofluid. *Materials Science and Engineering B.* 2007;139:141–148.

38. He Y, Jin Y, Chen H, Ding Y, Cang D, Lu H. Heat transfer and flow behaviour of aqueous suspensions of TiO2 nanoparticles (nanofluids) flowing upward through a vertical pipe. *International Journal of Heat and Mass Transfer.* 2007;50:2272–2281.

39. Patel HE, Sundararajan T, Das SK. An experimental investigation into the thermal conductivity enhancement in oxide and metallic nanofluids. *Journal of Nanoparticle Research.* 2010;12:1015–1031.

40. Xie H, Wang J, Xi T, Liu Y. Thermal conductivity of suspensions containing nanosized SiC particles. *International Journal of Thermophysics.* 2002;23:571–580.

41. Yongjin F, Boming Y, Peng X, Mingqing Z. The effective thermal conductivity of nanofluids based on the nanolayer and the aggregation of nanoparticles. *Journal of Physics D Applied Physics.* 2007;40:3164.

42. Hong KS, Hong T-K, Yang H-S. Thermal conductivity of Fe nanofluids depending on the cluster size of nanoparticles. *Applied Physics Letters.* 2006;88:031901.

43. Feng Y, Yu B, Feng K, Xu P, Zou M. Thermal conductivity of nanofluids and size distribution of nanoparticles by Monte Carlo simulations. *Journal of Nanoparticle Research.* 2008;10:1319–1328.
44. Prasher R, Song D, Wang J, Phelan P. Measurements of nanofluid viscosity and its implications for thermal applications. *Applied Physics Letters.* 2006;89:133108.
45. Huminic A, Huminic G, Fleaca C, Dumitrache F, Morjan I. Thermal conductivity, viscosity and surface tension of nanofluids based on FeC nanoparticles. *Powder Technology.* 2015;284:78–84.
46. Ding Y, Alias H, Wen D, Williams RA. Heat transfer of aqueous suspensions of carbon nanotubes (CNT nanofluids). *International Journal of Heat and Mass Transfer.* 2006;49:240–250.
47. Das PK, Mallik AK, Ganguly R, Santra AK. Synthesis and characterization of TiO2–water nanofluids with different surfactants. *International Communications in Heat and Mass Transfer.* 2016;75:341–348.
48. Lee S, Choi SUS, Li S, Eastman JA. Measuring thermal conductivity of fluids containing oxide nanoparticles. *Journal of Heat Transfer.* 1999;121:280–289.
49. Karthikeyan NR, Philip J, Raj B. Effect of clustering on the thermal conductivity of nanofluids. *Materials Chemistry and Physics.* 2008;109:50–55.
50. Zhu H, Zhang C, Liu S, Tang Y, Yin Y. Effects of nanoparticle clustering and alignment on thermal conductivities of Fe_3O_4 aqueous nanofluids. *Applied Physics Letters.* 2006;89:023123.
51. Cho T, Baek I, Lee J, Park S. Preparation of nanofluids containing suspended silver particles for enhancing fluid thermal conductivity of fludis. *Journal of Industrial and Engineering Chemistry.* 2005;11:400–406.
52. Liu M-S, Lin MC-C, Tsai CY, Wang C-C. Enhancement of thermal conductivity with Cu for nanofluids using chemical reduction method. *International Journal of Heat and Mass Transfer.* 2006;49:3028–3033.
53. Murshed SMS, Leong KC, Yang C. Enhanced thermal conductivity of TiO_2—water based nanofluids. *International Journal of Thermal Sciences.* 2005;44:367–373.
54. Kim HJ, Lee S-H, Lee J-H, Jang SP. Effect of particle shape on suspension stability and thermal conductivities of water-based bohemite alumina nanofluids. *Energy.* 2015;90:1290–1297.
55. Xuan Y, Li Q, Hu W. Aggregation structure and thermal conductivity of nanofluids. *AICHE Journal.* 2003;49:1038–1043.
56. Ruan B, Jacobi AM. Heat transfer characteristics of multiwall carbon nanotube suspensions (MWCNT nanofluids) in intertube falling-film flow. *International Journal of Heat and Mass Transfer.* 2012;55:3186–3195.
57. Chen L, Xie H, Li Y, Yu W. Nanofluids containing carbon nanotubes treated by mechanochemical reaction. *Thermochimica Acta.* 2008;477:21–24.
58. Pang C, Jung J-Y, Lee JW, Kang YT. Thermal conductivity measurement of methanol-based nanofluids with Al_2O_3 and SiO_2 nanoparticles. *International Journal of Heat and Mass Transfer.* 2012;55:5597-5602.
59. Hwang YJ, Ahn YC, Shin HS, Lee CG, Kim GT, Park HS et al. Investigation on characteristics of thermal conductivity enhancement of nanofluids. *Current Applied Physics.* 2006;6:1068–1071.
60. Younes H, Christensen G, Luan X, Hong H, Smith P. Effects of alignment, pH, surfactant, and solvent on heat transfer nanofluids containing Fe_2O_3 and CuO nanoparticles. *Journal of Applied Physics.* 2012;111:064308.
61. Das SK, Putra N, Roetzel W. Pool boiling characteristics of nano-fluids. *International Journal of Heat and Mass Transfer.* 2003;46:851–862.
62. Mintsa HA, Roy G, Nguyen CT, Doucet D. New temperature dependent thermal conductivity data for water-based nanofluids. *International Journal of Thermal Sciences.* 2009;48:363–371.
63. Sharshir SW, Elsheikh AH, Peng G, Yang N, El-Samadony MOA, Kabeel AE. Thermal performance and exergy analysis of solar stills—A review. *Renewable and Sustainable Energy Reviews.* 2017;73:521–544.
64. Park SR, Pandey AK, Tyagi VV, Tyagi SK. Energy and exergy analysis of typical renewable energy systems. *Renewable and Sustainable Energy Reviews.* 2014;30:105–123.
65. Saidur R, BoroumandJazi G, Mekhlif S, Jameel M. Exergy analysis of solar energy applications. *Renewable and Sustainable Energy Reviews.* 2012;16:350–356.
66. Lee J, Mudawar I. Assessment of the effectiveness of nanofluids for single-phase and two-phase heat transfer in micro-channels. *International Journal of Heat and Mass Transfer.* 2007;50:452–463.
67. Teng T-P, Yu C-C. Heat dissipation performance of MWCNTs nano-coolant for vehicle. *Experimental Thermal and Fluid Science.* 2013;49:22–30.
68. Bergman TL. Effect of reduced specific heats of nanofluids on single phase, laminar internal forced convection. *International Journal of Heat and Mass Transfer.* 2009;52:1240–1244.

69. Ho MX, Pan C. Optimal concentration of alumina nanoparticles in molten Hitec salt to maximize its specific heat capacity. *International Journal of Heat and Mass Transfer.* 2014;70:174–184.

70. Shin D, Banerjee D. Specific heat of nanofluids synthesized by dispersing alumina nanoparticles in alkali salt eutectic. *International Journal of Heat and Mass Transfer.* 2014;74:210–214.

71. Nieh H-M, Teng T-P, Yu C-C. Enhanced heat dissipation of a radiator using oxide nano-coolant. *International Journal of Thermal Sciences.* 2014;77:252–261.

72. Hentschke R. On the specific heat capacity enhancement in nanofluids. *Nanoscale Research Letters.* 2016;11:88.

73. Anoop KB, Sundararajan T, Das SK. Effect of particle size on the convective heat transfer in nanofluid in the developing region. *International Journal of Heat and Mass Transfer.* 2009;52:2189–2195.

74. Yang J-C, Li F-C, Zhou W-W, He Y-R, Jiang B-C. Experimental investigation on the thermal conductivity and shear viscosity of viscoelastic-fluid-based nanofluids. *International Journal of Heat and Mass Transfer.* 2012;55:3160–3166.

75. Sharifpur M, Adio SA, Meyer JP. Experimental investigation and model development for effective viscosity of Al_2O_3–glycerol nanofluids by using dimensional analysis and GMDH-NN methods. *International Communications in Heat and Mass Transfer.* 2015;68:208–219.

76. Choi SU, Eastman JA. *Enhancing Thermal Conductivity of Fluids with Nanoparticles.* Argonne National Lab, Argonne, IL; 1995.

77. Bashirnezhad K, Bazri S, Safaei MR, Goodarzi M, Dahari M, Mahian O et al. Viscosity of nanofluids: A review of recent experimental studies. *International Communications in Heat and Mass Transfer.* 2016;73:114–123.

78. Murshed SMS, Estellé P. A state of the art review on viscosity of nanofluids. *Renewable and Sustainable Energy Reviews.* 2017;76:1134–1152.

79. Mishra PC, Mukherjee S, Nayak SK, Panda A. A brief review on viscosity of nanofluids. *International Nano Letters.* 2014;4:109–120.

80. Iqbal M. *An Introduction to Solar Radiation.* Elsevier; 2012.

81. Saidur R, Meng TC, Said Z, Hasanuzzaman M, Kamyar A. Evaluation of the effect of nanofluid-based absorbers on direct solar collector. *International Journal of Heat and Mass Transfer.* 2012;55:5899–5907.

82. Sharshir SW, Peng G, Wu L, Essa FA, Kabeel AE, Yang N. The effects of flake graphite nanoparticles, phase change material, and film cooling on the solar still performance. *Applied Energy.* 2017;191:358–366.

83. Rativa D, Gómez-Malagón LA. Solar radiation absorption of nanofluids containing metallic nanoellipsoids. *Solar Energy.* 2015;118:419–425.

84. Kameya Y, Hanamura K. Enhancement of solar radiation absorption using nanoparticle suspension. *Solar Energy.* 2011;85:299–307.

85. Jana S, Salehi-Khojin A, Zhong W-H. Enhancement of fluid thermal conductivity by the addition of single and hybrid nano-additives. *Thermochimica Acta.* 2007;462:45–55.

86. Said Z, Saidur R, Rahim NA. Optical properties of metal oxides based nanofluids. *International Communications in Heat and Mass Transfer.* 2014;59:46–54.

87. Vijayaraghavan S, Ganapathisubbu S, Santosh Kumar C. Performance analysis of a spectrally selective concentrating direct absorption collector. *Solar Energy.* 2013;97:418–425.

88. Karami M, Akhavan-Behabadi MA, Raisee Dehkordi M, Delfani S. Thermo-optical properties of copper oxide nanofluids for direct absorption of solar radiation. *Solar Energy Materials and Solar Cells.* 2016;144:136–142.

89. Xuan Y, Duan H, Li Q. Enhancement of solar energy absorption using a plasmonic nanofluid based on TiO_2/Ag composite nanoparticles. *RSC Advances.* 2014;4:16206–16213.

90. Karami M, Akhavan Bahabadi MA, Delfani S, Ghozatloo A. A new application of carbon nanotubes nanofluid as working fluid of low-temperature direct absorption solar collector. *Solar Energy Materials and Solar Cells.* 2014;121:114–118.

91. Hordy N, Rabilloud D, Meunier J-L, Coulombe S. High temperature and long-term stability of carbon nanotube nanofluids for direct absorption solar thermal collectors. *Solar Energy.* 2014;105:82–90.

92. Mercatelli L, Sani E, Zaccanti G, Martelli F, Di Ninni P, Barison S et al. Absorption and scattering properties of carbon nanohorn-based nanofluids for direct sunlight absorbers. *Nanoscale Research Letters.* 2011;6:282.

93. Sani E, Barison S, Pagura C, Mercatelli L, Sansoni P, Fontani D et al. Carbon nanohorns-based nano-fluids as direct sunlight absorbers. *Optics Express.* 2010;18:5179–5187.

94. He Y, Vasiraju S, Que L. Hybrid nanomaterial-based nanofluids for micropower generation. *RSC Advances.* 2014;4:2433–2439.

95. Han D, Meng Z, Wu D, Zhang C, Zhu H. Thermal properties of carbon black aqueous nanofluids for solar absorption. *Nanoscale Research Letters.* 2011;6:457.

96. Gorji TB, Ranjbar AA, Mirzababaei SN. Optical properties of carboxyl functionalized carbon nanotube aqueous nanofluids as direct solar thermal energy absorbers. *Solar Energy.* 2015;119:332–342.

97. Lee S-H, Jang SP. Extinction coefficient of aqueous nanofluids containing multi-walled carbon nano-tubes. *International Journal of Heat and Mass Transfer.* 2013;67:930–935.

98. Mehrali M, Sadeghinezhad E, Tahan Latibari S, Mehrali M, Togun H, Zubir MNM et al. Preparation, characterization, viscosity, and thermal conductivity of nitrogen-doped graphene aqueous nanofluids. *Journal of Materials Science.* 2014;49:7156–7171.

99. Luo Z, Wang C, Wei W, Xiao G, Ni M. Performance improvement of a nanofluid solar collector based on direct absorption collection (DAC) concepts. *International Journal of Heat and Mass Transfer.* 2014;75:262–271.

100. Ladjevardi SM, Asnaghi A, Izadkhast PS, Kashani AH. Applicability of graphite nanofluids in direct solar energy absorption. *Solar Energy.* 2013;94:327–334.

101. Jia L, Chen Y, Lei S, Mo S, Luo X, Shao X. External electromagnetic field-aided freezing of CMC-modified graphene/water nanofluid. *Applied Energy.* 2016;162:1670–1677.

102. Wang H, Yang W, Cheng L, Guan C, Yan H. Chinese ink: High performance nanofluids for solar energy. *Solar Energy Materials and Solar Cells.* 2018;176:374–380.

103. Zhang L, Liu J, He G, Ye Z, Fang X, Zhang Z. Radiative properties of ionic liquid-based nanofluids for medium-to-high-temperature direct absorption solar collectors. *Solar Energy Materials and Solar Cells.* 2014;130:521–528.

104. Taylor RA, Phelan PE, Otanicar TP, Adrian R, Prasher R. Nanofluid optical property characterization: Towards efficient direct absorption solar collectors. *Nanoscale Research Letters.* 2011;6:225.

105. He Q, Wang S, Zeng S, Zheng Z. Experimental investigation on photothermal properties of nano-fluids for direct absorption solar thermal energy systems. *Energy Conversion and Management.* 2013;73:150–157.

106. Wu Y, Zhou L, Du X, Yang Y. Optical and thermal radiative properties of plasmonic nanofluids contain-ing core–shell composite nanoparticles for efficient photothermal conversion. *International Journal of Heat and Mass Transfer.* 2015;82:545–554.

107. Zhang H, Chen H-J, Du X, Lin G, Wen D. Dependence of photothermal conversion characteristics on different nanoparticle dispersions. *Journal of Nanoscience and Nanotechnology.* 2015;15:3055–3060.

108. Shende R, Sundara R. Nitrogen doped hybrid carbon based composite dispersed nanofluids as working fluid for low-temperature direct absorption solar collectors. *Solar Energy Materials and Solar Cells.* 2015;140:9–16.

109. Ghazi S, Sayigh A, Ip K. Dust effect on flat surfaces—A review paper. *Renewable and Sustainable Energy Reviews.* 2014;33:742–751.

110. Sarsam WS, Kazi SN, Badarudin A. A review of studies on using nanofluids in flat-plate solar collec-tors. *Solar Energy.* 2015;122:1245–1265.

111. Iranmanesh S, Ong HC, Ang BC, Sadeghinezhad E, Esmaeilzadeh A, Mehrali M. Thermal performance enhancement of an evacuated tube solar collector using graphene nanoplatelets nanofluid. *Journal of Cleaner Production.* 2017;162:121–129.

112. Gudekar AS, Jadhav AS, Panse SV, Joshi JB, Pandit AB. Cost effective design of compound parabolic collector for steam generation. *Solar Energy.* 2013;90:43–50.

113. Hafez AZ, Soliman A, El-Metwally KA, Ismail IM. Solar parabolic dish Stirling engine system design, simulation, and thermal analysis. *Energy Conversion and Management.* 2016;126:60–75.

114. Raj P, Subudhi S. A review of studies using nanofluids in flat-plate and direct absorption solar collectors. *Renewable and Sustainable Energy Reviews.* 2018;84:54–74.

115. Gorji TB, Ranjbar AA. A review on optical properties and application of nanofluids in direct absorption solar collectors (DASCs). *Renewable and Sustainable Energy Reviews.* 2017;72:10–32.

116. Hussein AK. Applications of nanotechnology to improve the performance of solar collectors—Recent advances and overview. *Renewable and Sustainable Energy Reviews.* 2016;62:767–792.

117. Tyagi H, Phelan P, Prasher R. Predicted efficiency of a low-temperature nanofluid-based direct absorption solar collector. *Journal of Solar Energy Engineering*. 2009;131:041004.
118. Otanicar T PP, Prasher R, Rosengarten G, Taylor R. Nanofluid-based direct absorption PP solar collector. *Journal of Renewable and Sustainable Energy*. 2010;2.
119. Gupta HK, Agrawal GD, Mathur J. An experimental investigation of a low temperature Al_2O_3-H_2O nanofluid based direct absorption solar collector. *Solar Energy*. 2015;118:390–396.
120. Parvin S, Nasrin R, Alim MA. Heat transfer and entropy generation through nanofluid filled direct absorption solar collector. *International Journal of Heat and Mass Transfer*. 2014;71:386–395.
121. Verma KL. Thermal performance evaluation of a direct absorption flat plate solar collector (DASC) using Al_2O_3–H_2O based nanofluids. *IOSR Journal of Mechanical and Civil Engineering*. 2013;6:29–35.
122. Kundan SP. Performance evaluation of a nanofluid (CuO–H_2O) based low flux solar collector. *International Journal of Engineering Research*. 1 2013;2:108–112.
123. Karami M, Akhavan-Bahabadi MA, Delfani S, Raisee M. Experimental investigation of CuO nanofluid-based direct absorption solar collector for residential applications. *Renewable and Sustainable Energy Reviews*. 2015;52:793–801.
124. Delfani S, Karami M, Behabadi MAA. Performance characteristics of a residential-type direct absorption solar collector using MWCNT nanofluid. *Renewable Energy*. 2016;87:754–764.
125. Vakili M, Hosseinalipour SM, Delfani S, Khosrojerdi S, Karami M. Experimental investigation of graphene nanoplatelets nanofluid-based volumetric solar collector for domestic hot water systems. *Solar Energy*. 2016;131:119–130.
126. Chen M, He Y, Zhu J, Wen D. Investigating the collector efficiency of silver nanofluids based direct absorption solar collectors. *Applied Energy*. 2016;181:65–74.
127. Bandarra Filho EP, Mendoza OSH, Beicker CLL, Menezes A, Wen D. Experimental investigation of a silver nanoparticle-based direct absorption solar thermal system. *Energy Conversion and Management*. 2014;84:261–267.
128. Khullar V, Tyagi H, Hordy N, Otanicar TP, Hewakuruppu Y, Modi P et al. Harvesting solar thermal energy through nanofluid-based volumetric absorption systems. *International Journal of Heat and Mass Transfer*. 2014;77:377–384.
129. Cregan V, Myers TG. Modelling the efficiency of a nanofluid direct absorption solar collector. *International Journal of Heat and Mass Transfer*. 2015;90:505–514.
130. Turkyilmazoglu M. Performance of direct absorption solar collector with nanofluid mixture. *Energy Conversion and Management*. 2016;114:1–10.
131. Gorji TB, Ranjbar AA. Geometry optimization of a nanofluid-based direct absorption solar collector using response surface methodology. *Solar Energy*. 2015;122:314–325.
132. Gorji TB, Ranjbar AA. A numerical and experimental investigation on the performance of a low-flux direct absorption solar collector (DASC) using graphite, magnetite and silver nanofluids. *Solar Energy*. 2016;135:493–505.
133. Lee S-H, Jang SP. Efficiency of a volumetric receiver using aqueous suspensions of multi-walled carbon nanotubes for absorbing solar thermal energy. *International Journal of Heat and Mass Transfer*. 2015;80:58–71.
134. Hatami M, Jing D. Evaluation of wavy direct absorption solar collector (DASC) performance using different nanofluids. *Journal of Molecular Liquids*. 2017;229:203–211.
135. Li Q, Zheng C, Mesgari S, Hewkuruppu YL, Hjerrild N, Crisostomo F et al. Experimental and numerical investigation of volumetric versus surface solar absorbers for a concentrated solar thermal collector. *Solar Energy*. 2016;136:349–364.
136. Taylor RA, Phelan PE, Otanicar TP, Walker CA, Nguyen M, Trimble S et al. . Applicability of nanofluids in high flux solar collectors. *Journal of Renewable and Sustainable Energy*. 2011;3.
137. Liu J, Ye Z, Zhang L, Fang X, Zhang Z. A combined numerical and experimental study on graphene/ionic liquid nanofluid based direct absorption solar collector. *Solar Energy Materials and Solar Cells*. 2015;136:177–186.
138. Toppin-Hector A, Singh H. Development of a nano-heat transfer fluid carrying direct absorbing receiver for concentrating solar collectors. *International Journal of Low-Carbon Technologies*. 2016;11:199–204.
139. Lenert A, Perez Zuniga YS, Wang EN. Nanofluid-based absorbers for high temperature direct solar collectors. 2010:499–508.

140. Lenert A, Wang EN. Optimization of nanofluid volumetric receivers for solar thermal energy conversion. *Solar Energy*. 2012;86:253–265.
141. Veeraragavan A, Lenert A, Yilbas B, Al-Dini S, Wang EN. Analytical model for the design of volumetric solar flow receivers. *International Journal of Heat and Mass Transfer*. 2012;55:556–564.
142. Devendiran DK, Amirtham VA. A review on preparation, characterization, properties and applications of nanofluids. *Renewable and Sustainable Energy Reviews*. 2016;60:21–40.
143. Dhinesh Kumar D, Valan Arasu A. A comprehensive review of preparation, characterization, properties and stability of hybrid nanofluids. *Renewable and Sustainable Energy Reviews*. 2018;81:1669–1689.
144. Gupta M, Singh V, Kumar R, Said Z. A review on thermophysical properties of nanofluids and heat transfer applications. *Renewable and Sustainable Energy Reviews*. 2017;74:638–670.
145. Fuskele V, Sarviya RM. Recent developments in nanoparticles synthesis, preparation and stability of nanofluids. *Materials Today Proceedings*. 2017;4:4049–4060.
146. Leong KY, Razali I, Ku Ahmad KZ, Ong HC, Ghazali MJ, Abdul Rahman MR. Thermal conductivity of an ethylene glycol/water-based nanofluid with copper-titanium dioxide nanoparticles: An experimental approach. *International Communications in Heat and Mass Transfer*. 2018;90:23–28.
147. Babita, Sharma SK, Gupta SM. Preparation and evaluation of stable nanofluids for heat transfer application: A review. *Experimental Thermal and Fluid Science*. 2016;79:202–212.
148. Bahiraei M, Rahmani R, Yaghoobi A, Khodabandeh E, Mashayekhi R, Amani M. Recent research contributions concerning use of nanofluids in heat exchangers: A critical review. *Applied Thermal Engineering*. 2018;133:137–159.
149. Gómez-Villarejo R, Martín EI, Navas J, Sánchez-Coronilla A, Aguilar T, Gallardo JJ et al. Ag-based nanofluidic system to enhance heat transfer fluids for concentrating solar power: Nano-level insights. *Applied Energy*. 2017;194:19–29.
150. Raja M, Vijayan R, Dineshkumar P, Venkatesan M. Review on nanofluids characterization, heat transfer characteristics and applications. *Renewable and Sustainable Energy Reviews*. 2016;64:163–173.
151. Esfahani MR, Languri EM, Nunna MR. Effect of particle size and viscosity on thermal conductivity enhancement of graphene oxide nanofluid. *International Communications in Heat and Mass Transfer*. 2016;76:308–315.

Section V

Nanofluids in Oil and Gas Industry and Carbon Sequestration

20

Application of Nanofluids in Enhanced Oil Recovery: A Systematic Literature Review and Organizing Framework

Majid Mohammadi, Mehdi Sedighi, and Abdolhossein Hemmati-Sarapardeh

CONTENTS

20.1 Introduction

The petroleum industry is currently dealing with the major troubles in increasing productivity because demand for oil increases daily, especially in developed as well as developing countries.

A typical illustration of world oil consumption is presented in Figure 20.1. According to Statista report [1], global demand for crude oil in 2010 was 86.4 million barrels per day. It is expected to rise to around 99 million barrels per day by 2018. It can be stated that the increase in consumption is a main problem in the present context. Meanwhile, the detection of new oil fields is extremely confined. Most of the reservoirs are actually mature and have a low production rate. Application of enhanced oil recovery (EOR) is therefore heavily based on the current economy, the nature of the reserve oil, and the price of crude oil.

EOR includes several strategies intended to decrease oil saturation as well as enhance oil production [2]. On the whole, oil production is put into three levels: primary, secondary, and tertiary (improved). The latter process is referred to as EOR (see Figure 20.2).

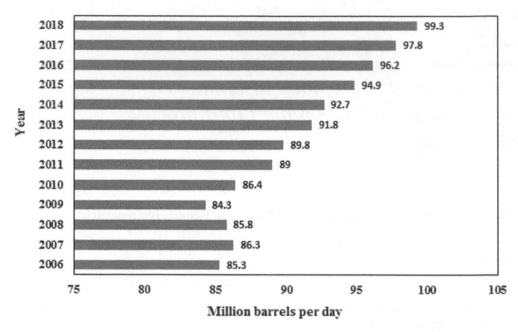

FIGURE 20.1   World consumption of oil. (From Statista. World consumption of oil, 2018. Available from: http://www. statista.com.)

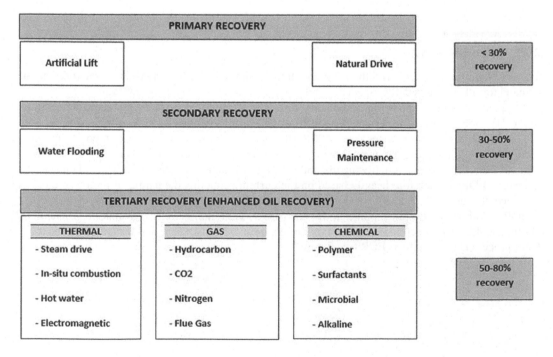

FIGURE 20.2   The different steps of oil recovery. (From Lee, W.J. and Wattenbarger, R.A. *Gas Reservoir Engineering.* Henry L. Doherty Memorial Fund of AIME, Richardson, Texas, 1996.)

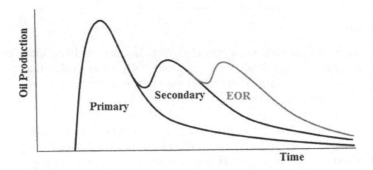

FIGURE 20.3 Additional oil production from EOR.

In other words, EOR is a technique which allows the extraction of an increased percentage of the original oil in place into a reservoir after the first two extraction stages, as indicated in Figure 20.3.

The EOR is employed when the primary and secondary techniques are no longer beneficial and typically about 60% to 80% of the oil is remained into the reservoir after primary and secondary oil recovery. The use of EOR can assist to recover additional oil. For gas wells, the primary production strategy is usually sufficient to produce about 80% of the reserves. For this reason, the EOR is required in the production of crude oil [3]. Normally, EOR is categorised into three different categories, as shown in Figure 20.2. Table 20.1 shows the important features of the available EOR methods.

These conventional EOR techniques face a number of key difficulties. For instance, in gas methods, the injected gas usually passes rapidly via injection well reservoirs to producing wells leading to a substantial amount of residual oil in the reservoirs due to the high mobility of the injected gas and oil [4,5]. In addition, chemical processes can be restricted by high chemical costs, feasible formation damage, as well as loss of chemicals [6–8]. Therefore, cost-effective, more efficient, and greener EOR types are absolutely necessary. Nanomaterials present new ways to deal with unsolved problems. In this chapter, the potential of nanoparticles (NPs) is investigated for EOR. In fact, some recent developments in this area are reviewed and evaluated.

TABLE 20.1

Major Characteristics of Available EOR Methods

EOR Technique	Working Principle	Mechanisms	When to Use
Thermal Injection	Injecting heat	• IFT decrease • Steam distillation • Oil expansion • Viscosity decrease • Gravity drainage	Heavy and extra heavy crude oils
Gas Injection	Injecting gas	• Pressure maintenance • Miscibility • Viscosity decrease • Oil expansion	After water injection development
Chemical Injection	Injecting chemical such as alkaline, polymers, surfactants	• IFT decrease • Wettability modification • Emulsification • Mobility control	After waterflood to extract residual oil; sandstone reservoir and less limestone

20.2 Nanofluids

Nanofluids are produced by the addition of NPs to fluids to intensify and ameliorate certain properties at low volume concentration. A particle is considered as NP when it is smaller than 100 nm [9,10]. For comparison, a human hair has a diameter of 80,000–100,000 nm in diameter.

The advantages of NPs in EOR methods:

Ultra-small size: Because of special small size, it is quite convenient to penetrate via porous media and transportation is simple. The common NPs, like CeO_2, Al_2O_3, MgO, TiO_2, ZrO_2, and SiO_2 are about 1–100 nm (see Figure 20.4), which is smaller compared to pore and throat sizes [11,12]. Therefore, they can readily flow via the porous media without a large decrease in permeability and getting trapped which enhances the oil recovery efficiency.

High surface to volume ratio: NPs have this property and therefore have great adsorption properties. Figure 20.5 illustrates this specific concept.

FIGURE 20.4   SEM image of different NPs (a) Al_2O_3, (b) SiO_2, (c) CeO_2, (d) TiO_2, (e) MgO, and (f) ZrO_2. (From Liu, Q. et al., *J. Petrol. Sci. Eng.*, 43, 75–86, 2004.)

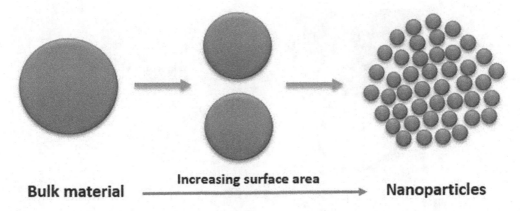

Bulk material —————— Increasing surface area ——————→ **Nanoparticles**

FIGURE 20.5 A picture of NPs with high surface to volume ratio.

Low costs and environmental friendliness: One of the concerns about the use of chemicals at the field level is the cost of injection. Since NPs price is generally more affordable than chemicals, these materials can be preferred for oil-field EORs. In addition, most NPs are sustainable materials compared to chemicals.

20.3 Nanofluid-EOR Mechanisms

The EOR mechanisms of NPs are not yet fully recognised. Various mechanisms including: decreasing the interfacial tension (IFT), altering the wettability, and lowering the viscosity are related to the recovery of trapped oil in reservoir. The study on NPs application in EOR has outlined on how these NPs affect the above properties and oil recovery.

20.3.1 Interfacial Tension

Reducing IFT in oil recovery is of a vital significance [13–17]. For conventional chemical flooding, IFT is a standard parameter for surfactant characterisation. The lowest oil-water interfacial tension through molecular surfactant can attain 10^{-3} to 10^{-2} mN/m [18,19]. Although nanofluid systems cannot reach low interfacial values, various researchers have attempted to decrease the IFT between oil and water with NPs [20–22]. It is suggested that NPs help to reduce IFT during EOR processes. The interfacial tension is usually evaluated by the pendant drop method [23] as well as spinning drop method [24].

Silicon oxide (SiO_2) NPs have been widely employed in EOR [25–28]. Lan et al. found the synergistic effect of silica NPs along with cationic surfactant on the IFT [29]. The cationic surfactant can adjust the surface of NPs from fully hydrophilic to partially hydrophobic, which supports the aggregation of NPs resulting in a decrease in IFT. It has been shown that ZrO_2 nanofluid substantially decreases the IFT [30]. Similarly, Moghadam and Azizian [31] identified that ZnO NPs could enhance the efficacy of surfactant in decreasing the IFT.

Furthermore, Al_2O_3 NPs were able to decrease IFT from 19 to 13 mN/m, while SiO_2 NPs can reduce the IFT to 16 mN/m [32]. In addition, carbon nanotubes (CNTs) have been shown to significantly reduce interfacial tension (Figure 20.6), indicating their superior potential as the EOR agent [33,34]. Table 20.2 represents the experimental results about IFT reduction by various nanofluids which have been reported in literature.

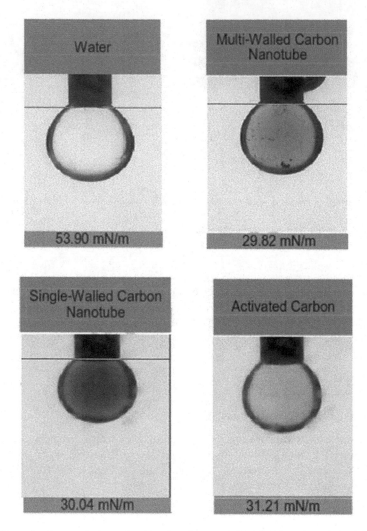

FIGURE 20.6   Effect of carbon nanotube and activated carbon on the interfacial tension. (From AfzaliTabar, M. et al., *J. Solid State Chem.*, 245, 164–173, 2017.)

20.3.2 Wettability Alteration

Wettability of a surface is related to solid-fluids and fluid-fluid interactions [43,44]. Wettability is an important parameter that regulates the distribution of fluids in the formation. Most recently, three main methods of measuring wettability are commonly employed by researchers: Amott test, the contact angle method, and the core displacement experiment [45–49]. Method of the contact angle is the most common method for determining wettability. Young equation is generally employed to discern the wettability type of the system using the methods of contact angle.

$$cos\theta = \frac{\sigma_{sw} - \sigma_{so}}{\sigma_{wo}}, \tag{20.1}$$

where θ is contact angle, σ_{sw}, σ_{so}, and σ_{wo} are interfacial tension between solid-water, solid-oil, and water-oil, respectively. According to the contact angle, the wettability of the rocks in 'oil water' system can be divided into three main types, i.e., water-wet ($\theta < 90°$), neutral-wet ($\theta = 90°$), and oil-wet ($\theta > 90$) [24] which are shown in Figure 20.7.

TABLE 20.2

Experimental Results about IFT Reduction by Nanofluids (T = 25°C)

NPs-Fluid	NPs Conc. (wt%)	IFT without NPs (mN/m)	IFT with NPs (mN/m)	Ref.
Al_2O_3-Propanol	0.3	38.5	2.25	[35]
Al_2O_3-Brine	0.05	19.2	12.8	[32]
Fe_2O_3-Propanol	0.3	38.5	2.75	[35]
SiO_2-Propanol	0.3	38.5	1.45	[35]
SiO_2-Ethanol	0.1	25	5	[36]
SiO_2-Water	0.1	13.62	10.69	[37]
SiO_2-Brine	0.05	19.2	16.9	[38]
SiO_2-Brine	0.1	21.7	4.5	[39]
SiO_2-Brine	0.5	21.7	5.2	[39]
SiO_2-Brine	1	21.7	5.8	[39]
SiO_2-Brine	1.5	21.7	6.1	[39]
SiO_2-Brine	2	21.7	6.3	[39]
SiO_2-Brine	0.05	19.2	17.5	[32]
SiO_2-Brine	0.25	15.58	13	[40]
ZrO_2-Water	0.01	16	3.1	[41]
ZrO_2-Water	0.001	51.4	37.2	[30]
ZrO_2-Water	0.01	51.4	37.2	[30]
ZrO_2-Water	0.1	51.4	36.8	[30]
Fe_3O_4@Chitosan-Brine	0.01	30	21.5	[42]
Fe_3O_4@Chitosan-Brine	0.02	30	22.5	[42]
Fe_3O_4@Chitosan-Brine	0.03	30	17.29	[42]
Multi-Walled CNT-Water	0.05	51.4	31.17	[34]
Multi-Walled CNT-Water	0.1	51.4	32.68	[34]
Multi-Walled CNT-Water	0.2	51.4	33.23	[34]
Multi-Walled CNT-Water	0.3	51.4	33.46	[34]
Multi-Walled CNT-Water	0.4	51.4	32.57	[34]
Multi-Walled CNT-Water	0.5	51.4	31.27	[34]
Multi-Walled CNT/SiO_2-Water	0.1	53.9	29.82	[33]
Single-Walled CNT/SiO_2-Water	0.1	53.9	30.04	[33]
Activated Carbon/SiO_2-Water	0.1	53.9	31.21	[33]

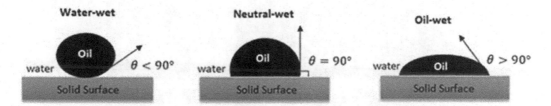

FIGURE 20.7 Wettability alteration in oil-water system.

Recent years, many research indicate that nanofluids have a major impact on altering the wettability, as shown in Figure 20.8. These studies have shown that NPs dispersion can change the wettability of oil-wet sandstones [45,50] and carbonates [51,52]. Consequently, wettability of the surface turns into more water-wet [45,51,53]. Al_2O_3, SiO_2, and ZrO_2 nanofluids had a significant impact on the modification of the properties of sandstone and carbonate rocks.

Hendraningrat et al. [54] found the effect of SiO_2 NPs on the wettability change on synthetic silica surface at various concentrations. Increasing the NP concentration decreases the contact angle and modifies

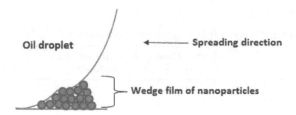

FIGURE 20.8   Wedge film of NPs displacing crude oil from a surface. (From Esfandyari Bayat, A. et al., *Energy & Fuels*, 28, 6255–6266, 2014; Nazari Moghaddam, R. et al., *Energy & Fuels* 29, 2111–2119, 2015.)

the wettability towards more water-wet. They suggested that smaller NPs can decrease the contact angle more than those with larger size [55]. Also, the change in wettability is a function of initial wettability, salinity, ionic composition, solid system, and the exposure time [56]. Roustaei and Bagherzadeh [57] experimentally examined the influence of SiO_2 NPs on the wettability of a carbonate rock. Al-Anssari et al. [58] identified that SiO_2 nanofluids induced wettability changes on oil-wet and mixed-wet carbonate substrates, in accordance with the results of Roustaei and Bagherzadeh [57]. Recently, some authors have attempted to analyse why NPs can cause changes in the wettability of rock surfaces through a variety of new characterisation techniques. For instance, Al-Anssari et al. [58] showed a surface modification of calcite with SEM images. It was realised that the distribution of NPs on the surface was homogeneous (see Figures 20.9 and 20.10). Contact angle data from various NPs are presented in Table 20.3.

20.3.3 Viscosity

As viscosity is an essential feature in the treatment and management of heavy crude oils, improving the mobility of oil in reservoir or surface conditions is currently an issue for the oil industry [63]. Mathematically, the mobility ratio is represented as follows:

$$M = \frac{k_{ri}\mu_o}{k_{ro}\mu_i},$$

(20.2)

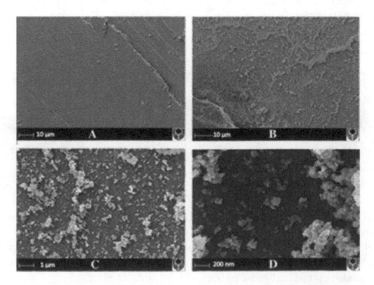

FIGURE 20.9   SEM images of a carbonate surface (a) before; (b) after nanofluid treatment; (c) high resolution; and (d) max-resolution. (From Al-Anssari, S. et al., *J. Colloid Interface Sci.*, 461, 435–442, 2016.)

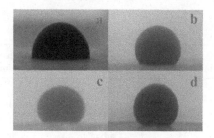

FIGURE 20.10 Image from oil droplet in the: (a) brine, (b) Al_2O_3 nanofluid, (c) TiO_2 nanofluid, and (d) SiO_2 nanofluid (T = 26°C). (From Esfandyari Bayat, A. et al., *Energy & Fuels*, 28, 6255–6266, 2014.)

TABLE 20.3

Experimental Results about Altering the Wettability by Nanofluids

NPs-Fluid	Temperature (°C)	Rock	Contact Angle, Before	Contact Angle, After	References
Al_2O_3-Propanol	25	Sandstone	134	90	[35]
Fe_2O_3-Propanol	25	Sandstone	134	98	[35]
SiO_2-Propanol	25	Sandstone	134	82	[35]
SiO_2-Water (Figure 20.10d)	26	Limestone	90	26	[59]
SiO_2-Water	40	Limestone	87	25	[59]
SiO_2-Water	50	Limestone	83	21	[59]
SiO_2-Water	60	Limestone	82	18	[59]
Al_2O_3-Water (Figure 20.10b)	26	Limestone	90	71	[59]
Al_2O_3-Water	40	Limestone	87	66	[59]
Al_2O_3-Water	50	Limestone	83	65	[59]
Al_2O_3-Water	60	Limestone	82	61	[59]
TiO_2-Water (Figure 20.10c)	26	Limestone	90	57	[59]
TiO_2-Water	40	Limestone	87	52	[59]
TiO_2-Water	50	Limestone	83	49	[59]
TiO_2-Water	60	Limestone	82	46	[59]
SiO_2-Water	25	Carbonate	122	24	[37]
SiO_2-Ethanol	23	Sandstone	135.5	66	[36]
ZrO_2-Acidic aqueous solution	25	Carbonate	140.2	59.7	[60]
Al_2O_3-Acidic aqueous solution	25	Carbonate	140.2	103.5	[60]
MgO-Acidic aqueous solution	25	Carbonate	140.2	98.2	[60]
TiO_2-Acidic aqueous solution	25	Carbonate	140.2	51.2	[60]
MWCNT-Paraffin	25	Carbonate	140.2	65.5	[60]
MWCNT-Paraffin/heptane	25	Carbonate	140.2	122.2	[60]
CeO_2-Aqueous solution	25	Carbonate	140.2	103.5	[60]
$CaCO_3$-Water and alcohol	25	Carbonate	140.2	58.2	[60]
SiO_2-Brine	25	Calcite	122	30	[58]
SiO_2-Brine	25	Calcite	143	63	[61]
SiO_2-Brine/PVP	25	Carbonate	151.7	83.5	[27,62]
TiO_2-Brine/PVP	25	Carbonate	151.7	60.7	[27,62]
80%TiO_2:20% SiO_2-Brine/PVP	25	Carbonate	151.7	37.4	[27,62]
SiO_2-Brine	25	Quartz	54	38.8	[32]
Al_2O_3-Brine/PVP	25	Quartz	54	28.6	[32]
TiO_2-Brine/PVP	25	Quartz	54	21.6	[32]

(*Continued*)

TABLE 20.3 (*Continued*)

Experimental Results about Altering the Wettability by Nanofluids

NPs-Fluid	Temperature (°C)	Rock	Contact Angle, Before	Contact Angle, After	References
Fe_3O_4@Chitosan-Brine	25	Carbonate	150	90	[42]
MWCNT/SiO_2-Water	25	Carbonate	>>100	60.5	[33]
SWCNT/SiO_2-Water	25	Carbonate	>>100	64.5	[33]
Activated Carbon/SiO_2-Water	25	Carbonate	>>100	75	[33]

where k_{ri} and k_{ro} stand for relative permeability of injected fluid and oil, respectively. μ_i and μ_o denote viscosity of injected fluid and oil, respectively. The mobility ratio decreases by reducing the viscosity of the oil phase or increasing the viscosity of injected fluids. Nanofluids can solve the mentioned problem because adding NPs to fluids (water, CO_2, or chemical) can enhance the viscosity of injected fluids [64–68].

Shah [69] presented that the viscosity of CO_2 nanofluids (1 wt% copper oxide NPs in CO_2) was 140 times higher than that of CO_2. In addition, Molnes et al. [70] identified that the shear viscosity of the water increased when cellulose NPs were dispersed in water. The nanofluids viscosity increases with raising NP concentration [71]. In addition, the nature of NPs influences nanofluids viscosity. At the similar concentration, the viscosity of the SiO_2 nanofluid exceeds that of the Al_2O_3 nanofluid [72]. Also, it has been reported that Fe_2O_3 NPs can increase the viscosity of the injected fluids [73,74].

The viscosity of heavy oil can be reduced using some metal oxide NPs even at high temperature [75]. The impact of NPs on the viscosity reduction, however, seems to lie more in the recovery of thermal process for high temperatures [75]. The percentage of viscosity reduction along with the optimal concentration of NPs highly depend on the nature of nano-metal oxide [75].

Taborda et al. [76] examined the use of SiO_2 nanofluid in EOR. Taborda et al. [76] described how the fluid should be made so that the NPs are entirely stable in the medium. The authors also presented that the reduction in viscosity is basically due to two factors: (a) the entraining fluid has a diluent effect and (b) the NPs can be dispersed through the crude oil. Taborda et al. [77] evaluated the influence of 8 nm sized SiO_2 NPs on the heavy oil viscosity. They identified that 1000 mg/L as the concentration of SiO_2 performed as the best viscosity-reducing agents. In addition, it was illustrated that as the particle size increases, the effect of NPs on the decrease of viscosity reduces, as shown in Figure 20.11.

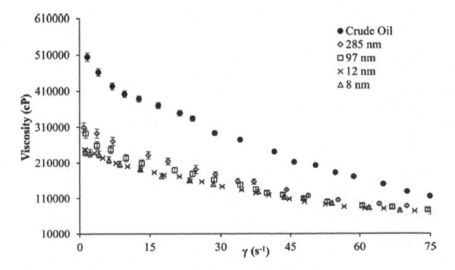

FIGURE 20.11 Changes of crude oil viscosity in the presence and absence of SiO_2 NPs of different sizes at 298 K and 1000 mg/L. (From Taborda, E.A. et al., *Energy & Fuels*, 31, 1329–1338, 2017.)

20.4 Nanofluid Flooding

Nanofluid flooding (also known as nanoflooding) is an innovative chemical EOR approach whereby nanofluids are injected into oil reservoirs to impact oil displacement or enhance injectivity [78]. A schematic of NPs injection into the oil reservoir is illustrated in Figure 20.12. Core flooding experiments are commonly employed in the oil industry to found the oil recovery at certain reservoir conditions. Data attained from the initial screening tests using viscosity, IFT, and wettability can minimise the number of core flooding tests. Ogolo et al. [66] injected nano metal oxides as an EOR agent at room temperature into a sandstone core sample. The outputs revealed a small increase in oil recovery compared to injection of distilled water only. Alternatively, when brine or ethanol was employed as dispersing agents, the recovery factor was efficiently decreased to an amount lower than that taken in the absence of NPs. Ehtesabi et al. [79–81] reported that water flooding tests with TiO_2 NPs could recover about 80% of the oil from oil-wet Berea sandstone. They explained this amount was only 49% in the absence of these NPs. The measurements performed by Ehtesabi et al. [79–81] showed that there is no significant change in the fluid viscosity or the interfacial tension of system when using TiO_2 NPs. Thus, the better recovery cannot be described by any of these two mechanisms. Contact angle measurements, however, demonstrated the ability of TiO_2 to modify the wettability of sandstone cores to water-wet. This change in wettability is due to the deposition of particles on the inner surfaces of the rock pores. The SEM image (Figure 20.13) confirmed the tendency of TiO_2 NPs to precipitate on the pore surfaces of the rock. Hence, this study identified the wettability change due to the deposition of NPs as the main mechanism for the extra oil recovery.

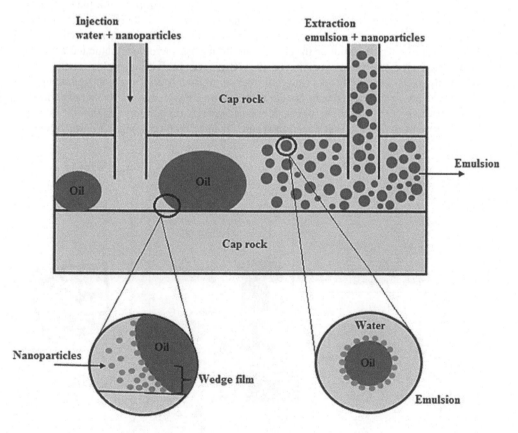

FIGURE 20.12 Schematic of NPs injection into the oil reservoir.

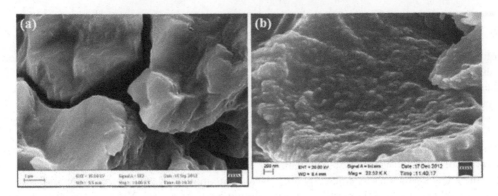

FIGURE 20.13 SEM image of core plug (a) before flooding test and (b) after flooding test. (From Ehtesabi, H et al., *Energy & Fuels*, 28, 423–430, 2013.)

In another study, Hendraningrat and Torsæter [32] employed Al_2O_3 nanofluid on sandstone cores with several wettability recovery levels and ended in the highest recovery for the intermediate wet, the lowest recovery for oil-wet, and an in-between recovery for the water-wet rocks.

Hendraningrat and Torsæter [32] documented that alumina was a good agent for EOR in a sandstone reservoir. In addition, Onyekonwu and Ogolo [82] investigated the ability of three polysilicon NPs in surfactant coreflood tests and found that these NPs yielded enhancement of surfactant performance. Moradi et al. [37] found corefloods representing that silica NPs were able to enhance water-alternating-gas (WAG) injection performance. Nguyen et. al. [83] employed silica/polymer nanocomposite for the EOR at the high temperature and brine concentration. The data displayed that the nanocomposites reduce IFT and viscosity.

Also, Latiff et al. [84] suggested to utilise ZnO together with non-invasive electromagnetic (EM). They examined the impact of ZnO nanofluids on the oil recovery from glass micromodel saturated with heavy oil. They efficiently recovered 26% of the remaining oil after 30 minutes. In addition, multi-metallic NPs can be transported via oil sands porous media into heavy oil reservoir as catalysts for heavy oil upgrading [85–89].

In addition to the above-mentioned core flooding experiments, the glass micromodel experiment is an important tool for studying microscopic oil recovery during nanofluid flooding [53,90–93]. A schematic of micromodel experimental setup is depicted in Figure 20.14. This type of glass micromodel makes it

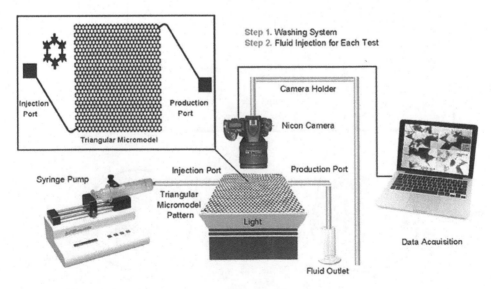

FIGURE 20.14 Schematic of micromodel experimental setup. (From Cheraghian, G. et al., *Ind. Eng. Chem. Res.*, 56, 8528–8534, 2017.)

TABLE 20.4

Experimental Results for Nanoflooding

NPs-Fluid	Oil Type	Core Type	Tem (°C)	Oil Recovery (%)	References
SiO$_2$-Brine[a]	Li oil[e]	Sa[b]	AT[d]	15	[94]
SiO$_2$-Brine[a]	Li oil	Sa	AT	12	[94]
SiO$_2$-Brinc[a]	Li oil	Sa	AT	6	[94]
SiO$_2$-Brine[a]	Li oil	Sa	AT	3	[94]
SiO$_2$-Brine[a]	Li oil	Sa	AT	3	[94]
SiO$_2$-Brine[a]	Li oil	Sa	AT	0	[94]
SiO$_2$-Brine	P oil[f]	Sa	AT	8	[95]
SiO$_2$-Water	Li oil	Sa	AT	11	[96]
Al$_2$O$_3$-Water	Li oil	Sa	AT	14	[96]
TiO$_2$-Water	Li oil	Sa	AT	5	[96]
CuO-Water	Li oil	Sa	AT	5	[96]
80%TiO$_2$:20% SiO$_2$/PVP-Brine	Li oil	Ca[c]	25	15	[27]
Nanoclay/polymer-Brine	He oil[g]	Sa	80	5.8	[97]
SiO$_2$/SDS-Water	He oil	Micromodel	AT	13	[93]
TiO$_2$/SDS-Water	He oil	Micromodel	AT	4.85	[98]
TiO$_2$-Water	Li oil	Ca	60	6.6	[59]
TiO$_2$-Water	Li oil	Ca	50	5.2	[59]
TiO$_2$-Water	Li oil	Ca	40	4.1	[59]
TiO$_2$-Water	Li oil	Ca	26	3	[59]
SiO$_2$/Polyacrylamide-Water	He oil	Micromodel	AT	10	[99]
SiO$_2$-Brine	Mo oil[h]	Micromodel	AT	22.6	[90]
NiO-Brine	Mo oil	Micromodel	AT	14.6	[90]
Fe$_3$O$_4$-Brine	Mo oil	Micromodel	AT	8.1	[90]
TiO$_2$-Brine	He oil	Sa	75	10–14	[80]
NiO-Brine	Me oil[i]	Ca	AT	7.59	[100]
Fe$_2$O$_3$-Brine	Me oil	Ca	AT	8.19	[100]
CuO-Brine	Me oil	Ca	AT	14.07	[100]
Al$_2$O$_3$-Ethanol	Me oil	Sa	AT	−0.9[j]	[66]
Al$_2$O$_3$-Water	Me oil	Sa	AT	12.5	[66]
Al$_2$O$_3$-Brine	Me oil	Sa	AT	5	[66]
Al$_2$O$_3$-Propanol	Li oil	Sa	AT	20.2	[35]
SiO$_2$-Propanol	Li oil	Sa	AT	22.5	[35]
MgO-Ethanol	Me oil	Sa	AT	−4.5[j]	[66]
MgO-Water	Me oil	Sa	AT	1.7	[66]
MgO-Brine	Me oil	Sa	AT	−2.5[j]	[66]
Fe$_2$O$_3$-Ethanol	Me oil	Sa	AT	−4.2[j]	[66]
Fe$_2$O$_3$-Water	Me oil	Sa	AT	9.2	[66]
Fe$_2$O$_3$-Brine	Me oil	Sa	AT	0	[66]
Fe$_2$O$_3$-Propanol	Li oil	Sa	AT	17.3	[35]
Ni$_2$O$_3$-Ethanol	Me oil	Sa	AT	−5[j]	[66]
Ni$_2$O$_3$-Water	Me oil	Sa	AT	2	[66]
Ni$_2$O$_3$-Brine	Me oil	Sa	AT	1.7	[66]
ZnO-Ethanol	Me oil	Sa	AT	−4.2[j]	[66]
ZnO-Water	Me oil	Sa	AT	3.3	[66]
ZnO-Brine	Me oil	Sa	AT	−4.2[j]	[66]
ZrO$_2$-Ethanol	Me oil	Sa	AT	−5[j]	[66]
ZrO$_2$-Water	Me oil	Sa	AT	4.2	[66]
ZrO$_2$-Brine	Me oil	Sa	AT	−3.3[j]	[66]

(Continued)

TABLE 20.4 (*Continued*)

Experimental Results for Nanoflooding

NPs-Fluid	Oil Type	Core Type	Tem (°C)	Oil Recovery (%)	References
SnO-Ethanol	Me oil	Sa	AT	−13.4[j]	[66]
SnO-Water	Me oil	Sa	AT	3.3	[66]
SnO-Brine	Me oil	Sa	AT	−3.3[j]	[66]
ZnO-Brine	Li oil	Sa	95	10.27	[101,102]
SiO$_2$-Brine	Me oil	Sa	50	4.5–10.3	[103]
MWCNT-Water	Li oil	glass bead	AT	12.2–18.6	[34]
Fe$_3$O$_4$@Chitosan-Brine	He oil	Ca	AT	10.8	[42]

[a] Different brine; [b] Sandstone; [c] Carbonate; [d] Ambient temperature; [e] Light Oil; [f] Paraffinic oil; [g] Heavy oil; [h] Model oil (asphaltene + toluene); [i] Medium crude oil; [j] The negative effect on oil recovery.

possible to continuously monitor the microscopic behaviour of oil and nanofluid two-phase flow and the distribution of fluid in porous media during the experiments. This is an important advantage for the detection of movement at nanofluid interfaces and for the investigation of the microscopic EOR of nanoflooding. Table 20.4 summarises a number of experimental details for the nanofluid flooding.

20.5 Conclusion

The use of NPs in enhanced oil recovery can improve upstream productivity. Various types of NPs used in chemical EOR are highlighted. The influence of NPs on viscosity, IFT, wettability, and oil recovery is outlined. This chapter presents that the conventional enhanced oil recovery techniques face many issues, some of which are properly addressed using NPs, as they have some unique properties. NPs are thus regarded as possible agents to EOR. The use of nanofluids can decrease the interfacial tension and viscosity, as well as altering the wettability of the rock to a more water-wet state.

Various NPs such as SiO$_2$, Ni$_2$O$_3$, Al$_2$O$_3$, ZnO, MgO, SnO, Fe$_3$O$_4$, Fe$_2$O$_3$, ZrO$_2$, and multi-walled carbon nanotube (MWCNT) were applied for EOR in laboratory. It was realised that above 3%–22% additional oil is recovered using nanofluids in sandstone and carbonate reservoirs. Among them, SiO$_2$ NPs had better performance. Nanotechnology is expected to be used permanently in the field applications in near future. Laboratory data suggest that NPs can be regarded as potential candidates for EOR process.

REFERENCES

1. Statista. World consumption of oil, 2018. Available from: http://www.statista.com.
2. Safari M, Mohammadi M, Sedighi M. Effect of neglecting geothermal gradient on calculated oil recovery. *Journal of Applied Geophysics* 2017;138:33–39.
3. Lee WJ, Wattenbarger RA. *Gas Reservoir Engineering*. 1996.
4. Ahmadi Y, Eshraghi SE, Bahrami P, Hasanbeygi M, Kazemzadeh Y, Vahedian A. Comprehensive Water–Alternating-Gas (WAG) injection study to evaluate the most effective method based on heavy oil recovery and asphaltene precipitation tests. *Journal of Petroleum Science and Engineering* 2015;133:123–129.
5. Sun X, Dong M, Zhang Y, Maini BB. Enhanced heavy oil recovery in thin reservoirs using foamy oil-assisted methane huff-n-puff method. *Fuel* 2015;159:962–973.
6. Chang H, Zhang Z, Wang Q, Xu Z, Guo Z, Sun H et al. Advances in polymer flooding and alkaline/surfactant/polymer processes as developed and applied in the People's Republic of China. *Journal of Petroleum Technology* 2006;58(2):84–89.
7. Vargo J, Turner J, Vergnani B, Pitts MJ, Wyatt K, Surkalo H et al. Alkaline-surfactant-polymer flooding of the Cambridge Minnelusa field. *SPE Rocky Mountain Regional Meeting*. Society of Petroleum Engineers; 1999:552–558.

8. Sofla SJD, Sharifi M, Sarapardeh AH. Toward mechanistic understanding of natural surfactant flooding in enhanced oil recovery processes:The role of salinity, surfactant concentration and rock type. *Journal of Molecular Liquids* 2016;222:632–639.

9. Krishnamoorti R. Extracting the benefits of nanotechnology for the oil industry. *Journal of Petroleum Technology* 2006;58(11):24–26.

10. Sedighi M, Mohammadi M, Sedighi M, Ghasemi M. Biobased cadaverine as a green template in the synthesis of NiO/ZSM-5 nanocomposites for removal of petroleum asphaltenes: Financial analysis, isotherms, and kinetics study. *Energy & Fuels* 2018.

11. Alomair OA, Matar KM, Alsaeed YH. Nanofluids application for heavy oil recovery. *SPE Asia Pacific Oil & Gas Conference and Exhibition.* Society of Petroleum Engineers 2014.

12. Sedighi M, Mohammadi M, Sedighi M. Green SAPO-5 supported NiO nanoparticles as a novel adsorbent for removal of petroleum asphaltenes: Financial assessment. *Journal of Petroleum Science and Engineering* 2018;171:1433–1442.

13. Liu Q, Dong M, Zhou W, Ayub M, Zhang Y, Huang S. Improved oil recovery by adsorption–desorption in chemical flooding. *Journal of Petroleum Science and Engineering* 2004;43(1–2):75–86.

14. Ali S, Thomas S. The promise and problems of enhanced oil recovery methods. *Journal of Canadian Petroleum Technology* 1996;35(7).

15. Alvarado V, Manrique E. *Enhanced Oil Recovery: Field Planning and Development Strategies.* Burlington, MA: Gulf Professional Publishing 2010.

16. Hemmati-Sarapardeh A, Ayatollahi S, Ghazanfari M-H, Masihi M. Experimental determination of interfacial tension and miscibility of the CO_2–crude oil system: Temperature, pressure, and composition effects. *Journal of Chemical & Engineering Data* 2013;59(1):61–69.

17. Moeini F, Hemmati-Sarapardeh A, Ghazanfari M-H, Masihi M, Ayatollahi S. Toward mechanistic understanding of heavy crude oil/brine interfacial tension: The roles of salinity, temperature and pressure. *Fluid Phase Equilibria* 2014;375:191–200.

18. Thomas S. Enhanced oil recovery: An overview. *Oil & Gas Science and Technology-Revue de l'IFP* 2008;63(1):9–19.

19. Sheng J. *Modern Chemical Enhanced Oil Recovery: Theory and Practice.* Burlington, MA: Gulf Professional Publishing 2010.

20. Glaser N, Adams DJ, Böker A, Krausch G. Janus particles at liquid–liquid interfaces. *Langmuir* 2006;22(12):5227–5229.

21. Xiang W, Zhao S, Song X, Fang S, Wang F, Zhong C et al. Amphiphilic nanosheet self-assembly at the water/oil interface: Computer simulations. *Physical Chemistry Chemical Physics* 2017;19(11):7576–7586.

22. Cheng Z, Mejia AF, Diaz A, Clearfield A, Mannan MS, Chang Y-W. Amphiphilic nanosheets and methods of making the same. U.S. Patent 9,586,983, issued March 7, 2017.

23. Suleimanov BA, Ismailov F, Veliyev E. Nanofluid for enhanced oil recovery. *Journal of Petroleum Science and Engineering* 2011;78(2):431–437.

24. Li S. An Experimental Investigation Of Enhanced Oil Recovery Mechanisms In Nanofluid Injection Process. PhD NTNU; 2016.

25. Youssif MI, El-Maghraby RM, Saleh SM, Elgibaly A. Silica nanofluid flooding for enhanced oil recovery in sandstone rocks. *Egyptian Journal of Petroleum* 2018;27(1):105–110.

26. Ramezanpour M, Siavashi M. Application of SiO_2–water nanofluid to enhance oil recovery. *Journal of Thermal Analysis and Calorimetry* 2018:1–16.

27. Mohammadi M, Dadvar M, Dabir B. Application of response surface methodology for optimization of the stability of asphaltene particles in crude oil by TiO_2/SiO_2 nanofluids under static and dynamic conditions. *Journal of Dispersion Science and Technology* 2018;39(3):431–442.

28. Ma H, Luo M, Dai LL. Influences of surfactant and nanoparticle assembly on effective interfacial tensions. *Physical Chemistry Chemical Physics* 2008;10(16):2207–2213.

29. Lan Q, Yang F, Zhang S, Liu S, Xu J, Sun D. Synergistic effect of silica nanoparticle and cetyltrimethyl ammonium bromide on the stabilization of O/W emulsions. *Colloids and Surfaces A: Physicochemical and Engineering Aspects* 2007;302(1–3):126–135.

30. Esmaeilzadeh P, Hosseinpour N, Bahramian A, Fakhroueian Z, Arya S. Effect of ZrO_2 nanoparticles on the interfacial behavior of surfactant solutions at air–water and n-heptane–water interfaces. *Fluid Phase Equilibria* 2014;361:289–295.

31. Moghadam TF, Azizian S. Effect of ZnO nanoparticles on the interfacial behavior of anionic surfactant at liquid/liquid interfaces. *Colloids and Surfaces A: Physicochemical and Engineering Aspects* 2014;457:333–339.
32. Hendraningrat L, Torsæter O. Metal oxide-based nanoparticles: Revealing their potential to enhance oil recovery in different wettability systems. *Applied Nanoscience* 2015;5(2):181–199.
33. AfzaliTabar M, Alaei M, Khojasteh RR, Motiee F, Rashidi A. Preference of multi-walled carbon nanotube (MWCNT) to single-walled carbon nanotube (SWCNT) and activated carbon for preparing silica nanohybrid pickering emulsion for chemical enhanced oil recovery (C-EOR). *Journal of Solid State Chemistry* 2017;245:164–173.
34. Soleimani H, Baig MK, Yahya N, Khodapanah L, Sabet M, Demiral BM et al. Impact of carbon nanotubes based nanofluid on oil recovery efficiency using core flooding. *Results in Physics* 2018;9:39–48.
35. Joonaki E, Ghanaatian S. The application of nanofluids for enhanced oil recovery: Effects on interfacial tension and coreflooding process. *Petroleum Science and Technology* 2014;32(21):2599–2607.
36. Roustaei A, Saffarzadeh S, Mohammadi M. An evaluation of modified silica nanoparticles' efficiency in enhancing oil recovery of light and intermediate oil reservoirs. *Egyptian Journal of Petroleum* 2013;22(3):427–433.
37. Moradi B, Pourafshary P, Jalali F, Mohammadi M, Emadi M. Experimental study of water-based nanofluid alternating gas injection as a novel enhanced oil-recovery method in oil-wet carbonate reservoirs. *Journal of Natural Gas Science and Engineering* 2015;27:64–73.
38. Hendraningrat L, Torsæter O. Effects of the initial rock wettability on silica-based nanofluid-enhanced oil recovery processes at reservoir temperatures. *Energy & Fuels* 2014;28(10):6228–6241.
39. Sun Q, Li Z, Li S, Jiang L, Wang J, Wang P. Utilization of surfactant-stabilized foam for enhanced oil recovery by adding nanoparticles. *Energy & Fuels* 2014;28(4):2384–2394.
40. Emadi S, Shadizadeh SR, Manshad AK, Rahimi AM, Mohammadi AH. Effect of nano silica particles on Interfacial Tension (IFT) and mobility control of natural surfactant (Cedr Extraction) solution in enhanced oil recovery process by nano-surfactant flooding. *Journal of Molecular Liquids* 2017;248:163–167.
41. Mohajeri M, Hemmati M, Shekarabi AS. An experimental study on using a nanosurfactant in an EOR process of heavy oil in a fractured micromodel. *Journal of Petroleum Science and Engineering* 2015;126:162–173.
42. Rezvani H, Riazi M, Tabaei M, Kazemzadeh Y, Sharifi M. Experimental investigation of interfacial properties in the EOR mechanisms by the novel synthesized Fe_3O_4@ Chitosan nanocomposites. *Colloids and Surfaces A: Physicochemical and Engineering Aspects* 2018;544:15–27.
43. Bera A, Ojha K, Kumar T, Mandal A. Mechanistic study of wettability alteration of quartz surface induced by nonionic surfactants and interaction between crude oil and quartz in the presence of sodium chloride salt. *Energy & Fuels* 2012;26(6):3634–3643.
44. Bera A, Mandal A, Kumar T. The effect of rock-crude oil-fluid interactions on wettability alteration of oil-wet sandstone in the presence of surfactants. *Petroleum Science and Technology* 2015;33(5):542–549.
45. Giraldo J, Benjumea P, Lopera S, Cortés FB, Ruiz MA. Wettability alteration of sandstone cores by alumina-based nanofluids. *Energy & Fuels* 2013;27(7):3659–3665.
46. Anderson WG. Wettability literature survey-part 6: The effects of wettability on waterflooding. *Journal of Petroleum Technology* 1987;39(12):1,605-1,22.
47. Amott E. Observations relating to the wettability of porous rock. SPE-1167-G 1959.
48. Anderson WG. Wettability literature survey-part 3: The effects of wettability on the electrical properties of porous media. *Journal of Petroleum Technology* 1986;38(12):1371–1378.
49. Teklu TW, Alameri W, Kazemi H, Graves RM. Contact angle measurements on conventional and unconventional reservoir cores. *Unconventional Resources Technology Conference*, San Antonio, TX, July 20–22, 2015. Society of Exploration Geophysicists, American Association of Petroleum Geologists, Society of Petroleum Engineers; 2015:2297–2311.
50. Ju B, Fan T. Experimental study and mathematical model of nanoparticle transport in porous media. *Powder Technology* 2009;192(2):195–202.
51. Karimi A, Fakhroueian Z, Bahramian A, Pour Khiabani N, Darabad JB, Azin R et al. Wettability alteration in carbonates using zirconium oxide nanofluids: EOR implications. *Energy & Fuels* 2012;26(2):1028–1036.

52. Karimi M, Mahmoodi M, Niazi A, Al-Wahaibi Y, Ayatollahi S. Investigating wettability alteration during MEOR process, a micro/macro scale analysis. *Colloids and Surfaces B: Biointerfaces* 2012;95:129–136.
53. Maghzi A, Mohammadi S, Ghazanfari MH, Kharrat R, Masihi M. Monitoring wettability alteration by silica nanoparticles during water flooding to heavy oils in five-spot systems: A pore-level investigation. *Experimental Thermal and Fluid Science* 2012;40:168–176.
54. Hendraningrat L, Li S, Torsæter O. A coreflood investigation of nanofluid enhanced oil recovery. *Journal of Petroleum Science and Engineering* 2013;111:128–138.
55. Hendraningrat L, Li S, Torsater O. A coreflood investigation of nanofluid enhanced oil recovery in low-medium permeability Berea sandstone. *SPE International Symposium on Oilfield Chemistry*. Society of Petroleum Engineers 2013.
56. Hendraningrat L, Torsæter O. Understanding fluid-fluid and fluid-rock interactions in the presence of hydrophilic nanoparticles at various conditions. *SPE Asia Pacific Oil & Gas Conference and Exhibition*. Society of Petroleum Engineers 2014.
57. Roustaei A, Bagherzadeh H. Experimental investigation of SiO_2 nanoparticles on enhanced oil recovery of carbonate reservoirs. *Journal of Petroleum Exploration and Production Technology* 2015;5(1):27–33.
58. Al-Anssari S, Barifcani A, Wang S, Maxim L, Iglauer S. Wettability alteration of oil-wet carbonate by silica nanofluid. *Journal of Colloid and Interface Science* 2016;461:435–442.
59. Esfandyari Bayat A, Junin R, Samsuri A, Piroozian A, Hokmabadi M. Impact of metal oxide nanoparticles on enhanced oil recovery from limestone media at several temperatures. *Energy & Fuels* 2014;28(10):6255–6266.
60. Nazari Moghaddam R, Bahramian A, Fakhroueian Z, Karimi A, Arya S. Comparative study of using nanoparticles for enhanced oil recovery: Wettability alteration of carbonate rocks. *Energy & Fuels* 2015;29(4):2111–2119.
61. Al-Anssari S, Arif M, Wang S, Barifcani A, Lebedev M, Iglauer S. Wettability of nanofluid-modified oil-wet calcite at reservoir conditions. *Fuel* 2018;211:405–414.
62. Mohammadi M, Dadvar M, Dabir B. TiO_2/SiO_2 nanofluids as novel inhibitors for the stability of asphaltene particles in crude oil: Mechanistic understanding, screening, modeling, and optimization. *Journal of Molecular Liquids* 2017;238:326–340.
63. Hart A. A review of technologies for transporting heavy crude oil and bitumen via pipelines. *Journal of Petroleum Exploration and Production Technology* 2014;4(3):327–336.
64. Li W, Zhu J-H, Qi J-H. Application of nano-nickel catalyst in the viscosity reduction of Liaohe extra-heavy oil by aqua-thermolysis. *Journal of Fuel Chemistry and Technology* 2007;35(2):176–180.
65. Chen Y, Wang Y, Lu J, Wu C. The viscosity reduction of nano-keggin-K3PMo12O40 in catalytic aqua-thermolysis of heavy oil. *Fuel* 2009;88(8):1426–1434.
66. Ogolo N, Olafuyi O, Onyekonwu M. Enhanced oil recovery using nanoparticles. *SPE Saudi Arabia Section Technical Symposium and Exhibition*. Society of Petroleum Engineers 2012.
67. Shokrlu YH, Babadagli T. Viscosity reduction of heavy oil/bitumen using micro-and nano-metal particles during aqueous and non-aqueous thermal applications. *Journal of Petroleum Science and Engineering* 2014;119:210–220.
68. Aristizábal-Fontal JE, Cortés FB, Franco CA. Viscosity reduction of extra heavy crude oil by magnetite nanoparticle-based ferrofluids. *Adsorption Science & Technology* 2018;36(1–2):23–45.
69. Shah RD. Application of nanoparticle saturated injectant gases for EOR of heavy oils. *SPE Annual Technical Conference and Exhibition*. Society of Petroleum Engineers 2009.
70. Molnes SN, Torrijos IP, Strand S, Paso KG, Syverud K. Sandstone injectivity and salt stability of cellulose nanocrystals (CNC) dispersions—Premises for use of CNC in enhanced oil recovery. *Industrial Crops and Products* 2016;93:152–160.
71. Hemmati-Sarapardeh A, Varamesh A, Husein MM, Karan K. On the evaluation of the viscosity of nanofluid systems: Modeling and data assessment. *Renewable and Sustainable Energy Reviews* 2018;81:313–329.
72. Salem Ragab AM, Hannora AE. A comparative investigation of nano particle effects for improved oil recovery–Experimental work. *SPE Kuwait Oil and Gas Show and Conference*. Society of Petroleum Engineers 2015.

73. Tarek M, El-Banbi AH. Comprehensive investigation of effects of nano-fluid mixtures to enhance oil recovery. *SPE North Africa Technical Conference and Exhibition*. Society of Petroleum Engineers 2015.

74. Afzal S, Nikookar M, Ehsani MR, Roayaei E. An experimental investigation of the catalytic effect of Fe_2O_3 nanoparticle on steam injection process of an Iranian reservoir. *Iranian Journal of Oil & Gas Science and Technology* 2014;3(2):27–36.

75. Afzal S, Ehsani MR, Nikookar M, Khodabandeh-Shahraki A, Roayaei E, Mohammadi AH. Reduction of heavy oil viscosity using nanoparticles in enhanced oil recovery process. In: Knight, A. (Ed.), *Enhanced Oil Recovery: Methods, Economic Benefits and Impacts on the Environment*. Hauppauge, NY: Nova Science Publishers; 2015.

76. Taborda EA, Franco CA, Lopera SH, Alvarado V, Cortés FB. Effect of nanoparticles/nanofluids on the rheology of heavy crude oil and its mobility on porous media at reservoir conditions. *Fuel* 2016;184:222–232.

77. Taborda EA, Franco CA, Ruiz MA, Alvarado V, Cortés FB. Experimental and theoretical study of viscosity reduction in heavy crude oils by addition of nanoparticles. *Energy & Fuels* 2017;31(2):1329–1338.

78. El-Diasty AI, Aly AM. Understanding the mechanism of nanoparticles applications in enhanced oil recovery. *SPE North Africa Technical Conference and Exhibition*. Society of Petroleum Engineers 2015.

79. Ehtesabi H, Ahadian M, Taghikhani V. Investigation of diffusion and deposition of TiO_2 nanoparticles in sandstone rocks for EOR application. *76th EAGE Conference and Exhibition* 2014.

80. Ehtesabi H, Ahadian MM, Taghikhani V. Enhanced heavy oil recovery using TiO_2 nanoparticles: Investigation of deposition during transport in core plug. *Energy & Fuels* 2014;29(1):1–8.

81. Ehtesabi H, Ahadian MM, Taghikhani V, Ghazanfari MH. Enhanced heavy oil recovery in sandstone cores using TiO_2 nanofluids. *Energy & Fuels* 2013;28(1):423–430.

82. Onyekonwu MO, Ogolo NA. Investigating the use of nanoparticles in enhancing oil recovery. *Nigeria Annual International Conference and Exhibition*. Society of Petroleum Engineers 2010.

83. Nguyen P-T, Do B-PH, Pham D-K, Nguyen Q-T, Dao D-QP, Nguyen H-A. Evaluation on the EOR potential capacity of the synthesized composite silica-core/polymer-shell nanoparticles blended with surfactant systems for the HPHT offshore reservoir conditions. *SPE International Oilfield Nanotechnology Conference and Exhibition*. Society of Petroleum Engineers 2012.

84. Latiff NRA, Yahya N, Zaid HM, Demiral B. Novel enhanced oil recovery method using dielectric zinc oxide nanoparticles activated by electromagnetic waves. *National Postgraduate Conference (NPC)*. IEEE; 2011:1–7.

85. Molina L, Javier H. Transport of catalytic particles immersed in fluid media through cylindrical geometries under heavy oil upgrading conditions. 2009;70(12).

86. Saidur R, Leong K, Mohammad H. A review on applications and challenges of nanofluids. *Renewable and Sustainable Energy Reviews* 2011;15(3):1646–1668.

87. Shokrlu YH, Babadagli T. Transportation and interaction of nano and micro size metal particles injected to improve thermal recovery of heavy-oil. *SPE Annual Technical Conference and Exhibition*. Society of Petroleum Engineers 2011.

88. Almao PP. In situ upgrading of bitumen and heavy oils via nanocatalysis. *The Canadian Journal of Chemical Engineering* 2012;90(2):320–329.

89. Hashemi R, Nassar NN, Almao PP. Nanoparticle technology for heavy oil in-situ upgrading and recovery enhancement: Opportunities and challenges. *Applied Energy* 2014;133:374–387.

90. Kazemzadeh Y, Eshraghi SE, Kazemi K, Sourani S, Mehrabi M, Ahmadi Y. Behavior of asphaltene adsorption onto the metal oxide nanoparticle surface and its effect on heavy oil recovery. *Industrial & Engineering Chemistry Research* 2015;54(1):233–239.

91. Li S, Hendraningrat L, Torsaeter O. Improved oil recovery by hydrophilic silica nanoparticles suspension: 2 phase flow experimental studies. *IPTC 2013: International Petroleum Technology Conference*. 2013.

92. Mohebbifar M, Ghazanfari MH, Vossoughi M. Experimental investigation of nano-biomaterial applications for heavy oil recovery in shaly porous models: A pore-level study. *Journal of Energy Resources Technology* 2015;137(1):014501.

93. Cheraghian G, Kiani S, Nassar NN, Alexander S, Barron AR. Silica nanoparticle enhancement in the efficiency of surfactant flooding of heavy oil in a glass micromodel. *Industrial & Engineering Chemistry Research* 2017;56(30):8528–8534.

94. Hendraningrat L, Torsæter O. A study of water chemistry extends the benefits of using silica-based nanoparticles on enhanced oil recovery. *Applied Nanoscience* 2016;6(1):83–95.
95. Torsater O, Engeset B, Hendraningrat L, Suwarno S. Improved oil recovery by nanofluids flooding: An experimental study. *SPE Kuwait International Petroleum Conference and Exhibition.* Society of Petroleum Engineers 2012.
96. Manan M, Farad S, Piroozian A, Esmail M. Effects of nanoparticle types on carbon dioxide foam flooding in enhanced oil recovery. *Petroleum Science and Technology* 2015;33(12):1286–1294.
97. Cheraghian G. Thermal resistance and application of nanoclay on polymer flooding in heavy oil recovery. *Petroleum Science and Technology* 2015;33(17–18):1580–1586.
98. Cheraghian G. Effects of titanium dioxide nanoparticles on the efficiency of surfactant flooding of heavy oil in a glass micromodel. *Petroleum Science and Technology* 2016;34(3):260–267.
99. Maghzi A, Mohebbi A, Kharrat R, Ghazanfari M. An experimental investigation of silica nanoparticles effect on the rheological behavior of polyacrylamide solution to enhance heavy oil recovery. *Petroleum Science and Technology* 2013;31(5):500–508.
100. Haroun MR, Alhassan S, Ansari AA, Al Kindy NAM, Abou Sayed N, Kareem A et al. Smart nano-EOR process for Abu Dhabi carbonate reservoirs. *Abu Dhabi International Petroleum Conference and Exhibition.* Society of Petroleum Engineers 2012.
101. Adil M, Lee K, Zaid HM, Latiff NRA, Alnarabiji MS. Experimental study on electromagnetic-assisted ZnO nanofluid flooding for enhanced oil recovery (EOR). *PLoS One* 2018;13(2):e0193518.
102. Alnarabiji MS, Yahya N, Nadeem S, Adil M, Baig MK, Ghanem OB et al. Nanofluid enhanced oil recovery using induced ZnO nanocrystals by electromagnetic energy: Viscosity increment. *Fuel* 2018;233:632–643.
103. Lu T, Li Z, Zhou Y, Zhang C. Enhanced oil recovery of low-permeability cores by SiO_2 nanofluid. *Energy & Fuels* 2017;31(5):5612–5621.

21

Application of Nanoparticle Suspension in Enhanced Oil Recovery

Dengwei Jing Jing, L. Sun, and M. Hatami

CONTENTS

21.1 Introductions

With the substantial decline of available oil resources all over the world, more and more attention is being paid to enhanced oil recovery (EOR) techniques [1]. However, in order to promote the application of these EOR technologies, many faced challenges must be emphasized. For instance, the sweep efficiency is still relatively low, the costs are high, and there are also some formation damages. Comparing with the traditional EOR methods, the applications of nanoparticles (NPs) are regarded by many literatures as potential alternative to tackle the challenges mentioned above. Now based on a lot of papers published recently, this chapter tries to provide an overview of the research of nanoparticles which are used often in the form of nanofluids concerning their application in EOR.

In nanofluid, not only water, but also oil and other liquid can be base fluid. In some cases even gas is a kind of base fluid [2]. In oil and gas industry, nanofluids composed by a single kind or various kinds of nanoparticles and base fluid (water or brine) are generally applied based on the practical needs of oil recovery enhancement. Obviously, it is necessary to understand the mechanisms of NPs in EOR process. Many studies have focused on the interactions of nanofluids with oil and rock and tried to suggest reasonable hypothesis. So far, the following mechanisms of nano-assisted EOR or key factors that affect the oil recovery process have been proposed. These mechanisms or key factors mainly include disjoining pressure, pore channels plugging, viscosity of injection fluids, interfacial tension, wettability of rock, and asphaltene precipitation. For the disjoining pressure, after the addition of nanoparticles, a film with wedge shape, also called "wedge film", is formed at the interface of rock surface and oil droplets. It has the function of separating the oil droplets with the reservoir, therefore, the efficiency of oil recovery can be improved compared with the traditional methods [3–4].

For the pore channel plugging, two mechanisms, i.e., mechanical entrapment and log-jamming can result in the plugging of pore channel [5]. If the diameter of nanoparticles is larger than that of pore channels, the mechanical entrapment may occur. But in fact, pore channels are in general in micrometer scale. It means that its diameter is thousand times larger than nanoparticles. Thus, it is not very hard for nanoparticles to pass through pore channels. The mechanical entrapment only plays a part in certain situations in the presence of some metal nanoparticles whose size is large [6–7]. Log-jamming mechanism works when a nanofluid flows from pores to throats where the change of the channel diameter and the resulting pressure difference can lead to the acceleration of the nanofluid's flow. The liquids molecules will flow faster than the nanoparticles due to their smaller size. Thus, the larger and slower nanoparticles will accumulate at the entrance of the throat of the pore leading to channel plugging, but it is not always disadvantageous. In some cases, the performance of nanofluid in oil recovery may be improved because of log-jamming. Mobility ratio of injected fluids is other important parameter in flooding processes. The mobility ratio can be expressed by the following equation [8]:

$$M = \frac{\lambda_i}{\lambda_o} = \frac{kr_i/\mu_i}{kr_o/\mu_o} = \frac{kr_i\mu_o}{kr_o\mu_i}. \tag{21.1}$$

In the equation above, the subscripts i and o refer to injected fluids and oil respectively, λ_i and λ_o the mobility of injected fluids and oil separately, kr_i and kr_o the relative permeability, and μ the effective viscosity. As indicated by Eq. 21.1, in the displacing processes, the mobility ratio can reach a high level because the viscosity of oil is larger than that of the injected fluids, such as water, CO_2, or other traditional chemicals. As a result, a high mobility ratio can result in viscous fingering of injected fluids within oil and low sweep efficiency. The mobility ratio can be decreased by either reducing viscosity of oil phase or increasing viscosity of injected fluids. However, these problems can be solved by addition of nanofluids because the effective viscosity of injected fluids could be increased with the addition of nanoparticles.

Rock and core can be seen as a kind of porous media and the interfacial tension (IFT) plays a key role in determining distribution and movement of fluids in it. Some mechanisms for nanofluid in EOR simply assumed that certain types of nanoparticles have potential to reduce IFT. The rock wettability is defined as the tendency of a fluid to adhere to the rock surface in competition with another immiscible fluid [9]. It is a key factor to control oil recovery because it can affect pressure of capillary, the saturation process of fluids, and the relative permeability of oil. Recent years, it has been confirmed by many researchers that nanofluids are promising to change the wettability of rock. At present, researchers use three main experimental methods to measure wettability, they are the contact angle method, the Amott test, and the core displacement test [10–12]. Among them, the most commonly used approach is the contact angle method. More details about the mechanisms mentioned above can be found in the following sections.

21.2 Application of Nanofluids in Liquid Displacement

21.2.1 Nanofluids for Liquid Displacement

The SiO_2 nanoparticles, which are the main components of sandstone, are most commonly used in the study. One of the reasons is their relatively low price and their easy access, another important reason is that their surface can be modified so that its chemical behavior can be controlled. The most common two factors that affect the performance of SiO_2 nanoparticles in EOR are their concentration and their size. Ragab [13] studied the effect of SiO_2 at different concentrations on medium crude oil recovery and found that at certain concentrations, SiO_2 nanoparticles can enhance oil recovery by 8.74%–13.88%. In Hendraningrat's work [14], SiO_2 particles with the average size of 21–40 nm were studied in flooding the core plugs with the permeability range from 9 to 400 mD, and it was found that at certain concentrations, the SiO_2 nanoparticles show an interesting potential for EOR in water-wet sandstone. They also studied the capability of nanofluids in the flooding of oil-wet cores. Their results show that the addition of nanoparticles in water can also have impact on the oil-wet core recovery in the process of flooding.

Besides the concentration and size, the type of base fluids, injection time, rock permeability, and temperature also have effect on nanofluids core flooding. In another work, Hendraningrat et al. studied these factors and found that a higher temperature and low injection rate are beneficial for oil recovery. Furthermore, they also found that when the injection time is secondary mode, the oil recovery efficiency is higher than that at tertiary mode. However, the rock permeability doesn't have an obvious effect on oil recovery, suggesting that nanofluids can enhance oil recovery at a large range of rock permeability [15].

Three possible mechanisms have been proposed to explain how SiO_2 nanoparticles can enhance oil recovery [3,4,17,18]. One is disjoining pressure mechanism. This pressure originated from both Brownian motion and the electrostatic repulsion between molecules [16]. As is shown in Figure 21.1 [17], when the SiO_2 nanoparticles are in a dispersing state, they have a tendency to rearrange in the wedge-shaped film in contact with oil phase, which act as discontinuous phase. The wedge-shaped film detaches the oil phase from the surface of rock, therefore, more oil can be recovered. Another is log-jamming mechanism. When the SiO_2 nanofluids passed the pore throat, their velocity will increase because the size of pore throats is much smaller than the pore body, and the pressure in this area is different than the others. Consequently, SiO_2 nanoparticles may fall behind and accumulate in the channel because their velocity is smaller than that of base fluids. Finally, they cause the pore channel to be blocked and the base fluids are forced to change their flow path. They may flow into those pores not invaded before and possibly containing oils [18]. The last mechanism to explain the effect of nanofluids in EOR is the wettability alteration mechanism. The addition of SiO_2, owing to the various designed surface properties, can change the wettability of rock, and the interfacial tension and the contact angle are supposed to be reduced [3,4]. Besides SiO_2 nanoparticles, some metal oxides, such as TiO_2, Al_2O_3, Fe_3O_4, ZnO, MgO, ZrO_2, and CeO_2 were also proved to be effective in enhancing oil recovery. For example, Ehtesabi et al. [19] measured the contact angle on the rock surface before and after the treatment of TiO_2 nanoparticles. It turned out that after the treatment of TiO_2 nanoparticles, the surface of rock changed from oil-wet to water-wet. However, when the concentration of TiO_2 surpass 0.01%, too much nanoparticles may cause block and plugging. Bayat et al. [20] also investigated the contact angle and interface tension at 26°C, 40°C, 50°C, and 60°C in the presence of different concentration of Al_2O_3, TiO_2, and SiO_2. They found that compared to SiO_2, Al_2O_3 and TiO_2 nanofluids demonstrated better performance in EOR regardless of temperature. Moreover, in their experiment, after flooded by Al_2O_3 and TiO_2 nanofluids at 50°C and 60°C, an obvious reduction in oil viscosity was observed. Hoseinet al. [21] conducted various experiments on carbonate

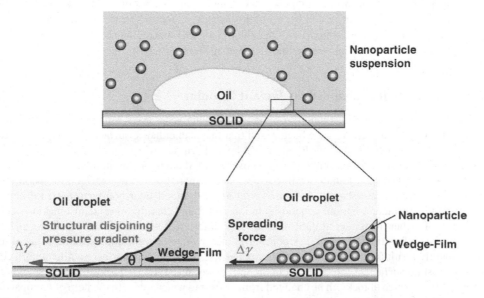

FIGURE 21.1 Nanoparticle structuring in the wedge-film resulting in structural disjoining pressure gradient at the wedge vertex drives the nanofluid spreading. (From Wasan, D. et al., *Curr. Opin. Colloid Interface Sci.*, 16(4), 344–349, 2011.)

sandpack flooding. They synthesized different concentrations of Fe_3O_4@Chitosan nanoparticles and measured contact angle, zeta potential, absorbance spectroscopy, viscosity, and interfacial tension. They are the first to study the possibility of Fe_3O_4@Chitosan in EOR and confirm that this is a good option for EOR because it can enhance the modification of the core-flooding system. Nwidee et al. [22] investigated the influence of ZrO_2 and NiO nanoparticles on the wetting preference of limestone formations whose wettability was assessed through scanning electron microscope (SEM), atomic force microscope (ATM) and contact angle. They tested the potentials of the nanoparticles to alter oil-wet calcite substrates water-wet at low nanoparticle concentrations (0.004–0.05 wt%). Moghaddam et al. [23] studied the impact of a series of nanofluids in flooding process on the surface of bonate rocks, including ZrO_2, $CaCO_3$, TiO_2, SiO_2, MgO, Al_2O_3, and CeO_2. Both their experiments and theoretical calculations prove that disjoining pressure is responsible for the enhanced oil recovery.

21.2.2 Nanoparticles Enhanced Polymer Flooding

In fact, many different materials have the potential in oil recovery enhancement. Among these materials, polymers have the ability to improve the mobility ratio because they can increase the viscosity of inject fluids while reduce the permeability of water [24]. Maghzi et al. [25] pointed out that nanoparticles have the potential to improve the performance of polymer flooding in the presence of salts. They saturated glass micromodel with heavy oil and then conducted polymer flooding experiment. To test the effect of silica nanoparticles, silica nanoparticles are dispersed in the polyacrylamide solutions with different salinities and then injected into the glass micromodel saturated with heavy oil. Viscosity measurement, continuously captured images during the displacement, and the microscopic monitoring were employed to analyze the results. The results confirmed the positive effect of polymer in viscosity increase during EOR, and the addition of nanoparticles in polymer can noticeably increase the viscosity.

The interfacial and rheological study for nanoparticle enhanced polymer flooding was conducted by Maurya et al. [26]. By characterizing surface activity and measuring rheological properties, they studied the usage of nanofluids in oil recovery from high temperature reservoir. Besides, they simulated the transport properties with STARS simulator CMG in order to figure out whether it is suitable for polymer flooding operations. Their results showed that the presence of polymer chains can help to improve the surface properties of silica nanoparticles and to reduce the interfacial tension.

One of the proper mechanisms of nanoparticles enhanced polymer flooding is that during the flooding process, the polymer may be degraded by the high shear rates, and this cause the decrease of viscosity [27]. However, the addition of nanoparticles has the function to control the rheological properties. That may be a reason why nanoparticles can enhance polymer flooding.

21.2.3 Nanoparticles Enhanced Surfactant Flooding

Surfactant has been studied and applied in flooding for many years and a lot of papers have been published in this area. Nanoparticles have been found that they can enhance surface flooding. The effect of nanoparticles in enhanced surfactant flooding has been discussed widely, and the following are some explanation of how surfactant can enhance oil recovery with the help of nanofluids.

Firstly, Munshi et al. [28] found that in the presence of nanofluids, the interfacial tension between water and oil can decrease sharply. Suleimanov et al. [29] added light non-ferrous metal nanoparticles and anionic surfactant into an aqueous solution and found that the interfacial tensions decreased by 70%–90%. Further, Ravera et al. [30] measured the interfacial tensions of the liquid/air and liquid/oil interfaces as a function of nanofluid's concentrations. At low concentrations, the interfacial tension decrease mainly result from the adsorption of nanoparticles on the liquid surface, while at high concentrations, almost no surfactant can be found in the liquids because they are removed by these nanofluids.

Secondly, the surface of rock can be changed from oil-wet to water-wet with the help of nanoparticles + surfactant. And this may be another explanation for the effect of nanoparticles in surfactant enhanced oil recovery [10,31].

Besides, the addition of nanoparticles in surfactant can also affect the rheological properties of the aqueous system [28]. Suleimanov et al. [29] found that after the addition of nanoparticles, the fluids change from Newtonian fluids into non-Newtonian and thus the viscosity increased by 2 times.

21.2.4 Nanoparticles Enhanced Emulsion Flooding

In recent years, nanotechnology has also been applied in emulsion flooding. Pei et al. [32] conducted an experiment to study the effect of nanoparticles in emulsion flooding. And the result shows the emulsion stability was improved, and the apparent viscosity of the emulsion increased after the treatment of nanoparticles and surfactant. Thus, they confirmed that the emulsion flooding stabilized with nanofluids and surfactant has great potential in heavy oil recovery after the rock was flooded by water. Their 2D homogeneous sandpack flooding test further indicates that after the emulsion flooding was stabilized by the nanoparticle-surfactant, both the sweep efficiency improved because the high permeable water channels were blocked and the displacement efficiency was improved because the trapped oil was mobilized.

21.3 Application of Nanofluids in Gas Displacement

Gas displacement has been widely applied in industry since 1970s [33]. However, there are many problems faced by gas displacement [34]. One of the most severe problems is the lack of acceptable sweep efficiency because the viscosity is obviously different between gas phase and oil phase. In order to solve the problems faced by gas injection above, surfactants are introduced in the flooding process because they are believed to have the effect of generating foams and reducing the velocity of gas phase. In some fields, the addition of surfactant has proved to be effective in increasing the sweep efficiency, but in other reservoirs with harsh environmental conditions, the performances of surfactant are far from satisfaction. In fact, high temperature, high salinity, and high adsorption of rock surface—all these can disturb the effect of surfactant. In such situation, nanofluids are introduced into the gas flooding process as a replacement of surfactant because they have the ability to produce highly stable gas-in-water foams. Even in harsh conditions, they can stay stable because nanoparticles themselves are chemically stable and on mineral surface, they have low retention.

Bulk- and bubble-scale stability experiments, macroscopic experiments in porous media (consolidated and unconsolidated), and visualization experiments in pore-scale are the three traditional methods to investigate the mobility control of foams and their application in EOR [35]. Similarly to the case in nanoparticle-assisted liquid flooding, SiO_2 is the most commonly used nanoparticle in such gas flooding processes. Yekeen et al. [36] studied the influence of SiO_2 and Al_2O_3 nanoparticles + surfactant on the foam stability and the kaolinite adsorptions under both static and dynamic conditions. They employed surface tension and two-phase titration methods to measure the adsorption and etched glasses micromodels which can make the pore scale experiments visualization to conduct foam stability experiment. On trough analyzing adsorption data with Langmuir, Freundlich, and Temkin adsorption foam analyzer and conducting bulk stability experiments with KRUSS dynamic foam analyzer, they reported a 75% decrease of surfactant adsorption on kaolinite when SiO_2 was added and a 38% decrease when Al_2O_3 was added. They also found that in the porous media, both SiO_2-surfactant and Al_2O_3-surfactant foams propagated well in the presence of oil. The presence of nanoparticles enhanced the interfacial elasticity of films and improved the microscopic displacement efficiency to almost 100%. Their investigation confirmed nanoparticles can improve foam stability and minimized surfactant adsorption.

Application of dimethyldichlorosilane (DMDCS) to modify the surface of SiO_2 nanoparticles and its effect on improving foam stability have also attracted some attentions. Sonn et al. [37] measured the decaying foam volume with time by a Foamscan. They found that the increase of DMDCS results in more hydrophobic SiO_2 nanoparticles through contact angle measurement, and the foam stability improved obviously in 1 wt% of unmodified SiO_2 nanoparticles in various concentrations of surfactant solutions. Besides contact angle and surface tension, the standard deviation of drop shapes is also measured. Vatanparast et al. [38] regarded the surfaces pressure as a function of drop surface area and investigated the adsorption process and surface structure. They not only found standard deviation (STD) trend is a useful tool for analyzing nanoparticles' adsorption and structure of surface, but also the most suitable mechanism to explain the improvement of surfactant adsorption at air/water surface: the repulsive electrostatic interactions between nanoparticles and surfactants.

The factors which are concerned to be related with the effect of nanoparticles in the EOR by gas flooding include temperature, pressure, the surface wettability of nanoparticles, the concentration of nanoparticles, the type of nanoparticles, and whether there is oil on the foam. Many researchers studied the effect of nanoparticles and surfactant on emulsification progress of gas-water under various concentrations [39–41]. Their results showed that the stability and viscosity of foam or emulsion increase with the concentration of nanoparticles. Alyousef et al. [42], however, reported a different trend. They confirmed that due to flocs formation, the addition of nanoparticles help the foam to be more stable, but the increase of concentration does not always lead to the increase of stability of foams. Only when the nanoparticles are at certain concentrations, they concluded, it can produce a more stable form in comparison with addition of surfactant alone. Their conclusions were also supported by the study from Yekeen et al. [43]. They also used a Hele-Shaw cell to study the effect of SiO_2 and Al_2O_3 nanoparticles and sodium dodecyl sulfate (SDS) on the stability of foam at bulk/bubble scales and under static/dynamic conditions. They found that the increase of nanoparticles concentration can result in the decrease of foam stability. When the concentration of nanoparticles reached 3 wt%, the largest amount of surfactant was adsorbed.

Hu et al. [44] studied the effects of nanoparticle size on foams stabilization. In their experiments, they choose three different sizes of SiO_2 (20, 100, and 500 nm) nanoparticles and two surfactants (hexadecyl trimethylammonium bromide [CTAB] and SDS) and tested their performance in improving foam stability. They pointed out that smaller SiO_2 nanoparticles show a more effective enhancement in stabilizing foam due to their low adhesion energy and high number concentration. SiC nanoparticles were also used as a novel material to enhance the stability of foams. Li et al. [45] conducted a series of experiments to test the effect of their new foaming apparatus and SiC nanoparticles. The result shows that the new SiC foaming system has ideal capacity to carrying fluids and to strengthen anti-condensate oil. The novel SiC foaming system has the potential to be used in deliquification of gas as well.

Mixture of positively charged surface-treated nanoparticles (S-AK) and anionic-nonionic surfactant sodium fatty alcohol polyoxyethylene ether sulfate (AES) mixture was used by Xu et al. to form foams of natural gas [46]. They observed the occurrence of synergy between S-AK and AES in the process of producing stable natural gas foams. They pointed out that this is the evidence which confirms that the mixture of the S-AK/AES can generate more stable foams than that of either single component. In the dispersion, when the monolayer of AES molecules is adsorbed on the S-AK surface, very stable natural gas foams will be obtained. The authors also tried to explain the mechanism of natural gas foams stabilization in the presence of surface-treated nanoparticles and AES as shown by Figure 21.2. In the

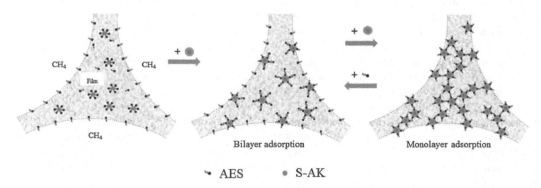

FIGURE 21.2 Stabilizing mechanism of the natural gas foams by S-AK/AES. (From Xu, L. et al. *Colloids Surf. A*, 550, 176–185, 2015.)

process of the adding of S-AK into the AES aqueous solution, two states called bilayer adsorption and monolayer adsorption appear. In the state of monolayer adsorption, AES molecules dragged a certain amount of S-AK nanoparticles into the interface of the CH_4 and aqueous phase. At the interface, the liquid loss of foam films is decreased by surface-treated nanoparticles. Meanwhile, the absorbed monolayer of surface-treated nanoparticles dispersed in the liquids phase becomes hydrophobic and gradually flocculates.

21.4 Other New Technologies Relating to Nanofluids Application in EOR

21.4.1 Mixture of Nanoparticles

Nowadays, in some experimental studies, the use of mixtures of nanofluids was investigated because it is believed that the mixture of nanoparticles can combine the advantages of various kinds of nanoparticles. Alomair et al. [47] tested the effect of nanofluid mixture on heavy oil flooding in Berea sandstone cores. The experimental results showed that when mixing nanofluids of SiO_2 and Al_2O_3 at 0.05 wt%, the best incremental oil recovery effect can be achieved compared to the other nanofluids. The enhanced oil recovery can be attributed to the mixture of nanofluids because it can prevent the occurrence of asphaltene precipitation and reduce viscosity and interfacial tension. In the following work, Tarek et al. [48] tested the effect of mixtures of nanofluids (40% Fe_2O_3 + 35% Al_2O_3 + 25% SiO_2), and the results confirmed its advantages over the application of a single nanofluid in EOR.

Tarek further studied the effect of concentrations of nanoparticle mixtures on the tertiary oil recovery on a high permeability core [49]. It was found that the effect of nanofluid mixture is related to the properties of the fluid and rock. They pointed out that it is the high packing capacity of nanofluids mixtures at the interface that might be responsible for the enhancement of oil recovery.

However, every coin has two sides. The use of nanofluids mixture isn't always better technique for EOR than single nanofluid flooding. This method also has its disadvantages. Basically, it requires a lot of time and technique in the complex process of the synthesis and characterization of nanofluid mixtures. And the flooding experiments using nanofluid mixture is more time-consuming and complicated. Besides time, the cost of nanofluid mixture flooding is often also higher than that of single nanofluid flooding.

21.4.2 Barium Acid Salt and Micro-gel Nano-spheres

The performance of barium acid salt and micro-gel nano-spheres are also studied in literature. Kanj et al. [50] conducted core-flooding experiments in the largest conventional oil field in the world: the giant Ghawar oil field. The effect of $BaSO_4$ nanofluids and $BaFe_{12}O_{19}$ nanofluids was studied comprehensively. Meanwhile, polyacrylamide micro-gel nano-spheres were synthesized and tested in a sandpack model during the process Zhuangxi heavy oil flooding by Wang et al. [51]. The experiments showed the good potential of nanospheres in the process of EOR. The combined experiment of laboratory-scale and field-scale of NPs application on the carbonate rocks from giant Ghawar oil field is conducted by Kanj et al. [52]. The nanoparticles studied are a certain kind of carbon based fluorescent NPs called A-Dots and were found to be a typical class of nanomaterial that are generally synthesized through a method of hydrothermal treatment. The recovery factor achieved using their nanoparticles was determined experimentally to be more than 96%.

21.4.3 Cellulose Nanocrystals

Cellulose nanocrystals have also been used in nano-assisted EOR experiments. Wei et al. [53] regarded nanocrystals as a novel surface-active and "green" material in EOR. They conducted rheological experiments and visual micromodel experiments and found that this kind of nanofluid is eco-friendly material and can gain promising future in the application of EOR. The injective capacity of cellulose nanocrystals (CNC) was studied in a sandstone core with high permeability by Molnes et al. [54]. Their results showed that CNC have positive effect on EOR in the water injecting process. Besides, cellulose nanocrystals were found to have the ability to stay stable after the temperature increased to as high as 140°C for 3 days, as was demonstrated by Heggset et al. [55].

21.4.4 Nanofluids to Control Fines Migration for Oil Recovery

Fine-scale particles transport in multiphase fluids saturated porous media is one of the major challenges in petroleum engineering and many other disciplines as well. Recently, the applications of nanoparticles to control fines migration have been proposed and investigated [56]. Yuan et al. presented series of analytical solutions to characterize the phenomenon of nanoparticle/fines transport with mutual interactions in porous media and confirmed the positive effects of nanoparticles treatment (both pre-flush and co-injection) to control fines migration [57]. They presented a comprehensive study to evaluate and optimize the effectiveness of nanofluids to both prevent fines migration and enhance oil recovery using different utilization approaches: nanofluids co-injection and pre-flush. Figure 21.3 shows the mutual interactions among nanoparticles, fines, and rock grains by which nanoparticles enhance fines attachment and control fines migration [58].

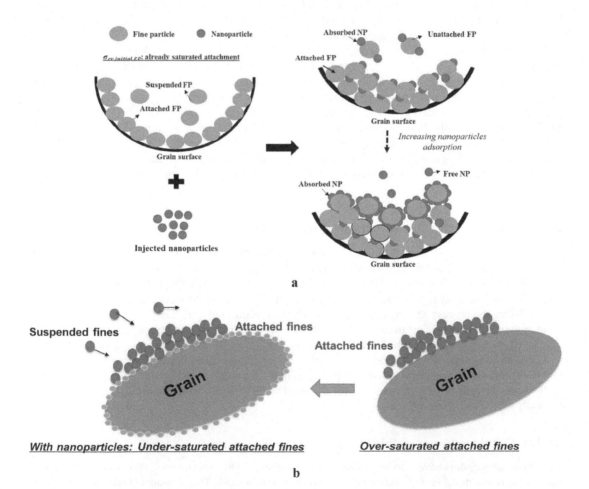

FIGURE 21.3 Mutual interactions among nanoparticles, fines, and rock grains by which nanoparticles enhance fines attachment and control fines migration. (a) Approach I: Co-injection of nanoparticles with fine suspension and (b) Approach II: Nanoparticle treated porous to fines injection. (From Yuan, B. et al., *Fuel*, 215, 474–483, 2018.)

21.5 Remarks and Perspectives

Although nanoparticles are confirmed to have the potential to be effective in many EOR processes by many research, there lacks large quantities of field scale applications and most of studies are limited to laboratory scale. If these nanoparticles are to be applied in practical oil and gas industry to enhance oil recovery, the following challenges must be overcome:

1. How to generate a stable suspension of nanoparticles has always been a severe technical problem. NPs in nanofluids always have the tendency to aggregate especially at the harsh reservoir conditions because the van der Waals interactions are strong in oil fields [57].

2. Currently, most of the attention is still paid on the process of nanofluid flooding. However, there is a lack of research on the mechanisms of the interaction between NPs, rock properties, and initial reservoir fluid during these processes, and the mechanism of other nano-assisted EOR methods as introduced above are not clearly understood.

3. More theoretical investigations and mathematical models are needed to understand the effect of nanoparticles in EOR besides experimental studies. EOR process in the presence of nanoparticles involves a series of very complex procedures. So it is not an easy job to have an accuracy calculation and comprehensive modeling.

4. Metal nanoparticles such as TiO_2, Al_2O_3, etc. have been frequently studied and their ability to enhance the effect in oil and gas flooding were proved. However, the advantages of these metal nanoparticles over silica nanoparticles still need to be clarified.

5. The application of mixture nanofluids as a novel technology to enhance oil recovery is still in its primary stage and more experiments should be carried out in order to understand its mechanism and promote its application in practical fields.

From point of view of research, more attention should be paid to the following areas: The nanofluid's stability should be improved, and the dispersion state of nanofluids should be under control. More theoretical works and mathematical models for nano-assisted EOR processes are needed to understand the fundamental mechanisms. More pilot-scale experiments and even field experiments need to be conducted. More functional nanoparticles and other advanced methods should be introduced in industries. Some novel technologies are to be developed to find an access to cheaper and environmental-friendly nanoparticles.

REFERENCES

1. Sun, X., Zhang, Y., Chen, G., and Gai, Z. (2017). Application of nanoparticles in enhanced oil recovery: A critical review of recent progress, *Energies*, 10, 345.
2. El-Diasty, A.I., and Ragab, A.M.S., Applications of nanotechnology in the oil & gas industry: Latest trends worldwide & future challenges in Egypt, *North Africa Technical Conference and Exhibition*, Society of Petroleum Engineers: Cairo, Egypt, 2013.
3. Mcelfresh, P.M., Olguin C., Ector D., The application of nanoparticle dispersions to remove paraffin and polymer filter cake damage, *Proceedings of the SPE International Symposium and Exhibition on Formation Damage Control*, Lafayette, LA, USA, February 15–17, 2012.
4. Aveyard, R., Binks, B.P., Clint, J.H. (2003). Emulsions stabilised solely by colloidal particles, *Advances in Colloid & Interface Science*, 100(2), 503–546.
5. Skauge, T., Spildo, K., Skauge, A., Nano-sized particles for EOR, *Proceedings of the SPE Improved Oil Recovery Symposium*, Tulsa, OK, April 24–28, 2010.
6. Hashemi, R., Nassar, N.N., Almao, P.P. (2013). Enhanced heavy oil recovery by in situ prepared ultra-dispersed multimetallic nanoparticles: A study of hot fluid flooding for Athabasca bitumen recovery, *Energy & Fuels*, 27(4), 2194–2201.
7. Mulgundmath, V.P., Tezel, F.H., and Saatcioglu, T. (2012). Adsorption and separation of CO_2/N_2 and CO_2/CH_4 by 13X zeolite, *Canadian Journal of Chemical Engineering*, 90(3), 730–738.

8. Lyons, W.C., and Plisga, G.J, *Standard Handbook of Petroleum and Natural Gas Engineering*, Gulf Professional Publishing: Houston, TX, 2011.
9. Si, L.V., and Bo, H.C. (2016). Chemical flooding in heavy-oil reservoirs: From technical investigation to optimization using response surface methodology, *Energies*, 9(9), 711.
10. Giraldo, J., Benjumea, P., Lopera, S. et al. (2013). Wettability alteration of sandstone cores by alumina-based nanofluids, *Energy & Fuels*, 27(7), 3659–3665.
11. Anderson, W.G. (1987). Wettability literature survey part 5: The effects of wettability on relative permeability, *Journal of Petroleum Technology*, 39, 1453–1468.
12. Anderson, W.G. (1986). Wettability literature survey part 1: Rock/oil/brine interactions and the effects of core handling on wettability, *Journal of Petroleum Technology*, 38, 1125–1144.
13. Ragab, A.M.S. (2015). A comparative investigation of nano particle effects for improved oil recovery–Experimental work, *The SPE Kuwait Oil & Gas Show and Conference*.
14. Hendraningrat, L., Li, S., and Torsæter, O. (2013). A coreflood investigation of nanofluid enhanced oil recovery, *Journal of Petroleum Science & Engineering*, 111(21), 128–138.
15. Hendraningrat, L., Li, S., and Torsaeter, O., Effect of some parameters influencing enhanced oil recovery process using silica nanoparticles: An experimental investigation. *SPE Reservoir Characterisation and Simulation Conference and Exhibition*, Abu Dhabi, September 16–18, 2013.
16. Chengara, A., Nikolov, A.D., and Wasan, D.T. et al. (2004). Spreading of nanofluids driven by the structural disjoining pressure gradient, *Journal of Colloid & Interface Science*, 280(1), 192–201.
17. Wasan, D., Nikolov, A., and Kondiparty, K (2011). The wetting and spreading of nanofluids on solids: Role of the structural disjoining pressure, *Current Opinion in Colloid & Interface Science*, 16(4), 344–349.
18. Youssif, M.I., El-Maghraby, R.M., Saleh, S.M., and Elgibaly, A. (2018). Silica nanofluid flooding for enhanced oil recovery in sandstone rocks, *Egyptian Journal of Petroleum*, 27, 105–110.
19. Ehtesabi, H., Ahadian, M.M., Taghikhani, V., and Ghazanfari, M.H. (2014). Enhanced heavy oil recovery in sandstone cores using TiO_2 nanofluids, *Energy & Fuels*, 28, 423–430.
20. Bayat, A.E., Junin, R., and Samsuri, A. et al. (2014). Impact of metal oxide nanoparticles on enhanced oil recovery from limestone media at several temperatures, *Energy & Fuels*, 28, 6255–6266.
21. Hosein, R., Masoud, R., Morteza, T., Yousef, K., and Mohammad, S. (2018). Experimental investigation of interfacial properties in the EOR mechanisms by the novel synthesized Fe_3O_4@ Chitosan nanocomposites, *Colloids & Surface A*, 544, 15–27.
22. Nwidee, L.N., Al-Anssari, S., Barifcani, A. et al. (2017). Nanoparticles influence on wetting behaviour of fractured limestone formation, *Journal of Petroleum Science & Engineering*, 149, 782–788.
23. Moghaddam, R.N., Bahramian, A., and Fakhroueian, Z. et al. (2015). Comparative study of using nanoparticles for enhanced oil Recovery: Wettability alteration of carbonate rocks, *Energy & Fuels*, 29(4), 2111–2119.
24. Shi, L., Ye, Z., and Zhang, Z. et al. (2010). Necessity and feasibility of improving the residual resistance factor of polymer flooding in heavy oil reservoirs, *Petroleum Science*, 7(2), 251–256.
25. Maghzi, A., Kharrat, R., and Mohebbi, A. et al. (2014). The impact of silica nanoparticles on the performance of polymer solution in presence of salts in polymer flooding for heavy oil recovery, *Fuel*, 123(5), 123–132.
26. Maurya, N.K., Kushwaha, P., and Mandal, A. (2017). Studies on interfacial and rheological properties of water soluble polymer grafted nanoparticle for application in enhanced oil recovery, *Journal of the Taiwan Institute of Chemical Engineers*, 70, 319–330.
27. Cheraghian, G., Nezhad, S.S.K., Kamari, M., Hemmati, M., Masihi, M., and Bazgir, S. (2015). Effect of nanoclay on improved rheology properties of polyacrylamide solutions used in enhanced oil recovery, *Journal of Petroleum Exploration and Production Technology*, 5, 189–196.
28. Munshi, A.M., Singh, V.N., Kumar, M., and Singh, J.P. (2008). Effect of nanoparticle size on sessile droplet contact angle, *Journal of Applied Physics*, 103, 084315.
29. Suleimanov, B.A., Ismailov, F.S., and Veliyev, E.F. (2011). Nanofluid for enhanced oil recovery, *Journal of Petroleum Science and Engineering*, 78, 431–437.
30. Ravera, F., Santini, E., and Loglio, G. et al. (2006). Effect of nanoparticles on the interfacial properties of liquid/liquid and liquid/air surface layers, *Journal of Physical Chemistry B*, 110(39), 19543–19551.

31. Mohajeri, M., Hemmati, M., and Shekarabi, A.S. (2015). An experimental study on using a nanosurfactant in an EOR process of heavy oil in a fractured micromodel, *Journal of Petroleum Science and Engineering*, 126, 162–173.

32. Pei, H., Shu, Z., Zhang, G., Ge, J., Jiang, P., Qin, Y., and Cao, X. (2018). Experimental study of nanoparticle and surfactant stabilized emulsion flooding to enhance heavy oil recovery, *Journal of Petroleum Science and Engineering*, 163, 476–483.

33. Chukwudeme, E.A., and Hamouda, A.A. (2009). Enhanced Oil Recovery (EOR) by miscible CO_2 and water flooding of asphaltenic and non-asphaltenic oils, *Energies*, 2(3), 714–737.

34. Cuthiell, D., Kissel, G., and Jackson, C. et al. (2006). Viscous fingering effects in solvent displacement of heavy oil, *Journal of Canadian Petroleum Technology*, 45(7), 29–39.

35. Yekeen, N., Manan, M.A., Idris, A.K., and Padmanabhan, E. (2018). A comprehensive review of experimental studies of nanoparticles-stabilized foam for enhanced oil recovery, *Journal of Petroleum Science & Engineering*, 164, 43–74.

36. Yekeen, N., Muhammad A.M., Ahmad, K.I., Ali, M.S., and Abdul, R.R. (2017). Experimental investigation of minimization in surfactant adsorption and improvement in surfactant-foam stability in presence of silicon dioxide and aluminum oxide nanoparticles, *Journal of Petroleum Science &Engineering*, 159, 115–134.

37. Sonn, J.S., Lee, J.Y., Jo, S.H., Yoon, I., Jung, C.H., and Lim, J.C. (2018). Effect of surface modification of silica nanoparticles by silane coupling agent on decontamination foam stability, *Annals of Nuclear Energy*, 114, 11–18.

38. Vatanparast, H., Samiee, A., Bahramian, A., and Javadi, A. (2017). Surface behavior of hydrophilic silica nanoparticle-SDS surfactant solutions: I. Effect of nanoparticle concentration on foamability and foam stability, *Colloids and Surfaces A*, 513, 430–441.

39. Farhadi, H., Riahi, S., Ayatollahi, S., and Ahmadi, H. (2016). Experimental study of nanoparticle-surfactant-stabilized CO_2 foam: Stability and mobility control, *Chemical Engineering Research and Design*, 111, 449–460.

40. Carn, F., Colin, A., and Pitois, O. et al. (2009). Foam drainage in the presence of nanoparticle-surfactant mixtures, *Langmuir*, 25(14), 7847–7856.

41. Worthen, A.J., Bryant, S.L., and Huh, C. et al. (2013). Carbon dioxide-in-water foams stabilized with nanoparticles and surfactant acting in synergy, *AIChE Journal*, 59(9), 3490–3501.

42. AlYousef, Z.A., Almobarky, M.A., and Schechter, D.S. (2018). The effect of nanoparticle aggregation on surfactant foam stability, *Journal of Colloid and Interface Science*, 511, 365–373.

43. Nurudeen, Y., Ahmad, K.I., Muhammad, A.M., Ali, M.S., Abdul, R.R., and Tan, X.K. (2017). Bulk and bubble-scale experimental studies of influence of nanoparticles on foam stability, *Chinese Journal of Chemical Engineering*, 25, 347–357.

44. Hu, N., Li, Y., Wu, Z., Lu, K., Huang, D., and Liu, W. (2018). Foams stabilization by silica nanoparticle with cationic and anionic surfactants in column flotation: Effects of particle size, *Journal of the Taiwan Institute of Chemical Engineers*, 88, 62–69.

45. Li, X.K., Xiong, Y., Chen, D.J., Zou, C.J. (2015). Utilization of nanoparticle-stabilized foam for gas well deliquification, *Colloids & Surfaces A*, 482, 378–385.

46. Xu, L., Rad, M.D., and Telmadarreie, A. et al. (2018). Synergy of surface-treated nanoparticle and anionic-nonionic surfactant on stabilization of natural gas foams, *Colloids and Surface A*, 550, 176–185.

47. Alomair, O.A., Matar, K.M., and Alsaeed, Y.H., Nanofluids application for heavy oil recovery, *Proceedings of the SPE Asia Pacific Oil & Gas Conference and Exhibition*, Adelaide, Australia, October 14–16, 2014.

48. Tarek, M., and El-Banbi, A.H., Comprehensive investigation of effects of nano-fluid mixtures to enhance oil recovery, *Proceedings of the SPE North Africa Technical Conference and Exhibition*, Cairo, Egypt, September 14–16, 2015.

49. Tarek, M., Investigating nano-fluid mixture effects to enhance oil recovery. *Proceedings of the SPE Annual Technical Conference and Exhibition*, Houston, TX, September 28–30, 2015.

50. Kanj, M.Y., Funk, J.J., and Al-Yousif, Z., Nanofluid coreflood experiments in the ARAB-D, *Proceedings of the SPE Saudi Arabia Section Technical Symposium*, Al-Khobar, Saudi Arabia, May 9–11, 2009.

51. Wang, L., Zhang, G., Ge, J.J., Li, G., Zhang, J.Q., and Ding, B., Preparation of microgel nanospheres and their application in EOR, *Proceedings of the International Oil and Gas Conference and Exhibition in China*, Beijing, China, June 8–10, 2010.
52. Kanj, M.Y., Rashid, M., and Giannelis, E., Industry first field trial of reservoir nanoagents, *Proceedings of the SPE Middle East Oil and Gas Show and Conference*, Manama, Bahrain, September 25–28, 2011.
53. Wei, B., Li, Q., Jin, F., Li, H., and Wang, C. (2016). The potential of a novel nanofluid in enhancing oil recovery, *Energy & Fuels*, 30, 2882–2891.
54. Molnes, S.N., Torrijos, I.P., and Strand, S. et al. (2016). Sandstone injectivity and salt stability of cellulose nanocrystals (CNC) dispersions—Premises for use of CNC in enhanced oil recovery, *Industrial Crops & Products*, 93, 152–160.
55. Heggset, E.B., Chinga-Carrasco, G., and Syverud, K. (2017). Temperature stability of nanocellulose dispersions, *Carbohydrate Polymers*, 157, 114–121.
56. Assef, Y., Arab, D., and Pourafshary, P. (2014). Application of nanofluid to control fines migration to improve the performance of low salinity water flooding and alkaline flooding, *Journal of Petroleum Science and Engineering*, 124, 331–340.
57. Yuan, B., and Moghanloo, R.G. (2017). Analytical modeling improved well performance by nanofluid pre-flush, *Fuel*, 202, 380–394.
58. Yuan, B., Moghanloo, R.G., and Wang, W.D., Using nanofluids to control fines migration for oil recovery: Nanofluids co-injection or nanofluids pre-flush? A comprehensive answer, *Fuel*, 215, 474–483, 2018.

22

Nanofluids: Applications and Its Future for Oil and Gas Industry

Ravi Shankar Kumar and Tushar Sharma

CONTENTS

22.1 Introduction

Continuous demand of oil and gas has led to a rapid decline of hydrocarbon reserves globally and posed a major challenge to the producers in regard to recovering the remaining (residual) oil and gas. Thus, improving oil recovery from the current reservoirs holds the key to meet the global energy demands. As such, the development of new technology (nanotechnologies) has become imperative to satisfy the ever growing and insatiable need for the energy. Nanofluid technology, as a part of nanotechnology, is a great area of interest where well-designed materials, i.e., nanoparticles governing the matter at nanoscale levels. These nanoparticles have higher surface area to volume ratio and unique properties, such as high adsorption potential, heat conductivity, superficial electrical conductivity, and higher surface energy due to the more exposed surfaces [1]. Nanofluid, which is a colloidal suspension of nanoparticles (1–100 nm) in base fluid such as water [2], oil [3], glycol [4], and polymeric solutions [5], offers promising solutions for various industrial applications including oil industry. Different nanoparticles can be used to design the nanofluids as per the requirement such as pharmaceuticals [6], surface engineering [7], polymer nanocomposite [8], foamability [9], drug delivery [10], and oilfield operations [11]. The oxides nanoparticles such as silica [12], titania [2], alumina [13], copper oxide [14], aluminium oxide [15], and zinc oxide [16] are found successful in oilfield industries. The successful application of nanofluids embraces some prerequisites such as stable suspension of nanoparticles in the base fluid, long term dispersion stability, and negligible agglomeration of nanoparticles in the nanofluids [16–17]. Therefore, a nanofluid meeting prerequisites made them superlative for petroleum industry in recent years. In our previous study [5], the stability and rheological properties of nanofluids stabilized by SiO_2 and TiO_2 nanoparticles have been investigated for oilfield applications, where very less amount of titanium nanoparticles (0.05 wt% and 0.1 wt%) provided noteworthy improvements in design of silica nanofluid. The various reports are available that confirm the addition of nanoparticle provided unique modifications in properties of base fluid which extended the applicability of nanofluids for oil recoveries [18], drilling and completion [19], exploration [20], fluid loss control [21], fine migration control [22–23], wellbore stability [24], interfacial tension reduction, and wettability alteration [25].

The nanofluid flooding as chemical injection method has proven the efficacy to boost the oil production [26]. From the literature, most of the investigations have shown the great interest to unlock the residual (remaining) oil from the reservoir using silica-based nanofluid than conventional enhanced oil recovery (EOR) agents. Lu et al. [27] conducted nanofluid flooding experiments using silica nanoparticles and received significant improvements in oil recovery. Furthermore, he suggested that the cyclic nanofluid injection can provide higher tertiary oil recovery than a continuous nanofluid injection scheme. Similarly, silica nanoparticles were used to enhance the efficacy of surfactant flooding in production of heavy oils [28]. Li et al. [29] investigated the effect of silica nanofluid on wettability alteration. The reduction in interfacial tension (IFT) and wettability alteration has been favorable to more oil recoveries [30]. Now, it is very clear that the efficacy of fluid which is used for IFT reduction and wettability alteration from oil-wet to water-wet can be improved by the addition of nanoparticles [31–33]. When nanoparticles interact with the reservoir oil in the pore space matrix of the rock surface, it has a tendency of self-arrangement in the form of wedge shape film [34]. The wedge shape film-like structure of nanoparticles improves the displacement of crude oil from rock surfaces. The use of stable nanofluid for wettability alteration is a comparatively novel approach. Several studies have been found which showed the impact of dispersed nanoparticles on wettability alteration from strongly oil-wet to water-wet for sandstone and carbonate reservoirs by a phenomenon called "adsorption". The adsorbed nanoparticles on the reservoir surfaces form a coating-like structure (wedge film), which authorize the rock surface to become more water-wet [15,31]. Many studies suggested that the combined effect of two or more wettability modifiers have more impact on wettability alteration and IFT reduction. The application of surfactant along with nanofluid is an example for more wettability alteration and toward water-wet. In this contexts, Sharma et al. [30] investigated a well stabilized silica nanofluid with or without surfactant for wettability alteration and reported shift toward strongly water-wet and IFT reduction which is favorable for more oil recoveries.

Thus, in order to recover more oil from the reservoir and meet the global energy demand, the nanofluid has been proven satisfactory results and enhances the oilfield production capabilities by controlling the key parameters such as flow behavior (control on viscosity), wettability alteration, IFT reduction, and thermal stability. Therefore, this study is fully devoted on the formulation methodology of stable nanofluid for the oil and gas industry by keeping wettability alteration, IFT reduction, and improved oil recovery as a primary objective.

22.2 Preparation of Nanofluids

There are mainly two methods for the preparation of nanofluids, i.e., one-step method [35] and two-step method [12]. However, two-step method is the common method for the preparation of nanofluids. In this method, the nanomaterials or nanoparticles are first produced in the form of dry powders (nano-sized powder) by chemical or physical methods, and then the nano-sized powders are dispersed into a base fluid with the help of intensive magnetic force agitation, ultrasonic agitation, high-shear mixing, and homogenization as shown in Figure 22.1a. Two-step method is the most economic and widely used to produce nanofluids for oil and gas industry due to the production of nanomaterial at industrial levels. In a stable nanofluid, nanoparticles are uniformly dispersed in the aqueous phase. Although, the dispersed nanoparticles may aggregate due to attractive forces such as van der Waals' force and surface tension, which are further responsible for sedimentation of nanoparticles [36]. The agglomeration of nanoparticles results in sedimentation, clogging of micro pores, and also reduction in thermal conductivity of nanofluid. Therefore, it is required to formulate the nanofluid in such a manner so that the agglomeration of nanoparticles in the nanofluid should be less which will further enhance the dispersion stability of the prepared nanofluids.

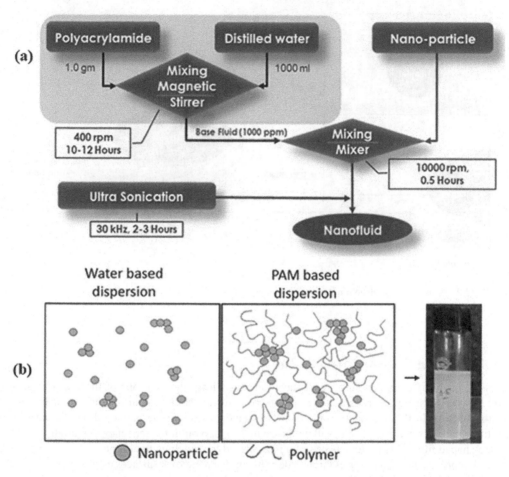

FIGURE 22.1 (a) Flowchart for the preparation of nanofluid using two-step method and (b) schematic showing influence of nanoparticles in nanofluid consist of PAM as dispersant.

Furthermore, it has been reported that a dispersing agent of high molecular weight polymer such as polyacrylamide (PAM) can be used as dispersion stabilizer by increasing the viscosity of base fluid [37]. PAM is a water soluble polymer which forms soft gel when hydrated and produces a highly viscous aqueous solution even using at low concentration resulting in enhanced dispersion stability (as shown schematic in Figure 22.1b) by decreasing gravitation forces on suspended particles [38] and provides additional steric repulsive forces between particles [39].

22.3 Characterization of Nanofluid

After preparation, it is required to characterize the nanofluid for dispersion stability. The characterization can be performed in several stages such as visual observations, rate of sedimentation (ROS), zeta potential and size measurement using dynamic light scattering (DLS) technique, dispersion stability using UV-vis spectroscopy, morphological structure and nanoparticle adsorption using field emission scanning electron microscopy (FESEM), settlement of nanoparticles in suspension using electrical conductivity measurements, and flow behavior using rheological measurements. These results confirm synthesis of stable and unstable nanofluid before its implementation in different applications.

In this section, all the practical aspects related to the characterization of nanofluid system will be thoroughly explained which will help to understand the importance of using nanofluid for oil and gas industry.

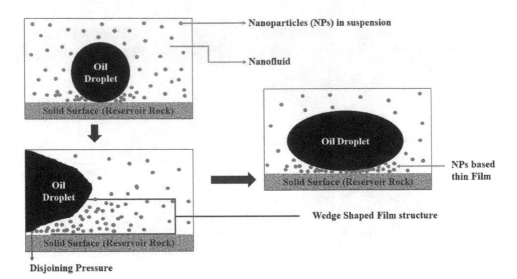

FIGURE 22.2 Schematic for disjoining pressure mechanism for nanofluid flooding in the aqueous phase using nanoparticle.

22.4 Disjoining Pressure Mechanism

In general, oil in hydrocarbon formations stays together with water, and one of these phases strongly adhere to surface. Nanoparticles attribute a special mechanism where nanoparticles affect the oil removal from the surface. In this mechanism, the dispersed nanoparticles in the nanofluid have great interest, when these particles come in contact with oil phase in the reservoir, it has tendency to rearrange and form a wedge-shaped film. This wedge-shaped film is subjected to detaching the oil phase from the rock surfaces and improves the oil recoveries through detachment from the rock surfaces as shown in Figure 22.2.

The disjoining pressure for a liquid film is well-defined as the additional pressure in the film relative to that in the bulk solution [40]. This additional pressure pushes the nanoparticles under the bank of oil to adhere to the surface strongly. It results in more water phase to adhere to the surface, and as a result oil leaves the surface to get mobilized.

22.5 Log-Jamming Mechanism

Log-jamming mechanism is caused by the nanoparticle which has smaller size than the pore throat of the reservoir. In this process, when injected nanofluid enters into the pore area through throat, it can reduce the size of low area at the throat, which can eventually create a pressure differential and support to the incremental velocity of nanofluid. In addition, the water passes through the pore is faster than the nanoparticles which leaves nanoparticles to accumulate at the entry point of the pores of the reservoir. Consequently, the accumulations of the nanoparticles assist to free the oil droplets from the pores and under temporary log-jamming effect, i.e., throat blockage disappear and started flowing with the aqueous phase which can further improve the performance of oil recovery. However, the gradual increment in the accumulation of the nanoparticles at the throat of the pore body leads to blockage of the pore or if the size of the nanoparticles is larger than the pore throat size, it can block the pore throat which can reduce the performance of oil recovery. This phenomenon is also known as "mechanical entrapment". Although the size of pore throat is in the range of microscale, more than thousand times larger than the nanoparticles size, few cases can show the tendency of nanoparticles agglomeration resulting into the formations of clusters of nanoparticles of larger than the size of pore throat. Thus, a rigorous study is required before the nanofluid is injected in reservoir pores to ensure the feasibility of EOR.

22.6 Nanoparticle-Assisted EOR Mechanism (Nanofluid Flooding)

Nanofluid-assisted EOR technique (also called nanoflooding) is a process to increase the recovery of residual oil from the reservoir by means of chemical injection process where nanofluid is expected to enhance the production of trapped crude oil. In this process, the nanoparticles of nanofluid penetrate into the reservoir to improve the displacement efficiency of oil by means of wettability alteration, reduction of IFT, and log-jamming effect. As per the literature, the most commonly used nanoparticle in the oil recovery application is silica nanoparticle. More than 95% of the silica nanoparticle contains silicon oxide which is the main component of the sandstone and makes nanofluid environmental friendly and comparatively cost effective, resulting silica nanoparticle to work as possible alternative for the application where other nanoparticles create contaminations and toxicity issues in oilfield.

A classical study on nanofluid use for oil recovery from porous and permeable sandpacks was presented by Sharma et al. [12], and the results were compared with the one of conventional oil recovery method. Nanofluid exhibited higher pressure drops during injection into sandpack than surfactant-polymer (SP) flood system, which resulted in a higher oil recovery. The trait of high oil recovery using nanofluids can also be linked with higher viscosity of nanofluid than SP solution. High viscous nanofluid may provide significant reduction in mobility ratio and consequently higher oil recovery from sandpacks is expected to be received. As we know that the smaller the size of nanoparticle, the higher the surface area and larger will be the interface of oil phase and nanoparticle, which is subjected to build a better reduction in IFT of oil phase leading to high recovery.

To accomplish flooding experiments using nanofluids, an experimental set-up called core flooding apparatus is required. This system consists of all essential component such as displacement pump, accumulators, connection tubing, valves, pressure gauge, sandpack holder, temperature controller, fluid collector, and heating jacket to perform nanofluid-based oil recovery experiments on porous and permeable sandstone reservoir core samples. A typical schematic of the discussed equipment set is shown in Figure 22.3 [41]. However, the actual mechanism behind higher oil recovery of nanofluids is still in discussion and not entirely clear particularly in oilfield jobs, but several researchers have quoted the

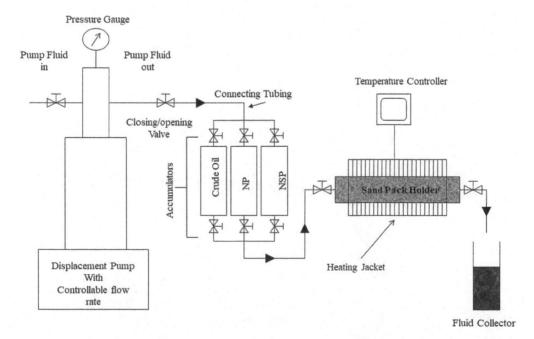

FIGURE 22.3 Schematic of core flooding apparatus used for flooding experiments. (Reprinted from *J. Petrol. Sci. Eng.*, 170, Goswami, R. et al., Effect of ionic strength on crude emulsification and EOR potential of micellar flood for oil recovery applications in high saline environment, 49–61, Copyright (2018), with permission from Elsevier.)

possible mechanism of oil recovery as disjoining pressure and log-jamming effect. Overall, in this section, the detailed discussion will illustrate the key parameters and the influence of nanoparticles on the performance of oil recovery.

22.7 Effect of Nanofluid on Mobility Ratio (Controllable Viscosity Profile)

The mobility ratio (M) is one of the important dimensionless numbers for EOR, which is defined as the ratio of mobility for displacing fluid (water, or chemical agents such as nanofluid) to the displaced fluid (crude oil), and mathematically, it can be written as [42]:

$$M = \frac{\lambda_w}{\lambda_o} = \frac{k_w / \mu_w}{k_o / \mu_o} = \frac{k_w / k_o}{\mu_w / \mu_o}, \tag{22.1}$$

where λ_w and λ_o are the water (injected fluid or displacing fluid) and oil (displaced fluid) mobilities, respectively, k_w and k_o are the relative permeabilities of water and oil, respectively, μ_w and μ_o are the effective viscosities of the water and oil. A high mobility ratio (M > 1) causes fingering effect that makes water to move further and bypass the trapped oil, resulting reduced oil recovery. Assuncao et al. [42] conducted displacement test using different polymer solutions ranges from 250 to 1550 ppm, and the results were compared for mobility ratio. On the basis of results, comparison was drawn with respect to water. In these findings, it has been clearly seen that the mobility ratio less than or equal to 1 is favorable for more oil recovery. Thus, the mobility ratio can be reduced by decreasing the viscosity of displaced fluid (oil) or by increasing the viscosity of displacing fluid or injected fluid. The viscosity of displacing fluid also increases in presence of nanoparticles [43]. Increased viscosity reduces the movement of water phase and maintains the mobility ratio with oil. The effect of nanoparticle inclusion on viscosity of conventional oil recovery method such as polymer (P) and SP is shown in the Figure 22.4 [12]. The viscosity of nanoparticle-polymer (NP) and

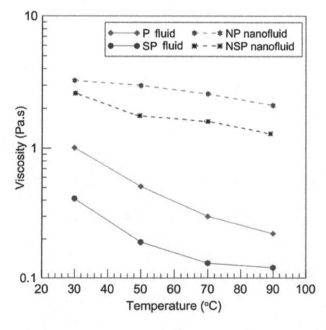

FIGURE 22.4 Effect of temperature on viscosity of fluid systems such as P, SP, NP, and NSP nanofluid. (Reprinted with permission from Sharma, T. et al., *Ind. Eng. Chem. Res.*, 55, 12387–12397, 2016. Copyright 2016 American Chemical Society.)

nanoparticle-surfactant-polymer (NSP) fluids significantly increased, and new fluids called nanofluids of P and SP with improved rheological properties against temperature were proposed for oil recovery applications. Thus, a proper designed nanofluid can provide best support on the aforesaid problem by decreasing the mobility ratio which can further increase the overall sweep efficiency resulting in more oil to be recovered [42].

22.8 Wettability Alteration Using Nanofluids

According to Craig [44], the propensity of a fluid to spread on a solid (rock) surface in the presence of other immiscible fluid is defined as wettability for that fluid. In the reservoir, the liquid phase can be water or oil, and the solid phase is the reservoir rock. Wettability of solid surface directly relates to the fluid-fluid and fluid-rock interactions and implies an interfacial tension. Thus, wettability is a result of interfacial tension between the fluid phases present and their individual adhesive attraction (electrostatic force) to the solid. Interfacial forces are related to one another by the famous Young's Law:

$$\cos\theta = \frac{\sigma_{sw} - \sigma_{so}}{\sigma_{wo}}, \tag{22.2}$$

where θ is contact angle and σ values are interfacial tension between the interfaces of solid-water (σ_{sw}), solid-oil (σ_{so}), and water-oil (σ_{wo}).

The scheme of interfacial forces acting on an oil droplet with water and rock is shown in the Figure 22.5.

Wettability is typically measured in terms of contact angle made by wetting phase on rock surface, and it is always measured from denser phase to lighter phase. The value of contact angle decides the wetting condition from the Young's Law, if $\theta < 90°$, implies water-wet and $\theta > 90°$, implies oil-wet, while $\theta \approx 90°$ is intermediate wet. Moreover, wettability works as a major factor in order to govern the rate of oil recovery by affecting the capillary pressure, relative permeability, and irreducible oil saturation. When nanoparticles are introduced, their presence promotes the surface adsorption during the disjoining pressure phenomena resulting in alteration in rock wetting toward water in case of water-wet formation. Thus, nanofluids make water-wet formation more water-wet which in turn results in oil phase removal from the surface. To understand the effect on wettability alteration with respect to oil viscosity, Maghzi et al. [32] conducted an experiment to monitor the change in wettability by silica nanoparticle during water flooding to heavy oils and found that SiO_2 nanoparticles-based nanofluid altered the wettability from oil-wet to water-wet. Similarly, zirconium oxide and silica oxide were dispersed in surfactant, alkali, and brine solution by Cao et al. [45], who found that the nanofluids showed great wettability alteration even under high pressure and high temperature (HPHT) reservoir conditions.

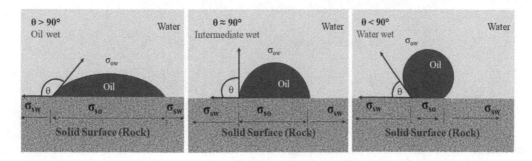

FIGURE 22.5 Surface interactions of crude oil in two phase fluid system with reservoir rock.

22.9 Interfacial Tension Reduction by Nanofluid

IFT is defined as a force that acts between oil and water, and it is normally measured in dyne/cm. IFT reduction allows one phase to penetrate into another by beating interfacial forces which may be for oil-water, oil-gas, and gas-water. Therefore, the entry of water into oil phase leads to oil bank deformation into smaller droplets, and mobilization of smaller droplets is more efficient than large size oil bank. Water alone cannot do IFT reduction, and hence, some surface active agents such as surfactant and nanoparticles make this possible. Surfactant micelles orient themselves around oil droplets and reduce IFT of the crude oil. Thus, IFT is an important parameter for the enhancement of oil production. Therefore, it is necessary to measure the IFT for oil and injected fluid in EOR techniques. From the literature, it has been found that some types of nanoparticles have great potential to lower the IFT for nanofluid flooding than conventional and surfactant [46–47,30]. Generally, IFT between crude oil and nanofluid is measured by pendant drop method, in which the IFT value is calculated by analyzing the entire shape of the oil droplet using accurate video processing and analysis software. Adel and Hannora [48] conducted experiments in order to compare the applicability of SiO_2 and Al_2O_3 on reduction for crude oil IFT, and in their findings, IFT was greatly reduced when either of the nanoparticles was used with brine. Moreover, the reduction of IFT was more for SiO_2 as compared to Al_2O_3. Therefore, SiO_2 nanoparticles pose more potential to be utilized as EOR agent. Similarly, another study also demonstrates the significant IFT reduction due to inclusion of nanoparticles is shown in the Figure 22.6 [12].

Similarly, Sharma et al. [30] prepared SiO_2 nanofluids with and without surfactant and tested them for IFT reduction of various paraffins to be applicable in EOR applications. For nanofluids IFT reduction results of different paraffins (n-decane, n-heptane, n-hexane, and n-pentane) under high temperature (reservoir equivalent), they reported nanofluids provided significant reduction in IFT values of paraffins, and their performance on IFT reduction of these oils even remained stable with increasing temperature. This improved behavior of nanofluids indicate that nanofluids as compared to conventional solutions may provide better result for high temperature condition of reservoir.

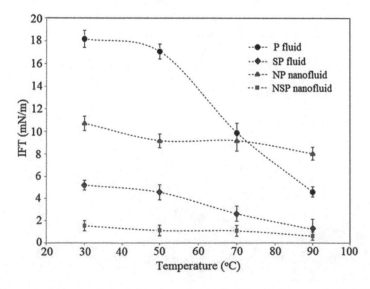

FIGURE 22.6 IFT results of crude oil for fluid system *viz.,* P, SP, NP, and NSP nanofluid at different temperature (Reprinted with permission from Sharma, T. et al., *Ind. Eng. Chem. Res.,* 55, 12387–12397, 2016. Copyright 2016 American Chemical Society.)

22.10 Other Advantages of Nanoparticles in Oilfield

Other than nanofluids, nanoparticles found interest in stabilizing emulsion for oil recovery applications at HPHT conditions. Emulsion is a solution of two immisicible phases where one phase is dispersed into the continuous phase of other. Oil-in-water (o/w) emulsions repersant oil dispersion in water phase and have been explored to improve the extent of oil recovery [49]. However, high temperature deforms the stability of o/w emulsions and made them unstable for operating conditions of oilfield. Hence, nanoparticle inclusion enhanced the stability of conventional o/w emulsions and these emulsions are called as Pickering emulsions. For Pickering emulsion, nanoparticle intends to adsorb at the surface of oil droplets and restrict the droplet destabilization through steric barrier around oil droplets. Strictly stabilized Pickering emulsion received more attention in oilfield practice due to their thermally stable nature and undeformed rheological properties [50,51]. Sharma et al. [51] prepared thermally stable Pickering o/w emulsion of SiO_2 nanoparticles, polyacrylamide, and sodium dodecyle sulfate. Emulsion without SiO_2 nanoparticles exhibited significant droplet coalesence with increasing temperature, while emulsion prepared using SiO_2 nanoparticles were coalesence stable and showed insignificant droplet destabilization with increasing temperature as shown in Figure 22.7. The trait of thermally stable nanoparticle also improved rheological properties of these emulsions, and as a result, Pickering emulsion exhibited higher viscosity and moduli at high temprature of 371 K, while viscosity of simple o/w emulsion significantly affected by high temperature [50]. It is clear from the figure that Pickering emulsion did not exhibit much different coalesence than simple o/w emulsion. In addition, the controlled viscosity profile of nanoemulsion can also control mobility ratio during flooding process, which is a key parameter to improve the oil recovery. Nanoemulsion are small enough to enter into the pore throats with less retention. Therefore, the aforesaid advantages make nanoemulsion more applicable for research as well as oil and gas industries.

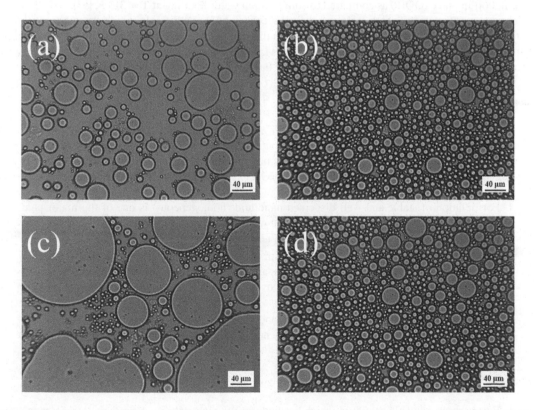

FIGURE 22.7 Microscopic images showing effect of temprature on thermal stability of conventional emulsion: (a) 298 K, (c) 371 K, and Pickering emulsion, (b) 298 K, and (d) 371 K.

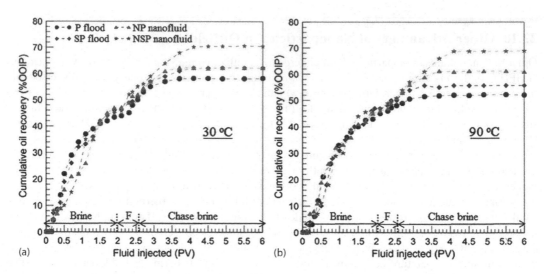

FIGURE 22.8 Cumulative oil recovery; (a) and (b) of fluid system (P, SP, NP, and NSP) as a function of fluid pore volume injected at 30°C and 90°C. Where F indicates the interval when P, SP, NP, and NSP nanofluid was injected. (Reprinted with permission from Sharma, T. et al., *Ind. Eng. Chem. Res.*, 55, 12387–12397, 2016, Copyright 2016 American Chemical Society.)

Silica nanoparticles are commonly used for the preparation of nanoemulsion due to their enhanced surface properties and flow behavior. Wettability alteration and IFT reduction can also be controlled by silica nanoparticles [2,52]. Pickering emulsion reported incremental oil recovery of more than 23% of original oil in place (OOIP) as compared to conventional water flooding at T = 313 K [53].

The temperature significantly affected the oil recovery of P & SP fluids, while the oil recovery of nanofluids marginally changed, which makes them thermally stable and applicable to be utilized at high temperature oil recovery applications. Nanofluid *viz.,* NP and NSP showed enhanced oil recovery as compared to conventional flooding system such as P and SP shown in Figure 22.8 [12].

22.11 Challenges and Future Scope

Nanoparticles have been proven potential surface active agent for EOR processes, but some of them are limited up to laboratory scale and simultaneously not fully approved for field applications. Numerous challenges still remain before nanoparticle-based EOR techniques can be used for real field application.

1. Formulation of stable and well dispersed nanofluid (homogeneous) is one of the major challenges for real field applications because the nanoparticles have tendency to aggregate which destabilizes the nanofluid system. Although, use of surfactant may stabilize the nanofluid [54].

2. Several researchers have been come up with the applicability of nanofluid for oil recovery applications. However, the mixture of nanoparticles still requires more laboratory investigations and modelling analysis for more oil recovery followed by cost effectiveness because mixing of two or more nanoparticles may increase the cost of nano-assisted EOR method.

3. The mechanism for nano-assisted EOR method, interactions with reservoir formation, and initial reservoir fluid properties are still clearly not understood, which require more evidence for the fundamentals.

The need of the hour is to formulate extensive study in the following research areas:

1. Formulation of stable nanofluid which has better dispersion stability, long term storage capacity, and potential nanofluid properties which supports the EOR recoveries for more oil.

2. Mixture of nanoparticles may increase the dispersion stability and offer better rheological properties followed by the necessary properties required for nano-assisted EOR techniques, which can help to increase the oil recoveries from the depleted reservoirs.

3. More feasible and robust authorizations should be required to understand the fundamentals of nano-assisted EOR techniques.

22.12 Conclusions

This chapter will help to understand the recent research and development in order to understand the applicability of nanoparticles for enhanced oil recovery. The conventional EOR chemicals/agents face numerous of challenges during oil recoveries applications, many of them are miraculously solved by nanoparticles by greatly inducing the key parameters like wettability alteration, IFT reduction, mobility control, controllable viscosity profile. In addition, introducing the surfactant and polymers in the nanofluid increases the sweep efficiency and adjusts the mobility ratio which are additional packages to the nanofluid system for EOR. Nanoparticles have great potential of forming emulsions, nanoemulsion for EOR applications. Silica nanoparticle is commonly used for nano-assisted enhanced oil recovery applications for sandstone/carbonate formations. Moreover, a homogeneous mixture of two or more nanoparticles may produce well dispersed and rheologically stable nanofluid which are more efficient than silica nanofluid for oil recoveries applications. Many parameters such as size, rate of sedimentation, viscosity of base fluid, effect of temperatures, and ionic strength of reservoir formations have great influence on the performance of nanofluid flooding. Polyacrylamide (PAM), hydrolyzed polyacrylamide (HPAM), and xanthan gum are some typical oilfield viscosifiers which are generally used for the enhancements of the viscosity of the base fluid which can further support the nanoparticles to be in suspension. Therefore, the simultaneous effect of nanoparticles, surfactant, polymers, and chemicals offers improved oil recoveries than the individual fluid system.

ACKNOWLEDGMENT

Department of Science and Technology, India (Grant: SB/S3/CE/057/2015) is also gratefully acknowledged for providing partial financial assistance for the work.

REFERENCES

1. Rovani, S., Santos, J. J., Corio, P. and Fungaro, A. D. (2018). Highly pure silica nanoparticles with high adsorption capacity obtained from sugarcane waste ash. *ACS Omega* 3, 2618–2627.
2. Bayat, A. E., Junin, R., Samsuri, A., Piroozian, A. and Hokmabadi, M. (2014). Impact of metal oxide nanoparticles on enhanced oil recovery from limestone media at several temperatures. *Energy & Fuels* 28, 6255–6266.
3. Ettefaghi, E., Ahmadi, H., Rashidi, A., Nouralishahi, A. and Mohtasebi, S. S. (2013). Preparation and thermal properties of oil-based nanofluid from multi-walled carbon nanotubes and engine oil as nanolubricant. *International Communications in Heat and Mass Transfer* 46, 142–147.
4. Xie, H., Yu, W. and Chen, W. (2010). MgO nanofluids: Higher thermal conductivity and lower viscosity among ethylene glycol-based nanofluids containing oxide nanoparticles. *Journal of Experimental Nanoscience* 5, 463–472.
5. Kumar, R. S. and Sharma, T. (2018). Stability and rheological properties of nanofluids stabilized by Sio_2 nanoparticles and SiO_2-TiO_2 nanocomposites for oilfield applications. *Colloids and Surfaces A: Physicochemical and Engineering Aspects* 539, 171–183.
6. Schmitt, M., Limage, S., Denoyel, R. and Antonia, M. (2017). Effect of SPAN80 on the structure of emulsified aqueous suspensions. *Colloids and Surfaces A: Physicochemical and Engineering Aspects* 521, 121–132.
7. Vafai, S., Wen, D. and Borca-Tasciuc, T. (2011). Nanofluid surface wettability through asymptotic contact angle. *Langmuir* 27, 2211–2218.

8. Narukulla, R., Ojha, U. and Sharma, T. (2017). Stable & re-dispersible polyacryloyl hydrazide–Ag nanocomposite Pickering emulsions. *Soft Matter* 13, 6118–6128.

9. Vatanparast, H., Samiee, A., Bahramian, A. and Javadi, A. (2017). Surface behavior of hydrophilic silica nanoparticle-SDS surfactant solutions: I. Effect of nanoparticle concentration on foamability and foam stability. *Colloids and Surfaces A: Physicochemical and Engineering Aspects* 513, 430–441.

10. Tang, H., Xiang, D., Wang, F., Mao, J., Tan, X. and Wang, Y. (2017). 5-ASA-loaded SiO_2 nanoparticles a novel drug delivery system targeting therapy on ulcerative colitis in mice. *Molecular Medicine Reports* 15, 1117–1122.

11. Ponmani, S., Nagarajan, R. and Sangwai, J. (2013). Applications of nanotechnology for upstream oil and gas industry. *Journal of Nano Research* 24, 7–15.

12. Sharma, T., Iglauer, S. and Sangwai, J. S. (2016). Silica nanofluids in an oilfield polymer polyacrylamide: Interfacial properties, wettability alteration and applications for chemical enhanced oil recovery. *Industrial & Engineering Chemistry Research* 55, 12387–12397.

13. Beck, M. P., Sun, T. and Teja, A. S. (2007). The thermal conductivity of alumina nanoparticles dispersed in ethylene glycol. *Fluid Phase Equilibria* 260, 275–278.

14. Srinivasan, A. and Shah, S. N. (2014). Surfactant-based fluids containing copper-oxide nanoparticles for heavy oil viscosity reduction, SPE-170800-MS.

15. Giraldo, J., Benjumea, P., Lopera, S., Cortes, F. B. and Ruiz, M. A. (2013). Wettability alteration of sandstone cores by alumina-based nanofluids. *Energy & Fuels* 27, 3659–3665.

16. William, J. K. M., Ponmani, S., Samuel, R., Nagarajan, R. and Sangwai, J. S. (2014). Effect of CuO and ZnO nanofluids in xanthan gum on thermal, electrical and high pressure rheology of water-based drilling fluids. *Journal of Petroleum Science and Engineering* 117, 15–27.

17. Xiu-tian-feng, E., Pan, L., Wang, F., Wang, L., Zhang, X. and Zou, J. J. (2016). Al-nanoparticle-containing nanofluid fuel: Synthesis, stability, properties, and propulsion performance. *Industrial & Engineering Chemistry Research* 55, 2738–2745.

18. Zheng, C., Cheng, Y., Wei, Q., Li, X. and Zhang, Z. (2017). Suspension of surface-modified nano-SiO_2 in partially hydrolyzed aqueous solution of polyacrylamide for enhanced oil recovery. *Colloids and Surfaces A: Physicochemical and Engineering Aspects* 524, 169–177.

19. Quintero, C. G., Noik, C., Dalmazzone, C. and Grossiord, J. L. (2008). Modelling and characterisation of diluted and concentrated water-in-crude oil emulsions: Comparison with classical behaviour. *Rheologica Acta* 47, 417–424.

20. Kapusta, S., Balzano, L. and Riele, P. (2012). Nanotechnology applications in oil and gas exploration and production. SPE, IPTC-15152-MS, In *International Petroleum Technology Conference*, Bangkok, Thailand, 7–9 February, 2012.

21. Javeri, S. M., Haindade, Z. W. and Jere, B. C. (2011). Mitigating loss control and differential sticking problems using silicon nanoparticles, SPE-145840-MS, In *SPE/IADC Middle East Drilling Technology Conference and Exhibition*, Muscat, Oman, 24–26 October, 2011.

22. Huang, T., Crews, J. B. and Willingham, J. R. (2008). Using nanoparticle technology to control formation fine migration, SPE-115384-MS, In *SPE Annual Technical Conference and Exhibition*, Denver, Colorado, 21–24 September, 2008.

23. Habibi, A., Ahmadi, M., Pourafshary, P. and Ayatollahi, S. (2014). Fines migration control in sandstone formation by improving silica surface zeta potential using a nanoparticle coating process. *Energy Sources, Part A: Recovery, Utilization, and Environmental Effects* 36, 2376–2382.

24. Riley, M., Stamatakis, E., Young, S., Hoelsher, K. P., Stefano, G. D., Ji, L., Guo, Q. and Friedheim, J. (2012). Wellbore stability in unconventional shale–The design of a nano-particle fluid, SPE-153729-MS, In *SPE Oil and Gas India Conference and Exhibition*, Mumbai, India, 28–30 March, 2012.

25. Roustaei, A., Saffarzadeh, S. and Mohammadi, M. (2013). An evaluation of modified silica nanoparticles' efficiency in enhancing oil recovery of light and intermediate oil reservoirs. *Egyptian Journal of Petroleum* 22, 427–433.

26. Peng, B., Zhang, L., Luo, J., Wang, P., Ding, B., Zeng, M. and Cheng, Z. (2017). A review of nanomaterials for nanofluid enhanced oil recovery. *RSC Advances* 7, 32246–32254.

27. Lu, T., Li, Z., Zhao, Y. and Zhang, C. (2017). Enhanced oil recovery of low-permeability cores by SiO_2 nanofluid. *Energy & Fuels* 31, 5612–5621.

28. Cheraghian, G. and Hendraningrat, L. (2016). A review on applications of nanotechnology in the enhanced oil recovery part B: Effects of nanoparticles on flooding. *International Nano Letters* 6, 1–10.

29. Li, R., Jiang, P., Gao, C., Huang, F., Xu, R. and Chen, X. (2017). Experimental investigation of silica-based nanofluid enhanced oil recovery: The effect of wettability alteration. *Energy & Fuels* 31, 188–197.

30. Sharma, T. and Sangwai, J. S. (2017). Silica nanofluids in polyacrylamide with and without surfactant: Viscosity, surface tension, and interfacial tension with liquid paraffin. *Journal of Petroleum Science and Engineering* 152, 575–585.

31. Karimi, A., Fakhroueian, Z., Bahramian, A., Khiabani, N. P., Darabad, J. B., Azin, R. and Arya, S. (2012). Wettability alteration in carbonates using zirconium oxide nanofluids: EOR implications. *Energy & Fuels* 26, 1028–1036.

32. Maghzi, A., Mohammadi, S., Ghazanfari, M. H., Kharrat, R. and Masihi, M. (2012). Monitoring wettability alteration by silica nanoparticles during water flooding to heavy oils in five-spot systems: A pore-level investigation. *Experimental Thermal and Fluid Science* 40, 168–176.

33. Moghaddam, R. N., Bahramian, A., Fakhroueian, Z., Karimi, A. and Arya, S. (2015). Comparative study of using nanoparticles for enhanced oil recovery: Wettability alteration of carbonate rocks. *Energy & Fuels* 29, 2111–2119.

34. McElfresh, P., Holcomb, D. and Ector, D. (2012). Application of nanofluid technology to improve recovery in oil and gas wells, SPE-154827-MS, In *SPE International Oilfield Nanotechnology Conference and Exhibition*, Noordwijk, The Netherlands, 12–14 June, 2012.

35. Zhu, H., Lin, Y. and Yin, Y. (2004). A novel one-step chemical method for preparation of copper nanofluids. *Journal of Colloid and Interface Science* 277, 100–103.

36. Wang, Z., Luo, Z. and Yan Y. (2018). Dispersion and sedimentation of titanium dioxide nanoparticles in freshwater algae and daphnia aquatic culture media in the presence of arsenate. *Journal of Experimental Nanoscience* 13, 119–129.

37. Wei, Y. and Huaqing, X. (2012). A review on nanofluids: Preparation stability mechanisms, and applications. *Journal of Nanomaterials* 435873, 1–17.

38. Yu, Y., Wang, J. and Parr, J. F. (2012). Preparation and properties of TiO_2/fumed silica composite photocatalytic materials. *Procedia Engineering* 27, 448–456.

39. Tang, T., Hui, C. Y. and Jagota, A. (2006). Adhesive contact driven by electrostatic forces. *Journal of Applied Physics* 99, 054906.

40. Chengara, A., Nikolov, A. D., Wasan, D. T., Trokhymchuk, A. and Henderson, D. (2004). Spreading of nanofluids driven by structural disjoining pressure gradient. *Journal of Colloid and Interface Science* 280, 192–201.

41. Goswami, R., Chaturvedi, K. R., Kumar, R. S., Chon, B. H. and Sharma, T. (2018). Effect of ionic strength on crude emulsification and EOR potential of micellar flood for oil recovery applications in high saline environment. *Journal of Petroleum Science and Engineering* 170, 49–61.

42. Assuncao, P. M., Rodrigues, L. M. N. R. and Romero, O. J. (2011). Effect of polymer injection on the mobility ratio and oil recovery, SPE-148875-MS, In *SPE Heavy Oil Conference and Exhibition*, Kuwait City, Kuwait, 12–14 December, 2011.

43. Zhang, T., Davidson, D., Bryant, S. L. and Huh, C. (2010). Nanoparticle-stabilized emulsions for applications in enhanced oil recovery, SPE-129885-MS, In *SPE Improved Oil Recovery Symposium*, Tulsa, Oklahoma, 24–28 April, 2010.

44. Craig FF. The reservoir engineering aspects of water flooding. Richardson: SPE Monograph Series 1971.

45. Cao, N., Mohammed, M. A. and Babadagli, T. (2015). Wettability alteration of heavy-oil/bitumen containing carbonates using solvents, high pH solutions and nano/ionic liquids. In *Offshore Technology Conference*; Rio de Janeiro, Brazil; October 27–29, 2015.

46. Hendraningrat, L. and Torsæter, O. (2014). Understanding fluid-fluid and fluid-rock interactions in the presence of hydrophilic nanoparticles at various conditions, SPE-171407-MS, In *SPE Asia Pacific Oil & Gas Conference and Exhibition*, Adelaide, Australia, 14–16 October, 2014.

47. Al-Anssari, S., Barifcani, A., Wang, S., Maxim, L., Iglauer, S. (2016). Wettability alteration of oil-wet carbonate by silica nanofluid, *Journal of Colloid and Interface Science* 461, 435–442.

48. Adel, M. S. R. and Hannora, A. E. (2015). A comparative investigation of nano particle effect for improved oil recovery—Experimental work, SPE-175395-MS, In *SPE Kuwait Oil and Gas Show and Conference*, Mishref, Kuwait, 11–14 October, 2015.

49. Maaref S., Ayatollahi, S., Rezaei, N. and Masihi, M. (2017). The effect of dispersed phase salinity on water-in-oil emulsion flow performance: A micromodel study. *Industrial & Engineering Chemistry Research* 56, 4549–4561.

50. Sharma, T., Kumar, G. S. and Sangwai, J. S. (2015). Comparative effectiveness of production performance of Pickering emulsion stabilized by nanoparticle-surfactant-polymer over surfactant polymer (SP) flooding for enhanced oil recovery for Brownfield reservoir. *Journal of Petroleum Science and Engineering* 129, 221–232.

51. Sharma, T., Kumar, G. S. and Sangwai, J. S. (2015). Viscoelastic properties of oil-in-water (o/w) Pickering emulsion stabilized by surfactant-polymer and nanoparticle-surfactant-polymer systems. *Industrial & Engineering Chemistry Research* 54, 1576–1584.

52. Huibers, B. M. J., Pales, A. R., Bai, L., Li, C., Mu, L., Ladner, D., Daigle, H. and Darnault, C. J. G. (2017). Wettability alteration of sandstones by silica nanoparticle dispersion in light and heavy crude oil. *Journal of Nanoparticle Research* 19, 323.

53. Sharma, T., Kumar, G. S. and Sangwai, J. S. (2014). Enhanced oil recovery using oil-in-water (o/w) emulsion stabilized by nanoparticle, surfactant and polymer in the presence of NaCl. *Geosystem Engineering* 17, 195–205.

54. Cao, H., Zhang, X., Ding, B., Wang, L. and Lu N. (2017). Synergistic action of TiO_2 particles and surfactants on the foamability and stabilization of aqueous foams. *RSC Advances* 7, 44972–44978.

23

CO$_2$ Capture via Nanofluids

Na Zhang, Jianchao Cai, Zhen Pan, Muftah H. El-Naas, Feng Chen, and Zhien Zhang

CONTENTS

23.1 Introduction

CO$_2$, as one of the major greenhouse gas (GHG) emissions, has attracted much attention in the world. Currently, a variety of CO$_2$ capture methods have been widely utilized, e.g., absorption [1], adsorption [2], membrane [3], chemical loop combustion [4], etc. Among them, chemical absorption is proved to be an effective and mature way of capturing CO$_2$ from emitted gas mixtures. This method provides high capture efficiency and mature techniques and equipment. To evaluate the CO$_2$ capture performance of the system, the chemical reaction process of the gas and liquid phases is the dominative factor. Thus, the solvent type is the main influencing factor of the capturing process [5–7]. Amine solutions are extensively used as the CO$_2$ capture solvents in laboratories and industries. However, some amine solutions or other solvents (e.g., NaOH, K$_2$CO$_3$) do not show a good mass transfer efficiency.

Adding nanoparticles into the solutions could significantly improve the gas mass transfer efficiency and capture efficiency. Thus, this book chapter introduces the reaction mechanisms between CO$_2$ and nanofluids. In addition, a variety of nanoparticles used for CO$_2$ capture are discussed and summarized. Finally, the conclusions and future prospects are given regarding the performance enhancement by nanoparticles.

23.2 Mechanisms between CO₂ and Nanofluids

Some scholars [8–10] have conducted in-depth research and discussion on the mechanism of enhanced gas absorption process by nanofluids in the previous work. At present, there are four main enhancement mechanisms that are generally recognized and focused on: the sweeping effect mechanism, the hydrodynamic effect mechanism, the bubble trapping mechanism, and the infiltration mechanism.

23.2.1 The Sweeping Effect Mechanism

The sweeping effect mechanism (Figure 23.1) was first proposed by Kars et al. [11] in 1979. They used a theoretical model to explain the presence of solid particles in a gas-liquid-solid three-phase system to enhance the rate of gas absorption in the liquid. In this mechanism, when the diameter of nanoparticle is smaller than the thickness of the mass transfer boundary layer, the particles can penetrate into the gas-liquid mass transfer boundary layer under the fluid by Brownian motion and stay in the mass transfer boundary for a period of time to adsorb a certain amount of gas molecules. After that, they pass through the mass transfer boundary layer and flow back to the fluid. Consequently, gas molecules are desorbed from the liquid and the nanoparticles are regenerated during the gas transport process. The process is a reciprocating cycle of movement through a continuous adsorption and dissolution. The gas concentration decreases resulting in forming a concentration gradient. Thus, the driving force of mass transfer is increased, and the absorption rate is also increased. However, in the study of Kluytmans et al. [12], it was found that the gas-liquid mass transfer coefficient did not increase with the increase of the concentration of activated carbon particles. Therefore, they claimed that the sweeping effect does not explain the phenomenon that solid particles can enhance gas-liquid mass transfer. Ruthiya et al. [13] also made a similar conclusion. The transport of solid particles between the gas-liquid interface and the liquid phase is not the main reason for solid particles enhancing the gas-liquid mass transfer.

23.2.2 The Hydrodynamic Effect Mechanism

The hydrodynamic effect mechanism, also called mixing of the gas-liquid boundary layer, is shown in Figure 23.2. The mechanism shows that the enhanced mass transfer is mainly caused by the fine particles at the near-gas-liquid interface passing through the disturbance to the mass transfer boundary layer and changing the hydrodynamic conditions in the mass transfer membrane [14,15]. The way in which particles change the convective mass transfer and concentration gradient is as follows: (1) the mass transfer membrane thickness can be reduced by collision of particles near the gas-liquid interface; (2) a particle diameter larger than the thickness of the mass transfer membrane will cause an impact on the entire liquid

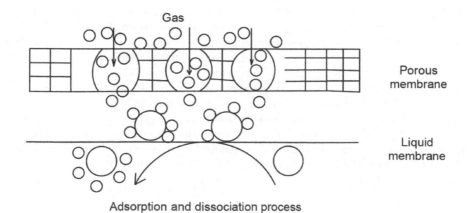

Adsorption and dissociation process

FIGURE 23.1   Schematic of the sweeping effect mechanism.

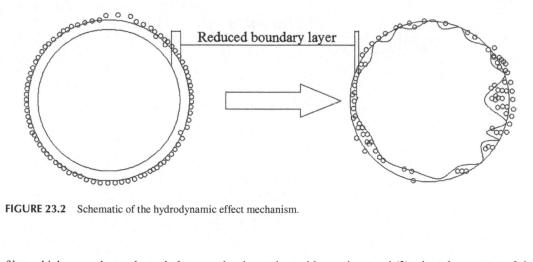

FIGURE 23.2   Schematic of the hydrodynamic effect mechanism.

film, which exacerbates the turbulence and reduces the residence time; and (3) when the content of the particles is large, the viscosity of the absorption system will change greatly, thereby affecting the mass transfer characteristics of the multiphase system. The liquid near the gas-liquid interface is disturbed by the Brownian motion of the nanoparticles, creating partial micro-convection and additional diffusion, which lead to the enhanced mass transfer [16,17].

23.2.3 The Bubble Trapping Mechanism

In inhibition of bubble polymerization mechanism, the solid particles are attached to the surface of the bubble or the dielectric in the solution is adsorbed to the surface of the bubble. Figure 23.3 shows the absorption mechanism of the bubble breaking effect. The bubble stiffness improves, the rate of bubble coalescence reduces, and the contact area of the gas-liquid interface increases. Finally, the mass transfer coefficient is enhanced [18,19].

23.2.4 The Infiltration Mechanism

The infiltration mechanism is applying the theory of infiltration to the diffusion in a heterogeneous fixed liquid membrane, including the diffusivity and solubility of the gas solute. The mechanism assumes that there is an average thickness of the liquid film at the gas-liquid interface, dividing the diffusion in the homogeneous phase of the interface into two parts [20]. In gas-liquid mass transfer process, the gas is first diffused into the continuous thin liquid layer, and then flows into the dispersed or continuous phase in the heterogeneous phase. Figure 23.4 depicts the infiltration mechanism effect.

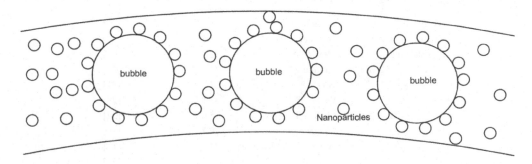

FIGURE 23.3   Schematic of the bubble trapping mechanism.

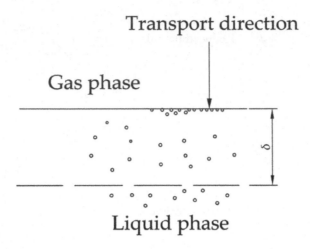

FIGURE 23.4   Schematic of the infiltration mechanism.

In some cases, the CO_2 absorption enhancement was induced by more than one mechanism. In the publications of Zhang et al. and Lee et al. [1,21], they believed that the enhanced CO_2 absorption could be explained by grazing effect, bubble breaking, and hydrodynamic effect mechanisms. In the CO_2 absorption process by nanoemulsion absorbents, the absorption was enhanced by grazing effect and hydrodynamic effect [22].

23.3 Enhancement via Different Nanoparticle Types

A large number of metallic and non-metallic nanofluids are used to capture CO_2. Table 23.1 lists the enhancement results of CO_2 absorption using different nanofluids.

TABLE 23.1

Common Use of Nanofluids for CO_2 Absorption Enhancement

Nanoparticles	Base Solution	Enhancement (%)	References
(0.01 vol%)Al_2O_3	Methanol	8.3	[23]
(0.01 vol%)Al_2O_3	Methanol	4.5	[24]
(0.01 vol%)Al_2O_3	Methanol	40	[25]
(0.01 vol%)Al_2O_3	NaCl	12.5	[26]
(0.05 wt%)Al_2O_3	DEA	33	[27]
(0.05 wt%)Al_2O_3	Water	3	[28]
(0.01 vol%)Al_2O_3	DI water	23.5	[21]
(0.39 vol%)Fe_3O_4	MDEA	92.8	[29]
(0.02 wt%)Fe_3O_4	Sulfinol-M	23.2	[30]
(0.02 wt%)Fe_3O_4	Water	24	[31]
(0.15 wt%)Fe_3O_4	Water	43.8	[28]
(0.01 vol%)SiO_2	Methanol	5.6	[24]
(0.05 vol%)SiO_2	Methanol	9.7	[19]
(0.021 wt%)SiO_2	Water	24	[18]
(0.01 vol%)SiO_2	DI water	13.1	[32]

(Continued)

TABLE 23.1 (Continued)

Common Use of Nanofluids for CO_2 Absorption Enhancement

Nanoparticles	Base Solution	Enhancement (%)	References
(0.01 vol%)SiO_2	DI water	23.5	[21]
(0.05 wt%)SiO_2	Water	25.9	[28]
(0.05 wt%)SiO_2	DEA	40	[27]
(0.01 vol%)TiO_2	Methanol	4.6	[33]
(0.08 wt%)TiO_2	MDEA	11.54	[34]
(0.02 wt%)CNT	MDEA	23	[31]
(0.01 wt%)CNT	Water	38	[28]
(0.05 wt%) CNT	Water	32	[35]
(40 mg/L) MWCNT	Water	36	[36]

23.3.1 Al_2O_3

Al_2O_3 nanoparticles are characterized by corrosion resistance, heat resistance, high strength, and high hardness. At present, Al_2O_3 nanoparticles have been widely studied in promoting the absorption of CO_2. Al_2O_3 fluids have good stability and are easy to prepare. The absorption of CO_2 is enhanced by the addition of Al_2O_3 nanoparticles in various solutions.

Pineda et al. [25] simulated the effect of Al_2O_3 nanoparticles on the absorption characteristics of CO_2 bubbles in methanol solution. The bubble size, ascending speed, and mass transfer rate were compared with previous experimental results. From the experimental and simulation results, it can be concluded that the mass transfer coefficient is increased by about 40% by adding 0.01 vol% of Al_2O_3 nanoparticles compared to the pure methanol absorbent. Kim et al. [16] also drew similar conclusions. Jung et al. [23] prepared Al_2O_3/methanol nanofluids with a nanoparticle concentration of 0.005 vol%–0.1 vol%. Based on their results, the maximum increase in CO_2 absorption (compared to pure methanol) was 8.3% at 0.01 vol% of Al_2O_3 nanoparticles. Yung et al. [37] studied the effect of Al_2O_3 nanoparticles on CO_2 absorption in methanol (base solution) and obtained similar results. Lee et al. [26] showed that Al_2O_3 nanoparticles and NaCl aqueous solution were combined into Al_2O_3/NaCl aqueous nanofluids to enhance the CO_2 absorption of the base fluid (NaCl aqueous solution). The absorption experiments were carried out in the bubble type absorber system. The particle concentration ranged from 0.005 vol% to 0.1 vol%. It is found that the optimum concentration of Al_2O_3 nanoparticles is 0.01 vol% in the NaCl aqueous solution-based nanofluids. It was found that the CO_2 solubility is enhanced to 11.0%, 12.5%, and 8.7% for 0.01 vol% Al_2O_3/NaCl aqueous solution at 30°C, 20°C, and 10°C, respectively. Meanwhile, the effect of nanoparticles on the mass transfer enhancement was more significant in the region of unsaturated state than that of the saturated state.

Lee et al. [21] found that when the Al_2O_3 nanoparticle concentration was 0.01 vol%, the CO_2 absorption rate increased by 23.5% compared to the base liquid (deionized [DI] water). In order to improve mass transfer based on the nature of nanoparticles, Taheri et al. [27] employed a wetted-wall column system in a laboratory scale. The results showed up to 33% improvement of CO_2 absorption at 0.05 wt% of Al_2O_3/diethanolamine (DEA) nanofluids. Peyravi et al. [28] studied the effect of Al_2O_3 nanoparticles on the mass transfer rate of CO_2 absorption process using a hollow fiber membrane contactor with distilled water as the base fluid. The results show that the absorption rate increased to 3.0% when the concentration of Al_2O_3 nanoparticles is 0.05 wt%.

23.3.2 Fe_3O_4

Magnetic fluids were first developed by Papell in the 1960s for high performance seals in space applications [38]. Fe_3O_4 nanoparticles have a tendency to aggregate in solution to reduce their surface energy due to their large surface area to volume ratio. In recent years, the use of these nanofluids to absorb CO_2 has received increasing attention due to its unique advantages.

Komati et al. [29] used a novel approach for increasing the mass transfer coefficient in gas/liquid mass transfer and applied it to the industrial system of CO_2 absorption. The approach used a ferrofluid additive to the liquid phase. To verify this effect, a CO_2/methyldiethanolamine (MDEA) system was tested by a wet-wall column. The results show that ferrofluids had an enhanced effect on mass transfer and the extent of which depends on the amount of added ferrofluid. The enhancement in mass transfer coefficient was 92.8% for a 50 vol% fluid (solid magnetite volume fraction of about 0.39%). These results had significance for the absorption of CO_2 in industrial absorbers using amine solutions.

Salimi et al. [39] reported that the addition of Fe_3O_4 nanoparticles to the solvent enhanced the mass transfer characteristics, which indicated a positive effect on CO_2 absorption. Mohammad et al. [40] studied the effect of Fe_3O_4-water nanofluids on the physical absorption of CO_2 under AC and DC magnetic fields. The results show that the mass transfer rate and solubility of CO_2 in magnetic nanofluids increased with the increase of magnetic field strength. Nabipour et al. [30] added 0.02 wt% Fe_3O_4 nanoparticles to the base solvent to increase the absorption rate, and the equilibrium solubility reached a maximum of 23.2%. Rahmatmand et al. [31] dispersed Fe_3O_4 nanoparticles in pure water to form nanofluids at concentrations of 0.02 wt%, 0.05 wt%, and 0.1 wt%. The results show that Fe_3O_4 was more effective at lower nanoparticle concentrations (0.02 wt%) and increases to 24%. Peyravi et al. [28] produced and tested Fe_3O_4/water nanofluids as new absorbents in hollow fiber membrane contactors to study the effect of nanoparticles on mass transfer rate during CO_2 absorption. The results show that the absorption rate was up to 43.8% with 0.15 wt% Fe_3O_4 nanoparticles.

23.3.3 SiO_2

The electrical insulating properties of SiO_2 nanoparticles suggest potential applications in electronic devices. However, based on SiO_2 nanofluids, it has the characteristics of easy preparation and good stability, so increasing the absorption of CO_2 has become a research hotspot.

Lee et al. [24] studied different levels of SiO_2/methanol nanofluids to remove CO_2 in bubble absorbers. When the SiO_2 nanofluid was 0.01 vol% at $-20°C$, the maximum enhancement of CO_2 absorption (compared to pure methanol) was 5.6%. Pineda et al. [19] prepared nanofluids by adding different amounts of SiO_2 nanoparticles to methanol solution and carried out CO_2 absorption test in a tray absorption tower. A comparative study was conducted in 0.005 vol%–0.1 vol% of the nanofluid, and the results showed that the CO_2 absorption rate was up to 9.7% at a nanofluid concentration of 0.05 vol%. SiO_2 promotes the separation of bubbles in the orifice. The movement of the fluid is more dynamic by the external force. The collision of the particles with the gas-liquid interface causes the bubbles to break into small-sized bubbles, which increases the interface area. This will promote the mass transfer of gas to the liquid, thereby increasing the CO_2 absorption rate. They also found that SiO_2 nanoparticles exhibited better CO_2 absorption than Al_2O_3 nanoparticles at the optimum nanofluids concentration, and the absorption rate was slightly higher by 0.3%.

Kim et al. [18] prepared a suspension of SiO_2 nanoparticles in water (nanoparticle dosage from 0.01 wt% to 0.04 wt%) for absorption of CO_2 in bubble-type absorbers. The results showed that the carbon dioxide removal rate was increased by 24% when the nanoparticle dose was 0.021 wt%. In the piperazine/K_2CO_3 absorbent, the average CO_2 absorption rate during the first 1 minute and total absorption were increased 11% and 12% by the addition of 0.021 wt% nanoparticles. Siad et al. [41] showed the relationship between gas consumption and time in the dissolution process of gas mixture in pure water and SiO_2 nanofluid through experimental research. It was found that the gas consumption of the SiO_2 nanofluids at a concentration of 0.1 wt%, 0.2 wt%, and 0.3 wt% was greater than that under pure water, and the consumption gradually increased with increasing concentration. Choi et al. [32] found that when the concentration of SiO_2 nanoparticles was 0.01 vol%, the CO_2 absorption rate increased by 13.1% compared with the base liquid (DI water). Lee et al. [21] obtained an increase in the absorption rate of CO_2 to 23.5% at the same concentration. Peyravi et al. [28] used SiO_2/water nanofluids as new absorbents in hollow fiber membrane contactors to study the effect of nanoparticles on mass transfer rate during CO_2 absorption. The results show that the absorption rate was up to 25.9% when the concentration of SiO_2 nanoparticles was 0.05 wt%.

Taheri et al. [27] studied the effect of nanofluids based on DEA on CO_2 absorption in a wet-wall absorber based on the properties of nanoparticles. The addition of SiO_2 nanoparticles to the DEA solution promoted the absorption of CO_2, and the results showed that the CO_2 absorption increased to 40% under 0.05 wt% of SiO_2/DEA nanofluid. Among other works, Park et al. [42–44] and Hwang et al. [29,45] studied the effect of SiO_2 nanoparticles on CO_2 absorption rates in aqueous solutions of DEA, MEA, and diisopropanolamine (DIPA). It was found that the absorption rate decreased as the concentration of the nanofluids increased due to the elasticity of the solution.

23.3.4 TiO₂

Pineda et al. [33] used methanol as the base fluid and found that TiO_2 nanoparticles have enhanced mass transfer effect. Compared with pure methanol, TiO_2 nanoparticles promoted the maximum increase in CO_2 absorption by 4.6%. Fang et al. [46] investigated the effect of different concentrations of TiO_2 nanoparticles on CO_2 absorption in aqueous solution by establishing a CO_2 bubble test bench. As the number of nanoparticles increased, the removal rate first increased and then decreased. As the nanoparticles increased, the degree of agglomeration of the nanoparticles increased, which weakened the Brownian motion and reduced the molecular diffusion coefficient and mass transfer rate. The increase in the content of TiO_2 nanoparticles and the increase in solution viscosity resulted in the loss of CO_2 adsorption capacity. This phenomenon can be explained by the adsorption of CO_2 by TiO_2 is difficult to remove quickly from the contact surface, and the particles lose their CO_2 adsorption capacity on the gas-liquid contact surface to form a flat surface, thus hindering the entry of CO_2 into the liquid, but generally greater than the efficiency of the blank solution. When the amount of nanoparticles added was 3.0 g/L, the CO_2 absorption rate reached a maximum of 2.5 mL/s. Similar phenomenon has been reported in previous work [47].

Li et al. [34] studied the effect of nanoparticles on the absorption of CO_2 bubbles in MDEA solution. The stability of TiO_2 nanofluids was maintained by mechanical agitation in the research. It can be found that the CO_2 absorption rate can be enhanced by 1.95%, 6.53%, 7.79%, and 11.54%, when the TiO_2 nanoparticles concentration is 0.05 wt%, 0.2 wt%, 0.4 wt%, and 0.8 wt%, respectively. The results show that the surface tensions, kinematic viscosities, and thermal conductivities all increased with increasing concentration of the nanoparticles.

Jiang et al. [48] also studied the effect of TiO_2 nanoparticles on enhanced CO_2 absorption. It is revealed that the TiO_2-MDEA nanofluid provides a higher enhancement factor than other nanoparticles. Since TiO_2 can absorb more CO_2, it increases the concentration gradient of CO_2 and eventually enhances absorption. In addition, TiO_2-MEA nanofluids exhibited better CO_2 absorption than TiO_2-MDEA nanofluids due to the faster chemical reaction rate between carbon dioxide and MEA. Similar results have been reported in other publications [46,47].

23.3.5 CNT

Since the establishment of carbon nanotubes (CNTs), scientists have begun to discover the excellent properties of these particles. Carbon nanotubes have shown great potential in the field of gas storage [30].

Darabi et al. [35] investigated numerically the CO_2 absorption performance in a hollow fiber membrane using CNT/water and SiO_2/water solutions. The rate of CO_2 absorption by CNT nanoparticles showed 16% higher than that by SiO_2 nanoparticles. This is attributed to a higher adsorption ability of CNT nanoparticles in comparison with SiO_2 nanoparticles. Lu et al. [49] claimed that both Al_2O_3 and CNT nanoparticles strengthened the absorption of CO_2 and CNT showed a significant enhancement. Amaris et al. [50] used CNTs to study the performance of NH_3/$LiNO_3$ tubular bubble absorbers and found that the presence of CNTs led to a significant increase in heat transfer and mass transfer rates. Rahmatmand et al. [31] obtained CNT nanofluids showing a good absorption performance at a low concentration of 0.02 wt%. It was also figured out that CNT nanoparticles were more effective for MDEA than DEA solutions and the CO_2 adsorption capacity increased up to 23%. Peyravi et al. [28] produced and tested CNT/water nanofluids as new absorbents in hollow fiber membrane contactors to study the effect of nanoparticles on mass transfer rate during CO_2 absorption. The results show that the absorption rate was up to 38.0% when the concentration of CNT nanoparticles is 0.01 wt%. Recently, Koronaki et al. [51] used CNTs to numerically study the absorption of CO_2. Their model considered the Brownian motion of suspended nanoparticles proposed by Nagy et al. [52]. It is concluded that the adsorption rate of CO_2 gas was higher than that of large particle size under the condition of small particle size nanoparticles.

Furthermore, the material of multi-walled carbon nanotubes (MWCNTs) has been reported due to its large surface area. Jorge et al. [36] studied amine-functionalized MWCNTs as absorbents for CO_2 capture. These functional groups not only enhanced the absorption of carbon dioxide, but also increased the hydrophilicity of multi-walled carbon nanotubes, allowing them to be suspended in water for at least 3 months at room temperature. When the concentration of MWCNT nanofluid was about 40 mg/L, the CO_2 absorption capacity was 36% higher than that of water. Nabipour et al. [30] studied the equilibrium adsorption of CO_2 on MWCNTs. Due to the strong interaction between multi-walled carbon nanotubes and CO_2, these particles may be a good choice to increase the absorption of CO_2 in Sulfinol-M solutions. Jorge et al. [36] used a nanofluid containing plasma-functionalized MWCNTs to absorb CO_2 in a bubble column. The CO_2 absorption capacity of this nanofluid was found to be significantly improved compared to water.

23.3.6 CuO

Fang et al. [46] studied the effect of nanoparticles on the CO_2 absorption by ammonia solutions. The enhancement factor of CuO nanofluids always hovered around 1. Nano-CuO was easily combined with ammonia to form copper-ammonium complex ions, which led to the loss of the influence of nanoparticles on the reaction. At the same time, the formation of copper-ammonium complex ions had no effect on the viscosity of the solution, so nano-CuO had no obvious enhancement or inhibition effect on the reaction. Kim et al. [53] studied the enhancement of NH_3/H_2O absorption by 5.32 times under the simultaneous action of chemical surfactants and CuO nanoparticles. CuO nanoparticles had been successfully dispersed into CO_2 using polydimethylsiloxane (PDMS) as a co-solvent by Shah [54]. In addition, CuO nanofluids can be effectively used as a coolant in shell-and-tube gas coolers to improve the performance of transcritical CO_2 refrigeration cycles [55].

23.3.7 Others

In addition to the above nanoparticles, Tang et al. [56] used a bubble reactor to study the strengthening effect of MgO particles on CO_2 absorption using anhydrous ethanol as the base liquid. The absorption rate in different nanofluid concentrations increased to a certain extent relative to pure anhydrous ethanol, and the absorption rate showed a significant upward trend with the increase of nanoparticle volume fraction. Jiang et al. [48] added MgO nanoparticles into MDEA solution and also proved that MgO promoted the absorption of CO_2. Haghtalab et al. [57] experimentally obtained that the absorption of CO_2 in ZnO/water-based nanofluids was greater than that in pure water. It is also believed that this type of nanoparticle had the high potential to enhance CO_2 absorption in water-based nanofluids. Yang et al. [37] also studied the effects of Fe_2O_3 and $ZnFe_2O_4$ nanoparticles on the absorption of CO_2 gas by ammonia. They obtained the best values for nanoparticles and surfactants. By adding Fe_2O_3 and $ZnFe_2O_4$ nanoparticles, the absorption rate of CO_2 was increased by 70% and 50%, respectively. In addition, NiO nanoparticles were found to have the promoting effect on CO_2 absorption [39]. Pang et al. [58] found that Ag nanoparticles could also enhance the gas absorption.

23.4 Conclusions and Prospects

Chemical absorption is regarded as an effective method for capturing CO_2. The liquid type is a determining factor to the capturing performance. The common solutions always show low CO_2 capture efficiency and high regeneration energy. By adding nanoparticles into the solutions could significantly improve the absorption efficiency and the gas-liquid mass transfer process. In this chapter, the detailed discussions about the latest developments in CO_2 absorption with nanoparticles were introduced, which is helpful to the choice of the nanofluids and their relevant properties. Thus, the main direction for nanofluids application in CO_2 capture area is to identify the solutions with high capture performance and less regeneration energy, which indicates a great potential in coping with greenhouse effect.

ACKNOWLEDGMENTS

The project was supported by Open Funds of Fujian Provincial Key Laboratory of Featured Materials in Biochemical Industry (No. FJKL_FMBI201704) and Fujian Province University Key Laboratory of Green Energy and Environment Catalysis (No. FJ-GEEC201702).

REFERENCES

1. Zhang, Z. et al. (2018). Progress in enhancement of CO_2 absorption by nanofluids: A mini review of mechanisms and current status. *Renewable Energy* 118, 527–535.
2. Liu, J. et al. (2012). Progress in adsorption-based CO_2 capture by metal-organic frameworks. *Chemical Society Reviews* 416, 2308–2322.
3. Zhang, Z. et al. (2018). Modeling of a CO_2-piperazine-membrane absorption system. *Chemical Engineering Research and Design* 131, 375–384.
4. Feng, H. et al. (2017). Particulate formation from a copper oxide-based oxygen carrier in chemical looping combustion for CO_2 capture. *Environmental Science & Technology* 514, 2482.
5. Li, H. and Zhang, Z. (2018). Mining the intrinsic trends of CO_2 solubility in blended solutions. *Journal of CO_2 Utilization* 26, 496–502.
6. Li, Y. et al. (2018). Carbon dioxide absorption from biogas by amino acid salt promoted potassium carbonate solutions in a hollow fiber membrane contactor: A numerical study. *Energy & Fuels* 32(3), 3637–3646.
7. Zhang, Z. (2016). Comparisons of various absorbent effects on carbon dioxide capture in membrane gas absorption (MGA) process. *Journal of Natural Gas Science & Engineering* 31, 589–595.
8. Das, P.K. (2017). A review based on the effect and mechanism of thermal conductivity of normal nanofluids and hybrid nanofluids. *Journal of Molecular Liquids* 240, 420–446.
9. Pang, C. et al. (2014). Heat conduction mechanism in nanofluids. *Journal of Mechanical Science & Technology* 287, 2925–2936.
10. Zhou, S.D. et al. (2014). Progress in research on enhancement of acid gas absorption by nanofluids. *Natural Gas Chemical Industry* 39, 83–87.
11. Kars, R.L., Best, R.J., and Drinkenburg, A.A.H. (1979). The sorption of propane in slurries of active carbon in water. *Chemical Engineering Journal* 172, 201–210.
12. Kluytmans, J.H.J. et al. (2003). Mass transfer in sparged and stirred reactors: Influence of carbon particles and electrolyte. *Chemical Engineering Science* 5820, 4719–4728.
13. Ruthiya, K.C., Kuster, B.F.M., and Schouten, J.C. (2010). Gas–liquid mass transfer enhancement in a surface aeration stirred slurry reactors. *Canadian Journal of Chemical Engineering* 813–814, 632–639.
14. Jamialahmadi, M. and Müller-Steinhagen, H. (2010). Effect of solid particles on gas hold-up in bubble columns. *Canadian Journal of Chemical Engineering* 691, 390–393.
15. Ozkan, O. et al. (2000). Effect of inert solid particles at low concentrations on gas–liquid mass transfer in mechanically agitated reactors. *Chemical Engineering Science* 5514, 2737–2740.
16. Kim, J.H., Jung, C.W., and Kang, Y.T. (2014). Mass transfer enhancement during CO_2 absorption process in methanol/Al_2O_3 nanofluids. *International Journal of Heat & Mass Transfer* 766, 484–491.
17. Yoon, S., Jin, T.C., and Yong, T.K. (2014). The particle hydrodynamic effect on the mass transfer in a buoyant CO_2-bubble through the experimental and computational studies. *International Journal of Heat & Mass Transfer* 739, 399–409.
18. Kang, H.U. (2008). Synthesis of silica nanofluid and application to CO_2 absorption. *Separation Science & Technology* 4311–4312, 3036–3055.
19. Pineda, I.T. et al. (2012). CO_2 absorption enhancement by methanol-based Al_2O_3 and SiO_2 nanofluids in a tray column absorber. *International Journal of Refrigeration* 355, 1402–1409.

20. Wei, S. et al. (2011). Mechanism and influence factors of nano-particles enhancing ammonia bubble absorption performance. *Journal of Chemical Engineering of Chinese Universities* 251, 30–36.
21. Lee, J.S., Lee, J.W., and Kang, Y.T. (2015). CO_2 absorption/regeneration enhancement in DI water with suspended nanoparticles for energy conversion application. *Applied Energy* 143, 119–129.
22. Jeong, M. et al. (2017). Mass transfer performance enhancement by nanoemulsion absorbents during CO_2 absorption process. *International Journal of Heat and Mass Transfer* 108, 680–690.
23. Jung, J.Y., Lee, J.W., and Kang, Y.T. (2012). CO_2 absorption characteristics of nanoparticle suspensions in methanol. *Journal of Mechanical Science & Technology* 268, 2285–2290.
24. Lee, J.W. et al. (2011). CO_2 bubble absorption enhancement in methanol-based nanofluids. *International Journal of Refrigeration* 348, 1727–1733.
25. Pineda, I.T., Kim, D., and Kang, Y.T. (2017). Mass transfer analysis for CO_2 bubble absorption in methanol/Al_2O_3 nanoabsorbents. *International Journal of Heat and Mass Transfer* 114, 1295–1303.
26. Lee, J.W. and Kang, Y.T. (2013). CO_2 absorption enhancement by Al_2O_3 nanoparticles in NaCl aqueous solution. *Energy* 535, 206–211.
27. Taheri, M. et al. (2016). Simultaneous absorption of carbon dioxide (CO_2) and hydrogen sulfide (H_2S) from CO_2–H_2S–CH_4 gas mixture using amine-based nanofluids in a wetted wall column. *Journal of Natural Gas Science & Engineering* 28, 410–417.
28. Peyravi, A., Keshavarz, P., and Mowla, D. (2015). Experimental investigation on the absorption enhancement of CO_2 by various nanofluids in hollow fiber membrane contactors. *Energy & Fuels* 29(12), 8135–8142.
29. Komati, S. and Suresh, A.K. (2008). CO_2 absorption into amine solutions: A novel strategy for intensification based on the addition of ferrofluids. *Journal of Chemical Technology & Biotechnology Biotechnology* 838, 1094–1100.
30. Nabipour, M., Keshavarz, P., and Raeissi, S. (2016). Experimental investigation on CO_2 absorption in Sulfinol-M based Fe_3O_4 and MWCNT nanofluids. *International Journal of Refrigeration* 73, 1–10.
31. Rahmatmand, B., Keshavarz, P., and Ayatollahi, S. (2016). Study of absorption enhancement of CO_2 by SiO_2, Al_2O_3, CNT, and Fe_3O_4 nanoparticles in water and amine solutions. *Journal of Chemical & Engineering Data* 614.
32. Choi, I.D., Lee, J.W., and Yong, T.K. (2015). CO_2 capture/separation control by SiO_2 nanoparticles and surfactants. *Separation Science & Technology* 505, 772–780.
33. Pineda, I.T., Chang, K.C., and Yong, T.K. (2014). CO_2 gas absorption by CH_3OH based nanofluids in an annular contactor at low rotational speeds. *International Journal of Greenhouse Gas Control* 234, 105–112.
34. Li, S.H., Ding, Y., and Zhang, X.S. (2013). Enhancement on CO_2 bubble absorption in MDEA solution by TiO_2 nanoparticles. *Advanced Materials Research* 631–632, 127–134.
35. Darabi, M., Rahimi, M., and Dehkordi, A.M. (2017). Gas absorption enhancement in hollow fiber membrane contactors using nanofluids: Modeling and simulation. *Chemical Engineering & Processing Process Intensification* 119, 7–15.
36. Jorge, L., Coulombe, S., and Girard-Lauriault, P.L. (2016). Nanofluids containing MWCNTs coated with nitrogen–rich plasma polymer films for CO_2 absorption in aqueous medium. *Plasma Processes & Polymers* 1211, 1311–1321.
37. Yang, L. et al. (2011). Experimental study on enhancement of ammonia–water falling film absorption by adding nano-particles. *International Journal of Refrigeration* 343, 640–647.
38. Beckman, K.J. (1999). Preparation and properties of an aqueous ferrofluid. *Journal of Chemical Education* 767, 943–948.
39. Salimi, J., Haghshenasfard, M., and Etemad, S.G. (2015). CO_2 absorption in nanofluids in a randomly packed column equipped with magnetic field. *Heat & Mass Transfer* 515, 621–629.
40. Karimi Darvanjooghi, M.H., Pahlevaninezhad, M., Abdollahi, A., and Davoodi, S.M. (2017). Investigation of the effect of magnetic field on mass transfer parameters of CO_2 absorption using Fe_3O_4-water nanofluid. *AIChE Journal* 63, 2176–2186.
41. Samer, S. et al. (2016). A study on the influence of nanofluids on gas hydrate formation kinetics and their potential: Application to the CO_2 capture process. *Journal of Natural Gas Science & Engineering* 32, 95–108.
42. Park, S.W. et al. (2008). Absorption of carbon dioxide into aqueous colloidal silica solution with diisopropanolamine. *Journal of Industrial & Engineering Chemistry* 142, 166–174.

43. Park, S.W., Choi, B.S., and Lee, J.W. (2006). Effect of elasticity of aqueous colloidal silica solution on chemical absorption of carbon dioxide with 2-amino-2-methyl-1-propanol. *Korea-Australia Rheology Journal* 183, 133–141.

44. Park, S.W. et al. (2006). Absorption of carbon dioxide into aqueous colloidal silica solution. *Separation Science & Technology* 4114, 3265–3278.

45. Hwang, B.J. et al. (2009). Absorption of carbon dioxide into aqueous colloidal silica solution with different sizes of silica particles containing monoethanolamine. *Korean Journal of Chemical Engineering* 263, 775–782.

46. Fang, L. et al. (2017). Experimental study on enhancement of bubble absorption of gaseous CO_2 with nanofluids in ammonia. *Journal of Harbin Institute of Technology* 242, 80–86.

47. Dagaonkar, M.V. et al. (2003). The application of fine TiO_2 particles for enhanced gas absorption. *Chemical Engineering Journal* 921, 151–159.

48. Jiang, J. et al. (2014). Experimental study of CO_2 absorption in aqueous MEA and MDEA solutions enhanced by nanoparticles. *International Journal of Greenhouse Gas Control* 29, 135–141.

49. Sumin, L.U. et al. (2013). Experimental and theoretical studies of CO_2 absorption enhancement by nano-Al_2O_3 and carbon nanotube particles. *Chinese Journal of Chemical Engineering* 219, 983–990.

50. Amaris, C., Bourouis, M., and Vallès, M. (2014). Passive intensification of the ammonia absorption process with $NH_3/LiNO_3$ using carbon nanotubes and advanced surfaces in a tubular bubble absorber. *Energy* 688, 519–528.

51. Koronaki, I.P., Nitsas, M.T., and Vallianos, C.A. (2016). Enhancement of carbon dioxide absorption using carbon nanotubes–A numerical approach. *Applied Thermal Engineering* 99, 1246–1253.

52. Nagy, E., Feczkó, T., and Koroknai, B. (2007). Enhancement of oxygen mass transfer rate in the presence of nanosized particles. *Chemical Engineering Science* 6224, 7391–7398.

53. Kim, J.K., Jung, J.Y., and Yong, T.K. (2007). Absorption performance enhancement by nano-particles and chemical surfactants in binary nanofluids. *International Journal of Refrigeration* 301, 50–57.

54. Shah, R.D. (2009). Application of nanoparticle saturated injectant gases for EOR of heavy oils. *SPE Annual Technical Conference and Exhibition*, New Oriean, Louisiana, USA, 4–7 October 2009.

55. Sarkar, J. (2011). Performance of nanofluid-cooled shell and tube gas cooler in transcritical CO_2 refrigeration systems. *Applied Thermal Engineering* 3114–3115, 2541–2548.

56. Tang, Z.L., Peng, L.M., and Zhang, S.Y. (2012). Experiment on enhancement of bubble absorption of gaseous CO_2 with nanofluids. *Journal of Tianjin University* 456, 534–539.

57. Haghtalab, A., Mohammadi, M., and Fakhroueian, Z. (2015). Absorption and solubility measurement of CO_2 in water-based ZnO and SiO_2 nanofluids. *Fluid Phase Equilibria* 392, 33–42.

58. Pang, C. et al. (2012). Mass transfer enhancement by binary nanofluids (NH_3/H_2O+Ag nanoparticles) for bubble absorption process. *International Journal of Refrigeration* 358, 2240–2247.

Index

Note: Page numbers in italic and bold refer to figures and tables, respectively.